MARINE MAMMALS

BIOLOGY AND CONSERVATION

MARINE MAMMALS
BIOLOGY AND CONSERVATION

Edited by

Peter G. H. Evans
Sea Watch Foundation
Oxford, England

and

Juan Antonio Raga
University of Valencia
Valencia, Spain

Kluwer Academic / Plenum Publishers
New York, Boston, Dordrecht, London, Moscow

Library of Congress Cataloging-in-Publication Data

Marine mammals: biology and conservation/edited by Peter G.H. Evans and Juan
Antonio Raga.
    p.   cm.
    Includes bibliographical references and index.
    ISBN 0-306-46572-8—ISBN 0-306-46573-6 (pbk.)
      1. Marine mammals.   2. Wildlife conservation.   I. Evans, Peter G.
H. II. Raga, Juan
  Antonio.

    QL713.2 .M354 2001
    599.5—dc21

                            2001041359

GENERALITAT
VALENCIANA
**CONSELLERIA DE
MEDIO AMBIENTE**

VNIVERSITAT
ID ǑVALÈNCIA

MINISTERIO
DE MEDIO AMBIENTE

Cover photo: Atlantic Spotted Dolphins underwater—B. Würsig.
Back cover photos from left top: Mating herd of Florida manatees—Florida Fish and Wildlife
Conservation Commission, Sea otter breaking open a clam—P. Chanin,
and Two adult male New Zealand fur seals fighting—I. Stirling,
Spine photos: Adult male walrus—I. Stirling and Polar bear on ice—I. Stirling.

Cover design by Susanne Van Duyne (Trade Design Group).

ISBN 0-306-46572-8 (hardback)
ISBN 0-306-46573-6 (paperback)

©2001 Kluwer Academic/Plenum Publishers, New York
233 Spring Street, New York, New York 10013

http://www.wkap.nl/

10 9 8 7 6 5 4 3 2 1

A C.I.P. record for this book is available from the Library of Congress

Printed in the United States of America

Preface

Interest in marine mammals has increased dramatically in the last few decades, as evidenced by the number of books, scientific papers, and conferences devoted to these animals. Nowadays, a conference on marine mammals can attract between one and two thousand scientists from around the world. This upsurge of interest has resulted in a body of knowledge which, in many cases, has identified major conservation problems facing particular species. At the same time, this knowledge and the associated activities of environmental organisations have served to introduce marine mammals to a receptive public, to the extent that they are now perceived by many as the living icons of biodiversity conservation.

Much of the impetus for the current interest in marine mammal conservation comes from "Save the Whale" campaigns started in the 1960s by environmental groups around the world, in response to declining whale populations after over-exploitation by humans. This public pressure led to an international moratorium on whaling recommended in 1972 by the United Nations Conference on the Human Environment in Stockholm, Sweden, and eventually adopted by the International Whaling Commission ten years later. This moratorium largely holds sway to this day, and further protective measures have included the delimitation of extensive areas of the Indian Ocean (1979) and Southern Ocean (1994) as whale sanctuaries.

The United States was one of the first countries to introduce specific legislation for marine mammals under the Marine Mammal Protection Act in 1972. This not only significantly advanced conservation action in that country, but also stimulated research through the availability of more substantial resources for scientists. Other countries such as Australia and New Zealand have followed with their own legislative measures.

In the last three decades of the twentieth century, various global conventions for environmental protection have been established, including the MARPOL Agreement (1973/78) for control of marine pollution; UNCLOS (1982) - the UN Convention on the Law of the Sea; and the Earth Summit (1992) - Convention on Biological Diversity.

In recent years, conservation problems facing marine species have driven several coastal nations to develop protective measures. Specific regional agreements that have been established in Europe include OSPAR - the Oslo Paris Convention for the Protection of the Marine Environment of the North-East Atlantic (1992), and, specifically for cetaceans, the "Agreement on the Conservation of Small Cetaceans of the Baltic and North Sea" (ASCOBANS) (1992), and the "Agreement on the Conservation of Cetaceans of the Black Sea, Mediterranean Sea and Contiguous Atlantic

Area" (ACCOBAMS) (1996), within the Bonn Convention - the Convention on Conservation of Migratory Species of Wild Animals (1979).

Marine mammal science has developed especially in North America. This is reflected in the formation in 1981 of the Society of Marine Mammalogy (MMS) in that continent, with members drawn particularly from the United States. Most text books on marine mammals emanate from the USA. Ninety percent of contributors to the recent Smithsonian text "Biology of Marine Mammals" and its companion volume "Conservation and Management of Marine Mammals" also come from the United States. Largely due to lack of resources, the study of marine mammals in Europe has tended to fall behind North America.

However, as a result of the increased interest in scientific and conservation aspects of marine mammals in Europe, and the need to disseminate this knowledge in the region, the European Cetacean Society (ECS) was founded in 1987. This society has progressively consolidated as an organisation interested in fostering cetacean (and other marine mammal) research. The ever increasing and enthusiastic attendance of students at ECS annual meetings raised the need to provide them with specialised up to date information on marine mammals that is usually absent from most university courses. For this reason, and with sponsorship from the International University Menendez Pelayo, the First European Seminar on Marine Mammals: Biology and Conservation was organised in 1996 in Valencia (Spain). After this course, the possibility of producing a book for undergraduate and graduate students was considered, and this eventually materialised in 1998, following contributions to the Second European Seminar.

During the 1990s there has also been growing interest in marine mammals from other parts of the world, particularly South America, Asia, and Australasia. In 1996, for example, a new society called Sociedad Latino Americana de Especialistas en Mamiferos Aquaticos (SOLAMAC) was founded, bringing together marine mammalogists from throughout that subcontinent, and a society catering for the Pacific region is in the progress of being formed. We have tried to take account of this in the contributions to this volume.

Our book is aimed particularly at advanced undergraduate and graduate students, although it should also be of value to teachers and indeed anyone with a special interest in marine mammal science. We have tried to provide a wide range of views and perceptions on the biology and conservation of marine mammals, but with emphasis upon applied aspects of research given the pressing conservation problems that marine mammals face at present.

Within the confines of a single book, it is impossible to cover extensively every subject within marine mammal science. We have therefore been selective, opting away from a more comprehensive but necessarily less detailed book on the subject (for which there is already an excellent text by Annalisa Berta and James L. Sumich entitled "Marine Mammals: Evolutionary Biology", published by Academic Press). Our emphasis has been on evolution, ecology, behaviour, and various aspects of marine mammal health and conservation, not only because of the needs for applied research in those subject areas in this modern world but also because these have proved to be of particular interest to students. We only briefly touch on aspects of anatomy and physiology, for example. For people specialising in those areas, we recommend the relevant chapters in the recent book "Biology of Marine Mammals", edited by John E. Reynolds III and Sentiel A. Rommel, and published by the Smithsonian Institution, USA.

Our book is structured around seventeen chapters, grouped into five sections. Each section has a short introduction that attempts to provide an overview of that theme, filling in some gaps, and identifying key literature references.

The first section is on *Life History and Ecology* and includes chapters under the following headings: "Life history strategies of marine mammals" by Peter Evans and Ian Stirling; "How persistent are marine mammal habitats in an ocean of variability? - habitat use, home range and site fidelity in marine mammals" by Arne Bjørge; and "Ecological aspects of reproduction of marine mammals" by Christina Lockyer.

The second section on *Sensory Systems and Behaviour* consists of three chapters dealing with "Sound and cetaceans" by Jonathan Gordon and Peter Tyack; "Behavioural ecology of cetaceans" by James Boran, Peter Evans and Martin Rosen; and "New perspectives on the behavioural ecology of pinnipeds" by Humberto Luis Cappozzo.

The third section concerns *Survey and Study Techniques*, including also three chapters: "The assessment of marine mammal population size and status" by Philip Hammond; "Acoustic techniques for studying cetaceans" by Jonathan Gordon and Peter Tyack; and "Applications of molecular data in cetacean taxonomy and population genetics with special emphasis on defining species boundaries" by Michel Milinkovitch, Rick LeDuc, Ralph Tiedemann and Andrew Dizon.

Under the fourth section on *Health, Parasites, and Diseases*, are three chapters: "Marine mammal health: holding the balance in an ever-changing sea" by Joseph Geraci and Valerie Lounsbury; "Living together: the parasites of marine mammals" by Javier Aznar, Mercedes Fernandez, Juan Balbuena, and Juan Antonio Raga; and "Mass mortalities in marine mammals" by Mariano Domingo, Seamus Kennedy and Marie Van Bressem.

The fifth and last section focuses upon *Conservation and Management*, and features five chapters on "Marine mammals and ecosystems: ecological and economic interactions" by Enrique Crespo and Martín Hall; "Environmentalists, fishermen, cetaceans and fish: is there a balance and can science help to find it?" by Martín Hall and Gregory Donovan; "Organohalogenated contaminants in marine mammals" by Ailsa Hall; "Cetaceans and humans: influences of noise" by Bernd Würsig and Peter Evans; and "Global climate change and marine mammals" by Bernd Würsig, Randall Reeves and J.G. Ortega-Ortiz.

We owe a great debt of gratitude to the panel of almost thirty scientists, drawn from nine countries, who contributed chapters on the specific topics within their field of interest.

Each chapter has been reviewed additionally by at least two recognised authorities in its respective field, and for their incisive and helpful comments, we thank: John R. Baker, Juan Antonio Balbueno, Arne Bjørge, John Calambokidis, Philip J. Clapham, Christopher W. Clark, Justin G. Cooke, James D. Darling, Douglas P. DeMaster, Karen L. Forney, Roger L. Gentry, Joseph R. Geraci, David I. Gibson, Eric P. Hoberg, A. Rus Hoelzel, Aleta A. Hohn, Vincent Janik, Paul Jepson, Robert D. Kenney, Robin J. Law, Burney J. Le Boeuf, David J. Marcogliese, Simon P. Northridge, Gianni Pavan, William F. Perrin, Andrew J. Read, Peter J.H. Reijnders, W. John Richardson, Randall S. Wells, Koen van Waerebeek, Graham A.J. Worthy, and Bernd Würsig.

Photographs to accompany this book were kindly provided by A. Aguilar, I. Birks, A. Bjørge, J.-M. Bompar, J. Boran, L. Cappozzo, P. Chanin, E. Crespo, P. Evans, M. Fedak, J. Geraci, D. Glockner, J.A. Gomez, J. Gordon, B. Haase, A. Hall, M. Hall, T. Henningsen, J. Heyning, B. Hicks, S. Kraus, F. Larsen, V. Lounsbury, J. Moncrieff, P. Morris, K. Norris Library, R. Pitman, J.A. Raga, F. Ritter, W. Rossiter, M. Scheer, E. Secchi, T. Similå, C. Smeenk, I. Stirling, N. Tregenza, F. Trujillo, P. Tyack, T. Walmsley, B. Wilson, J. Wang, S. Wright, B. Würsig, K. Young, and E. Zúñiga.

Support for the publication of this book was generously provided by the Conselleria de Medio Ambiente of Generalitat Valenciana, Ministerio de Medio Ambiente of Spain, and the University of Valencia.

Finally, we would like to thank Bridget Eichoff for her valuable assistance in the formatting of chapters, and Joanna Lawrence of Kluwer Academic/Plenum Publishers for her continuous help and encouragement during this book's production.

Peter G.H. Evans and Juan Antonio Raga

Contents

Contents

A. Introduction - Life History and Ecology

Throughout this book, emphasis is placed upon the many adaptive features to an aquatic mode of life exhibited by marine mammals. These may be viewed in terms of anatomy and physiology, life history and ecology, sensory systems and social organisation. In this introductory section, we briefly review each of the main groups, indicating some of the special anatomical and physiological features they possess. For more details of morphology, distribution, and identification, see the systematic texts in Harrison and Ridgway (1981a, b, 1985, 1989, 1994, 1999).

Marine mammals comprise two entirely aquatic orders, Cetacea (whales, dolphins and porpoises) with 14 families, and Sirenia (dugongs and manatees) with two families; and five families within the largely terrestrial order, Carnivora, with representatives that spend at least a portion of their lives in the sea (see pages 577-581 for complete list of species).

There are somewhere between 79 and 84 species of Cetacea (Jefferson *et al.*, 1993; Heyning and Perrin, 1994; Rice, 1998) (the exact number recognised varies according to systematist, and a further four species have recently been assigned specific status by some authorities - see Rice, 1998). These are divided into two sub-orders, Mysticeti or baleen whales (11-12 species) and Odontoceti or toothed whales which include also the dolphins and porpoises (68-72 species). Their ancestry can be traced to the order Artiodactyla (the even-toed ungulates) although there is lively debate at present concerning their origins and systematics (see, for example, Graur and Higgins, 1994; Milinkovitch, 1997; Heyning, 1999; Milinkovitch *et al.*, this volume). The evidence, as reviewed in Chapter 9, points most persuasively to the Cetacea being simply highly derived artiodactyls, and best placed as a sub-order within Artiodactyla (Gatesy *et al*, 1999).

Both anatomically and physiologically, cetaceans show several adaptations to living in an aquatic environment. They have streamlined, torpedo or spindle shaped bodies, and reduced appendages (no external ears, reproductive organs or hind limbs - two small openings on the side of the head lead to the hearing organs; the penis of the male is tucked within muscular folds and the teats of the female concealed within slits either side of the genital area; and hind limbs are reduced to traces of the bony skeleton as vestigial remains of the pelvic girdle and in some cases of the femur).

Cetaceans are large or very large (from 1.2 m to 30 m in length; 30 kg to 150 tonnes in weight), and they include the largest mammal to have ever lived, the blue whale. Large size is facilitated by the buoyancy provided by the aqueous medium in which they live, and by their thermoregulatory abilities.

Unlike other mammals, cetaceans have largely dispensed with hair (otherwise normally reduced to the snout), but have a thick insulating layer of subdermal fat called blubber. The two forelimbs have a skeletal structure like a human arm but with the 'fingers' contained within a common integument and flattened to form a pair of horizontal paddle-shaped flippers used for steering and stability with an additional tactile function. Most (but not all) have a dorsal fin or ridge made of fibrous and fatty material, which may help provide stability as well as aid thermoregulation. And all have a boneless horizontal tail fluke, powered by two muscle masses, for propulsion through the water. The skull has become telescoped so that both upper and lower jaws extend well beyond the entrance to the nasal passages or nares, and the one to two blowholes have migrated to the top of the head where they allow air exchange whilst the animal is moving at the surface. The seven neck vertebrae have become compressed and in most species are fused together (although in river dolphins, white whales and the rorquals, the vertebrae remain separate for greater flexibility). The bones of the skeleton are porous and oil-filled. Other internal adaptations include a stomach comprising multiple compartments and a specialised larynx. Although most mammals produce sound by vibrating vocal cords in the larynx, the site of sound production in odontocetes at least, appears to be the nasal plug and elaborate nasal sac system.

Some species (*e.g.* sperm whale and beaked whales) can dive to great depths (possibly up to 3,000 m) and remain below the surface for periods exceeding an hour. Like all mammals, they are air breathing, and generally must hold their breath on diving, the lungs, blood and muscles all serving as sites for oxygen storage.

The two sub-orders of Cetacea are distinguished from one another in many ways. First, the mysticetes are generally much larger (generally more than 10 m in length), and their feeding apparatus has been modified such that they have fringed plates of keratin or baleen which are used to filter organisms such as plankton and small fish; odontocetes, on the other hand, are mostly less than 10 m length (the sperm whale being a notable exception) and they have jaws often extended as a beak-like snout behind which the forehead rises in a rounded curve or "melon", and they possess teeth (although in females of the family Ziphiidae these do not erupt through the gums). Mysticetes have a symmetrical skull with two external nostrils or blowholes whereas odontocetes have only one (the two nasal passages joining below the surface).

Five species of Sirenia exist, split into two families: Trichechidae or manatees which inhabit waters of varying salinity close to shores fringing the tropical or subtropical Atlantic (though the Amazonian manatee is entirely freshwater); and Dugongidae now represented only by the dugong which lives in a marine environment in the tropical and subtropical Indo-Pacific (a second member, the Steller's sea cow is now extinct) (Reynolds and Odell, 1991; Jefferson *et al.*, 1993).

Sirenians also evolved from ungulates; their nearest relatives are elephants. They are entirely aquatic, living in shallow waters and feeding mainly upon plant matter, sea grasses and the like. This herbivorous diet requires a specialised digestive system resembling a horse or elephant. They are large animals (3-4 m long), fusiform in shape, and lacking hind limbs or a dorsal fin, but with flexible paddle-like pectoral flippers and a powerful broad tail fluke. Their nostrils are located on top of the muzzle or at the tip, below which are short robust vibrissae or hairs. The sparse, fine hair covering a thick, tough skin gives them a naked appearance. They have small eyes which can be closed by a sphincter rather than the usual eyelids that other mammals possess. Unlike cetaceans, they have heavy, dense bones, and they have mammary glands in an axillary rather than the more usual abdominal position in other marine mammals.

Within the order Carnivora are five families which have marine representatives. Three are families within the sub-order Pinnipedia - Otariidae, the eared seals (16 species); Phocidae, the earless or true seals (19 species); and Odobenidae which includes a single representative, the walrus; one is the family Ursidae, or bears, which includes the polar bear; and the fifth is the Mustelidae (weasels, otters, mink, *etc.*) with two species - the marine otter of Chile and Peru, and the sea otter of Pacific North America and Russia (Stirling, 1988; Riedman, 1990; Jefferson *et al.*, 1993).

Of carnivores, the pinnipeds spend the greatest amount of time in water, and show the greatest anatomical and physiological adaptations to an aquatic mode of life. They have a spindle-shaped body, insulation in the form of fur or blubber or both, pectoral flippers for movement over land or ice, and a pelvic flipper which, in the phocids, are used for underwater propulsion. Like most marine mammals, pinnipeds are generally of large size (2-5 m in length, and up to 5,000 kg in weight), and, though cumbersome on land where they may haul out particularly during the breeding season, at sea they are excellent swimmers, some phocids diving to depths of 500 m or more, and making seasonal migrations that can be in excess of 12,000 km.

The Otariidae have small external ears, giving them the common name of eared seals, and they possess dense fur comprising a sparse covering of long, coarse guard hairs under which is a thick layer of hairs that traps air for better insulation. The family is divided into the sea lions and fur seals, both groups of which have the ability to walk using their hind flippers which can be tucked under the body by rotating the pelvis. The large front flippers are used for propulsion underwater. Otariids live mainly in temperate and subtropical regions. Males are usually much larger than females.

The Phocidae or 'true seals' have no external ears; their pelvis cannot rotate so they are only able to move with difficulty on land (or ice), and their small front flippers are used more for steering than propulsion which instead is performed by movement of the tail flipper. They have only thin fur which can become wetted, and so they rely upon their blubber for insulation. They are more streamlined than otariids and tend to be excellent swimmers and divers. With the exception of monk seals, most live in temperate to polar regions. Males and females are usually of similar size (with the notable exception of the northern and southern elephant seals).

The Odobenidae comprises a single species, the walrus (subdivided geographically into three subspecies). It is immediately recognisable by its large tusks which are extruded down from the jaws, its large front flippers and tail flippers, thick blubber, and almost naked, rather loose skin. Like otariids, it can rotate its pelvis and move (though not very easily) on land/ice using its tail flippers. Both these and the front flippers can be used for underwater propulsion. Like phocids, it has no external ears. Walruses have a circumpolar distribution, and males are larger than females.

For many people, marine mammals comprise the orders Cetacea, and Sirenia, and sub-order, Pinnipedia. However, two other carnivore families have representatives that are dependent in some way upon the sea. The polar bear is the largest member of the family Ursidae (reaching a length of 2.5 m). It has a distinctive white pelage, small ears and head, long neck and a more streamlined body than its relatives, lacking a shoulder hump. It propels itself through water with its forelimbs, letting its hindlimbs drag behind. It inhabits circumpolar regions of the northern hemisphere, and is often associated with ice from which it may hunt, and carve out a den.

Finally, there are two species of mustelid, the sea otter living beside the North Pacific, and the marine otter of South America. The sea otter (1.5 m long and 45 kg in weight) is much the larger of the two. Like other otters, sea otters have a small head and small ears, a streamlined body for swimming and a horizontally flattened rudder-like tail. Characteristic features include extremely dense underfur with sparse guard hairs, a loose flap of skin below the chin that can be used to store food or tools; posterior cheek teeth lacking cutting edges; retractile claws on the front feet only; and flattened hind feet for propulsion in the water. They also lack any functional anal glands which in other mustelids are used extensively for scent marking.

Our first three chapters explore the nature of the above adaptive features of marine mammals in terms of their life history and ecology. In Chapter 1, Peter Evans and Ian Stirling review life history strategies of marine mammals, with special emphasis upon cetaceans and pinnipeds. They point not just to many of the specialisations for a mammal imposed by an aquatic existence, but also to some of the parallels that exist with terrestrial mammals and other taxa by application of general evolutionary theory. In the second chapter, Arne Bjørge examines habitat use, home range and site fidelity in marine mammals by posing the question "How persistent are marine mammal habitats in an ocean of variability?" He argues that an understanding of the influences of spatial and temporal variability is extremely important if humans are to properly manage habitats upon they so often have an impact. Then, in Chapter 3, Christina Lockyer considers reproduction in marine mammals from an ecological perspective. She demonstrates how reproductive habits and behaviour are influenced by environment. Today, those effects are accelerated by human influences - chemical contamination, prey depletion, disturbance, and, indirectly, through climate change. Although several examples are given of the different patterns of reproduction exhibited by marine mammals, the concluding message is that our knowledge remains too dependent upon a limited number of detailed studies drawn from a few species.

REFERENCES

Gatesy, J., Milinkovitch, M.C., Waddell, V., and Stanhope, M. (1999) Stability of Cladistic Relationships between Cetacea and Higher-Level Artiodactyl Taxa. *Systematic Biology*, **48**, 6-20.

Graur, D. and Higgins, D.G. (1994) Molecular evidence for the inclusion of cetaceans within the order Artiodactyla. *Molecular Biology and Evolution*, **11**, 357-364.

Heyning, J.E. (1999) Whale Origins - Conquering the Seas. *Science*, **283**, 943.

Heyning, J.E. and Perrin, W.F. (1994) Evidence for two species of common dolphins (genus *Delphinus*) from the eastern North Pacific. *Contributions in Science, Natural History Museum of Los Angeles County, CA*, **442**, 1-35.

Jefferson, T., Leatherwood, S., and Webber, M.A. (1993) *Marine Mammals of the World*. FAO Species Identification Guide. Food and Agriculture Organisation of the United Nations, Rome.

Milinkovitch, M.C. (1997) The Phylogeny of Whales: a Molecular Approach. In: *Molecular Genetics of Marine Mammals*, (Ed. by A.E. Dizon, S.J. Chivers, and W.F. Perrin), pp. 317-338. Special Publication Number 3; The Society for Marine Mammology.

Reynolds, J.E. III, and Odell, D.K. (1991) *Manatees and Dugongs*. Facts on File, New York, NY.

Reynolds, J.E. III, and Rommel, S.A. (Eds.) (1999) *Biology of Marine Mammals*. Smithsonian Institution Press, Washington DC. 578pp.

Rice, D.W. (1998) Marine mammals of the world: systematics and distribution. *Society Marine Mammology Special Publication*, **4**, 231pp.

Ridgway, S.H. and Harrison, R.J. (Eds.) (1981a) *Handbook of Marine Mammals, Vol. 1: The Walrus, Sea Lions, Fur Seals, and Sea Otter*. Academic Press, London & New York.

Ridgway, S.H. and Harrison, R.J. (Eds.) (1981b) *Handbook of Marine Mammals, Vol. 2: Seals.* Academic Press, London & New York.

Ridgway, S.H. and Harrison, R.J. (Eds.) (1985) *Handbook of Marine Mammals, Vol. 3: The Sirenians and Baleen Whales.* Academic Press, London & New York.

Ridgway, S.H. and Harrison, R.J. (Eds.) (1989) *Handbook of Marine Mammals, Vol. 4: River Dolphins and the Larger Toothed Whales.* Academic Press, London & New York.

Ridgway, S.H. and Harrison, R.J. (Eds.) (1994) *Handbook of Marine Mammals, Vol. 5: The First Book of Dolphins.* Academic Press, London & New York.

Ridgway, S.H. and Harrison, R.J. (Eds.) (1999) *Handbook of Marine Mammals, Vol. 6: The Second Book of Dolphins.* Academic Press, London & New York.

Riedman, M. (1990) *The Pinnipeds: Seals, Sea Lions, and Walruses.* University of California Press, Berkeley, CA.

Stirling, I. (1988) *Polar Bears.* University of Michigan Press, Ann Arbor, MI.

Chapter 1

Life History Strategies of Marine Mammals

[1]PETER G.H. EVANS and [2]IAN STIRLING
[1]Department of Zoology, University of Oxford, South Parks Road, Oxford OX1 3PS, UK.
E-mail: peter.evans@zoo.ox.ac.uk;
[2]Canadian Wildlife Service, 5320 122 St., Edmonton, AB, T6H 3S5, Canada.
E-mail: ian.stirling@ec.gc.ca

1.   INTRODUCTION

The evolution of the life history strategies of marine mammals comprise a remarkable suite of adaptations to the particular set of environmental pressures that these mammals face, living as they do largely in an aquatic medium.

An underlying constraint faced by cetaceans, pinnipeds, sirenians and sea otters appears to be the inability of rearing more than a single young at a time. All species of marine mammal (except the polar bear) normally give birth to a single offspring even if twin foetuses or blastocysts may occasionally occur. Presumably single births are necessary because a mother is unable to produce sufficient rich milk to nourish more than one offspring to a size sufficient to have a high enough probability of postweaning survival. As a result, reproductive rates are of necessity low. To increase survivorship of newborns, cetaceans, otariids, odobenids, polar bears and sea otters, invest in a variable, but generally protracted, amount of maternal care. The phocids by contrast, have a strategy of compensating for a short period of maternal care by transferring a large amount of fat to the newborn pup through the milk of the mother which enables it to survive for a protracted period while it learns to be self-sufficient.

After intensive nurturing for an extended period in the womb (usually 8-16 months), the young of marine mammals are suckled over several months

Marine Mammals: Biology and Conservation, edited by
Evans and Raga, Kluwer Academic/Plenum Publishers, 2002

and, in some cases, for up to 2-3 years, depending on environmental circumstances. The long period of maternal care is followed by slow physical and sexual maturation which further limits the frequency at which marine mammals can reproduce. However, low reproductive rates are offset by high survivorship and hence relatively long life spans. If adult survival is markedly reduced, as can occur as a direct consequence of human exploitation or accidental capture in fishing gear, populations may take a long time to recover, even if various life history parameters change in a density-dependent manner. This has important implications on the management and conservation of marine mammal populations. These various life history traits will be explored further below. In general, the breeding biology of land-breeding pinnipeds, sea otters, or polar bears is easier to study than that of most other marine mammals and so we have more detailed knowledge of members of these groups.

2. NUMBER OF OFFSPRING

All species of whales and dolphins, sirenians, and sea otters, without exception, typically produce a single young: they differ from most other placental mammals which have litters of more than one young. Although female cetaceans possess paired mammary glands, twin foetuses have only very rarely been recorded (0.6 per cent in humpback to 2.3 per cent in sei whales - Gambell, 1968; Kimura, 1957; Ohsumi, 1977). Almost invariably one of these apparently dies.

Twin births in pinnipeds are also extremely rare (Spotte, 1982), although histological or tissue examination of female reproductive tracts have revealed the existence of twin foetuses or blastocysts in thirteen species representing all the major families (King, 1983); live births of twins have also been occasionally witnessed (see, for example, Peterson and Reeder, 1966; Bell, 1979), and unborn triplets in a ringed seal have even been reported (Kumlien, 1879). Unlike cetaceans, there have been a few cases of twin seal pups surviving to weaning (Bester and Kerley, 1983; Doidge, 1987) although their subsequent fate was unknown. However, despite such exceptions, both female seals and cetaceans typically are limited to the production of a single offspring at any one time, and this automatically limits the intrinsic rate of increase possible for a marine mammal population.

Sirenians also usually bear single offspring although twins have frequently been confirmed for both captive and free-living Florida manatees (4% recorded in one carcass study - Marmontel, 1995), and twin foetuses have been occasionally reported in the dugong (Boyd *et al.*, 1999; Marmontel, 1995; Odell *et al.*, 1995; Rathbun *et al.*, 1995).

In contrast, polar bears commonly have litters of 1-3 cubs, with two being most common and up to four having been recorded rarely (Ramsay and Stirling, 1988).

3.   GESTATION & DELAYED IMPLANTATION

The chronology of reproductive events, in the interval between fertilisation and birth is variable between different groups. Most species of seals (plus sea otters and polar bears) have evolved a mechanism called delayed implantation where, after a female mates and her egg is fertilised, the embryo stops growing for a period at the early blastocyst stage and remains in a state of suspended animation sometimes for several months, a state defined by Sinha *et al.* (1966) as unimplanted pregnancy. Growth of the embryo, commonly referred to as the period of gestation, is the period between when the fertilised egg implants in the uterine wall and begins full foetal development through to the time it is born, a state defined by Sinha *et al.* (1966) as implanted pregnancy.

Delayed implantation enables seals and polar bears to give birth when environmental conditions are favourable, and facilitates the synchrony of both pupping and mating within a restricted period at a particular time each year. This is particularly important for several species of seals that aggregate in large concentrations to give birth to their young in a relatively short period of time before returning to the sea and becoming widely dispersed through much of the rest of the year. Some of the more highly sexually dimorphic species such as northern elephant seals and Steller sea lions may also feed in different areas during at least part of the time they are at sea. Although sea otters also exhibit delayed implantation (Kenyon, 1981), pupping rates are constant throughout the year in California (Riedman *et al.*, 1994). Pups are also born throughout the year in Alaska, although most are born in spring and summer (Kenyon, 1981).

The length of the active gestation period partly determines the duration of the inactive phase when the embryo remains unattached to the uterine wall. In most seals that give birth annually, the total gestation period is roughly 10.5-11.8 months. Otariid seal species (comprising the fur seals and sea lions) that breed annually have a longer period between mating and birth of the pup - consistently around 11.8 months, since mating usually takes place one week or so after giving birth, while the period of actual foetal growth is generally between 7.8 and 8.5 months, following a delay in implantation of about 3.5-4.0 months (see review in Riedman, 1990). Phocid seals, which have a relatively short period of pup dependency (4-40+ days), have a gestation period of about 11 months in most species; the females of most species mate at the end of the lactation period and then give birth approximately one year later. In ringed and Weddell seals, however, which both have a weaning time of about 6 weeks, mating takes place before the pup is weaned (Smith, 1987; Kaufman *et al.*, 1975). There is a tendency for those species with longer periods of pup dependency to have shorter gestation periods (9-10 months) and those with shorter lactation periods to have a gestation of 11.3-11.5 months. The delay in implantation ranges from 1 to 4 months, averaging about 2.5 months (see review in Riedman, 1990).

Walruses have the longest total gestation period of any seal - 15-16 months, with a delay in implantation of 4-5 months. The walrus calf is nursed for two or more years (the mother does not mate again for over a year after giving birth), which results in the longer interval between pregnancies in that species (Fay, 1982).

The gestation period of manatees is not accurately known, but in both captive and free-ranging Florida manatees, it is estimated at 12-14 months (Odell *et al.*, 1995; Rathbun *et al.*, 1995), and in dugongs at 13.9 months, though this is based upon a small sample size (Marsh, 1995).

In the case of the polar bear, mating takes place in about April-May and implantation follows about mid-September to mid-October, with birth of the cubs about two months later, at which time the females are ensconced in maternity dens, usually in snow (Derocher *et al.*, 1992). The young, about 0.6 kg at birth, are highly altricial so the den functions as an external womb to provide protection from the environment until they are large enough to be able to follow their mothers onto the sea ice (Ramsay and Dunbrach, 1986).

Cetaceans do not show delayed implantation although, like some seals, their gestation period frequently is confined to a period of 10-12 months, and is linked to specific seasons. Baleen whales commonly seasonally migrate over long distances from equatorial breeding grounds to cold temperate and polar feeding grounds (Mackintosh, 1965; Lockyer, 1984, this volume). Their gestation periods range from 10 to 13 months (mean = 11.6, median = 12, mode = 12 months) (Fig. 1). Toothed whales and dolphins are more variable in the length of the gestation period (10-17 months), and this appears to depend upon their diet. Those with predominantly cephalopod diets (sperm whales, beaked whales, pilot whales & Risso's dolphin) have longer periods of gestation - ranging from 11 to 17 months (mean = 14.0, median = 14, mode = 12-16 months). Those with a predominantly fish diet (common dolphin, striped dolphin, bottlenose dolphin, *Lagenorhynchus* spp., phocoenids, boto, baiji, tucuxi, and franciscana) have gestation periods ranging from 10 to 12 months (mean = 10.9, median = 11, mode = 10 & 11 months) (Fig. 1). Other species (killer & false killer whales, melon-headed whale, beluga, narwhal, spotted & spinner dolphins) where neither fish nor cephalopods predominate in the diet were omitted from the analysis.

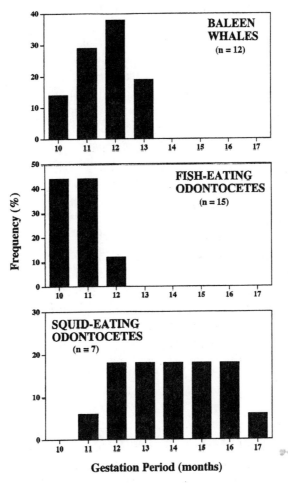

Figure 1. Gestation Periods of Cetaceans (data derived from Appendix; some species have estimates of gestation period that range across months).

Cephalopods are known generally to have lower energy contents than most fish and plankton (see references in Evans 1987, 1990). The calorific values of plankton such as euphausiids (krill), and many fish are around 4-10 kJoules per gram body weight, whereas the corresponding values for squid are about 3-3.5 kJoules per gram body weight. Furthermore, cephalopods may be more difficult to capture than either plankton or most species of fish. Diet may therefore have an important effect on the rate at which the foetus can grow in the embyro, thus lengthening the gestation period for those species nourished on a lower energy diet.

Although cetaceans do not undergo delayed implantation, the growth of the embryo is often relatively slow during the earlier developmental stages, followed by a rapid acceleration in growth towards the end of gestation. This is most clearly exemplified by the baleen whales. In the case of the blue whale, for example, the foetus in the latter stages of development grows at an average rate of 34 kg per day, faster than any other mammal known. Despite such rapid growth in whales leading to the birth of large offspring, in relative terms they remain small compared with other mammals (see Read & Harvey, 1989). Seals, on the other hand, are born at a larger relative size than most other mammals. Presumably the need for the neonate whale to be precocial and as large as possible to cope with the austere environment of the sea is counterbalanced by the time constraint upon the amount of growth possible by a whale foetus within the seasonal cycle.

Before leaving this topic, we should note the uncertainties that inevitably surround estimations of both the gestation period and neonatal size. For many marine mammal species, these are necessarily crude, based in the case of the former upon estimates of seasonal changes in foetal body lengths, and in the case of the latter, upon comparisons between the size of the largest foetus and smallest free-swimming neonate. The resulting estimates furthermore may be biased if data cover only short periods which include breeding and/or calving seasons, or if the breeding season is long compared with the gestation time (Kasuya, 1972, 1995; Martin and Rothery, 1993; Sergeant, 1962). Precise gestation periods are known only for a few species in captivity where pregnancy has been determined by progesterone analysis (Asper *et al.*, 1992).

4.   LACTATION & WEANING

When the young are born they must grow quickly to overcome the unfavourable energy balance caused by being small in a relatively cold aquatic environment. All marine mammals produce milk that is extremely rich, ranging from 30-60 percent fat. In seals, the fat content is highest (40-50%) in phocids which have a short lactation period (lasting between 4 days and 1.5 months), and lower (20-35%) in otariids with a longer lactation period ranging from 4 months to as long as 3 years, and in walruses (c. 30%) with a lactation period of 2-3 years (see review in Riedman, 1990, and case studies in Trillmich and Ono, 1991). In cetaceans, this milk may yield between 15,500 kJoules per kilogram body weight (in the beluga) and 18,000 kJoules per kilogram body weight (in large rorqual whales) (Gaskin, 1982). Marine mammal milk also contains relatively large amounts of protein (5-15 per cent or more), which is important for rapid tissue growth. By comparison, human and cow's milk contain only about 2-4 percent fat and 1-3 percent protein (Riedman, 1990). On the other hand, the milk of seals and cetaceans contains little or no lactose, or sugar, whereas that of terrestrial mammals typically contains 3-5 percent sugar, and human milk has a sugar content of 6-8 percent (see review in Riedman, 1990). In polar bears, the fat content of the mothers' milk declines from about 36% when the females first leave their maternity dens in spring to about 21%, 18 months later, when accompanied by yearlings (Derocher *et al.*, 1993).

The variation in lactation periods between different seal taxa reflects the different maternal strategies imposed by their respective ecologies (Bonner, 1984; Oftedal *et al.*, 1987). Female phocids have short lactation periods during which the pups ingest only milk, which is particularly rich in fat. The content of fat in those species that fast, such as the northern elephant seal, increases from 12 percent at birth to a plateau of 55 percent during the last week of lactation, thus enabling the pups to grow rapidly (Riedman and Ortiz, 1979). The mothers fast and remain with the pups throughout this period, and weaning occurs abruptly. In contrast, otariid seals have a longer period of nursing, leaving their pup to feed on some solid food during the latter stages of lactation, and suckling with milk of lower fat content. Their pups grow more slowly and weaning is more prolonged. Walrus pups continue to be nursed for one or two years, or longer, after reaching nutritional independence, accompanying their mothers on feeding trips. Both otariid seals and walruses have relatively small-sized pups compared with phocids which are also more precocial.

Amongst phocid seals, those that breed on unstable pack ice have the shortest and most intensive lactation periods of all. The hooded seal has a lactation period of 4 days, the shortest known for any mammal (Bowen *et al.*, 1985, 1987). Harp seals have nursing periods of 10-12 days (Kovacs and Lavigne, 1986), and bearded seals 24 days (Lydersen, 1998). The eight species of pinnipeds that breed either on land or on landfast ice have long lactation periods by comparison - ranging from about three to six weeks (see review in Riedman, 1990).

Like seals, cetaceans show great variation in the lactation period, both between species and, within species, between individuals/populations. Baleen whales generally have the shortest lactation periods, ranging from 4 to 12 months (mean = 7.7 months, median = 7 months, mode = 6 months) (Fig. 2). Odontocetes, on the other hand, possess lactation periods typically varying between 6 months and 3-4 years (mean = 18.8 months, median = 19 months, mode = 18 months). The precise length is rarely known since the presence of a suckling juvenile of a particular age does not necessarily infer that its mother has been suckling continuously over the entire period. Thus long-finned and short-finned pilot whales aged 12-14 years have been found with lactose in their stomachs, indicating at least some intermittent suckling, but these could involve either allo-mothers (*i.e.* adult females other than the parent who may engage in some maternal care), or the real mother whilst suckling a new offspring. In contrast, adoptive nursing is rare in otariid seals, but does occur to some degree in phocids such as the Hawaiian monk seal, the grey seal, and the Weddell seal (Stirling, 1975a; Boness, 1990).

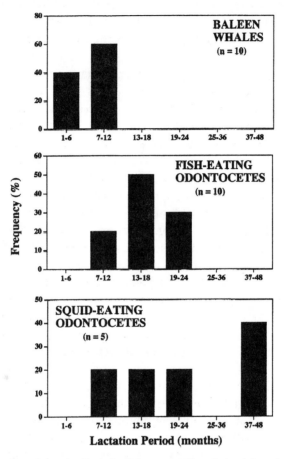

Figure 2.   Lactation Periods of Cetaceans (data derived from Appendix).

As with the gestation periods of cetaceans, it appears that those odontocetes feeding primarily upon cephalopods have longer lactation periods (mean = 27.0 months, median = 24 months, mode = no particular month) compared with those feeding predominantly upon fish (mean = 12.8 months, median = 13 months, mode = 18 months) (Fig. 2).

The young of manatees and dugongs remain with their mother for at least 12 months, and in some cases (notably the calves of young females) up to 2 years before weaning, although calves begin eating sea grasses shortly after birth (manatees - Rathbun *et al.*, 1995; Reid *et al.*, 1995; dugong - Marsh *et al.*, 1984; Marsh, 1995).

Sea otters nurse their young for approximately six months but the time to weaning can vary between four and nine months (Riedman *et al.*, 1994). In most parts of the Arctic, polar bears nurse their young for about 2.5 years, though in Western Hudson Bay up to 40% of females in some years may successfully wean their cubs at only 1.5 years of age (Ramsay and Stirling, 1988).

As with mammals in general, the period of lactation appears to be much more flexible within a species than the gestation period. On the other hand, it is much more difficult to estimate the lactation period and some of the variation observed may relate to this. Three methods are generally used. The first and most common method takes the gestation period of a species and multiplies this by the proportion of the sample that is lactating divided by the proportion that is pregnant (including some which may be both pregnant and lactating) (Perrin & Reilly, 1984). This is based on the assumption that the length of time spent in lactation will be directly proportional to the proportion of mature lactating females sampled. There are various potential biases related mainly to sampling (but also to the possibility of overlooking animals in early pregnancy). If, for example, there is some segregation of animals of particular status this could give a misleading overall estimate.

The second method estimates the age at weaning by calculating the age of the assumed oldest suckling calf in a sample containing a lactating female (Perrin & Reilly, 1984). This method has various biases associated with variation in growth rate of different calves and the fact that the calf that is suckling for the longest time may continue to do so for an unknown length of time. Furthermore, the average age of weaning will not necessarily tell one the lactation period, because it takes no account of the higher chance of some lactating females having lost their calves as the lactation period progresses. It is also the case that many marine mammal species have a prolonged weaning period, often taking some solid foods within a short time of birth, and intermittent suckling may continue for an extended period that can last for years.

The final method determines the presence of lactose (a constituent of milk) in the stomach of a juvenile by a direct chemical test (Perrin & Reilly, 1984). This also estimates the age at weaning rather than the lactation period and may be affected by bias if there is differential mortality between young that continue suckling for a long time compared with those that are weaned earlier. As with the previous method, some calves may continue suckling intermittently after the main shift to solid food, and this may involve mothers that are now suckling another, younger, calf of their own.

5.   GROWTH RATES

The foetus in baleen whales grows faster than in any other mammal, accelerating particularly in the latter stages of pregnancy (see Laws, 1959a, b; Mikhalev, 1980; Lockyer, 1981). This accelerated growth generally coincides with arrival on the feeding grounds. Rapid growth follows birth and continues for several months, nourished by the mother's fat-rich milk. During those first few months, the suckled young may increase its weight by five to eight times.

Small and medium-sized odontocetes (with gestation periods of one year or less) show similarly rapid linear foetal growth rates, whereas a few medium or large-sized toothed whales (sperm whale, short-finned pilot whale, and killer whale) have much lower foetal growth rates associated with relatively long gestation periods (exceeding one year) (Kasuya, 1995). Kasuya (1995) considered that the main distinction between the two groups related to the seasonal constraints in food availability imposed upon the former compared with the latter group. A greater sample size is needed to test this hypothesis further, and it would be interesting to see whether variation in foetal growth rates is better explained by seasonality of food resources or the type of food resource itself and its calorific content.

Although growth rates may not be as fast as for baleen whales, odontocete neonates show rapid growth whilst suckling in the first year of life, continuing though at a slower rate until a plateau is reached more or less around the period when sexual maturity is reached (Perrin and Reilly, 1984; Best *et al.*, 1984; Read *et al.*, 1993; Stewart, 1994; Duffield *et al.*, 1995; Hohn *et al.*, 1996). As with other breeding parameters, following birth, juvenile growth rates are lower in those odontocetes with a predominantly cephalopod diet compared with those feeding primarily upon fish (Evans, 1987).

Seals likewise grow rapidly until weaning starts. In phocid seals, a pup may gain an average of 23 percent of its body weight per day (Bonner, 1984). Hooded seal pups, with a mean suckling period of only 4 days, almost double their weight over the first four days, from a birth weight of 22 kilograms to 42.6 kilograms (Bowen *et al.*, 1985). During the 10- to 12-day lactation period of the harp seal, a pup may more than triple its birth weight (from 10.8 to 34.4 kilograms, a mean rate of 2.5 kg/day) (Stewart and Lavigne, 1980). Northern elephant seal pups quadruple their birth weight, from an average of 34 kilograms at birth to 136 kilograms at weaning during the one-month lactation period (Le Boeuf and Ortiz, 1977). Female phocids fast or feed very little during lactation, living instead from their own fat reserves. Over the lactation period, whilst the fat content of the mother's milk increases, the water content declines, allowing the mother to conserve water during fasting (Riedman and Ortiz, 1979). Following weaning, the pups of some phocids, such as elephant seals, may fast for up to three months, during which time they can lose a considerable amount of weight while surviving off their fat stores. The enormous transfer of energy from mother to pup is exemplified in the northern elephant seal, where the mother can transfer 138 kilograms of milk (equivalent to 604,000 kJoules) during the four-week nursing period, with a mean rate of energy transfer equivalent to 21,600 kJoules per day (five times the metabolic requirements of the pup) (Ortiz *et al.*, 1984). Costa *et al.* (1986) has shown that by the end of lactation, a mother has lost 42 percent of her initial body weight, while her pup has gained about 55 percent of the mass lost by its mother. Most of the female's weight loss is fat rather than muscle tissue, with milk production accounting for 60 percent of her energy expenditure during the 26.5 day lactation period (Riedman, 1990).

Figure 3.   The southern right whale nurses its calf for 6-7 months but the two may remain together for much longer. (Photo: B. Würsig)

Female otariid seals generally feed while nursing (fasting for no more than one or two days at a time, except for a period of several days immediately following birth), and they have a relatively protracted lactation period during which the pups gain weight slowly. The pup may also begin to eat solid food during the latter part of the nursing period and are therefore weaned more gradually than phocid pups. Otariid pups gain weight at a rate of between 45 grams per day (in the New Zealand fur seal - Mattlin, 1981, 1987) to 76-90 grams per day (in the Antarctic fur seal - Doidge *et al.*, 1984). During the first month of life, a California sea lion pup has an average daily energy intake of 2,296 kJoules (Oftedal *et al.*, 1985, 1987).

The relatively long period of maternal care (Fig. 3) provided by most marine mammals allows the young to grow rapidly and thus quickly attain the large absolute size that is essential for survival in the generally cold environment in which most species live for at least part of their lives. Differences in maternal strategies between groups of species reflect patterns in space and time of food availability, and maybe to some extent to exposure to predation.

6.   SEXUAL MATURITY

Age at maturity has been variously defined as the age at which a male or female is physiologically capable of breeding (physiological maturity) and the age at which it is actually successful (social maturity) - in the case of the female, this is taken as equivalent to the age of first birth. In the marine mammal literature, the former definition has generally been used (see IWC reproduction workshop - Perrin *et al.*, 1984). Sexual maturity in females is estimated by the presence of evidence for at least one ovulation (with or without pregnancy) - either a corpus luteum or a corpus albicans. The corpus luteum is the endocrine gland which develops from the cellular components of the ovarian follicle after ovulation (Harrison & Weir, 1977); while the corpus albicans is a regressing or regressed corpus luteum (whether that corpus luteum is from pregnancy or not). In the event that pregnancy does not follow ovulation, no corpus albicans forms. It is much more difficult to estimate sexual maturity in males. The increased weight of the testis tends to be used as the criterion for sexual maturity for practical reasons, but, ideally, histological evidence (spermatogenesis and the presence of sperm), and measurements of the diameter of the testis tubule, should also be collected (see, for example, studies by Hohn *et al.*, 1985 of spotted dolphin and by Desportes *et al.*, 1993 of long-finned pilot whale).

Marine mammals typically show sexual bimaturism, where one sex matures sooner than the other (see Appendix). In most odontocetes and all seal species, females reach sexual maturity more quickly than males. In baleen whales, there is little difference between the sexes, and where a difference has been detected, it is males that reach sexual maturity earlier.

The breeding structure of a species influences the difference in age at maturity between males and females. In highly polygynous species (such as all the otariid seals, the walrus, certain phocids such as elephant seals and grey seals, and several odontocete species particularly sperm whale, pilot whales, and killer whale) where males compete directly with one another for females, males tend to delay maturity, grow larger, and gain more experience before attempting to compete effectively for mates (Evans, 1987; Riedman, 1990; see also Chapters 5-6).

On the other hand, in species with promiscuous mating where males do not control access to females, one expects the opposite trend. Because females gain fecundity with size at a higher rate than males, one expects the males to be smaller and younger at maturity than the females (see Bell, 1980; Stearns, 1992). In pair-forming species, there is little or no competition among males for mates following pair-formation (Wiley, 1974). An exception to this generalization is the Antarctic ice-breeding phocids, in which the females are larger than the males, including the Weddell seal which is strongly polygynous and defends underwater territories adjacent to breathing holes (Stirling, 1983).

Most pinniped males do not become "socially mature" until several years (usually at least 3-4 years) after reaching sexual maturity. Although a young male may be capable of breeding at a certain age, he is rarely able to copulate successfully with a female or compete effectively with the dominant or territorial bull until he is older. Grey seal males typically reach sexual maturity at 6 years, but most do not become active breeders until 12-18 years of age. Walruses generally reach sexual maturity at 9-10 years, but most males probably do not actually begin to mate until 13-16 years (Fay, 1982). Adolescent five-year old northern elephant seal males are excluded from mating activities until they are at least 4-5 year older (Le Boeuf, 1974). The same applies to some cetaceans. For example, puberty in male sperm whales begins at 7-11 years (at a body length of 8.7-10.3 metres) and testis growth continues until a body length of 13.7 to 14.0 metres is reached, at about 25-28 years (Best *et al.*, 1984). At this point, testes begin to grow more rapidly, reaching their maximum weight of 8-18 kilograms in physically mature animals. As frequently occurs with sexually dimorphic species with strongly polygynous mating systems, there is a second spurt in growth of body size which coincides with this accelerated testis growth, and at least in some cases, the acquisition of a harem of females (Best *et al.*, 1984).

Age at maturity is pivotal to the evolution of life history strategies, for fitness is often more sensitive to changes in this trait than to changes in any other (Stearns, 1992). With maturation, selection pressures and trade-offs change dramatically. The demographic pressure to mature early must be balanced by trade-offs with other fitness components to explain delayed maturity. Species that delay maturity tend to be large, long-lived and to have a few, large offspring. An optimal age and size at maturity are attained where the benefits and costs of maturation at different ages and sizes balance at a stable equilibrium point or along a reaction norm.

The costs of earlier maturation are the benefits of later maturation. The principal benefit to early maturation is demographic. Simply because they spend less time as juveniles, early maturing individuals have a higher probability of surviving to maturity. Individuals that mature earlier also have higher fitness because their offspring are born earlier and start reproducing sooner (Cole, 1954; Lewontin, 1965; Hamilton, 1966). The magnitude of this benefit depends on the type of life history. It is weaker in organisms that delay maturity. Compared to other life history traits, changes in age at maturity and in juvenile survival have large impacts on fitness across a wide range of types of life histories.

Of the many factors that could act to delay maturity, the two generally important ones are that delayed maturity leads to higher initial fecundity, and, if the quality of the offspring produced or of parental care provided is higher, then delaying maturity will normally reduce the early, and usually high, mortality rate of juveniles.

When the reproductive fitness of all species of mammals, large and small are considered in a comparative sense, it is apparent that the benefits of earlier maturation include shorter generation time and a higher rate of survival to maturity because the juvenile period is shorter (Stearns, 1992). In general, the benefits of delayed maturation are higher initial fecundity because of the greater body size permitted by a longer period of growth, lower instaneous juvenile mortality rates, and higher overall lifetime fecundity. After the effects of size and phylogeny have been controlled, there is a strong pattern for mammals that delay maturity to have long lives and low fecundities. Since cetaceans and pinnipeds normally produce but one offspring, regardless of the size or age of first breeding of the female, the principal reproductive goal for a female is to grow to a large enough size before first giving birth to give her offspring as high a probability of survival as possible. In pinnipeds, females must reach 87% of adult body size before breeding (Laws, 1959a, b) so that age alone is less of a determining factor for successful reproduction than size. Thus, at high densities when resources become limiting, growth of females is slowed and the age of first reproduction is delayed. In the case of those whales without social structure, females mature later and larger than males because they continue to gain appreciable fecundity after males have grown into the region of diminishing returns. Where males compete with each other to control access to females, and become highly polygynous, such as the elephant seals, males mature later and become larger than females.

The fact that age and size at maturity can respond rapidly to natural selection accounts for the wide variation in these parameters observed within a species both between populations and within a population between individuals. Amongst seals, most females tend to give birth by a certain age (5-6 years, for example), some females may mate for the first time at 3-4 years, whilst others may postpone initial breeding until several years later (*e.g.* Carrick *et al.*, 1962). A good example is the walrus, where although most ovulate for the first time at 5-6 years, some females initially mate at 4 years, while others delay until they are as old as 12 years. The majority of males become sexually mature at 9-10 years, although a few are fertile at 7-8 years (Mansfield, 1958; Krylov, 1966; Fay, 1982). Age at sexual maturity varies between individuals in sirenians, ranging from 2-11 years in manatees (Hernandez *et al.*, 1995; Marmontel, 1995; Rathbun *et al.*, 1995; Reid *et al.*, 1995), and 9-18 years in dugongs (Marsh *et al.*, 1984; Marsh, 1986), with no obvious difference between the sexes, although male Florida manatees mature at a much earlier age than male dugongs.

Many cetacean species have similarly wide variations in the age at which sexual maturity is reached. In the gray whale, this may vary from 5-11 years in males, and 8-12 years in females (Jones *et al.*, 1984); in the fin whale this varies from 6-12 years in both males and females (Lockyer, 1984; Sigurjónsson, 1995); and in the minke whale, males are sexually mature at 3-6 years and females at 5-7 years (Lockyer, 1984; Sigurjónsson, 1995). This has been less well studied for odontocete species, but those for which extensive data exist often show similar levels of variation. Some of the variability observed (particularly within geographical areas) appears to be the consequence of density dependent factors, populations exposed at different time periods to different competitive pressures (presumably related to variation in availability of food resources) frequently reaching sexual maturity at different average ages. A good example of this is the change in average age at maturity in fin whales off Iceland from 10-12 years in the early 1950's to 7-8 years around 1980, increasing again to around 10 years by 1990 (Sigurjónsson, 1995).

In the case of polar bears, the age of first breeding also appears to be related to the overall level of biological productivity of the ecosystem and its influence on the density of the ringed seals that form their primary food (Stirling and Øritsland, 1995). In the eastern Beaufort Sea, where biological productivity appears to be lower than in areas such as Baffin Bay, most female polar bears mate for the first time at the age of five, and produce their first cubs a year later. In most other areas of the Arctic, females breed and produce cubs for the first time one year-of-age earlier.

7.   SIZE DIMORPHISM & SEX RATIOS

Sex-ratio theory predicts that a parent should vary the amount of maternal investment in relation to its offspring's sex, investing more in the sex with the highest variance in reproductive success (Trivers, 1972, 1985, Trivers and Willard, 1973, Maynard Smith, 1980, Charnov, 1982). The assumption is that those offspring that attain a greater body size will be healthier and thus likely to produce more young in their lifetime. In highly polygynous species, males show the highest variance: most produce few or no offspring whilst a few may sire very many offspring. Among polygynous and sexually dimorphic species, a female should produce a male offspring when she is in good condition since she will then be best able to provide it with sufficient maternal care for it to be successful reproductively, compared with others. If she is in poor physical condition (or is of subordinate social status), she may be better to produce female offspring since these will have less trouble finding a mate and are therefore more likely to reproduce irrespective of their size or dominance status. Evidence to support this hypothesis exists for a number of vertebrates, although there are many cases which appear to contradict it (see Clutton-Brock and Iason 1986, Armitage, 1987, for reviews; also Clapham 1996). So far as we know, no-one has demonstrated any support for this hypothesis from marine mammals, to date at least; however, it would be worth testing this experimentally.

For most polygynous and sexually dimorphic seals, male pups weigh more at birth and at weaning, and take in more milk than female pups (Riedman, 1990). They also grow faster than females during the lactation period (Kovacs and Lavigne, 1986). This difference between male and female young is most pronounced among the highly polygynous species such as the northern elephant seal and the grey seal, where selection should favour weaning the largest male possible. Northern elephant seal male pups are nursed a full day longer than female pups, milk intake by male pups is 61 percent higher than the female pups, and the males are more likely to steal milk from other mothers (Reiter *et al.*, 1978; Costa and Gentry, 1986). Grey seal mothers invest 10 percent more energy in raising male pups compared with female pups (nearly all of this occurring postpartum). More males were born to large mothers than to small mothers, and mothers of male pups were 8 kilograms heavier than those with female pups (Anderson and Fedak, 1985). In all polygynous otariids that have been studied, male pups weigh more at birth, grow at a faster rate, or ingest more milk than female pups (see review in Riedman, 1990).

For phocid seals that are not highly polygynous, the difference in maternal investment for males and females is less pronounced. Male and female harp seals, for example, are the same size at birth, grow at similar rates, and are close to the same weight at weaning (Kovacs, 1987). Amongst sirenians, there is little sexual size dimorphism, although female dugongs may grow slightly larger than males (Marsh, 1980).

There are exceptions to the sex-ratio theory amongst a few highly polygynous seals. Among California seal lions, maternal care continues longer for female than male offspring, although this may be caused by female juveniles deriving greater benefits by remaining with their mothers near the natal site where they can learn the skills of mothering and become familiar with local distribution of food resources upon which adult females depend more heavily (Francis and Heath, 1985). Early in the lactation period, females did not nurse pups of one sex longer than pups of the other sex during their time onshore (Boness *et al.*, 1985), nor did these workers find any sex bias towards females amongst juveniles sampled. However, the estimated milk intake was 19 percent higher for male pups than in female pups though not as a percentage of body weight (Oftedal *et al.*, 1987).

Maternal investment in relation to offspring sex has not been investigated in anything like the same detail in cetaceans compared with seals. Seger and Trivers (1983) found a significant excess of male foetuses in a number of baleen whale species (most marked in the humpback and blue whale). Female foetuses were slightly larger than male ones for all five species examined (as they are throughout life), and they appear to grow slightly faster in all species except the minke whale. There is also some evidence for a preponderance of males early in gestation but that this then declines to approach parity prior to birth. Humpback and blue whales with the highest excess of male foetuses also show the highest rates of differential mortality before birth.

Within a whale species, there are also differences between large and small mothers. In humpback, blue and minke whales, larger females would be predicted to produce more sons than smaller females whereas larger fin and sei whales would be more likely to produce daughters (Seger and Trivers, 1983). Sex-ratio theory interprets this finding as selection for parents equalising their investment in the production of male and female offspring. More male foetuses are conceived than female ones but these use less energy by growing more slowly, and they suffer a higher mortality so reducing the number of males during the gestation period. This should apply only if mothers that lose a foetus then have a higher chance of reproducing successfully next time. Although intuitive, this has not actually been demonstrated as yet. A similar sex-ratio puzzle, demonstrated from both grey and Weddell seals, is that while the overall sex ratio is even, there is a significant trend for males to be born earlier than females (Coulson and Hickling, 1961; Stirling, 1971).

For species that may conceive in successive years (presumably bearing a smaller reproductive cost), this reproductive compensation is less likely to be effective. We may, therefore, not expect it to operate in sei, fin and minke whales, with maximum pregnancy rates ranging between 0.6 and 0.85. In those cases, larger females, arriving in polar seas after smaller ones, have less opportunity to compensate for lost foetuses. Thus large females equalise their investment in the sexes at an effective sex ratio at conception nearer 1:1 than smaller females who equalise their investment by producing relatively more sons. On the other hand, in humpback and blue whales, with pregnancy rates around 0.5, such reproductive compensation may not apply. Larger females arrive in polar seas before smaller ones and it is they that would equalise investment in the sexes by having more sons. Wiley and Clapham (1993) recently found a correlation between maternal condition and offspring sex ratio in the humpback whale. Mature females with longer interbirth intervals (and thus likely to be in better condition having less recently incurred the costs of lactation) were significantly more likely to produce males than females. The skew towards males was among the strongest found for any mammal, and yet the overall sex ratio in the population appears to be 1:1, both overall and among different age classes (Clapham *et al.*, 1995). As yet, we have no way of explaining this apparent paradox. Presumably either there is differential mortality between the sexes early in life that compensates for the male sex-ratio bias at the foetal stage, or there is some differential dispersal as yet undetected.

One way of evaluating adaptive variation in the sex ratio of offspring is to concentrate upon its consequence (reproductive success) rather than the cause (such as maternal condition), since it is upon this that selection acts (Armitage, 1987). Examination of other taxa has indicated that females produce the sex with the higher probability of reproductive success irrespective of the costs associated with the production of that sex (Armitage, 1987). The bias towards one sex or another also varies with current social and environmental conditions.

Like seals, polygynous cetacean species show differences in growth rates between the sexes, with males growing faster than females, ultimately attaining greater body size. This is most pronounced in the sperm whale where a physically mature male is around 30-40 percent longer and some 100-140 percent heavier than the female (Best *et al.*, 1984). In most social odontocetes, in fact, males are larger than females, presumably because sexual selection favours large size in males where there is some direct competition for access to females. In those odontocetes with less social structure, such as the river dolphins and some phocoenids like the harbour porpoise, and in baleen whales, females are larger than males. Ralls (1976) speculated that reversed sexual size dimorphism might be that larger females make more efficient mothers, being better able to offset the energetic costs associated with rapid foetal growth and lactation. In delphinids, the ratio of males to females appears to decline with age (see Perrin and Reilly, 1984), consistent with the theory that mortality is greater in the larger sex (Ralls *et al.*, 1980).

8.   MATING SYSTEMS

Marine mammal mating systems have been the subject of many reviews (see for example Bartholemew, 1970; Stirling, 1975, 1983; Jouventin and Cornet, 1980; Le Boeuf, 1986 for seals; Brownell and Ralls, 1986; Evans, 1987, Aguilar and Monzon, 1990 for cetaceans; and Orians, 1969, Emlen and Oring, 1977, Vehrencamp and Bradbury, 1984, Kenagy and Trombulak, 1986, Clutton-Brock, 1989 for general reviews).

A major problem with a discussion of the evolution of mating systems in marine mammals lies in the ability to accurately determine parentage. Without genetic evidence, it is rarely possible to be certain whether males successfully inseminate one or several females. Most evidence to date has to come from direct behavioural observation, which necessarily is better for terrestrially mating seal species than cetaceans. New studies using molecular genetic techniques are now underway with many species and promise to provide exciting new insights in the near future.

For most species of mammals, females provide all parental care, nursing and protection of the young; males simply contribute their sperm. This asymmetry of investment has important consequences on the type of mating system that is likely to evolve, favouring for the most part some form of polygyny. If males are freed from the burden of parental care, they can best maximise their reproductive success by fertilising as many females as possible. Males therefore compete for females whereas, because female cetaceans and pinnipeds can only have one offspring per year, there is a significant benefit to being choosy about their mates. All seals and at least the majority of cetaceans fall into this category.

The mating systems of seals range from those species which are highly polygynous and sexually dimorphic (in which males are much larger than females) through those showing some element of promiscuity and only a small amount of sexual dimorphism, to a few which are monogamous, or at least serially monogamous, one of which (the hooded seal) shows a large degree of sexual dimorphism while in the other (the crabeater seal), females tend to be slightly larger than males overall (Laws 1958; see also review in Stirling, 1983).

The spatial and temporal distribution of breeding habitat also determines the particular mating system likely to evolve since it has a strong influence on the distribution of receptive females. The form and amount of breeding habitat available to a species may also vary throughout its range, resulting in the development of intraspecific differences in mating behaviour between populations or groups. For example, in different locations, the grey seal breeds on ice, long smooth sandy beaches, or in rocky coves, resulting in significant differences in their social organisation (*e.g.* Fogden, 1971; Anderson *et al.*, 1975; Boness and James, 1979).

All seals give birth to their young on either land or ice but they can be divided into three categories according to the habitat on which they mate: (1) those that mate on land; (2) those that mate on or under floating pack ice; and (3) those that mate beneath landfast ice. Of the 33 species of seals extant, 20 mate on land and most of the remainder mate in the water beneath ice attached to land, with the exception of the crabeater seal which appears to mate on the surface of the pack ice to reduce the risk from marine predators (Siniff *et al.*, 1979).

Figure 4.   A male crabeater seal mounting a female on the surface of the pack ice. (Photo: I. Stirling)

Eighteen of the 20 species of land-breeding seals are highly polygynous, strikingly sexually dimorphic, and mate in moderate-sized to very large colonies. In some of the sexually dimorphic species, males have developed secondary sexual characteristics such as the pendulous nose and frontal chest shield of the elephant seal or the balloon-like inflatable nasal sac of the hooded seal. Highly or moderately polygynous species that mate on land include all sea lions and fur seals, northern and southern elephant seals, and the grey seals. When oestrus females are gregarious and congregate at the same sites on land each year to give birth to their pups and to mate, it becomes possible for polygyny to evolve because males need only to control access by competitors in a two dimensional environment. In such cases, competition for access to females for mating becomes intense and it is a successful strategy for males to become larger because they have a substrate on which they can grip and push while fighting opponents. In this circumstance, they can become dominant enough to be able to restrict access to variable numbers of females (in extreme cases, 15-20 or more) during a breeding season. For species such as the northern fur seal, where females show a strong degree of fidelity to specific sites for parturition in subsequent years, a territorial social system develops in which males compete to occupy those specific terrestrial areas to which females will return to have their pups each year and subsequently mate. In species such as elephant seals, where females are also highly gregarious but may move about a certain amount on a beach to which they show fidelity, males do not defend individual territories but instead move locally to remain with the female groups in order to be present when they ovulate and mate, while excluding other males from their presence. In that circumstance, a dominance hierarchy develops. In general, the more polygynous a species is, the more sexually dimorphic they become.

Harbour seals and monk seals are coastal species that haul out on land throughout the year but mainly mate in water (Venables and Venables, 1957; Johanos *et al.*, 1994) and they appear to be promiscuous.

Most ice-breeding seals on the other hand tend to be widely distributed at low densities and do not group together in large masses and food is generally available. The exceptions to this generalisation are harp and hooded seals which aggregate in large relatively dense patches near the edge of the drifting pack. Both these species are highly migratory and travel long distances to do most of their feeding (Sergeant, 1965; Bowen *et al.*, 1987). However, even in harp or hooded seal patches, the females are sufficiently spaced that males are unable to monopolise more than one female at a time and the short breeding season that characterises the species breeding in the relatively unstable pack ice further imposes a constraint on availability of numbers of receptive females. Aquatic copulation further inhibits males from being able to control access to groups of females and so 11 of the 13 species of ice-breeding seals appear to have evolved a degree of promiscuity and, in the case of crabeater and hooded seals, what has been termed serial monogamy.

In most of the ice-breeding seals the sexes are similar in size and appearance although, in general, males tend to be slightly larger than females (except for hooded seals, in which the males are much larger than the females). Except for the walrus, all of the ice breeders are phocid seals. The two ice breeders in which moderate polygyny has evolved are the Weddell seal and the walrus, though the circumstances through which they developed polygyny appear to be different. Walruses originally occupied large terrestrial haul-outs, much as the otariids do today, and at one time were numerous and taxonomically diverse in the North Pacific, there being no less than 7 genera only 4 million years ago (Repenning, 1976). Thus, it appears that the strong degree of sociality, polygyny, and sexual dimorphism exhibited today is probably a phylogenetic legacy from an earlier time when they mated on land.

Only two species of seals, the ringed seal of the Arctic and the Weddell seal of the Antarctic, mate in the water beneath the landfast ice. Both species maintain their own breathing holes and so, theoretically, have the opportunity to share maintenance of these holes and aggregate to some degree on the ice to give milk to their young, before returning to the water to mate. Although several Weddell seals regularly haul out on the ice around the same breathing hole, this option is not open to ringed seals because it leaves the individuals too vulnerable to predation by polar bears (Stirling, 1977). If a bear gets close enough to a group of seals to charge, not all will be able to escape down the breathing hole quickly enough. In contrast, the Weddell seal which mates in the normally three dimensional aquatic environment in the Antarctic, does so under landfast ice where there are a limited number of self-maintained breathing holes. Thus, because all males must surface to breathe at these few holes in the ice near where the oestrus females will eventually enter the water to mate, territorial males are able to control access to breathing holes, exclude competitors, and maintain a polygynous mating system in much the same way their terrestrially mating counterparts do. For reasons that are not understood, despite the fact that well developed polygyny has evolved in the Weddell seal, males are still slightly smaller than females, as they are in the other antarctic ice-breeding phocids (Stirling, 1983).

For cetaceans, our limited knowledge of their behaviour during the breeding season has frequently forced us to do no more than guess at the mating system, inferred on the basis of evolutionary theory. Baleen whales are notable for the seasonal cycles of migration that they exhibit between high latitude summer feeding and low latitude winter breeding. They tend to live singly or in small loose groups, perhaps because their large overall energy demands which cannot sustain greater numbers over an extended period on the localised plankton swarms upon which they generally feed.

Figure 5. Gray whale mating behaviour. (Photo: T. Walmsley)

These plankton concentrations may be relatively stable during a summer season, and so could usefully be defended, although they might not support more than a few individuals. However, the formation of even temporary harems seems unlikely since most mating occurs when there is little feeding, and therefore few opportunities for a male to maintain a group of females in a polygynous mating system. Monogamy tends to occur where the long-term association of the male parent with the female and young is likely to increase the chance of his offspring surviving. For this to operate, food should be a relatively scarce resource which can be provided to the young by the male. In the case of baleen whales, where for several months the food is provided by the mother alone (in the form of milk), this is not likely to occur. Furthermore, the food resource needs to be sufficient to sustain both male and female (and their young) throughout the year unless site fidelity is so strong that both adults return to exactly the same area from one year to the next. If these sites are not very discrete so enabling pairs to readily come into contact with one another, then a promiscuous mating system is more likely to occur.

Observations of sexual activity (primarily made in coastal waters) suggest that mating in baleen whales generally occurs in multi-male groups.

Females of southern right, bowhead and gray whales apparently commonly copulate with more than one male, and interactions between males are generally not very aggressive. This has been interpreted as reflecting sperm competition (sperm-producing males attempting to displace or dilute the sperm of others within the female), and has been related to the large testes and long penises that those three species (particularly the right whale) possess (Brownell & Ralls, 1986). Larger testes are required to produce the large numbers of sperm needed under those conditions, and the longer penises presumably help to deliver the sperm closer to the ova. On the other hand, the mating systems of humpback whales (and maybe other balaenopterid whales and the pygmy right whale) appear not to be selected in this way. Humpback males compete primarily by monopolising females, preventing other males from copulating with them; females commonly copulate with only one male, male-male interactions are sometimes highly aggressive, and males have relatively small testes and shorter penises. Females are widely and unpredictably distributed on the breeding grounds, occurring either alone or in small unstable groups. This means that males cannot defend either resource-based territories or groups of females, and therefore compete for single rather than multiple females. Possibly to help achieve this, male humpbacks also produce a song that appears to serve as a reproductive advertisement display perhaps to attract potential mates, and is likened to a form of lekking (similar to the female-defence polygyny of male walruses in the water adjacent to the pack ice in the Arctic - Sjare and Stirling, 1996), although there is no evidence for the existence of male territories nor of any spatial structuring within singing areas (Mobley and Herman, 1985; Clapham, 1996). Other pelagic balaenopterid whales may also use loud vocalisations for reproductive advertisement. Clapham (1996), describing the mating system of the humpback, considers that it incorporates most of the elements of male dominance polygyny and all of the characteristics of a classical lek except for rigid spatial structuring. For this, he coined the new term 'floating lek' to incorporate the feature of movement of displaying males.

We have seen that males should maximise the number of mates whereas females should maximise parental care and mate choice. Female dispersion is influenced primarily by distribution of food resources whereas male dispersion is determined mainly by female dispersion. In few cases, can single cetacean males monopolise several females so that extreme harem polygyny is rarely likely to evolve. Amongst odontocetes, one species that appears to have many of the features of this mating system (for example extreme sexual size dimorphism, long-lasting social groupings) is the sperm whale. Female sperm whales and their offspring form long-lasting familial units which feed together in structured groups and temporary aggregations

(Whitehead, 1993). Large mature males (which do not seem to form coalitions) spend periods of weeks or months within a concentration, roving between the groups of females, though generally spending only a few hours with each. Differences in dispersion of female groups affect whether males remain within a group or move between groups to obtain mates. The matrilinear and polygynous social structure of sperm whales is seen also in some other medium-sized odontocetes - notably killer whale, short-finned pilot whale, and long-finned pilot whale (and possibly also false killer whale) (see Boran *et al.*, this volume; also Kasuya, 1986, 1995). In those species, males appear to bond strongly with their mothers only leaving them either during temporary fusions of different pods for mating purposes or to travel temporarily with other pods with receptive females (these two strategies promoting outbreeding). On the other hand, the typical delphinids (such as bottlenose, spotted, and striped dolphins) live in large loose social networks where males may maintain social contacts in their natal areas (strong male-male bonds often forming) but also range widely to gain access to unrelated females (see Wells *et al.*, 1987). In this way, a mating system based upon either promiscuity or slight polygny is likely to be the norm. As has been indicated for larger odontocetes like the sperm whale, paternity studies upon the bottlenose dolphins of Sarasota, Florida, have shown that males do not become fathers until they are in their mid-twenties, long after they have become sexually mature (Duffield *et al.*, 1991).

Compared with terrestrial mammals, cetaceans have larger testes relative to their body size (Aguilar and Monzon, 1992). Members of the families Delphinidae, Phocoenidae, Kogiidae, and Balaenidae have larger testes than those predicted, whilst the Ziphiidae and the Pontoporidae have relatively smaller testes than expected. Those families with proportionately larger testes may rely more intensely on sperm competition wheras for those with smaller testes, males may compete for mates through individual aggression. For some species, both systems of male competition may exist. The lack of a relationship between sexual dimorphism and relative testis size tends to argue in favour of this.

Sirenians are thought to mate promiscuously (Rathbun *et al.*, 1995). For most of the year, they are solitary, but during mating, groups of males may compete for sexual access to the female by vigorously pushing each other out of the way. Dugongs may show more intense physical competition for females than manatees (Preen, 1989), and frequently bear scars presumed to be from the tusks of other males (Preen, 1989), though in Shark Bay, Australia, they appear to lek with males defending territories in which they display to approaching females (Anderson, 1997).

Sea otters are polygynous. Males are up to 50% larger than females and defend aquatic territories which females appear to be free to move between (Kenyon, 1988; Monson *et al.*, 2000).

Polar bears are also polygynous and males are twice the size of females but they do not defend territories (Ramsay and Stirling, 1988). Since most females wean their cubs at 2.5 years of age, the most frequent interval of availability for mating is every three years. Athough the sex ratio of males to females in undisturbed situations, is even, the long intervals between breeding of females creates a functional sex ratio of animals available for mating of 3 males: 1 female, which, in practice, is likely to be higher because older males out compete younger males. Ovulation in polar bears is induced so individual adult males attempt to defend a breeding female from competitors by moving her away from areas where other males are present if possible, and by fighting if necessary, for up to about two weeks before mating is completed, after which he seeks another female with which to repeat the process. The results of an analysis of maternal investment and factors affecting offspring size suggested that female polar bears do not normally invest differentially between offspring of different sexes except with triplet litters in spring (Derocher and Stirling, 1998).

9.   MORTALITY RATES & LIFE SPANS

Marine mammals are generally characterised by low annual mortality and long life spans (Fig. 6).   Amongst mammals in general, life span is correlated with body size, larger species living longer than smaller ones (Harvey *et al.*, 1989). In marine mammals, this rule generally holds only within families.

The average life span for most seals is estimated to be from 15 to 25 years (King, 1983). Phocid seals appear to live longer than otariids. The average life span for otariids (including both wild and captive animals) is around 17-18 years, whereas for phocids it is about 25 years. The maximum life span recorded for a pinniped is 46 years for a wild grey seal (Bonner, 1971).

Sirenians are also long-lived with manatees recorded at ages of nearly 60 years and dugongs more than 70 years (Marsh, 1980; Boyd *et al.*, 1999).

Maximum longevity of male and female sea otters in the wild is about 15 and 20 years respectively (Riedman and Estes, 1990; Monson *et al.*, 2000). Male and female polar bears in the wild live to 29 and 32 years respectively while in zoos they may live up to 40 years or more (Stirling, 1988).

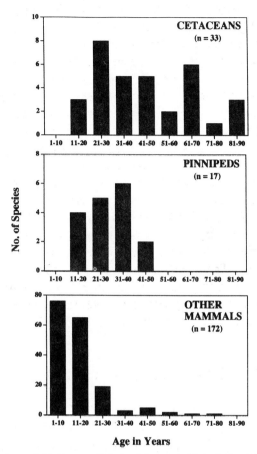

Figure 6. Maximum Life Spans of Marine Mammals (data derived from Appendix).

Data on average lifespans for cetaceans is sparse, and for most species, only the maximum lifespan has been recorded in the literature. Amongst baleen whales, there is some broad positive relationship with body size (Fig. 7) - the minke whale has a maximum recorded life span of 57 years, humpback whale 48 years, sei whale 60-70 years, fin and blue whales between 80 and 94 years, and bowheads possibly in excess of 100 years (Perrin *et al.*, 1984; Evans, 1987; Martin, 1990; George *et al.*, 1998). These reflect their corresponding annual mortality rate estimates: 0.09-0.10 for minke whale, 0.06-0.08 for sei whale, 0.04-0.05 for fin and blue whale. Amongst odontocetes, maximum life spans vary with taxonomic group: Phocoenidae - 15-24 years; Pontoporiidae - 16 years; Platanistidae & Iniidae - 28 years; Monodontidae - 25-50 years; Delphinidae - 18-90 years; Ziphiidae - 30-71 years; Kogiidae - 22 years; and Physeteridae - 65-70 years (Martin, 1980; Perrin *et al.*, 1984; Hohn *et al.*, 1989; Olesiuk *et al.*, 1990;

Bloch *et al.*, 1993; Kasuya and Tai, 1993; Heide-Jørgensen and Teilmann, 1994; Da Silva, 1994; Kasuya, 1995; Lockyer, 1995; see Appendix for details).

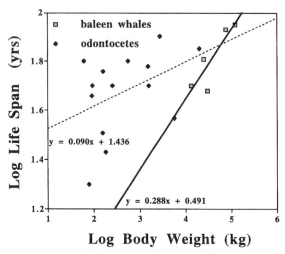

Figure 7.   Log-log Plot of Life Spans vs Body Weight for Cetaceans .

Female seals tend to live longer than males, possibly because males have to compete, often in aggressive encounters, for mating access to females (Riedman, 1990). Le Boeuf and Reiter (1988) found that about 77 percent of tagged northern elephant seals die before reaching their breeding age of 5 years. As many as 86-93 percent of males are unable to survive to 8-9 years of age, the time when males are socially mature and able to compete for females (Le Boeuf, 1981).

Males of polygynous cetacean species thought to actively compete for mate access, also show indications of some differential mortality with females surviving longer than males (Ralls *et al.*, 1980). Annual mortality rates of male sperm whales, for example, average 0.06-0.08 whereas females average 0.05-0.07. Likewise, the maximum lifespan of male killer whales is 60 years but 90 years for females (Olesiuk *et al.*, 1990). Killer whales show the typical mammalian U-shaped mortality curve: neonatal mortality is 43 percent, juvenile mortality 0.5-4.0 percent, and adult mortality 3.9 percent for males and 0.5-2.1 percent for females (Olesiuk *et al.*, 1990). In some dolphins of the genus *Stenella*, adult mortality is 4-7 per cent higher in males than in females (Kasuya, 1976), and about 4 percent higher (averaged over all age classes) in male short-finned pilot whales (Kasuya and Marsh, 1984). Amongst those baleen whales, that do not exhibit any pronounced polygyny, mortality rates do not show the same bias against males. In the fin whale, for example, average annual mortality is 0.035 for males and 0.045 for females, and it is also higher in female gray whales (0.088) than in males (0.080) (Reilly, 1984).

In mammals, juvenile mortality tends to be much higher than adult mortality. Gray whales show a much higher juvenile mortality rate (10 percent per annum) compared with adults (5.5 percent) (Reilly, 1984), and the same appears to be true for short-finned pilot whales (Kasuya and Marsh, 1984), killer whales (Olesiuk *et al.*, 1990), and bottlenose dolphins (Wells and Scott, 1990). As yet, there is little information from other cetacean species, although in some cases (*e.g.* sei whale and sperm whale), differences in mortality rates between adults and juveniles have not been detected. In the case of the short-finned pilot whale, mortality rate appears to increase after 28 years in the male, but not until 46 years in the female (Kasuya and Marsh, 1984).

Figure 8.   Harp seal pup. (Photo: T. Walmsley)

10.   BIRTH RATES & CALVING INTERVALS

The long period of maternal care exhibited by many marine mammal species results in relatively long calving intervals for individuals and correspondingly low birth rates for populations. This applies to nearly all cetaceans and to some seals, although in most seals birth rates are relatively high (80-90 percent or more).

Most female seals and sea otters give birth annually although in walruses, some fur seals, and some sea lions, the period between pupping may be three years or more, especially during periods when environmental conditions are poor (*e.g.* Trillmich and Ono, 1990). In sirenians, the breeding interval can vary greatly, ranging from 2.5-5 years in Florida manatees (Marmontel, 1995; Odell *et al.*, 1995; Rathbun *et al.*, 1995; Reid *et al.*, 1995) to 2.5-c.7 years in dugong (Marsh, 1995).

In cetaceans, only a few species are known to give birth annually (*e.g.* minke whale - Lockyer, 1984; harbour porpoise - Read and Hohn, 1995; and *Kogia* spp. - Plön, 1999), and 2-3 years is more common (see Appendix). Females of some species, such as sperm whale, killer whale, long-finned and short-finned pilot whales, may breed even less frequently, at intervals ranging from 3-9 years. Such low breeding frequency coupled with late maturation means that the total lifetime reproductive output of a female is never very high, with maxima probably typically in the order of 5-25 offspring for cetaceans, and 10-12 offspring for seals.

For a number of cetacean species, a further constraint upon the number of young a female may produce in her lifetime is that, beyond a certain age, she may cease reproducing at all. In toothed whales with a matrilineal and polygynous social structure, such as sperm whale, killer whale, short-finned and long-finned pilot whales, the pregnancy rate declines with increasing age, and conception ceases 18-27 years before the age of maximum longevity, although females may remain receptive to mating attempts (Kasuya, 1995). Even in the smaller delphinids, for example spotted dolphins, reproductive rates may decrease with age (Myrick *et al.*, 1986).

There may also be an element of reproductive suppression with females showing variability in calving intervals that could be related to dominance (Marsh and Kasuya, 1986). Reproductive senescence, though rare in cetaceans, has been found in individuals (see Myrick *et al.*, 1986), and is a common part of the life cycle in pilot whales (Marsh and Kasuya, 1984). Here, there is a shift in maternal investment from calf bearing to calf rearing, and post-reproductive females may serve an important function in taking care of young whilst the parent is feeding. Some short-finned pilot whales have a post-reproductive life span of 20-30 years (the mean is 14 years), and some of these appear to lactate for extended periods (Kasuya and Marsh, 1984), thus providing opportunities for fostering of the young of other females. This may not only increase reproductive success of female kin, but also enhance pod stability (Kasuya *et al.*, 1993). This female "terminal investment" may be timed to reduce the chance of dying before her last calf reaches maturity (which could be 20 years for a male killer whale, for example).

Fostering behaviour has been reported to occur in several species of pinnipeds but much more frequently in phocids than otariids (Stirling, 1975a; Boness *et al.*, 1992; Perry *et al.*, 1988). It is possible that fostering may help the survival of a small number of pups, especially in the case of the Hawaiian monk seal (Boness *et al.*, 1998) but in most circumstances, the nursing appears to be given to non kin (*e.g.* Boness *et al.*, 1992; Perry *et al.*, 1988) and it is unknown how such behaviour relates to the longer term survival and reproductive success of the individuals affected. There has been speculation about possible benefits of fostering including gaining of extra mothering experience for young, inexperienced females (as has been found in northern elephant seals - Riedman, 1990), and maybe help in inducing ovulation for those foster mothers that have lost their own pup (Le Boeuf *et al.*, 1972) but these remain largely untested hypotheses to date. Lunn *et al.* (2000) also confirmed genetically that a small amount of cub adoption occurs with polar bears in the wild, but its possible significance is not understood.

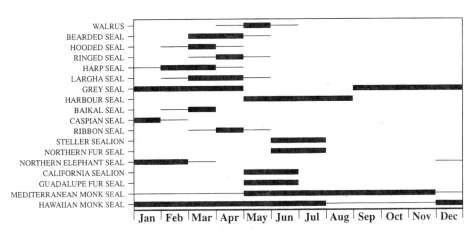

Figure 9.   Breeding Seasons of Northern Hemisphere Pinnipeds
(data derived from Appendix) .

11.   SEASONALITY OF PARTURITION

The seasons play a crucial role in determining the life cycles of most marine mammal species. Living as they do in a mainly tropical environment, manatees and dugongs are only diffusely seasonal, with any peak in calving (usually spring/early summer) tending to be linked to high river levels, peak production of algae/plants upon which they browse, or high water temperatures which are energetically more favourable to the young (Boyd *et al.*, 1999). Most seals give birth during the spring and summer months of the particular hemisphere they inhabit. For those breeding in Arctic and subarctic areas, the pupping season begins in spring when ice conditions are most favourable for rearing a pup, and food availability is increasing (Fig. 9). Some seal species, however, do not conform to this pattern. The northern elephant seal breeds in winter from December to early March, and the grey seal breeds during the autumn or winter (September to early April) depending on whether it is resident in the eastern Atlantic, western Atlantic, or the Baltic Sea. Similarly, the harbour seal on the western coast of North America pups between May and September at different locations (Bigg, 1969).   Several reasons have been put forward to account for these exceptions (Riedman, 1990), but as yet none provides a cogent explanation.

Pupping seasons can also vary in length both between and within species. Those populations and species with prolonged pupping seasons (for example monk seals and Australian sea lions) live in areas characterised by mild and less seasonal climatic variation. However, residence in a less seasonal climate does not provide a complete explanation for a prolonged pupping season. In South and Western Australia, Australian sea lions breed on many of the same islands as do New Zealand fur seals which have a very distinct spring pupping and mating season (Stirling, 1983). A similarly curious exception is the sea otter. In both California and Alaska, female otters may ovulate, breed, and give birth throughout the year (Riedman *et al.*, 1994), while seals and sea lions in the same area are highly seasonal breeders.

In the Arctic, polar bears give birth over a relatively short period in mid-winter, between about early December and early January. The females and cubs leave their maternity dens for the sea ice about late February to early March in the southern end of their range in Hudson Bay and up to a month later in more northerly latitudes in the High Arctic.

Seals with pupping seasons of intermediate length (for example harbour seals, grey seals, elephant seals, and many otariids) live in temperate regions or areas of stable ice, while those with the most compressed seasons (most of the arctic phocids) live in areas of unstable pack ice of the high latitudes. Phocids with the longest weaning periods, Weddell and ringed seals, have their pups on the stable landfast ice of high latitude polar regions (Stirling, 1983).

Seasonality of breeding in cetaceans is likewise related to climatic seasonality which influences availability of food. In baleen whales and those species breeding at high latitudes, the season of births is relatively restricted (Fig. 10a) whereas for those living mainly in lower latitudes, breeding seasons are frequently protracted (Fig. 10b). The summer months in temperate and polar regions experience an explosion of food production. In those areas, the smaller odontocete species give birth during this period of food abundance. This is particularly the case for phocoenids which seem to be strictly seasonal breeders (see, for example, Kasuya, 1978; Read and Hohn, 1995; Hohn *et al.*, 1996). On the other hand, the larger odontocetes and all baleen whales except the tropical Bryde's whale tend to give birth outside the summer period, some of these undertaking long-distance migrations from high-latitude feeding grounds to calving areas in low latitudes.

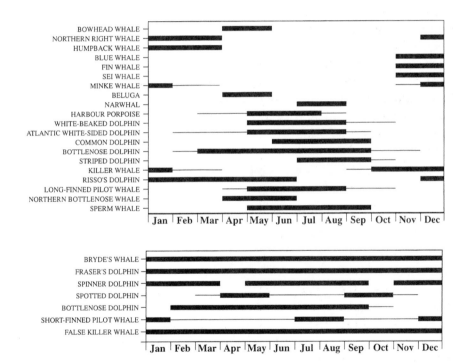

Figure 10. Breeding Seasons of cetaceans at a) high latitudes (top panel) and b) low latitudes (lower panel) (data derived from Appendix) .

Delphinids in tropical waters calve in general during any month of the year, although timing and length of the peak season may vary between populations. Spotted dolphins living offshore in the eastern tropical Pacific have a distinct calving peak between March and May in northern populations and two indistinct peaks centred around April and October respectively in southern populations (Barlow, 1984). Breeding peaks are not necessarily synchronised between species with similar ecologies and sharing the same habitat (see, for example, Barlow, 1984; Shirakihara *et al.*, 1993; Kasuya, 1995). Urian *et al.* (1996) have shown how if one considers bottlenose dolphins collectively along the US coast, it seems that they have either diffusely seasonal calving or no calving season. But if one examines the situation area by area, within each area calving is seasonal though different between areas. Furthermore, when a bottlenose dolphin is taken from the wild and placed in captivity, it retains the same ovulation/calving season as the wild population from which it came.

The causes of bimodal births such as may be found in populations of several odontocete species are also not known. Since they are found mainly, though not exclusively, in species with gestation periods exceeding 12 months, the unlinking of attachment to an annual cycle means that individuals may shift their breeding season in one of two directions, in relation to food and mate availability and their own body condition. It is clear that we still do not understand some of the factors influencing breeding seasonality.

12.   MIGRATION PATTERNS

The use of telemetry (particularly by satellite) has dramatically increased our knowledge of long-distance movements of several marine mammal species. Many pinnipeds make long-distance seasonal migrations to breeding rookeries or warm-water pupping grounds. Northern elephant seals currently hold the record for the longest annual migration of any mammal, making two long-distance migrations between breeding and moulting sites on the southern California Channel Islands and offshore foraging areas in the North Pacific, a round-trip of 18,000-21,000 km (Stewart and DeLong, 1995). For many seals, migrations are associated not only with the movement to favourable feeding grounds but also with the seasonal ice drift. The migratory movements of many arctic pinnipeds are closely linked to annual movements of the ice pack The Pacific walrus, migrates south in the autumn from the Chuckchi and northern Bering seas as the ice advances and north again in spring as the ice recedes, an annual migration that can amount to up to 3,000 kilometres. By comparison, some populations of Atlantic walruses winter in polynyas and migrate very little (Stirling, 1997) The harp seal has a similar annual migration with the seasons, with recorded movements of as much as 5,000 kilometres (Sergeant, 1965).

Most baleen whale species feeding in plankton-rich polar seas during the summer months undertake very extensive annual migrations towards the equator where they give birth in temperate to tropical waters. During the six to eight months spent away from their feeding grounds, they consume little food, living instead off the large fat reserves they have accumulated in their blubber during the previous season's feeding. Some whale species migrate very great distances (a one-way journey of as much as 8,000 km has been recorded for one individually identifiable humpback - Stone *et al.*, 1990). Timing and speed of migration varies between age and reproductive cohorts.

In the California gray whale, for example, (which can make a round trip of some 20,000 km between Alaska and the Gulf of California), the southwards migration is led by females in the late stages of pregnancy, apparently travelling singly and at a greater speed than the rest (Braham, 1984; Herzing and Mate, 1984; Swartz, 1986). Next come the recently impregnated females who have weaned their calves the previous summer. Then come immature females and adult males, and finally the immature males. These travel usually in groups of two but groups of up to eleven may occur. They tend to travel more slowly, particularly towards the end of the migration. The return migration is led by the newly pregnant females, followed by adult males and non-breeding females, then immatures of both sexes, and finally the females with their newborn calves. Presumably the newly pregnant females move north as soon as they can to make the maximum use of the time in Arctic seas, building up food reserves to nourish their developing foetus. They then move south as quickly as possible to give birth to their calves.

In nearly all other baleen whale species, newly pregnant females are the first to migrate to feeding grounds, but, unlike gray whales, they are usually the last to leave, possibly due to a need for extra energy stores in these species, for their shorter gestation periods.

The length of the migration can vary greatly within a species, both within and between populations. Juvenile blue, fin and sei whales tend not to penetrate such high latitudes as adults, whilst female minke whales are almost absent from higher latitudes of the Antarctic (Best, 1982; Lockyer, 1981; Lockyer and Brown, 1981). Humpback whales from the Gulf of Maine travel some 4,600 km to the West Indies and back (Clapham and Mattila, 1988); there, they mix with whales from, among other places, West Greenland, whose return migratory journey will be at least 2,000 km longer (Clapham, 1996).

Amongst odontocetes, only the sperm whale shows extensive latitudinal migrations, and in this case it is only the male that does so. Females and calves remain throughout the year in low latitudes but some adult and a few pubertal males migrate to cold temperate and polar seas during summer

Marine Mammals: Biology and Conservation

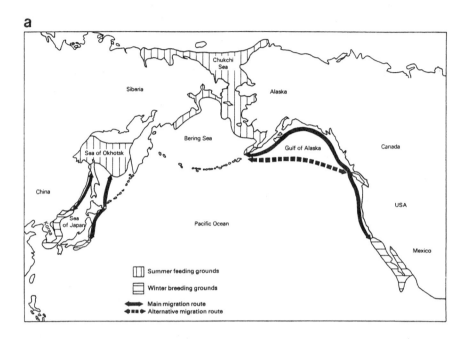

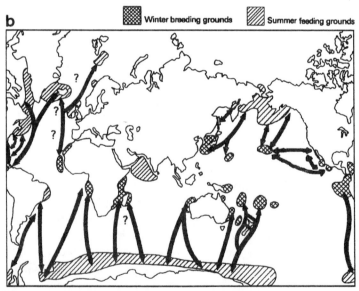

Figure 11. Migration routes for a) Gray Whale (top panel) and b) Humpback Whale (lower panel) (after Evans, 1987, with modifications; derived from Jones *et al.*, 1984, for gray whale; and from Winn & Winn, 1985; Baker *et al.*, 1986, 1990; Calambokidis *et al.*, 1996, for humpback whale).

months, a minority remaining at least at intermediate latitudes into the winter. Other odontocete species, though not making substantial latitudinal migrations, may nevertheless travel great distances. Belugas, for example, have been recorded (by satellite radio telemetry) travelling over 3,000 km (often to depths exceeding 600 metres under heavy pack ice with only small occasional leads) (Martin *et al.*, 1993; Smith and Martin, 1994), and it is likely that as further species are more closely investigated, similar great distances of movement will be recorded.

Polar bears are facultative migrants and their movements are largely determined by patterns of break-up and freeze-up of the sea ice and by individual variation in annual patterns (Garner *et al.*, 1980; Amstrup *et al.*, 2000). The size of home ranges may vary from a few hundred square kilometres to over 30,000.

13.   CONSERVATION ISSUES

Compared with many other mammals, marine mammals for a large part face few predators. Although some marine mammals are heavily preyed upon by sharks, leopard seals, and polar bears (*e.g.* Siniff and Bengtson, 1977; Stirling and Øritsland, 1995), the most notable predator is man. One of the most profound changes to an ecosytem occurred in the Southern Ocean when in a period of about one hundred years, numbers of all the great whale species (first humpback and right whales, then blue, fin, and finally sei whales) were hunted until numbers had declined so much that worldwide pressure forced whaling nations to adopt a moratium on their commercial exploitation. In this region, all rorquals tend to feed mainly upon krill, and there is considerable overlap in food preference, feeding area and timing of feeding (Nemoto, 1959, 1962). Laws (1977a, b) and Mackintosh (1973) have calculated that the overall reduction of baleen whale stocks from about 43 million tonnes in the 1900s to about 6.6 million tonnes today has made available a 'surplus' of about 153 million tonnes of krill. It has been further estimated that the total whale density in the Antarctic summer feeding area has declined from a biomass of 2.6 gram per square metre to 0.4 gram per square metre. These changes to the food web of the Southern Ocean have had important effects on a variety of biological parameters of the whales themselves, highlighting those which show greatest plasticity in the facing of changing environmental pressures.

The reduced abundance of the large blue, fin and humpback whales has had an effect upon the smaller sei and minke whales as well as upon the larger whales themselves (Gambell, 1973, 1976; Kawamura, 1978; Laws, 1977a, b; Lockyer, 1979, 1984). About double the proportion (50-55 percent) of mature non-lactating female blue whales are now pregnant compared with the rate before the 1930s, implying that the interval between successive births has been halved. Similar changes have taken place in fin and sei whales, and presently a very high proportion of minke whale females appear to be pregnant. Furthermore, declines in the mean age of sexual maturity have been observed in various species - from 10 to 6 years in fin, 11.4 to 7 years in sei, and 14 to 6 years in minke whales, although this finding has been the subject of some controversy, and estimated changes in age at first birth would be better measures than age at sexual maturity - De Master, 1984; Lockyer, 1984. Despite some areas of uncertainty surrounding those calculations, direct evidence of declines in age at maturity has been shown for fin whale stocks in the North Atlantic (Sigurjónsson 1995), North Pacific and Antarctic (Ohsumi 1986), and for minke whales in the Antarctic (Lockyer, 1984).

Despite apparent declines in the age at which various baleen whale species become sexually mature, the size at which they reach sexual maturity has remained constant, for fin and sei whales at about 90 percent maximum length. This implies an increased annual food intake of 20-25 percent to cover the extra energetic costs (Lockyer, 1978, 1981). Minke whales would require a similar increase of 27 percent to attain sexual maturity eight years earlier. Although many uncertainties exist in all these calculations, they lie within the realms of possibility given the amount of krill thought to be released by the reduction of the great whales.

Changes in pregnancy rates, individual growth rate and attainment of sexual maturity appear to correlate with the history and pattern of exploitation, being greatest in those areas with the longest history of exploitation (Lockyer, 1984). Together these findings represent strong evidence that the reduction in density of various whale species has released food to the remaining individuals of that and related species. Similar density-dependent effects have been found in the sperm whale, which showed a decline in calving interval in South African populations from 6 to 5.2 years between 1962-65, 1967, and 1973-75 (Best *et al.*, 1984); spinner dolphins in the eastern tropical Pacific, reaching sexual maturity in five instead of the usual six years in the most heavily exploited population (Perrin and Henderson, 1984; Perrin and Reilly, 1984); and striped dolphins in the western North Pacific, where pregnancy rate has increased and lactation period decreased (Kasuya, 1985). Even small-scale removals for live-capture of killer whales has been shown to increase reproductive rates of local populations (Olesiuk *et al.*, 1990).

Similar changes in life history parameters in response to exploitation have been found in several seal species. Northern fur seals (York, 1979), harp seals (Sergeant, 1973), and southern elephant seals (Carrick *et al.*,

1962) began breeding at an earlier age and their population growth rates increased after their numbers were reduced by commercial harvesting. When exploited colonies of southern elephant seals were compared with unexploited colonies at Signy Island and Macquarie Island, breeding age was found to vary greatly with the density of the colony (Carrick *et al.*, 1962). The average age of breeding bulls was 10 years (with a maximum of 20 years) at Signy Island and 6 years at Macquarie Island, whereas males reached sexual maturity at 4 years at South Georgia (Laws, 1953, 1956). At the unexploited high-density colony at Macquarie Island, the average age of primiparous females was 6 years whilst at the exploited South Georgia colony most females reached sexual maturity at age 2, and gave birth at age 3, and all 3-year-old females were found to be pregnant (Carrick *et al.*, 1962; Laws, 1953, 1956). At Macquarie Island, only 33 percent of four-year old females were pregnant, and one-quarter postponed pregnancy until they were 7 years old. Seals on Macquarie Island also grew more slowly than those on South Georgia.

An environment with abundant food resources and breeding opportunities has not yet reached carrying capacity, and its population is still growing and expanding into new areas. In those circumstances, the age at sexual maturity tends to decline, and younger individuals of both sexes may emigrate. In a saturated environment, food resources are limited and there is little suitable breeding space or few reproductive opportunities available to young animals.

Although marine mammals with their long life spans have shown strong resilience to centuries of human exploitation, very few species actually going extinct, their relatively low intrinsic rates of increase through comparatively long maturation and low reproductive rates, prevent them from rapidly responding in a compensatory manner to any marked increase in mortality. This makes them particularly vulnerable to over-exploitation by humans and to other sources of "unnatural" mortality such as incidental capture in fishing gear. Their wide dispersion may offer some protection particularly when reduced to low densities, but on the other hand, this in turn may hinder recovery rates once additional mortality has ceased. Despite a quarter of a century of protection, southern hemisphere blue whale populations show little sign of recovery.

Sea otters, like many other marine mammals, were over harvested and have since been recovering through much of their range, though the annual rates of population increase are markedly different in the California population (9%) and those in Alaska, British Columbia, and Washington (17-20%) (Riedman *et al.*, 1994). All the reasons are not fully understood but it appears that higher preweaning mortality in the California population may be the principal cause.

Polar bears were heavily harvested throughout much of their circumpolar range, especially after high prices for hides and accessibility by oversnow machines, aircraft, and ships for hunting increased in the 1960s (Prestrud and Stirling, 1994). In response, the nations responsible for polar bear conservation signed the Agreement on Conservation of Polar Bears in 1973, which provided the framework for international cooperation in research and management of this species (Stirling, 1983; Appendix I). The world population is currently estimated to number between 25-30,000 animals but the state of knowledge and status of individual populations is variable (IUCN Polar Bear Specialists Group, 1998).

Sirenians also face a variety of human pressures which impact upon their reproductive rates and survivorship. Collisions with vessels have become a serious threat leading to increased mortality in Florida manatees (Reynolds, 1995). Both dugongs and manatees may delay reproduction (both the age at first reproduction, and calving intervals) under certain conditions. Prolonged dry seasons (which can be exacerbated by large-scale deforestation) may affect food supply in the case of Amazonian manatees (Best, 1983); a major die-back of sea grasses in Torres Strait affected reproductive activity in dugongs in the mid-1970s (Marsh, 1986) whilst a large-scale loss of sea grass following two floods and a cyclone in Hervey Bay, southern Queensland in 1992 caused a marked reduction in dugong fecundity (Marsh, 1995; Boyd *et al.*, 1999). The great plasticity in life history parameters of sirenians can have positive effects in that following population reduction, increased testicular activity and pregnancy rates, and increased proportions of juveniles have been observed in populations of Florida manatees (Marmontel, 1995) and dugongs (Marsh, 1995). Manatees reach sexual maturity earlier and have a higher calving rate than dugongs so their intrinsic rates of increase are higher. Nonetheless, with their generally low reproductive rates, both groups are specially vulnerable to the widespread increasing human pressures upon their habitat, particularly if adult survival should decline.

14.   CONCLUSIONS

We have seen that most marine mammals share a number of life history traits: they mature late (often after an extended period of parental, usually maternal, care), bear only single offspring at any time, usually at infrequent intervals so that annual rates of fecundity are low; on the other hand, they have low natural mortality rates and are therefore relatively long-lived. Inter- and intra-specific variation in several life history parameters (for example the length of parental care including gestation and lactation, growth and maturation rates, natality and mortality rates) can usually be related to the availability and dispersion of important resources (food and potential mates), which in turn determine social organisation and mating system. For a number of species, the seasonality of food resources has further resulted in the evolution of annual cycles of breeding and migration.

A number of interesting life history questions remain to be answered. Why, for example, do baleen whales have no parental care and no reproductive senescence whereas the odontocete whales have parental care and strong reproductive senescence (Promislow, 1991)? Why do many baleen whale species bother to migrate to equatorial regions where they are forced to fast whilst giving birth and nursing their young (Gaskin 1982; Evans, 1987; Corkeron and Connor, 1999)? Why do various populations of seal and cetacean species give birth in different seasons often several months apart despite living under apparently similar conditions, sometimes even in the same geographical region (Riedman, 1990; Kasuya, 1995)? Similarly, why do aspects of the social organisation of some seal species vary significantly in different ecological circumstances (Stirling, 1983)? Answers to these questions are but a few of the many intriguing challenges awaiting the evolutionary biologist of the future.

REFERENCES

Aguilar, A. and Monzon, F. (1992) Interspecific variation of testis size in cetaceans: a clue to reproductive behaviour? *European Research on Cetaceans*, **6**, 162-164.

Amstrup, S.C., Durner, G.M., Stirling, I., Lunn, N., and Messier, F. (2000) Movements and distribution of polar bears in the Beaufort Sea. *Canadian Journal of Zoology*, **78**, 948-966.

Anderson, P.K. (1997) Shark Bay dugongs in summer. 1. Lekking. *Behaviour*, **134**, 433-462.

Anderson, S.S. and Fedak, M. (1985) Grey seal males: Energetic and behavioural links between size and sexual success. *Animal Behaviour*, **33**, 829-838.

Anderson, S.S., Burton, R.W. and Summers, C.F. (1975) Behaviour of grey seals (*Halichoerus grypus*) during a breeding season at North Rona. *Journal of Zoology* (London), **177**, 179-195.

Armitage, K.B. (1987) Do female yellow-bellied marmots adjust the sex ratios of their offspring? *American Naturalist*, **129**, 501-519.

Asper, E.D., Andrews, B.F., Antrim, J.E. and Young, W.G. (1992) Establishing and maintaining successful breeding programs for whales and dolphins in a zoological environment. *IBI Reports* (Kamogawa, Japan), 1992, **3**, 71-84.

Baker, C.S., Herman, L.M., Perry, A., Lawton, W.E., Straley, J.M., Wolman, A.A., Kaufman, G.D., Winn, H.E., Hall, J.D., Reinke, J.M. and Ostman, J. (1986) Migratory movement and population structure of humpback whales (*Megaptera novaeangliae*) in the central and eastern North Pacific. *Marine Ecology Progress Series*, **31**, 105-119.

Baker, C.S., Palumbi, S.R., Lambertsen, R.H., Weinrich, M.T., Calambokidis, J. and O'Brien, S. (1990) Influence of seasonal migration on geographic distribution of mitochondrial DNA haplotypes in humpback whales. *Nature* (London), **344**, 238-240.

Barlow, J. (1984) Reproductive seasonality in pelagic dolphins (*Stenella* spp.): implications for measuring rates. *Report of the International Whaling Commission* (special issue **6**), 191-198.

Bartholomew, G.A. (1970) A model for the evolution of pinniped polygyny. *Evolution*, **24**, 546-559.

Bell, G. (1980) The costs of reproduction and their consequences. *American Naturalist*, **116**, 45-76.

Bell, R.D. (1979) *Progress of fur seal pups mortality programme, 2V15 Bird Island 1978/79.* British Antarctic Survey, Madingley Road, Cambridge CB3 0ET, UK.

Best, P.B. (1982) Seasonal abundance, feeding, reproduction, age and growth in minke whales off Durban (with incidental observations from the Antarctic). *Report of the International Whaling Commission*, **32**, 759-786.

Best, P.B., Canham, P.A.S. and Macleod, N. (1984) Patterns of reproduction in sperm whales, *Physeter macrocephalus*. *Report of the International Whaling Commission* (special issue **6**), 51-79.

Best, R.C. (1983) Apparent dry-season fasting in Amazonian manatees, (Mammalia; Sirenia). Biotropica, 15, 61-64.

Bester, M.N. and Kerley, G.I.H. (1983) Rearing of twin pups to weaning by subantarctic fur seal, *Arctocephalus tropicalis* female. *South African Wildlife Research*, **13**, 86-87.

Bigg, M.A. (1969) The harbour seal in British Columbia. *Bulletin Fisheries Research Board of Canada*, **172**, 1- 33.

Bloch, D., Lockyer, C. and Zachariassen, M. (1993) Age and growth parameters of the long-finned pilot whale off the Faroe Islands. *Report of the International Whaling Commission* (special issue **14**), 163-208.

Boness, D.J. (1990) Fostering behavior in Hawaiian monk seals: is there a reproductive cost? *Behavioural Ecology and Sociobiology*, **27**, 113-122.

Boness, D.J., Craig, M.P., Honigman, L., and Austin, S. (1998) Fostering behavior and the effect of female density in Hawaiian monk seals, *Monachus schauinslandi*. *Journal of Mammalogy*, **79**, 1060-1069.

Boness, D.J. and James, H. (1979) Reproductive behavior of the grey seal (*Halichoerus grypus*) on Sable Island, Nova Scotia. *Journal of Zoology* (London), **188**, 477-500.

Boness, D.J., Bowen, D., Iverson, S.J., and Oftedal, O.T. (1992) Influence of storms and maternal size on mother-pup separations and fostering in the harbor seal, *Phoca vitulina*. *Canadian Journal of Zoology*, **70**, 1640-1644.

Boness, D.J., Dabek, L., Ono, K. and Oftedal, O.T. (1985) Female attendance behavior in California sea lions. In: *Proceedings of the Sixth Biennial Conference on the Biology of Marine Mammals*, Nov 22-26, Vancouver, British Columbia.

Bonner, W.N. (1971) An aged seal (*Halichoerus grypus*). *Journal of Zoology* (London), **164**, 261-262.

Bonner, W.N. (1984) Lactation strategies in pinnipeds: Problems for a marine mammalian group. *Symposium of the Zoological Society of London*, **51**, 253-272.

Bowen, W.D., Boness, D.J. and Oftedal, O.T. (1987) Mass transfer from mother to pup and subsequent mass loss by the weaned pup in the hooded seal, *Cystophora cristata*. *Canadian Journal of Zoology*, **65**, 1-8.

Bowen, W.D., Oftedal, O.T. and Boness, D.J. (1985) Birth to weaning in four days. Remarkable growth rate in the hooded seal, *Cystophora cristata*. *Canadian Journal of Zoology*, **63**, 2841-2846.

Bowen, W.D., Myers, R.A., and Hay, K. (1987) Abundance estimation of a dispersed, dynamic population: Hooded seals (*Cystophora cristata*) in the Northwest Atlantic. *Canadian Journal of Fisheries & Aquatic Sciences*, **44**, 282-295.

Boyd, I.L., Lockyer, C.H. and Marsh, H. (1999) Reproduction in marine mammals, Chapter 8. In: *Marine Mammals, Volume 1*. (Ed. by J.E Reynolds, III and S.A.Rommel), pp. 218-286. Smithsonian Institution Press, Washington D.C.

Braham, H.W. (1984) Distribution and migration of gray whales in Alaska. In: *The Gray Whale Eschrichtius robustus* (Ed. by M.L. Jones, S.L. Swartz and S. Leatherwood), pp. 249-266. Academic Press, Orlando, FL.

Brownell, R.L. and Ralls, K. (1986) Potential for sperm competition in baleen whales. *Report of the International Whaling Commission* (special issue **8**), 97-112.

Calambokidis, J., Steiger, G.H., Evenson, J.R., Flynn, K.R., Balcomb, K.C., Claridge, D.E., Waite, J.M., Darling, J.D., Ellis, G. and Green, G.A. (1996) Interchange and isolation of humpback whales off California and other North Pacific feeding grounds. *Marine Mammal Science*, **12**, 215-226.

Carrick, R., Csordas, S.E. and Ingham, S.E. (1962) Studies on the southern elephant seal, *Mirounga leonina* (L.), part 4: Breeding and development. *Commonwealth Scientific and Industrial Research Organization (C.S.I.R.O.) Wildlife Research*, **7**, 161-197.

Charnov, E.L. (1982) *The theory of sex allocation*. Princeton University Press, Princeton.

Clapham, P.J. (1996) The social and reproductive biology of Humpback Whales: an ecological perspective. *Mammal Review*, **26**, 27-49.

Clapham, P.J. and Mattila, D.K. (1988) Observations of migratory transits of two humpback whales. *Marine Mammal Science*, **4**, 59-62.

Clapham, P.J., Bérubé, M.C. and Mattila, D.K. (1995) Sex ratio of the Gulf of Maine humpback whale population. *Marine Mammal Science*, **11**, 227-231.

Clutton-Brock, T.H. (1989) Mammalian mating systems. *Proceedings of the Royal Society of London B*, **236**, 339-372.

Clutton-Brock, T.H. and Iason, G.R. (1986) Sex ratio variation in mammals. *Quarterly Review of Biology*, **61**, 339-374.

Cole, L.C. (1954) The population consequences of life history phenomena. *Quarterly Review of Biology*, **29**, 103-137.

Corkeron, P.J. and Connor, R.C. (1999) Why do baleen whales migrate? *Marine Mammal Science*, **15**, 1228-1245.

Costa, D.P., Le Boeuf, B.J., Huntley, A.C. and Ortiz, C.L. (1986) The energetics of lactation in the northern elephant seal. *Journal of Zoology* (London), **209**, 21-33.

Costa, D.P. and Gentry, R.L. (1986) Free-ranging energetics of northern fur seal. In: *Fur seals: Maternal strategies on land and at sea* (Ed. by R.L. Gentry and G.I. Kooyman), pp. 79-101. Princeton University Press, Princeton.

Coulson, J.C. and Hickling, G. (1961) Variation in the secondary sex ratio of the grey seal, *Halichoerus grypus* (Fabr.) during the breeding season. *Nature*, (London), **190**, 281.

Da Silva, V.M.F. (1994) *Aspects of the biology of the Amazonian dolphins genus Inia and Sotalia fluviatalis*. Ph.D. Thesis, University of Cambridge. 327pp.

De Master, D.P. (1984) Review of techniques used to estimate the average age at attainment of sexual maturity in marine mammals. *Report of the International Whaling Commission* (special issue **6**), 175-179.

Derocher, A.E. and Stirling, I. (1998) Geographic variation in growth of polar bears (*Ursus maritimus*). *Journal of Zoology* (London), **245**, 65-72.

Derocher, A.E., Andriashek, D., and Arnould, J.P.Y. (1993) Aspects of milk composition and lactation in polar bears. *Canadian Journal of Zoology*, **70**, 561-566.

Derocher, A.E., I. Stirling, and Andriashek, D. (1992) Pregnancy rates and serum progesterone levels of polar bears in western Hudson Bay. *Canadian Journal of Zoology*, **70**, 561-566.

Doidge, D.W. (1987) Rearing of twin offspring to weaning in Antarctic fur seals, *Arctocephalus gazella*. In: *Status, biology and ecology of fur seals* (Ed. by J. Croxall and R.L. Gentry), pp. 107-111. Proceedings of an International Symposium and Workshop, Cambridge, England, Apr 23-27, 1984. NOAA Technical Report NMFS 51.

Doidge, D.W., Croxall, J.P. and Ricketts, C. (1984) Growth rates of Antarctic fur seal *Arctocephalus gazella*. pups at South Georgia. *Journal of Zoology* (London), **203**, 87-93.

Duffield, D.A. and Wells, R.S. (1991) The combined application of chromosome, protein and molecular data for the investigation of social unit structure and dynamics in *Tursiops truncatus*. In: *Genetic Ecology of Whales and Dolphins* (Ed. by A.R. Hoelzel). *Report of the International Whaling Commission* (special issue **13**), 155-169.

Duffield, D.A., Odell, D.A., McBain, J.F., and Andrews, B. 1995. Killer whale (*Orcinus orca*) reproduction at sea world. *Zoo Biology*, **14**, 417-430.

Emlen, S.T. and Oring, L.W. (1977) Ecology, sexual selection, and the evolution of mating systems. *Science*, **197**, 215-223.

Estes, J.A. (1979) Exploitation of marine mammals: r-selection of K-strategists? *Journal of Fisheries Research Board of Canada*, **36**, 1009-1017.

Evans, P.G.H. (1987) *The Natural History of Whales & Dolphins*. Academic Press, London. 343pp.

Evans, P.G.H. (1990) European cetaceans and seabirds in an oceanographic context. *Lutra*, **33**, 95-125.

Fay, F.H. (1982) Ecology and biology of the Pacific walrus, *Odobenus rosmarus divergens* Illiger. North American Fauna, no. 74. U.S. Dept. of the Interior, Fish and Wildlife Service, Washington D.C.

Fogden, S.C.L. (1971) Mother-young behavior at grey seal breeding beaches. *Journal of Zoology* (London), **164**, 61-92.

Francis, J.M. & Heath, C.B. (1985) Duration of maternal care in the California sea lion - bias by sex. In: *Proceedings of the Sixth Biennial Conference on the Biology of Marine Mammals*, Nov 22-26, Vancouver, British Columbia.

Gambell, R. (1968) Seasonal cycles and reproduction in sei whales of the Southern Hemisphere. *Discovery Reports*, **35**, 31-134.

Gambell, R. (1973) Some effects of exploitation on reproduction in whales. *Journal of Reproductive Fertility, Supplement*, **19**, 533-553.

Gambell, R. (1976) Population biology and management of whales. In: *Applied Biology, Vol. 1* (Ed. by T.H. Coaker), pp. 247-343. Academic Press, New York and London.

Garner, G. W., Knick, S.T., and Douglas, D.C. 1990. Seasonal movements of adult female polar bears in the Bering and Chukchi Seas. *International Conference on Bear Research and Management*, **8**, 219-226.

Gaskin. D.E. (1982) *The Ecology of Whales and Dolphins*. Heinemann, London.

George, J.C., Bada, J., Zeh, J., Scott, L., Brown, S.E. and O'Hara, T. (1998) Preliminary age estimates of bowhead whales via aspartic acid racemisation. Paper SC/50/AS10, IWC Scientific Committee, April 1998, Oman.

Hamilton, W.D. (1966) The moulding of senescence by natural selection. *Journal of Theoretical Biology*, **12**, 12-45.

Harvey, P.H., Read, A.F. and Promislow, D.E.L. (1989) Life history variation in placental mammals: unifying the data with theory. *Oxford Surveys of Evolutionary Biology*, **6**, 13-32.

Heide-Jørgensen, M.P. and Teilmann, J. (1994) Growth, reproduction, age structure and feeding habits of white whales (*Delphinapterus leucas*) in West Greenland waters. *Meddelelser om Grønland, Bioscience*, **39**, 195-212.

Hernandez, P., Reynolds, J.E., Marsh, H. and Marmontel, M. (1995) Age and seasonality in spermatogenesis of Florida manatees. In: *Population Biology of the Florida Manatee* (Ed. by T.J. O'Shea, B.B. Ackermann and H.F. Percival), pp. 84-97. U.S. Department of the Interior, National Biological Service, Information and Technology Report 1.

Herzing, D.L. and Mate, B. (1984) Gray whale migrations along the Oregon coast, 1978-1981. In: *The Gray Whale Eschrichtius robustus* (Ed. by M.L. Jones, S.L. Swartz and S. Leatherwood), pp. 289-307. Academic Press, Orlando, FL.

Hohn, A.A., Chivers, S.J. and Barlow, J. (1985) Reproductive maturity and seasonality of male spotted dolphins, *Stenella attenuata*, in the eastern tropical Pacific. *Marine Mammal Science*, **1** (4), 273-293.

Hohn, A.A., Read, A.J., Fernandez, S., Vidal, O. and Findley, L. (1996) Life history of the vaquita, *Phocoena sinus* (Phocoenidae, Cetacea). *Journal of Zoology* (London), **239**, 235-251.

Hohn, A.A., Scott, M.D., Wells, R.S., Sweeney, J.C. and Irvine, A.B. (1989) Growth layers in teeth from known-age, free-ranging bottlenose dolphins. *Marine Mammal Science,* **5**, 315-342.

Horn, H.S. and Rubenstein, D.I. (1978) Behavioural Adaptations and Life History. Pp. 279-298. In: *Behavioural Ecology. An Evolutionary Approach.* (Ed. by J.R. Krebs & N.B. Davies), pp. 279-298. Blackwell, Oxford.

IUCN Polar Bear Specialist Group (1998) *Worldwide Status of the Polar Bear.* Pp. 23-44, In: Proceedings of the 12th Working Meeting of the IUCN Polar Bears Specialists, January 1997. Oslo, Norway.

Johanos, T.C., Becker, B.L., and Ragen, T.J. (1994) Annual reproductive cycle of the female Hawaiian monk seal (*Monachus schauinslandi*). *Marine Mammal Science*, **10**, 13-30.

Jones, M.L., Swartz, S.L. and Leatherwood, S. (Eds.) (1984) *The Gray Whale* Eschrichtius robustus. Academic Press, Orlando, Florida.

Jouventin, P. and Cornet, A. (1980) The sociobiology of pinnipeds. In: *Advances in the study of behavior*, vol. 2, pp. 121-141. Academic Press, New York.

Kasuya, T. (1972) Growth and reproduction of *Stenella coeruleoalba* based on the age determination by means of dentinal growth layers. *Scientific Report of the Whales Research Institute, Tokyo*, **24**, 57-79.

Kasuya, T. (1976) Reconsideration of life history parameters of the spotted and striped dolphins based on cemental layers. *Scientific Report of the Whales Research Institute, Tokyo*, **28**, 73-106.

Kasuya, T. (1985) Effect of exploitation on reproductive parameters of the spotted and striped dolphins off the Pacific coast of Japan. *Scientific Report of the Whales Research Institute, Tokyo,*, **30**, 1-63.

Kasuya, T. (1986) Distribution and behaviour of Baird's beaked whales off the Pacific coast of Japan. *Scientific Report of the Whales Research Institute, Tokyo*, **37**, 61-83.

Kasuya, T. (1995) Overview of cetacean life histories: an essay in their evolution. In *Whales, Seals, Fish and Man* (Ed. by A.S. Blix, L. Walløe, and Ø. Ulltang), pp. 481-497. Developments in Marine Biology, 4. Elsevier, Amsterdam.

Kasuya, T. and Marsh, H. (1984) Life history and reproductive biology of the short-finned pilot whale, *Globicephala macrorhynchus*, off the Pacific coast of Japan. *Report of the International Whaling Commission* (special issue **6**), 259-310.

Kasuya, T. and Tai, S. (1993) Life history of short-finned pilot whale stocks off Japan and a description of the fishery. *Report of the International Whaling Commission* (special issue **14**), 439-473.

Kasuya, T., Marsh, H. and Amino, A. (1993) Non-reproductive mating in short-finned pilot whales. *Report of the International Whaling Commission* (special issue **14**), 426-437.

Kaufman, G.W., Siniff, D.B., and Reichle. (1975) Colony behaviour of Weddell seals, *Leptonychotes weddelli*, at Hutton Cliffs, Antarctica. *Rapports P.-v Reunion Conseil International Exploration de la Mer*, **169**, 228-226.

Kawamura, A. (1978) An interim consideration on a possible interspecific relation in southern baleen whales from the viewpoint of their food habits. *Scientific Report of the Whales Research Institute, Tokyo*, **28**, 411-420.

Kenagy, G.J. and Trombulak, S.C. (1986) Size and function of mammalian testes in relation to body size. *Journal of Mammalogy*, **67**(1), 1-22.

Kenyon, K.W. 1981. Sea Otter. Pp. 209-223 In: *Handbook of Marine Mammals Vol. 1: Walrus, Sea Lions, Fur Seals and Sea Otter* (Ed. by S.H. Ridgway and R.J. Harrison). Academic Press, London.

Kimura, S. (1957) The twinning in southern fin whales. *Scientific Report of the Whales Research Institute, Tokyo*, **12**, 103-125.

King, J.E. (1983) *Seals of the world*. British Museum of Natural History, London.

Kovacs, K.M. (1987) Maternal behavior and early behavioral ontogeny of harp seals, *Phoca groenlandica*. *Animal Behaviour*, **35**, 844-855.

Kovacs, K.M. and Lavigne, D.M. (1986) Maternal investment and neonatal growth in phocid seals. *Journal of Animal Ecology*, **55**, 1035-1051.

Krylov, V.I. (1966) The sexual maturation of Pacific walrus females. *Zoolicheskii zhurnal* (Moscow), **45**, 919-927.

Kumlien, L. (1879) . Contributions to the natural history of Arctic America, made in connection with the Howgate Polar Expedition, 1877-78. *Bulletin of the U.S. National Museum*, **15**, 1-179.

Laws, R.M. (1953) The elephant seal (*Mirounga leonina* Linn.), part 1: Growth and age. *Falkland Islands Dependencies Survey, London. Sci. Rep.* No. 8, 1-62.

Laws, R.M. (1953) The elephant seal (*Mirounga leonina* Linn.), part 1: General social and reproductive behaviour. *Falkland Islands Dependencies Survey, London. Sci. Rep.* No. 13, 1-88.

Laws. R.M. (1958) Growth rates and ages of crabeater seals (*Lobodon carcinophagus*), Jacquinot and Pucherhan. *Proceedings of the Zoological Society of London*, **130**, 275-288.

Laws, R.M. (1959a) Accelerated growth in seals, with special reference to the Phocidae. *Norsk Hvalfangsttttid*, **48**, 425-452.

Laws, R.M. (1959b) The foetal growth rates of whales with special reference to the fin whale, *Balaenoptera physalus* Linn. *Discovery Reports*, **29**, 281-308.

Laws, R.M. (1977a) The significance of vertebrates in the Antarctic marine ecosystem. In *Adaptations within Antarctic Ecosystems* (Ed. by G.A. Llano), pp. 411-438. Third Symposium on Antarctic Ecology, Smithsonian Institution, Washington D.C..

Laws, R.M. (1977b) Seals and whales of the southern ocean. *Philosophical Transactions of the Royal Society of London B*, **279**, 81-96.

Le Boeuf, B.J. (1974) Male-male competition and reproductive success in elephant seals. *American Zoologist*, **14**, 163-176.

Le Boeuf, B.J. (1981) Elephant seals. In *The natural history of A–o Nuevo* (Ed. by B.J. Le Boeuf and S. Kaza), pp. 326-374. Boxwood Press, Pacific Grove, California.

Le Boeuf, B.J. (1986) Sexual strategies of seals and walruses. *New Scientist*, **1491**, 36-39.

Le Boeuf, B.J. and Ortiz, C.L. (1977) Composition of elephant seal milk. *Journal of Mammalogy*, **58**, 683-685.

Le Boeuf, B.J. and Reiter, J. (1988) Lifetime reproductive success in northern elephant seals. In *Reproductive success* (Ed. by T.H. Clutton-Brock), pp. 344-362. University of Chicago Press, Chicago.

Le Boeuf, B.J., Whiting, R.J. and Grant, R.F. (1972) Perinatal behavior of northern elephant seal females and their young. *Behaviour*, **43**, 121-156.

Lewontin, R.C. (1965) Selection for colonizing ability. In: *The genetics of colonizing species* (Ed. by H.G. Baker and G.L. Stebbins), pp. 77-94. Academic Press, New York.

Lockyer, C. (1978) A theoretical approach to the balance between growth and food consumption in fin and sei whales, with special reference to the female reproductive cycle. *Report of the International Whaling Commission*, **28**, 243-249.

Lockyer, C. (1979) Changes in a growth parameter associated with exploitation of southern fin and sei whales. *Report of the International Whaling Commission*, **29**, 191-196.

Lockyer, C. (1981) Growth and energy budgets of large baleen whales from the southern hemisphere. *Food and Agriculture Organisation of the United Nations Fisheries Series*, **5**, 379-487.

Lockyer, C. (1984) Review of baleen whale (Mysticeti) reproduction and implications for management. *Report of the International Whaling Commission* (special issue **6**), 27-50.

Lockyer, C. (1995) Investigation of aspects of the life history of the harbour porpoise, *Phocoena phocoena*, in British waters. *Report of the International Whaling Commission* (special issue **15**), 189-197.

Lockyer, C. and Brown, S.G. (1981) The migration of whales. In: *Animal Migration* (Ed. by D.J. Aidley), pp. 105-137. Society for Experimental Biology Seminar Series 13. Cambridge University Press, Cambridge.

Lunn, N.J., Paetkau, D., Calvert, W., Atkinson, S., Taylor, M., and Strobeck, C. (2000) Cub adoption by polar bears (*Ursus maritimus*): determining relatedness with microsatellite markers. *Journal of Zoology* (London), **251**, 23-30.

Lydersen, C. (1998) *Behaviour and energetics of ice-breeding North Atlantic phocid seals during the lactation period.* Doctor of Philosophy Thesis. University of Oslo, Oslo, Norway. 88pp.

Mackintosh, N.A. (1965) *The stocks of whales*. Buckland Foundation, Aberdeen. 232pp.

Mackintosh, N.A. (1973) Distribution of post-larval krill in the Antarctic. *Discovery Reports*, **36**, 95-156.

Mansfield, A.W. (1958) The biology of the Atlantic walrus, *Odobenus rosmarus rosmarus* (Linnaeus) in the eastern Canadian Arctic. *Fisheries Research Board Canada*, Manuscr. Rep. Ser. (Biol.) **653**.

Marmontel, M. (1995) Age and reproduction in female Florida manatees. In: *Population Biology of the Florida Manatee* (Ed. by T.J. O'Shea, B. B. Ackermann, and H. F. Percival), pp. 98-119. U.S. Department of the Interior, National Biological Service, Information and Technology Report 1.

Marsh, H. (1980). Age determination of the dugong *(Dugong dugon* (Müller)) in northern Australia and its biological implications. *Report of the International Whaling Commission* (special issue **3**), 181-201.

Marsh, H. (1986) The status of the dugong in Torres Strait. In: *Torres Strait Fisheries Seminar, Port Moresby* (Ed. by A. K. Haines, G. C. Williams and D. Coates) pp. 53-76. Australian Government Publishing Service, Canberra.

Marsh, H. (1995) The life history, pattern of breeding and population dynamics of the dugong. In: *Population Biology of the Florida Manatee* (Ed. by T.J. O'Shea, B.B. Ackermann and H.F. Percival) pp. 75-83. U.S. Department of the Interior, National Biological Service, Information and Technology Report 1.

Marsh, H. and Kasuya, T. (1986) Evidence for reproductive senescence in female cetaceans. Pp. 57-74. In: *Behaviour of Whales in relation to Management.* (Ed. by G. Donovan). *Report of the International Whaling Commission* (special issue **8**). International Whaling Commission, Cambridge. 282 pp.

Marsh, H., Heinsohn, G.E. and Marsh, L.M. (1984) Breeding cycle, life history and population dynamics of the dugong, *Dugong dugon.* (Sirenia, Dugongidae). *Australian Journal of Zoology*, **32**, 767-785.

Martin, A.R. (1980) An examination of sperm whale age and length data from the 1949-78 Icelandic catch. *Report of the International Whaling Commission,* **30**, 227-231.

Martin, A.R. (1990) *Whales and dolphins.* Salamander Books Ltd, London. 192pp.

Martin, A. and Rothery, P. (1993) Reproductive parameters of female long-finned pilot whale *(Globicephala melas)* around the Faroe Islands. *Report of the International Whaling Commission* (Special Issue **14**), 263-304.

Martin, A.R., Smith, T.C. and Cox, O.P. (1993) Studying the behaviour and movements of high Arctic belugas with satellite telemetry. *Symposia of the Zoological Society (London)*, **66**, 195-210.

Mattlin, R.H. (1981) Pup growth of the New Zealand fur seal, *Arctocephalus forsteri* on the Open Bay Islands, New Zealand. *Journal of Zoology* (London), **193**, 305-314.

Mattlin, R.H. (1987) New Zealand fur seal, *Arctocephalus forsteri* , within the New Zealand region. In: *Status, biology and ecology of fur seals* (Ed. by J.P. Croxall and R.L. Gentry), pp. 49-51. Proceedings of an International Symposium and Workshop, Cambridge, England, Apr 23-27, 1984. NOAA Technical Report NMFS 51.

Maynard Smith, J. (1980) A new theory of sexual investment. *Behavioral Ecology and Sociobiology*, **7**, 241-251.

Mikhalev, Yu. A. (1980) General regularities in prenatal growth in whales and some aspects of their reproductive biology. *Report of the International Whaling Commission*, **30**, 249-253.

Mobley, J.R. Jr and Herman, L.M. (1985) Transcience of social affiliations among humpback whales (Megaptera novaeangliae) on the Hawaiian wintering grounds. *Canadian Journal of Zoology*, **63**, 762-772.

Monson, D. H., Estes, J. A., Bodkin, J.L., and Siniff, D.B. (2000) Life history plasticity and population regulation in sea otters. *Oikos*, **90** (3), 457-468.

Myrick, A., Hohn, A.A., Barlow, J. and Sloan, (1986) Reproductive biology of female spotted dolphins, Stenella attenuata, from the eastern tropical Pacific. *Fisheries Bulletin*, **84** (2), 247-259.

Nemoto, T. (1959) Food of baleen whales with reference to whale movements. *Scientific Report of the Whales Research Institute, Tokyo*, **14**, 149-290.

Nemoto, T. (1962) Food of baleen whales collected in recent Japanese Antarctic whaling expeditions. *Scientific Report of the Whales Research Institute, Tokyo*, **16**, 89-103.

Odell, D. K., G. D. Bossart, M. T. Lowe, and Hopkins, T. D. (1995) Reproduction of the West Indian manatee in captivity. In: *Population Biology of the Florida Manatee* (Ed. by T.J. O'Shea, B.B. Ackermann, and H.F. Percival), pp. 192- 93. U.S. Department of the Interior, National Biological Service, Information and Technology Report 1.

Oftedal, O.T., Boness, D.J. and Tedman, R.A. (1987) The behavior, physiology, and anatomy of lactation in Pinnipedia. In: *Current Mammalogy*, vol. 1 (Ed. by H.H. Genoways), pp. 175-246. Plenum Press, New York.

Oftedal, O.T., Iverson, S.J. and Boness, D.J. (1985) Energy intake in relation to growth rate in pups of the California sea lion, *Zalophus californianus. Proceedings of the Sixth Biennial Conference on the Biology of Marine Mammals,* Nov 22-26, Vancouver, British Columbia.

Ohsumi, J.S. (1977) Bryde's whales in the pelagic whaling ground of the North Pacific. *Report of the International Whaling Commission* (special issue 1), 140-150.

Ohsumi, J.S. (1986) Yearly change in age and body length at sexual maturity of a fin whale stock in the eastern North Pacific. *Scientific Report of the Whales Research Institute, Tokyo*, 37, 1-16.

Olesiuk, P.K., Bigg, M.A. and Ellis, G.M. (1990) Life history and population dynamics of resident killer whales (*Orcinus orca*) in the coastal waters of British Columbia and Washington State. *Report of the International Whaling Commission* (special issue 12), 209-243.

Orians, G.H. (1969) On the evolution of mating systems in birds and mammals. *American Naturalist*, 103, 589-603.

Ortiz, C.L., Le Boeuf, B.J. and Costa, D.P. (1984) Milk intake of elephant seal pups: An index of parental investment. *American Naturalist*, 124, 416-422.

O'Shea, T.J., Ackerman, B.B. and Percival, H.F. (eds.) (1995) *Population Biology of the Florida Manatee.* Natl Biological Service: Information and Technology Report No. 1.

Perrin, W.F., Brownell, R.L. Jr. and DeMaster, D.P. (1984) *Reproduction in Whales, Dolphins and Porpoises. Report of the International Whaling Commission* (special issue 6). International Whaling Commission, Cambridge. 495pp.

Perrin, W.F. and Henderson, J.R. (1984) Growth and reproductive rates in two populations of spinner dolphins, *Stenella longirostris*, with different histories of exploitation. *Report of the International Whaling Commission* (special issue 6), 417-430.

Perrin, W.F. and Reilly, S.P. (1984) Reproductive parameters of dolphins and small whales of the family Delphinidae. *Reports of the International Whaling Commission* (special issue 6), 97-133.

Perry, E.A., Boness, D.J., and Fleischer, R.C. (1998) DNA fingerprinting evidence of nonfilial nursing in grey seals. *Molecular Ecology*, 7, 81-85.

Peterson, R.S. and Reeder, W.G. (1966) Multiple births in the northern fur seal. *Zür Saugetierk.*, 31, 52-56.

Plön, S. (1999) The fast lae revisited: life history strategies of *Kogia* from Southern Africa. *Abstract, Proceedings of Thirteenth Biennial Conference of Marine Mammals*, 28 Nov - 3 Dec 1999, Maui, Hawaii.

Preen, A. (1989) Observations of mating behavior in dugongs. *Marine Mammal Science*, 5, 382-386.

Prestrud, P. and Stirling, I. The International Polar Bear Agreement and the current status of polar bear conservation. *Aquatic Mammals*, 20,1-12.

Promislow, D.E.L. (1991) Senescence in natural populations of mammals: A comparative study. *Evolution*, 45, 1869-1887.

Ralls, K. (1976) Mammals in which females are larger than males. *Quarterly Review of Biology*, **51**, 245-276.

Ralls, K. Brownell, R.L. Jr and Ballou, J. (1980) Differential mortality by sex and age in mammals, with special reference to the sperm whale. *Report of the International Whaling Commission* (special issue **2**), 233-243.

Ramsay, M.A., and Dunbrach, R.L. (1986) Physiological constraints on life history phenomena: the example of small bear cubs at birth. *American Naturalist*, **127**, 735-743.

Ramsay, M.A., and Stirling, I. (1988) Reproductive biology and ecology of female polar bears (*Ursus maritimus*). *Journal of Zoology*, London, **214**, 601-634.

Rathbun, G.B., Reid, J.P., Bonde, R.K. and Powell, J.A. (1995). Reproduction in free-ranging Florida manatees. In: *Population Biology of the Florida Manatee* (Ed. by T.J. O'Shea, B.B. Ackermann and H.F. Percival), pp. 135-156. U.S. Department of the Interior, National Biological Service, Information and Technology Report 1.

Read, A.F. and Harvey, P. (1989) Life history differences among the eutherian radiations. *Journal of Zoology* (London), **219**, 329-353.

Read, A.J. and Hohn, A.A. (1995) Life in the fast lane: the life history of harbor porpoises from the Gulf of Maine. *Marine Mammal Science*, **11**, 423-440.

Read, A.J., Wells, R.S., Hohn, A.A. and Scott, M.D. (1993) Patterns of growth in wild bottlenose dolphins, *Tursiops truncatus*. *Journal of Zoology* (London), **231**, 107-123.

Reid, J.P., Bonde, R.K. and O'Shea, T.J. (1995) Reproduction and mortality of radio-tagged and recognizable manatees on the Atlantic Coast of Florida. In: *Population Biology of the Florida Manatee* (Ed. by T.J. O'Shea, B.B. Ackermann and H.F. Percival), pp. 171-191. U.S. Department of the Interior, National Biological Service, Information and Technology Report 1.

Reilly, S.B. (1984) Observed and published rates of increase in gray whales, *Eschrichtius robustus*. *Report of the International Whaling Commission* (special issue **6**), 389-399.

Reiter, J., Stinson, N.L. and Le Boeuf, B.J. (1978) Northern elephant seal development: The transition from weaning to nutritional independence. *Behavioural Ecology and Sociobiology*, **3**, 337-367.

Repenning, C.A. (1976) Adaptive evolution of sea lions and walruses. *Systematic Zoology*, **25**, 375-390.

Reynolds, J.E. III (1995) Florida manatee population biology: Research progress, infrastructure and applications of conservation and management. In: *Population Biology of the Florida Manatee* (Ed. by T.J. O'Shea, B. B. Ackermann, and H. F. Percival), pp. 6-12. U.S. Department of the Interior, National Biological Service, Information and Technology Report 1.

Reynolds, J.E. III and Odell, D.K. (1991) *Manatees and Dugongs*. Facts on File, Inc., New York, NY.

Ridgway, S.H. and Harrison, R.J. (Eds.) (1981a) *Handbook of Marine Mammals, Vol. 1: The Walrus, Sea Lions, Fur Seals, and Sea Otter*. Academic Press, London & New York.

Ridgway, S.H. and Harrison, R.J. (Eds.) (1981b) *Handbook of Marine Mammals, Vol. 2: Seals*. Academic Press, London & New York.

Ridgway, S.H. and Harrison, R.J. (Eds.) (1985) *Handbook of Marine Mammals, Vol. 3: The Sirenians and Baleen Whales*. Academic Press, London & New York.

Ridgway, S.H. and Harrison, R.J. (Eds.) (1989) *Handbook of Marine Mammals, Vol. 4: River Dolphins and the Larger Toothed Whales*. Academic Press, London & New York.

Ridgway, S.H. and Harrison, R.J. (Eds.) (1994) *Handbook of Marine Mammals, Vol. 5: The First Book of Dolphins*. Academic Press, London & New York.

Ridgway, S.H. and Harrison, R.J. (Eds.) (1999) *Handbook of Marine Mammals, Vol. 6: The Second Book of Dolphins*. Academic Press, London & New York.

Riedman, M. (1990) *The Pinnipeds. Seals, Sea Lions and Walruses*. University of California Press, Los Angeles. 439pp.

Riedman, M.L. and Estes, J.A. (1990) The sea otter (*Enhydra lutris*): behavior, ecology, and natural history. *Biological Report*, **90**(14), U.S. Fish and Wildlife Service. 126pp.

Riedman, M. and Ortiz, C.L. (1979) Changes in milk composition during lactation in the northern elephant seal. *Physiological Zoology*, **52**, 240-249.

Riedman, M., Estes, J.A., Staedler, M.M., Giles, A.A., and Carlson, D.R. (1994) Breeding patterns and reproductive success of California sea otters. *Journal of Wildlife Management*, **58**, 391-399.

Seger, J. and Trivers, R.L. (1983) Sex ratios of whales before birth. *Abstract, Proceedings of Fifth Biennial Conference of Marine Mammals*, 27 Nov - 1 Dec 1983, Boston, Mass.

Sergeant, D.E. (1962) The biology of the pilot or pothead whale *Globicephala melaena* (Traill) in Newfoundland waters. *Bulletin of the Fisheries Research Board of Canada*, **132**, 1-84.

Sergeant, D.E. (1965) Migrations of harp seals *Pagophilus groenlandicus* (Erxleben) in the northwest Atlantic. *Journal of the Fisheries Research Board of Canada*, **23**, 433-464.

Sergeant, D.E. (1973) Biology of white whales (*Delphinapterus leucas*) in Western Hudson Bay. *Journal of the Fisheries Research Board of Canada*, **30**, 1065-1090.

Shirakihara, M., Takemura, A. and Shirakihara, K. (1993) Age, growth and reproduction of the finless porpoise, *Neophocoena phocaenoides*) in the coastal waters of western Kyushu, Japan. *Marine Mammal Science.*, **9**, 392-406.

Sigurjónsson, J. (1995) On the life history and autecology of North Atlantic rorquals. In: *Whales, Seals, Fish and Man* (Ed. by A.S. Blix, L. Walløe, and Ø. Ulltang), pp. 425-441. Developments in Marine Biology, 4. Elsevier, Amsterdam.

Sinha, A.A., Conaway, C.H., and Kenyon, K.W. (1966) Reproduction in the female sea otter. *Journal of Wildlife Management*, **30**, 121-130.

Siniff, D.B. and Bengtson, J.L. (1977) Observations and hypotheses concerning the interactions among crabeater seals, leopard seals, and killer whales. *Journal of Mammalogy*, **58**, 414-416.

Siniff, D.B., Stirling, I., Bengtson, J.L., and Reichle, R. (1979) Social and reproductive behavior of crabeater seals (*Lobodon carcinophagus*). *Canadian Journal of Zoology*, **57**, 2243-2255.

Sjare, B. and Stirling, I. (1996) The breeding behavior of Atlantic walruses, *Odobenus rosmarus rosmarus*, in the Canadian High Arctic. *Canadian Journal of Zoology*, **74**, 897-911.

Smith, T.G. (1987) The ringed seal, *Phoca hispida*, of the Canadian Western Arctic. *Canadian Bulletin of Fisheries and Aquatic Sciences*, **216**, 1-81.

Smith, T.G. and Martin, A.R. (1994) Distribution and movement of belugas, *Delphinapterus leucas*, in the Canadian High Arctic. *Canadian Journal of Fisheries and Aquatic Sciences*, **51**, 1653-1663.

Spotte, S. (1982) The incidence of twinning in pinnipeds. *Canadian Journal of Zoology*, **60**, 2226-2233.

Stearns, S.C. (1992) *The Evolution of Life Histories*. Oxford University Press, Oxford.

Stewart, R.E.A. (1994) Size-at-age relationships as discriminators of white whale (*Delphinapterus leucas*) stocks in the eastern Canadian arctic. *Meddelelser om Grønland, Bioscience*, **39**, 217-235.

Stewart, B.S. and DeLong, R.L. (1995) Double migrations of the northern elephant seal, *Mirounga angustirostris*. *Journal of Mammalogy*, **76**, 179-194.

Stewart, R.E.A. and Lavigne, D.M. (1980) Neonatal growth of northwest Atlantic harp seals, *Pagophilus groenlandicus*. *Journal of Mammalogy*, **61**, 670-680.

Stirling, I. (1971) Variation in the sex ratio of newborn Weddell seals during the pupping season. *Journal of Mammalogy*, **52**, 842-844.

Stirling, I. (1975a) Adoptive suckling in pinnipeds. *Journal of Australian Mammalogy*, **1**, 389-391.

Stirling, I. (1975b) Factors affecting the evolution of social behavior in the Pinnipedia. *Rapport P.-v. Reunion Conseil International Exploration de la Mer*, **169**, 205-212.

Stirling, I. (1977) Adaptations of Weddell and ringed seals to exploit polar fast ice habitat in the presence or absence of land predators. In: *Adaptations within Antarctic ecosystems* (Ed. by G.A. Llano), pp. 741-748. Proceedings of the Third SCAR Symposium on Antarctic Biology, Aug 26-30, 1974. Smithsonian Institution, Washington D.C..

Stirling, I. (1983) The evolution of mating systems in pinnipeds. In: *Recent advances in the study of mammalian behavior* (Ed. by J.F. Eisenburg and D.G. Kleiman), pp. 489-527. Special Publication, American Society of Mammalogy, No. 7.

Stirling, I. (1988) *Polar Bears*. University of Michigan Press, Ann Arbor, MI.

Stirling, I. (1997) The importance of polynyas, ice edges, and leads to marine mammals and birds. *Journal of Marine Systems*, **10**, 9-21.

Stirling, I. and Øritsland, N.A. (1995) Relationships between estimates of ringed seal and polar bear populations in the Canadian Arctic. *Canadian Journal of Fisheries and Aquatic Sciences*, **52**, 2594-2612.

Stone, G.S., Florez-Gonzalez, L. and Katona, S. (1990) Whale migration record. *Nature, London*, **346**, 705.

Swartz, S.L. (1986) Gray whale migratory, social and breeding behavior. *Report of the International Whaling Commission* (special issue **8**), 207-229.

Trillmich, F. and Ono, K.A. (Eds.) (1991) *Pinnipeds and El Niño: Responses to Environmental Stress*. Springer-Verlag, Berlin. 293pp.

Trivers, R.L. (1972) Parental investment and sexual selection. In *Sexual selection and the descent of man, 1871-1971*. (Ed. by B. Campbell), pp. 136-179. Aldine, Chicago.

Trivers, R.L. (1985) *Social evolution*. Benjamin-Cummings, Menlo Park, California.

Trivers, R.L. and Willard, D.E. (1973) Natural selection of parental ability to vary the sex ratio of offspring. *Science*, **179**, 90-92.

Urian, K.W., Duffield, D.A., Read, A.J., Wells, R.S. and Shell, R.D. (1996) Seasonality of reproduction in bottlenose dolphins, *Tursiops truncatus*. *Journal of Mammalogy*, **77**, 394-403.

Vehrencamp, S.L. and Bradbury, J.W. (1984) Mating Systems and Ecology. In: *Behavioural Ecology. An Evolutionary Approach*. (Ed. by J.R. Krebs and N.B. Davies), pp. 251-278. Blackwell, Oxford.

Venables, U.M. and Venables, S.V. (1957) Mating behaviour of the seal *Phoca vitulina* in Shetland. *Proceedings of the Zoological Society of London*, **128**, 387-396.

Wells, R.S., Scott, M.D. and Irvine, A.B. (1987) The social structure of free-ranging bottlenose dolphins. In: *Current Mammalogy* (Ed. by H.H. Genoway), pp. 247-305. Plenum Press, New York.

Whitehead, H. (1993) Breeding behaviour of sperm whales. In: *Abstracts, Tenth Biennial Conference on the Biology of Marine Mammals*, 11-15 Nov 1995, Galveston, Texas, p. 10.

Wiley, R.H. (1974) Evolution of social organization and life history patterns among grouse: Tetraonidae. *Quarterly Review of Biology*, **49**, 201-227.

Wiley, D.N. and Clapham, P.J. (1993) Does maternal condition affect the sex ratio of offspring in humpback whales? *Animal. Behaviour*, **46**, 321-324.

York, A.E. (1979) Analysis of pregnancy rates in female fur seals in the combined United States-Canada pelagic collections, 1958-74. In: *Preliminary analysis of pelagic fur seal data collected by the United States and Canada during 1958-74* (Ed. by H. Kajimura, R.H. Landers, M.A. Perez, A.E. York and M.A. Bigg), pp. 50-122. Unpubl. MS. Natl Marine Mammal Lab., U.S. Natl Marine Fisheries Service, Seattle, Wash.

APPENDIX. LIFE HISTORY PARAMETERS FOR MARINE MAMMALS

	Gestation Period (mo.)	Peak Month(s) of Births	Lactation Period (d/mo)	Inter-birth Interval (yrs)	Age at Sex. Matur. (yr) Male	Female	Maximum Life Span (yr) Male	Female
PINNIPEDIA								
Otariidae								
Arctocephalus australis	8 (+4)	Nov-Dec	8-24 mo	1-2+	7	2-3		
A. forsteri	8 (+4)	Nov-Dec	10-12 mo	1	10-12	2-5		
A. galapagoensis	8 (+4)	Sept-Oct	18-36 mo	2+		3-5?		
A. gazella	8 (+3-4)	Dec	4 mo	1	3-4	3-4	13+	23
A. philippi		Nov-Dec						
A. pusillus	8 (+3-4)	Nov-Dec	8-18 mo	1-2	3-6	3-5	20**	
A. townsendi	8 (+3-4)	May-June	8-11 mo	1				
A. tropicalis	8 (+3-4)	Nov-Dec	9-11 mo	1	3-4	4-6		
Callorhinus ursinus	8 (+3-4)	June-July	4-5 mo	1	5	3-7	30	
Eumetopias jubatus	8 (+3.5)	June-July	11-36+ mo	2+	2-7	3-8		
Neophoca cinerea	8 (+3-4)	June, Oct-Dec	14-17 mo	2+		3?		
Otaria byronia	8 (+3-4)	Jan	6-24+ mo	1-2+	5-6	3-4	20	
Phocarctos hookeri	8 (+3-4)	Dec-Jan	6-12+ mo	1			30	
Zalophus californianus	8 (+3?)	May-June	6-12+ mo	1	4-5	4-5		
Odobenidae								
Odobenus rosmarus	10-11 (+4-5)	May	18-24 mo	2-4	10-14	5-10	30-40	
Phocidae								
Hydrurga leptonyx	9.5 (+1.5)	Nov-Dec	?30 d	1	4-6	2-7		
Leptonychotes weddellii	8.5 (+1.5)	Sept-Nov	c. 56 d	1	3-6	2-6		
Lobodon carcinophagus	5 (+6)	Oct-Nov	14-28 d	1	2-6	2-6		
Mirounga angustirostris	7 (+4)	Jan-Feb	22-29 d	1	5	2-6	20	20
M. leonina	7-8 (+4)	Oct-Nov	20-25 d	1	4-6	2-7		
Monachus monachus		Extended	100-120+ d					20
M. schauinslandi		Dec-July	35-42 d			5-6+	30	
Ommatophoca rossii	8-9 (+2-3)	Nov-Dec	c. 28 d	1	2-7	3-5	?12	
Cystophora cristata	8 (+3-4)	Mar	3-12 d	1	4-6	2-9	c. 35	
Erignathus barbatus	9.5 (+2)	Mar-Apr	24 d	1	6-7	3-6	31	
Halichoerus grypus	8 (+3-4)	Jan-Apr, Sept-Dec	16-21 d	1	6-8	3-5	41**	46
Phoca fasciata	7-8 (+3-4)	(Mar)Apr(May)	21-28 d	1	3-5	2-4	c. 30	
P. groenlandica	9 (+2.5)	Feb-Mar	10-12 d	1	3-5	3-7	30+	
P. hispida	7-8 (+3-4)	Apr-May	35-49 d	1	5-7	3-7		43
P. largha	? (+1.5-3?)	Mar-Apr	28 d	1	3-6	2-5	30+	
P. vitulina	8-9.5 (+1.5-3)	Apr-Sept	21-42 d	1	3-7	2-7	30+	
Pusa caspica	7-8 (+3-4)	Jan(Feb)	28-35 d	1	6	4-6		
Pusa sibirica		(Feb)Mar	68 d	1	4	3-6		

CARNIVORA								
Ursidae								
Ursus maritimus	2-4 (+4-5)	Nov-Dec	30 mo	2-4	6-10	4-5	29	32
Mustelidae								
Enhydra lutris	4 (+2-3)	Extended	4-7 mo	1(-2)	6-7	4	15	20
SIRENIA, the Sea Cows								
Dugongidae								
Dugong dugon	13.9	Aug-Jan	18+ mo	2-7	9-15	9-18	73	
Trichechidae								
T. manatus	12-14	Mar-Aug	12-24 mo	(1)2-5	2-11	2.5-7	59	
CETACEA								
MYSTICETI, the Baleen Whales								
Balaenidae								
Balaena mysticetus	c. 12-13	Apr-May	5-6 mo			4	? 100+	
Eubalaena australis	10	May-Aug	10-12 mo	3-4	10	10	60+	
E. glacialis	12-13	Dec-Mar		3-5				
Eschrichtidae								
Eschrichtius robustus	11-13	Jan-Feb	7 mo	2	5-11	8-12	40+	
Balaenopteridae								
Balaenoptera acutorostrata	10-11	Dec-Jan, May-June	4-6 mo	1-2	3-6(8)	5-7(8)	40-50(57)	
B. borealis	11-13	Nov-Dec, June	6-9 mo	2(-3)	<10	6-12	60-70	
B. edeni	12	Extended	? 6 mo	2+	9-13	8-11		
B. musculus	11	Nov-Dec?, May	7 mo	2(-3)	5-10	5-10	80-90	
B. physalus	11-12	Nov-Dec, May	(6)-7 mo	2(-3)	6-12	6-12	85-90(94)	
Megaptera novaeangliae	10-12	Jan-Feb, July-Aug	10-12 mo	1-2	(2)5-11	(2)5-11	48	
Neobalaenidae								
Caperea marginata	c. 12	? Extended	? 5-6 mo					
ODONTOCETI, the Toothed Whales								
Physeteridae								
Physeter macrocephalus	(14)15-16(17)	May-Sept, Feb-Apr	24(19-42) mo	3-15	18-19	7-12	65-70	
Kogiidae								
Kogia breviceps	11	?, Mar-Aug	12 mo	1	4-5	4-5	16	22
K. simus				1	4-5	4-5	21.5	
Ziphiidae								
Berardius bairdii	17	Mar-Apr			8-10	8-10	71	39
Hyperoodon ampullatus	12+	Apr-Jun	12+ mo	2	7-11	11	37+	27+
Ziphius cavirostris							36+	30

Monodontidae								
Delphinapterus leucas	(11)13-15	Apr-May, July-Aug	20-24+ mo	c. 3	6-7	4-7		25-30
Monodon monoceros	14.5	July-Aug	20 mo		5-8	11-13		50
Delphinidae								
Cephalorhynchus commersonii		Oct-Mar						18
Delphinus delphis	10.5	June-Sept,	c. 19 mo	2.6	5-6	5-6		30-35
Globicephala macrorhynchus	c.12-16	Dec-Jan, July-Aug	(24-34)42 mo	6-9	5-7	6	45.5	62.5
G. melas	c.12-16	May-Aug(Sept)	44 mo	3.3	16	8-9	50	60+
Grampus griseus	c.13-14	Dec-June			12-14	6-8	13+	17+
Lagenodelphis hosei		?Extended			3-4	3-4		
Lagenorhynchus acutus	c.11	May-Aug	c. 18 mo	2-3	7-8	7-8	22	27
L. albirostris	c.11	May-Aug			7-11	6-12		
L. obliquidens	10-12	May-Aug			7-10	7-9		46
L. obscurus	11-13	Aug-Oct, Nov-Feb	12-18 mo	2-3	4-6/?7-10	10		35-36
Orcaella brevirostris		?July-Dec						
Orcinus orca	(12-15)17	(Sept)Oct-Jan(Mar)	12+(36) mo	3.0-8.3	14-16	8-10	60	90
Peponocephala electra	12	?Extended			3-7	4-12		47
Pseudorca crassidens	15.5	Extended	18-24 mo		11	11	58	63
Sotalia fluviatilis	10	Oct-Nov	18 mo					
Stenella attenuata	11.5	Extended (*spr/aut)	20 mo	2.4	12	9	40	46
S. coeruleoalba	12	Extended (*sum/win)	12-18 mo	3-4	9	8-14		30-35(58)
S. longirostris	10-11	Extended (*spr/aut)	11-34 mo	3	7	5	19	23
Steno bredanensis					14	10	32	30
Tursiops truncatus	12	(Feb)Mar-Sept(Nov)	18-20 mo	2-6	8-15	5-13	41+	52
Phocoenidae								
Neophocaena phocaenoides		Apr-May						
Phocoena dioptrica	11	? July-Sept	6-15 mo					
P. phocoena	10-11	(Mar)May-July(Aug)	4-8 mo	1-2	3-4	3-5	24	24
Phocoenoides dalli	11	Aug-Sept	24(6-42) mo	3	5-6	3-4		
Platanistidae								
Platanista gangetica		Extended						28
Iniidae								
Inia geoffrensis	10	May-July					18+	28(35)
Pontoporidae								
Lipotes vexillifer	10	Mar-Apr						16
Pontoporia blainvillei	10-11	Nov-Dec	8-9 mo	2	2-3	2-3		13

NOTES ? = data uncertain   * = evidence for bimodal birth peaks, e.g. in spring & autumn, or summer & winter   ** = in captivity
(-) = outlying data
Gestation period = developmental gestation; for pinnipeds, polar bear & sea otter, an additional period of delayed implantation is put in parentheses
Many species show geographical variation in breeding seasonality; in the case of cetaceans, where two periods are given, the first usually refers to northern hemisphere and the second to the southern hemisphere. Ranges in Ages at Sexual Maturity usually reflect spatial or temporal variations.

SOURCES Main sources (containing more detailed references) are: Riedman, 1990 (pinnipeds); Perrin *et al.*, 1984 (cetaceans); Reynolds & Odell, 1991, O'Shea *et al.*, 1995 (sirenians); Stirling, 1988 (polar bear); Riedman & Estes, 1990 (sea otter); Ridgway & Harrison, 1981a,b, 1985, 1989, 1994, 1999, Boyd *et al.*, 1999 (marine mammals). Often, more detailed studies have modified previous estimates of life history parameters, and this has been taken into account here.

Chapter 2

How Persistent are Marine Mammal Habitats in an Ocean of Variability?
Habitat use, home range and site fidelity in marine mammals

ARNE BJØRGE
Institute of Marine Research, P.O. Box 1870 Nordnes, 5817 Bergen, Norway
E-mail: Arne.Bjorge@imr.no

1.   INTRODUCTION

Marine mammals are highly mobile and capable of travelling long distances. Some species do undergo long seasonal migrations, mainly north-south movements between breeding and foraging grounds. Other species may follow their prey species in an offshore-onshore seasonal migration pattern. Some species, despite their capacity for migration, are resident in relatively small areas throughout the year. These species are dependent upon the continuous availability of food within their local habitats. In some cases, displacements or shifts in distribution of marine mammals occur which are not related to annual migrations, and these events may reflect changes in the quality or availability of their preferred habitats.

The habitat (or habitats) of a marine mammal population (or species) is contained within the range of that particular population. The range may be illustrated by drawing a line which encompasses the home ranges of all individuals of that population. However, within this boundary, there are areas which are frequently used (preferred habitats) and other areas where animals rarely occur.

Knowledge of movements of marine mammals is a prerequisite for estimating their home ranges. However, this information alone is not sufficient for an in-depth understanding of their habitat use. We also need data on the temporal and spatial distribution of their different activities, *e.g.* breeding and foraging in order to identify breeding and foraging habitats.

Marine Mammals: Biology and Conservation, edited by
Evans and Raga, Kluwer Academic/Plenum Publishers, 2002

Better understanding of the marine mammals and their roles in the ecosystem requires knowledge of the geographical distribution of their foraging activity and the properties of their foraging habitats. In this chapter, emphasis will be put on foraging activity and foraging habitats. However, the same principles may be applied to identify and describe breeding activity and breeding habitats.

Many oceanic mammals feed on pelagic prey species. Their foraging habitats may be related to temporal oceanic, climatic or biological phenomenon, such as frontal zones, gyres, upwelling areas or the distribution in space and time of pelagic prey populations. In general, the geographic distribution of pelagic habitats often change within and between seasons. The foraging habitats of mammals feeding on benthic communities are often determined by bathymetry and substrate in addition to availability of benthic prey organisms. Benthic prey populations may also change in space and time. However, in general, the changes in benthic communities often occur on a longer time scale than the relatively rapid fluctuations in pelagic communities.

Fixed coordinates are frequently used to define foraging grounds of marine mammals. While fixed coordinates often are relevant for describing foraging habitats (and estimating home range) of benthic feeders, fixed coordinates (and estimators based on these) may not fully reveal the properties of the foraging habitats of pelagic feeders unless they are considered at appropriate temporal and spatial scales.

In this chapter, I will explore some of the aspects of marine mammal habitat use, including habitat preferences, home range, migration patterns and site fidelity. I will also review some cases of changes in marine mammal habitats due to environmental processes and anthropogenic effects. Further, I will discuss the problems of choosing the best possible spatial and temporal scales when planning a project. The examples are taken from international scientific literature, but with emphasis on the Northeast Atlantic and the European Arctic.

The review is not intended to be exhaustive, and students interested in planning their own projects related to marine mammal habitat use, are recommended to survey relevant literature. However, this chapter may serve as a source of ideas and key words to initiate such literature surveys. The rest of the chapter is referenced with citations to recent literature as well as some of the classical works in animal behaviour. Finally, some examples of hypotheses to be tested and methods to be used are taken from my own work on harbour seals *Phoca vitulina* at the Norwegian coast.

2. MARINE MAMMAL HABITATS AND HABITAT DEGRADATION

2.1 Characteristics of Marine Mammal Habitats

Habitat is defined as the locality or environment in which a plant or animal lives (Lawrence, 1989). There is a rich scientific literature describing studies of habitats and habitat use in terrestrial animals. Habitats of most terrestrial mammals are at the ground level and in principle two-dimensional. These habitats may often be delineated by fixed ground coordinates, and characterised by ground or surface moisture, angle of inclination, vegetation, *etc.* Although seasonal and inter-annual changes in vegetation, moisture or snow coverage normally occur, the landscape is a relatively stable configuration at the time-scale of a mammal's life span.

Marine mammals live in a volume of water, *i.e.* a three-dimensional environment. The marine environment constitutes a continuum (virtually the opposite of island habitats). However, marine mammals are not evenly distributed across the world's oceans. A set of more specific factors is therefore important in determining the distribution of the various species. Each locality is associated with some favourable (*e.g.* high density of prey) and some disadvantageous (*e.g.* high risk of encounters with predators) characteristics with regard to suitability as habitat for a species. Preferred habitats are often environments possessing an optimal balance between favourable and disadvantageous characteristics.

The favourable factors may be grouped into two broad categories: environmental factors that are suitable for breeding, and factors suitable for nourishment of the animals. Therefore, the home range of a marine mammal is often divided into a breeding habitat and a foraging habitat. When these habitats are not overlapping, they may be linked by a migration route. The California gray whale *Eschrichtius robustus* is an example of a species with geographically separated breeding and foraging habitats linked by a well-defined migration route. In other species, breeding and foraging occur in the same general area and no seasonal migrations are observed, *i.e.* the foraging habitat and the breeding habitat are overlapping such as in some of the pelagic dolphins.

2.2   Habitat Changes and Anthropogenic Effects

The oceans are in continuous change: this includes long-term changes due to tectonic movements of the continents and changes in sea level (*e.g.* the closure of the Central American Atlantic-Pacific connection at the geological time scale), and rapid changes due to oceanographic fluctuations or oscillations with subsequent effects on primary and secondary production (*e.g.* the El Niño Southern Oscillation, ENSO, at the scale of years or decades, and formations of eddies at the scale of days or weeks). In this ocean of variability, marine mammal habitats are also changing at different temporal scales.

Habitats formed by eddies, thermoclines, and fronts may shift from one locality to another several or many times during the life span of a marine mammal. Pelagic inhabitants in such habitats often change geographic distribution to remain within their preferred habitat, and large shifts in distribution may occur without significant effects on population size.

Habitats determined by geomorphologic factors (*e.g.* depth, available haul-out sites, *etc.*) are relatively stable over time with regard to location. When food availability changes, the inhabitants may be forced to move to another area which may possess less favourable physiographic properties, or they may stay within their former range. In order to equalise for changes in carrying capacity, adjustments in growth and fecundity may occur with subsequent effects on the population's size. The stronger site fidelity a population possesses regarding a geographical site (*e.g.* haul-out site), the more it may experience the effects from changes in local food availability.

Habitat degradation may be described as a shift in the characteristics of an area from favourable factors to increased disadvantageous factors. Human activities frequently influence marine mammal habitats, and often in a negative direction (cf. ICES, 1998). In addition to reduced quality of the habitat, *e.g.* by chemical or acoustic pollution, or over fishing of prey populations, anthropogenic effects might also include habitat loss (the habitat is completely removed, or the size or availability of the preferred habitat is reduced).

Marine mammal habitats formed by sea ice are sensitive to global or regional changes in temperature (see Tynan and DeMaster, 1997). In the European Arctic, special food webs of importance to marine mammals are dependent on ice-edge-related primary production and associated amphipod and fish fauna (Syvertsen, 1991; Gulliksen and Lønne, 1989; Lønne and Gulliksen, 1989). In addition, pinnipeds are dependent on ice as a platform for resting, parturition and lactation. Global warming may reduce or even eliminate suitable pinniped ice habitats (*e.g.* for seals resident in the Baltic).

2.2.1   Chemical Pollution and Habitat-related Health Deficiency

Chemical pollution is transferred to marine mammals primarily through their diet. In particular, populations foraging in coastal habitats may suffer health deficiencies due to local concentration of contaminants in the food web.

Three species of seals are resident in the Baltic: grey seals *Halichoerus grypus*, harbour seals, and the endemic subspecies of ringed seal *Phoca hispida botnica*. In the 1970s, lesions of the reproductive system, such as stenosis and occlusion of the uterus, were described in seals from Swedish and Finnish waters (Helle, 1980; Helle *et al.*, 1976a, 1976b; Olsson *et al.*, 1994; Bergman, 1997). These lesions have been attributed to high PCB and DDT levels in the seals. Between 1977 and 1986, about 42% of 4-year-old female Baltic grey seals that were examined showed stenosis or occlusions of the uterus, whereas in 1987-1996 only 11% had those lesions. The pregnancy rate during the first period was only 17% whereas in the second period it increased to 60% (Bergman, 1997).

Several authors have described loss of bone substance on the skull and asymmetry of the skull for seal species from the North and Baltic Seas (Stede and Stede, 1990; Bergman *et al.*, 1992; Mortensen *et al.*, 1992; Olsson *et al.*, 1994). Swedish investigations suggested that lesions of the skeletal system were related to hyperplasia of the adrenal cortex, caused by a high load of organochlorines (Bergman and Olsson, 1986; Bergman, 1997).

A higher prevalence of tumours, such as leiomyoma, was also described for the Baltic grey seal (Bergman, 1997). About 53% of females investigated between 1977 and 1986 showed leiomyomas, whereas between 1987 and 1996, 44% carried these benign tumours, which in some cases may hamper reproductive success.

These deficiencies in health status and reproduction were not recorded to the same degree in seals from neighbouring Atlantic waters, and it has been suggested that the improved health status of Baltic seals observed recently, are caused by a declining contaminant burden of the Baltic biota (ICES, 1997).

Investigations of beluga *Delphinapterus leucas* resident in the St. Lawrence estuary in North America revealed high prevalence of tumours, including several types of cancers (Martineau *et al.*, 1985; De Guise *et al.*, 1994a). In total, 28 of the 75 (37%) tumours, and 13 of 27 cases of cancer (47%), reported in cetaceans world-wide have been found in this single geographically isolated and genetically closed beluga population (Geraci *et al.*, 1987; De Guise *et al.*, 1994a; IWC, 1999; Martineau *et al.*, 1999). Except for eight beluga with gastric papillomas caused by papillomaviruses

(De Guise *et al.*, 1994b), the etiology of the beluga tumours remains unclear. It has been suggested that this high prevalence was due to carcinogenic compounds and decreased resistance to the tumour development (De Guise *et al.*, 1995). High levels of toxic compounds, such as polychlorinated biphenyls (PCBs) and polycyclic aromatic hydrocarbons (PAHs) are found in the St. Lawrence beluga (Martineau *et al.*, 1987). Some PAHs are among the most potent carcinogens, acting as initiators whereas other compounds, such as PCBs are recognised as promoters for the induction of tumours in initiated cells. The prevalence of tumours is higher, and the reproductive performance of this population is lower, than in the Canadian Arctic populations of beluga (Beland *et al.*, 1993; De Guise *et al.*, 1995). However, despite these deficiencies, recent surveys suggest recovery of the beluga population in the St. Lawrence estuary after hunting was closed in 1979 (Kingsley, 1998).

2.2.2   Acoustic Pollution and Displacement from Preferred Habitats

Change in behaviour is frequently expected to be a typical marine mammal response to acoustic disturbance (ICES, 1998); however, this has rarely been examined critically. One of the very few experiments testing for effects of anthropogenic sound on marine mammal behaviour in open oceanic waters was conducted in relation to the ATOC sound source at Pioneer Seamount, 85 km west of San Francisco. (The Acoustic Thermometry of Ocean Climate project is designed to repeatedly measure the speed of sound in the ocean over time in order to determine whether the oceans, which are the main heat sink of the planet, are warming.) For that experiment, 20-minute sequences of high-intensity (195 dB re 1μPa@1m), low-frequency (mainly 60-90 Hz) sound pulses were transmitted at depths of around 900 m every four hours for more than 24-hours.

Aerial surveys were conducted in 1995-97 in an 80 by 80 km box centred around the sound source. Experimental surveys (within 20 hours after the end of a 24-hour cycle of sound transmission) were compared with control surveys (flown at least 48 hours after the end of a sound transmission cycle). A total of 22,117 marine mammals were sighted during 34,095 km of control effort, and 23,068 marine mammals were sighted during 34,808 km of experimental effort. No significant differences in numbers of marine mammals of any species were detected between control and experimental surveys. However, there were significant differences in how two cetacean species, humpback whales *Megaptera novaeangliae* and sperm whales *Physeter macrocephalus*, were distributed in relation to the sound source. Both species were on average further from the sound source during experimental periods (Calambokidis *et al.*, 1998).

Two types of acoustic devices are developed to reduce interactions between marine mammals and the fishing industry. Deterrents are designed either to divert marine mammals from nets in order to reduce entanglement, or to keep seals away from fish farms to reduce or prevent damage to the farmed stock. Alarms (pingers) are constructed to make fishnets more detectable to marine mammals to avoid incidental mortality. In coastal waters, deterrents placed on fish farms have been shown to be effective for a period of weeks in keeping seals away, but the seals may habituate to the sounds, which may even become an attractant (Reeves *et al.*, 1996). Jefferson and Curry (1996) reviewed the various attempts up to 1995 to reduce interaction with fisheries, and they concluded that alarms greatly reduced the interaction between baleen whales and fish traps, but so far similar success was not demonstrated with small cetaceans and gillnet fisheries. Woodward and Goodson (1994) suggested that higher frequencies are more aversive and that the strongest avoidance is in response to increasing signal bandwidth. Kastelein *et al.* (1995) found distinctly stronger porpoise reactions to alarms whose signals included strong harmonics than to those with no harmonics. Recent experiments have demonstrated significant reduction of incidental mortality of harbour porpoises in nets with acoustic alarms (Gearin *et al.*, 1998). The possible problems that may appear if acoustic alarms are evolving towards deterrents, and they are deployed on fishing gear in a commercial fishery with high gear density over relatively large areas, are not yet fully explored. Displacement of marine mammals from habitats with abundant prey populations may be an unwanted effect of an effort to reduce incidental mortality.

3. EXAMPLES OF MARINE MAMMAL HABITATS AND HABITAT USE

3.1 Oceanic Pelagic Habitats

3.1.1 Pilot Whales and Temperature-related Habitats

Pilot whales, genus *Globicephala*, are primarily oceanic, occurring over deep oceans, but often entering shallow, coastal waters. In the Northern Hemisphere, the long-finned pilot whale *G. melas* occurs only in the Atlantic. In the Faroe Islands there is a long history of traditional drive harvest of long-finned pilot whales. Catch statistics are available for almost three hundred years. Information at this time scale is useful for comparison with environmental information, *e.g.* long-term climatic changes. The catch records show large-scale fluctuations with two peaks, in 1710-1730 and 1830-1850, and a high level of catches since 1935. Within the recent period, there are several narrower peaks (Zachariassen, 1993). The two large peaks correlate well with periods with warmer climate (Bloch *et al.*, 1990), and the narrower peaks within the recent period correlate well with short-term fluctuations in North Atlantic climate and higher water temperatures at the Faroe Islands (Zachariassen, 1993). However, it is not evident that the temperature influences the occurrence of pilot whales directly. A positive correlation also exists between the occurrence of the pilot whales and their main prey, the pelagic squid *Todarodes sagittatus* (Bloch *et al.*, 1990) which may be influenced by temperature directly, or indirectly through effects on hydrography or productivity. It seems likely that the occurrence of pilot whales at the Faroe Islands may be determined by prey availability and not by the temperature boundary of their own "habitat". Martin (1990) writes that *G. melas* is probably not a migratory species in the typical sense, but that the whales more likely go to habitats where prey occurs and remain there as long as adequate prey stocks persist.

3.2   The Shelf Edge and Slope

3.2.1   Risso's Dolphin, a Sub-tropical Slope Dweller

Risso's dolphins *Grampus griseus* occur in all deep oceanic tropical and warm temperate waters, including most of the Gulf of Mexico. On a large geographical scale it therefore would be correct to list the Gulf of Mexico as a Risso's dolphin habitat (see «known range» and habitat description in Martin, 1990). Baumgartner (1997) studied the distribution of Risso's dolphins in the northern Gulf of Mexico on a much finer geographical scale based on a total of 9,102 km of shipboard surveys and 72,300 km of aerial surveys. His analyses showed a significantly non-random distribution of Risso's dolphins with respect to both depth and depth gradient. Gulf of Mexico Risso's dolphins utilise a core habitat with depths of 350-975 m and depth gradients greater than 24 m per 1.1 km, and they were never sighted landward of the 250-m isobath. The core habitat is the steep upper section of the continental slope, including only 2% of the total surface area of the Gulf of Mexico. Within this narrow habitat, sighting rates were nearly five and six times higher than the average for shipboard and aerial surveys, respectively.

It is, however, unlikely that physiography attracts the dolphins directly. Baumgarter (1997) discussed several plausible trophic reasons why the Risso's dolphins prefer steep sections of the upper continental slope. The shelf edge is often associated with oceanic fronts and high biological production, and in the north-eastern Gulf the shelf waters interact with the Loop Current. Baumgartner (1997) suggested that the shortfin squid *Illex illecebrocus*, known to be abundant in these waters, may be the primary prey species for Risso's dolphins and a major determinant of its distribution in the Gulf of Mexico.

The upper continental slope and shelf break region is a transition zone between two distinct ecosystems, and an alternative hypothesis is that the Risso's dolphins may take advantage of a wide variety of cephalopod prey. A predator situated along this wind and current driven boundary is in a strategic position to benefit from advection of prey species into its vicinity. As fronts move on or off the shelf and upper continental slope, the Risso's dolphins may move with them to continue to take advantage of prey aggregations. Once the formations dissipate the dolphins may move back to the core habitat to meet the next shelf-break front (Baumgartner, 1997). To investigate this hypothesis, a suitable geographic scale and survey frequency are required in addition to simultaneous recording of relevant oceanographic data.

Figure 1. A pod of Risso's dolphins forages for octopus and cuttlefish on a 30-50 m slope in West Scotland (Photo: P.G.H. Evans).

3.2.2   Ice-breeding Hooded Seals Travel Far to Forage at the Shelf Break

The hooded seal *Cystophora cristata* is a North Atlantic pinniped (Fig. 2). There are three separate breeding grounds of this ice- breeding species: at Newfoundland, in the Davis Strait, and off Jan Mayen in the East Greenland Sea. Outside the spring breeding and mid summer moulting seasons, hooded seals are believed to be pelagic and range widely. Very little was known about foraging movements and foraging habitats. However, in a recent study, Folkow and Blix (1995) deployed satellite-linked transmitters on post-moult hooded seals at the resting sites on ice floes off East Greenland, which made it possible to follow individual seals over long distances.

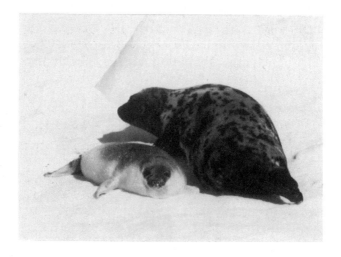

Figure 2.   Female hooded seal with pup (Photo: I. Stirling) .

The seals dispersed widely and shelf-break and slope waters from Svalbard to the west of the British Isles were visited (Fig. 3). Several of the tagged seals regularly crossed the Northeast Atlantic on 3-7 weeks excursions to alter between resting sites on the ice floes and foraging grounds at the continental slope off north-western Europe (Folkow and Blix, 1995; 1999).

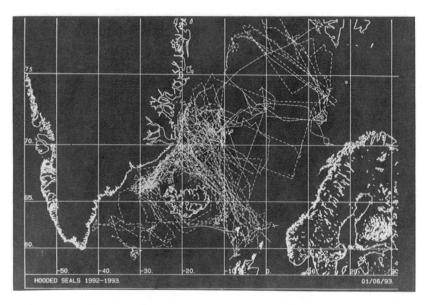

Figure 3.   Tracks of satellite-tagged hooded seals between the ice floes off East Greenland and the continental slope off Western Europe (Courtesy of L. Folkow, Dept of Arctic Biology, University of Tromsø).

Folkow *et al.* (1996) showed that hooded seals from the Greenland Sea stock spent 15% of their time in the waters around the Faroe Islands and off the west coast of the British Isles. These seals foraged at depth of 100 - 300 m off the Faroes in early autumn. A shift towards deeper dives (300-600m) off the Hebrides was observed during winter months. These shelf-edge and upper-slope foraging grounds seem to be determined by the bathymetry and possibly linked to the occurrence of certain prey species in these areas and depths.

The blue whiting *Micromesistius poutassou* spawns in April-May in waters west of the British Isles at 300-600 m depth and is found in large numbers at 100-300 m at the continental edge during the rest of the year (Bailey, 1982). The blue whiting displays diurnal vertical migration from 100-200 m during daylight hours to 300-500 m depth at dusk (Bailey, 1982). Folkow and Blix (1995) pointed out that for those seals that migrated towards the Faroes and British Isles, the seasonal distribution of the foraging excursions and the diurnal and seasonal variations in dive depths match the known distribution of the blue whiting quite well. However, to verify this plausible blue whiting hypothesis, further information may be required, *e.g.* on stomach contents of hooded seals foraging in this area.

3.3   Shelf Waters

3.3.1   North-east Atlantic Minke Whales

Minke whales *Balaenoptera acutorostrata* (Fig. 4) are thought to migrate between high latitudes in summer and lower latitudes in winter. However, migration routes are not well documented. During the summer season, the Northeast Atlantic (NEA) stock (IWC definition) of minke whales forages in shelf waters off the coasts of north-western Europe from the North Sea along the Norwegian coast to the Barents Sea. The majority of the NEA stock of minke whales forages in the Barents Sea. Biological production in the Barents Sea is high, about 4 million TJyr-1 (Sakshaug *et al.*, 1994). (This is equivalent to about 10 times the total annual production of electricity in Norway.) This very high production is caused by the 24-hour daylight in summer, relatively shallow waters, and the nutrient-rich mixing zone (the Polar Front) between Atlantic and Arctic waters.

Figure 4.   Minke whale feeding amongst seabirds (mainly kittiwakes).
(Photo: P.G.H. Evans)

The Atlantic water has a salinity of >3.5% and a temperature of >3°C, while Arctic water has a salinity and temperature of 3.43-3.48% and <0°C, respectively. The influx of Atlantic water into the Barents Sea fluctuates, and the location of the Polar Front fluctuates accordingly. These shifts in location of the Polar Front cause changes in the spatial distribution of primary and secondary production. The geographical position of the Polar Front and the water temperature of the Barents Sea influence the species composition of the primary and secondary production with subsequent effects on distribution and year-class strength of fish species foraging on the secondary production. In years with large influx of Atlantic waters into the Barents Sea, the Polar Front has an easterly and northerly distribution, and the conditions are favourable for the establishment of strong year-classes of boreal fish species such as herring *Clupea harengus* and cod *Gadus morhua*. In years of reduced influx of Atlantic waters, the Polar Front has a westerly distribution, and Arctic fish species are favoured, *e.g.* polar cod *Boreogadus saida* and capelin *Mallotus villosus*. Pelagic schooling species such as herring and capelin forage on zooplankton and turn the production of lower trophic levels into food available for piscivorous species in higher trophic levels, *e.g.* seabirds, mammals and large fish.

When comparing "producers" and "consumers" in the Barents Sea, we find that crustaceans, in particular copepods, are very important in the lower trophic levels with an annual production 450,000 TJyr^{-1}. In an "average" year, the production of capelin is 15,000 TJyr^{-1} (the abundance of the short-lived capelin fluctuates 1,000-fold in the Barents Sea), and the production of cod is 7,000 TJyr^{-1}. The consumption of some of the key consumers is 25,000 TJyr^{-1} for cod, 16,000 TJyr^{-1} for whales and 3,800 TJyr^{-1} for seals (Sakshaug *et al.*, 1994).

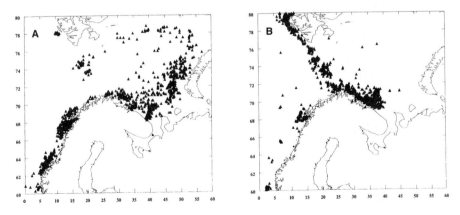

Figure 5.　　Shifts in distribution of foraging minke whales based on positions of harvested minke whales in a) 1952 and b) 1980.

Large between-year shifts in distribution of foraging minke whales have been observed within the overall Barents Sea habitat (Fig. 5). There are also between-season and within-season switches in the diet of minke whales in the Barents Sea; krill and capelin may alternate as the most predominant prey species in the northern part, while herring is dominating the diet in the southern Barents Sea in some years (Haug *et al.*, 1995). These switches between prey species and shifts in spatial distribution are believed to be related to the oceanographic conditions and species composition of the prey communities. These shifts in foraging habitats and diet are not likely to be detected unless the temporal scale of the research allows for years encompassing the full range of influx of Atlantic waters into the Barents Sea to be included in the investigations.

3.4   Benthic and Near Shore Habitats, Ice Edges and Polynyas

3.4.1   Walrus, the Giant of the Arctic Shallows

Walruses *Odobenus rosmarus* have a disjunct circumpolar distribution and occupy a narrow ecological niche (Fig. 6). Three important factors in determining their habitats are: 1) the availability of large areas of shallow (80 m or less) waters with suitable bottom substrate to support a productive bivalve community; 2) the reliable presence of open water over rich feeding areas; and 3) the presence of haul-out platforms (mostly ice floes during winter and spring and terrestrial sites during summer and autumn (Born *et al.*, 1995).

Walruses are able to crack ice up to 20 cm; however, when ice grows thicker, they are forced to areas where leads and polynyas are numerous and the ice is thick enough to support their weight (Born *et al.*, 1995; Burns *et al.*, 1981; Fay, 1982). A strong preference for hauling out on floes of first-year ice with areas ranging between 50 and 400 square metres has been observed (Wartzok and Ray, 1980, 1981).

Soft parts of bivalve molluscs constitute about 95% (by weight) of walrus food (Fay *et al.*, 1977; Fay and Stoker, 1982, Wiig *et al.*, 1993). The narrow ecological niche and dependence on certain habitats made the walrus in the Northeast Atlantic very vulnerable to exploitation. In the Norwegian Arctic (as in the rest of the western Eurasian Arctic), walruses suffered a severe over-exploitation beginning in about 1600. This exploitation was initiated by the Basques (primarily whaling) and continued with expeditions from Holland, Great Britain, Russia (primarily walrus hunting) and also Norwegians from the late 1700s onwards (Fig. 7).   Fedosev (1976) suggested a pre-exploited population of the Barents Sea and Kara Sea region in excess of 70 to 80 thousand animals. Three and a half centuries of

Figure 6.   Portrait of a young walrus at Svalbard. This highly specialised pinniped occupies a narrow ecological niche and is vulnerable to changes in its habitats. (Photo: A Bjørge)

exploitation had practically exterminated walrus from the Norwegian Arctic when the species became protected in 1952 (1956 in the Soviet Union). About 1970, walruses started to reappear at Svalbard, possibly as dispersal from remnant groups of walruses at Franz Josef Land. Today, walruses are distributed around the northern and eastern parts of the Svalbard archipelago, where aerial surveys in 1992-93 showed a minimum of 723 animals hauled out (Gjertz and Wiig, 1995). These were predominantly males. The females and calves were believed to reside at Franz Josef Land. Assuming a 1:1 sex ratio, the total population of the Svalbard - Franz Josef Land region now numbers about 2,000 animals.

Recent tagging of walrus at Svalbard using satellite-linked transmitters confirmed that the Svalbard - Franz Josef Land walruses constitute one population and that males frequently move between areas and utilise the shallows at Svalbard for foraging (Wiig *et al.*, 1993). Franz Josef Land therefore may serve as the main breeding habitat and foraging grounds for females of the same population.

The narrow ecological niche also makes walruses vulnerable to threats other than exploitation. The increasing tourist industry is of concern for this recovering but still very small population. New commercial activities may have detrimental effects on walrus habitats. At Svalbard there were rich concentrations of scallops (*Chlamys islandica*) between 20 and 70 m depth.

Figure 7.   Bones and skulls of walruses on the Moffen Island at Svalbard. Note that all skulls are without tusks. The island is a memorial site for the over-exploitation of walruses during the seventeenth and eighteenth centuries. Removal of bones is illegal. The island is protected as a nature reserve, and human access is prohibited. The walruses are now recolonising this area. (Photo: A. Bjørge) .

Beginning in the early 1980s, a commercial fishery for scallops developed. The trawlers operated three dredges, each up to 5 metres wide, in the deeper part of the scallop banks. The most efficient trawlers harvested up to 50 tons of living scallops per day. This fishery ploughed important parts of the foraging habitat of the walrus and may have caused significant mortality to the main prey of walrus, the soft-shelled bivalves. The scallop fishery collapsed after the 1987 season.

All major walrus habitats at Svalbard are now permanently protected as nature reserves; fishing, dredging and mineral and oil exploitation are prohibited, while human access or motorised traffic at the haul-out sites are either prohibited or subject to regulations. However, as human use of the Arctic increases and changes, further investigations on the habitat requirements of walrus are needed to ensure continued conservation of this specialised pinniped.

3.4.2 Harbour Seal, Resident and Littoral

The harbour seal has typically been regarded as non-migratory and littoral in distribution (Bigg, 1981), and as exhibiting a diurnal haul-out pattern (Stewart, 1984; Roen and Bjørge, 1995). Recent studies howev·r (*e.g.* Thompson, 1989; Thompson and Miller, 1990; Thompson *et al.*, 1991), have revealed seasonal and inter-annual movements between a few haul-out sites and between foraging locations. Thompson (1993) nevertheless concluded that harbour seals in Scottish waters are resident in the same geographical area throughout the year. Their diurnal haul-out patterns and limited travelling speed, indicate that harbour seals forage within a few kilometres of their haul out sites. This is supported by studies of foraging movements of radio-tagged seals where most foraging activity was less than 50 km from haul-out sites (Stewart *et al.*, 1989; Thompson and Miller, 1990; Thompson *et al.*, 1991; Thompson, 1993). Therefore, a typical harbour seal habitat should provide suitable haul-out sites, shelter during the parturition and lactation periods, and sufficient food within reach of the haul-out sites to sustain the population throughout the year.

The littoral and non-migratory habits of the harbour seals make them accessible for behavioural studies, and in the last part of this chapter I will use harbour seals as example to present some ideas that may encourage students in planning their own projects.

Figure 8. Harbour seal haul-out in the Shetland Islands, Scotland (Photo: P.G.H. Evans) .

4.   PLANNING A STUDENT PROJECT ON MARINE MAMMAL BEHAVIOUR AND HABITAT USE

When students start planning a project, the starting point is often an idea of which species and what aspects of behaviour they would like to investigate. The first step is to narrow down the scope of the project to one (or more) hypothesis that is possible to verify or reject, or to a well-defined question to answer. Give due consideration to the spatial and temporal scales for your sampling, and make sure you choose a sampling method or technique that gives consistent data throughout the period and area of investigation (see Pribil and Picman, 1997), and which is likely to provide sufficient accuracy in your results. In the initial phase, you should also consider the nature and amount of data that are feasible to obtain within the time frame and financial budget available for your project, and then choose statistical methods and tests relevant for that particular type of data. Then you need to consider the amount of data, number of replicates, *etc.*, that may be required to obtain sufficient statistical power and precision of your estimates.

Define and describe precisely the types of activity or activity patterns you are planning to record. For most behavioural studies you need to be able to identify and observe one individual conducting one specific activity or a sequence of specific activities. Individual recognition may be important for your study, and there is a variety of different methods available ranging from direct visual observation of natural markings to indirect methods such as audio-recognition, biopsy sampling, tagging, branding or deployment of data-loggers or satellite-linked radio tags (see Read, 1995).

4.1   Examples From Studies of Harbour Seals at the Norwegian Coast

The Norwegian coast is topographically very complex. The mainland coast is indented by a large number of fjords, and some fjord basins are more than a thousand metres deep. The outer coast is fringed by a large number of islands and islets, and countless inter-tidal rocks, constituting large archipelagos reaching far off the mainland coast. Harbour seals haul out on inter-tidal rocks in these archipelagos, and the complex topography makes a diversity of habitat types, in terms of depth, substrate, vegetation and prey communities, available close to the haul-out sites. For a number of years, my colleagues and I have studied harbour seal habitat use at the Norwegian coast (Bekkby and Bjørge, 1998a; Bjørge, 1991; Bjørge *et al.*, 1994, 1995; Braaten *et al.*, 1998; Roen and Bjørge, 1995).

Although haul-out bouts were recorded both during day and night and at high and low tide, there was a tendency for seals to haul out more frequently during the day and at low tide (Roen and Bjørge, 1995). Their at-sea behaviour was monitored by a combination of VHF-radio and ultrasonic telemetry (Bjørge *et al.*, 1995). After a trip to sea, seals regularly returned to the same haul-out rock or to adjacent rocks in the same area (Fig. 9a). The radio-tagged seals typically hauled out among other seals on inter-tidal rocks in clusters of small islets. They were social on land, but solitary when foraging at sea. Individual seals utilised different foraging habitats, and they repeatedly returned to the same or same type of habitat throughout the period of tracking. Most of the seals used 100 -200 m deep basins with soft substrate for foraging; other seals used shallows with kelp forest, sandy bays and slopes between the kelp zone and soft substrate basins. All radio-tagged seals foraged at or close to the sea floor. The utilisation of different foraging habitats may be a result of intra-specific competition for foraging sites and benthic fish resources in close proximity to haul-out sites. When large schools of pelagic fish, *e.g.* herring, entered the area, the harbour seal diet switched to pelagic fish until the abundant resource moved out of the area (Bjørge, 1993; but see also studies of changing habitat quality and squirrel populations by Lurz *et al.*, 1997).

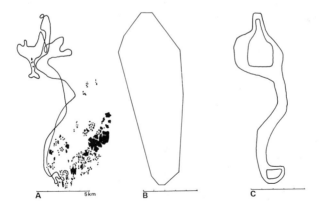

Figure 9.   a) A foraging trip of a radio-tagged adult male harbour seal. This seal repeatedly travelled about 20 km from his haul-out site to forage at some offshore shallow rocks. Between foraging trips he returned to the same haul-out site. (This illustration is composed to demonstrate a typical foraging trip of this particular seal. It is based on discontinuous track lines from more than one trip, recorded when tracking the tagged seal with a hydrophone mounted onboard an inflatable boat.) b) An illustration of home range of this male harbour seal presented as minimum convex polygon. c) A schematic illustration of the same home range if estimated by the harmonic mean method (*e.g.* 90% and 60% isopleths).

During the breeding season, mature males held underwater "territories", vocalising whilst performing a distinct type of dives. These dives were defined as display dives, and the display sites or "territories" were mainly in the deeper channels penetrating into the shallow water plateau between the haul-out sites which were used by females with pups and juvenile seals (Bjørge *et al.*, 1995).

In our studies of VHF-radio-tagged harbour seals (Bjørge *et al.*, 1995; Bekkby and Bjørge, 1998a), we divided the activity into 1) hauling out (including both resting and lactation); 2) transit swimming, 3) foraging, 4) display diving (males only); and 5) other (including swimming at or close to the surface around haul-out sites). We considered that continuous VHF signals for 8 minutes or more indicated that seals were hauled out. When the seals were at sea, we used dive profiles to characterise their behaviour. Travelling to and from the foraging grounds, the seals moved in typical V-shaped dives. Usually we observed several consecutive V-shaped dives in a constant direction, and we defined these dives as transit dives. These dives did not always reach to the sea floor.

When seals ceased directed travel at sea, a different type of dive was observed. The dive profile was U-shaped, the dives usually reached the sea floor, and the swimming direction frequently changed during and between dives. We defined these dives as foraging dives. The stomach temperature was monitored and used as further support in identification of foraging activity (Bekkby and Bjørge, 1998b; Bjørge *et al.*, 1995). Significant drops in temperature were interpreted as ingestion of food. On a few occasions, in particular when seals foraged in areas with complex topography, transit and foraging type dives were mingled. Durations of transit and foraging dives were not significantly different, and the average dive duration was 3.3 minutes (s.d. 1.9 minutes). The longest dive recorded was 14.3 minutes. For both transit and foraging dives the swimming speed was typically between 1.1 and 1.6 m/sec (but see Braaten *et al.*, 1998).

4.2   Home Range Estimation

Definition: Home range is here defined as the area one animal utilises for its normal activities related to foraging, mating and (in females) lactation. This definition is modified after Burt (1943).

Examples and Methods: A subject of interest is to characterise the home range of an animal under the assumption that the animal is not migrating at the time. The description of home range may include computation of: the centre of activity (Hayne, 1949); the distribution of activity radii (Tester and Siniff, 1965) and an index of the magnitude of the home range (Mohr, 1947; Sanderson, 1966). Home range has conventionally been estimated as the area of the minimum convex polygon (Fig. 9b) that could be established by drawing lines through distal plots of location for an animal (Mohr, 1947). However, two other approaches have been used to estimate home ranges: bivariate normal ellipses (*e.g.* Dunn and Gibson, 1977) and nonparametric methods based on grids (*e.g.* Dixon and Chapman, 1980). One problem in the minimum convex polygon method is that occasional movements may result in significant and sometimes dramatic changes in the size and shape of the area, and the described home range no longer reflects the regular activity of the animal.

When using tracking data from radio-tagged animals, subsequent locations are likely not to be independent. Most of the home range estimators in use (except the Dunn-Gibson estimator) require that location data are independent (White and Garrott, 1990). The precision of an estimated home range will therefore diminish when the time span between locations used for the estimate decreases.

Neither the elliptical nor convex polygon method allows home ranges with concave shapes. A non-parametric home range estimator based on the harmonic mean of the area distribution was suggested by Dixon and Chapman (1980). This method allows home ranges with two or more centra of activity, which is useful for most marine mammals where there is a migration between breeding and foraging grounds (Fig. 9c). The method has no restrictions on the contours of the home range. The probability of encountering the animal at intersections of an arbitrary grid is estimated. Contours with equal probability (isopleths of activity) are established by harmonic mean of the distances from the grid node to the observed tracking locations. Grid nodes located at centres of activity have the shortest distances and the lowest values of harmonic mean (White and Garrott, 1990).

However, the harmonic mean is sensitive to the positioning of the grid net and size of grid cells. The kernel method (Worton, 1989) is a modified harmonic mean method which is less grid-dependent and produces more consistent results than the harmonic mean.

Both the harmonic mean and the kernel methods are available in software packages, *e.g.* "Ranges V" distributed by Institute of Terrestrial Ecology, UK (Kenward and Hodder, 1996). Recently, the United States Geological Survey launched "Animal Movement" as an extension to the Geographical Information System (GIS) software ArcView (by Environmental Systems Research Institute Inc.) This extension facilitates the combination of animal tracking data and geographical data, and improves possibilities for straightforward comparison of home ranges and activity centres with habitat maps.

White and Garrott (1990) pointed out that a weakness in all non-parametric methods (and most other home range estimators) is that they produce estimates without confidence intervals. This makes comparison between home ranges difficult.

4.3   Habitat Preferences

Definition: If an animal spends more time in one habitat type than expected by chance alone, this is defined as preference for that particular habitat. In some populations, *e.g.* as shown for harbour seals, individual animals may have preferences for different habitats.

Examples and Methods: To estimate habitat preference, you may need to know the relative size of each habitat type available to the animal. This is often measured on a habitat map. The size of a habitat type can be measured using a planimeter (White and Garrott, 1990). More efficient and accurate is the use of digitised maps and commercially available GIS programmes in combination with statistical packages (see Staus, 1998). For our studies of harbour seals we used ArcView from Environmental Systems Research Institute Inc. Such programmes are usually based on two different methods: grid cell or polygon. In the first method, the maps are divided into grid cells and the dominant habitat type for that cell is coded as the habitat type of that cell. The sum of cell areas for each habitat type then provides the measurement of the area of that habitat type. The accuracy of this method depends on the scale at which the grid cells are defined. In the polygon method, a sequence of x-y coordinates is used to represent the habitat boundary. The line segments are then connected to form polygons for each portion of habitat. The estimated area of a habitat type is then the sum of areas of polygons for that particular habitat type (White and Garrott, 1990).

Hypothesis and tests: In our study of harbour seals we tested for habitat preference, with the null hypothesis that harbour seals have no preference for foraging habitat (*i.e.* there is no significant difference between habitat used and habitat available). This hypothesis may be tested using a chi square test for goodness of fit between used and available habitat types. If the null hypothesis that the animals have no habitat preference is rejected (*i.e.* there is a preference), the next step may be to test for differences between animals. A chi square test of independence may be used for this purpose (White and Garrott, 1990). If sufficient data are accumulated, you may proceed and test whether habitat preferences are dependent upon time of year, age, sex or reproductive status.

4.4 Site Fidelity

Definition: Site fidelity is defined as the tendency of an animal to remain in an area over an extended period, or to return to an area previously occupied (White and Garrott, 1990).

Examples and Methods: Site fidelity may be related to the total home range of an individual, or it may be related to part of the range, *e.g.* the foraging grounds of an animal. In the case of harbour seals at the Norwegian coast, we divided the activity into, among other things, haul-out, foraging and display (for adult males in breeding season only). If sufficient data are accumulated, site fidelity may therefore be tested for haul-out sites, foraging grounds, and display sites, *etc.*

Hypothesis and Tests: To test for site fidelity, you may test if the coordinates of the site change over time, *e.g.* by a multivariate method such as the Hotelling's T2 test. To test for differences between sex or age groups, a Mann Whitney test on x and y values may be used to test for differences between groups. You may wish to test for increasing site fidelity with age as animals mature and eventually establish territories. If data on temporal and spatial distribution of potential prey species are available, site fidelity for foraging grounds, or rather change in foraging grounds, may be used to discuss changes in foraging behaviour in relation to prey availability. Information on site fidelity may also be used in assessing impact human disturbance and studies of habitat degradation. However, when considering site fidelity, you need to pay particular attention to your spatial and temporal scales.

5. SYNOPSIS

The ability of mankind to impact and change the global environment is accelerating. Some of the most severe detrimental anthropogenic impacts are observed in the coastal zone. Estuaries are turned into industrial harbours, wetlands are drained for agricultural purposes, and our coastal waters are used as recipients for sewage and pollutants. Many of our marine mammal populations are resident in coastal waters and their habitats may be severely degraded, directly or indirectly, by human activity. For future conservation of marine mammals, we need further information on the habitat requirements of these animals, and better knowledge on how they respond to changes in their habitats.

Our perception of marine mammal habitats and the persistence of these habitats is dependent upon the temporal and spatial scales by which we are considering the habitats. The temporal and spatial aspects are therefore of vital importance both when you are planning a project, and when you evaluate the significance of your findings.

From a management point of view we may wish to know whether an observed change is an anthropogenic effect, or caused by natural fluctuations or oscillations. However, it is too late for documenting baseline data, if studies of habitat use and habitat requirements are initiated when a change is observed. Without the baseline, it is difficult to interpret the nature of and dynamics behind observed changes. Baseline studies and long term research or monitoring are critically important for understanding the persistence of marine mammal habitats in an ocean of variability.

ACKNOWLEDGEMENTS

This chapter has been based upon a lecture for the Second European Seminar on Marine Mammals at the Universidad Internacional Menéndez Pelayo, Valencia to whom I am grateful for inviting me to contribute.

Robert D. Kenney, University of Rhode Island, provided valuable comments to the content and comprehensive suggestions for linguistic improvements. Lars Folkow, University of Tromsø, and Trine Bekkby, Norwegian Institute for Nature Research, commented on a draft version.

Data and results on harbour seal behaviour originate from collaborative projects by the Norwegian Institute for Nature Research, Oslo, Norway, the Sea Mammal Research Unit, St Andrews, UK, and Institute of Marine Research, Bergen, Norway.

REFERENCES

Bailey, R.S. (1982) The population biology of blue whiting. *Advances in Maine Biology,* **19,** 257-355.

Baumgartner, M.F. (1997) The distribution of Risso's dolphin (*Grampus griseus*) with respect to the physiography of the northern Gulf of Mexico. *Marine Mammal Science,* **13**(4), 614-638.

Bekkby, T. and Bjørge, A. (1998a) Behaviour of pups and adult female harbour seals *Phoca vitulina* during late lactation and first post weaning period. World Marine Mammal Science Conference, January 1998 [Abstract]

Bekkby, T. and Bjørge, A. (1998b) Variation in stomach temperature as indicator of meal size in harbour seal, *Phoca vitulina. Marine Mammal Science,* **14**(3), 627-637.

Béland, P., De Guise, S., Girard, C., Lagace, A., Martineau, D., Michaud, R., Muir, D.C.G., Norstrom, R.J., Pellitier, E., Ray, S. and Shugart, L.R. (1993) Toxic compounds and health and reproductive effects in St. Lawrence beluga whales. *Journal of Great Lakes Research,* **19**, 766-775.

Bergman, A. (1997) Trends of disease complex in Baltic grey seals (*Halichoerus grypus*) from 1977 to 1996: improved gynecological health but still high prevalence of fatal intestinal wounds. WP 19 presented at the ICES Working Group on seals and small cetaceans in European seas, April 1997 (unpublished).

Bergman, A., and Olsson, M. (1986) Pathology of Baltic grey seal and ringed seal females with special reference to adrenocortical hyperplasia: Is environmental pollution the cause of a widely distributed disease syndrome? *Finnish Game Research,* **44**, 47-62.

Bergman, A., Olsson, M., and Reiland, S. (1992) Skull-bone lesions in the Baltic grey seal (*Halichoerus grypus*). *Ambio,* **21**(8), 517-519.

Bigg M.A. (1981) Harbour seal *Phoca vitulina* Linnaeus, 1758 and *Phoca largha* Pallas, 1811. In: *Handbook of Marine Mammals.* Volume 2: Seals. (Ed. by S.H. Ridgway and R.J. Harrison), pp. 1-27. Academic Press, London.

Bjørge A. (1991) Status of the harbour seal *Phoca vitulina* L. in Norway. *Biological Conservation,* **58**, 229-238.

Bjørge, A. (1993) *The harbour seal, Phoca vitulina L., in Norway and the role of science in management.* Dr.scient thesis. Institute of Fishery and marine Biology. University of Bergen.

Bjørge, A., Steen, H. and Stenseth, N.C. (1994) The effect of stochasticity in birth and survival on small populations of the harbour seal *Phoca vitulina* L. *Sarsia,* **79**, 151-5.

Bjørge, A., Thompson, D., Hammond, P.S., Fedak, M.A., Bryant, E., Aarefjord, H., Roen, R. and Olsen, M. (1995) Habitat use and diving behaviour of harbour seals in a coastal archipelago in Norway. In: *Whales, seals, fish and man.* (Ed. by A.S. Blix, L. Walløe and Ø. Ulltang), pp. 211-223. Elsevier Science, Amsterdam.

Bloch, D., Hoydal, K., Joensen, J.S. and Zachariassen, P. (1990) The Faroese catch of long-finned pilot whales. Bias shown of the 280 year time series. *Journal of North Atlantic Studies,* **2** (1-2), 45-46.

Born, E.W., Gjertz, I. and Reeves, R.R. (1995) *Population assessment of Atlantic walrus.* Norwegian Institute of Polar Research, Meddelelser no. 138, 1-100.

Braaten, B.E.P., Bjørge, A., Thompson, D., and Bryant, E. (1998) Strategies for optimising energy expenditure and energy intake during foraging dives of free ranging harbour seals. Paper ICES CM.1998/CC1 presented to ICES ASC, September 1998 (unpublished), 1-22.

Brown, F. and Mate, B.R. (1983) Abundance, movements, and feeding habits of harbor seals, *Phoca vitulina*, at Netarts and Tillamook Bays, Oregon. *Fishery Bulletin,* **81**, 291-301.

Burns, J.J., Shapiro, L.H. and Fay, F.H. (1981) Ice as marine mammal habitat in the Bering Sea. In: *The Eastern Bering Sea Shelf: Oceanography and Resources.* Vol. 2. (Ed. by D.W. Hood and J.A. Calder), pp. 781-797. University of Washington Press, Seattle.

Burt, W.H. (1943) Territoriality and home range concepts as applied to mammals. *Journal of Mammalogy,* **24**, 346-352.

Calambokidis, J., Chandler, T.E., Costa, D.P., Clark, C.W., and Whitehead, H. (1998) Effects of the ATOC sound source on the distribution of marine mammals observed from aerial surveys off Central California. The World Marine Mammal Science Conference, January 1998. [Abstract]

De Guise, S., Lagacé, A., and Béland, P. (1994a) Tumors in St. Lawrence beluga whales (*Delphinapterus leucas*). *Veterinary Pathology,* **31**(4), 444-449.

De Guise, S., Lagacé, A., and Béland, P. (1994b) Gastric papillomas in eight St. Lawrence beluga whales (*Delphinapterus leucas*). *Journal of Veterinary Diagnostic Investigations*, **6**, 385-388.

De Guise, S., Martineau, D., Béland, P., and Fournier, M. (1995a) Possible mechanisms of action of environmental contaminants on St. Lawrence beluga whales (*Delphinapterus leucas*). *Environmental Health Perspectives*, **103** (Suppl. 4), 73-77.

Dixon, K.R. and Chapman, J.A. (1980) Harmonic mean measures of animal activity areas. *Ecology*, **61**, 1040-1044.

Dunn, J.E. and Gibson, P.S. (1977) Analysis of radio telemetry data in studies of home range. *Biometrics*, **33**, 85-101.

Fay, F.H. (1982) Ecology and biology of the Pacific walrus, *Odobenus rosmarus divergens* Illiger. North American Fauna 74. US Dept of the Interior, Fish and Wildlife Service, 1-279.

Fay, F.H., Fedwer, H.M. and Stoker, S.W. (1977) *An estimate of the impact of the Pacific walrus population on its food resources in the Bering Sea.* PB 273 505, Nat. Tech. Inform. Serv. Springfield, 1-38.

Fay, F.H. and Stoker, S.W. (1982) Analysis of reproductive organs and stomach contents from walruses taken in the Alaskan native harvest, spring 1980. Report presented to US Fish and Wildlife Service. Anchorage, 1- 86.

Fedosev, G.A. (1976) Giants of the polar seas. *Priroda*, 1976 (8), 76-83.

Folkow, L.P. and Blix, A.S. (1995) Distribution and diving behaviour of hooded seals. In: *Whales, seals, fish and man*. (Ed. by A.S. Blix, L. Walløe and Ø. Ulltang), pp. 193-202. Elsevier Science, Amsterdam.

Folkow, L.P. and Blix, A.S. (1999) Diving behaviour of hooded seals (*Cystophora cristata*) in the Greenland and Norwegian seas. *Polar Biology*, **22**, 61-74.

Folkow, L., Mårtensson, P.-E. and Blix, A.S. (1996) Annual distribution of hooded seals (*Cystophora cristata*) in the Greenland and Norwegian Seas. *Polar Bio*logy, **16**, 179-189.

Gearin, P.J., Gosho; M.E., Laake, J.L. and DeLong, R.L. (1998) Evaluation of effectiveness of pingers to reduce incidental entanglement of harbour porpoise in a set gillnet fishery. World Marine Mammal Science Conference, January 1998 [Abstract]

Geraci, J.R., Palmer, N.C., and St. Aubin, D.J. (1987) Tumours in cetaceans: analysis and new findings. *Canadian Journal of Fisheries and Aquatic Sciences*, **4**, 1289-1300.

Gjertz, I. and Wiig, Ø. (1995) The number of walrus (*Odobenus rosmarus rosmarus*) in Svalbard in summer. *Polar Biology*, **15**, 557-530.

Gulliksen, B. and Lønne, O.J. (1989) Distribution, abundance and ecological importance of marine sympagic fauna in the Arctic. *Raport proces-verbal de la Reunion Conseil International por l'Exploration de la Mer*. **188**, 133-138.

Haug, T., Gjøsæter, H., Lindstrøm, U., Nilssen, K.T. and Røttingen, I. (1995) Spatial and temporal variations in northeast Atlantic minke whale (*Balaenoptera acutorostrata*) feeding habits. In: *Whales, seals, fish and man*. (Ed. by A.S. Blix, L. Walløe and Ø. Ulltang), pp. 225-40. Elsevier Science, Amsterdam.

Hayne, D.W. (1949) Calculations of size of home range. *Journal of Mammalogy*, **39**, 190-206.

Helle, E., Olsson, M., and Jensen, S. (1976a) PCB levels correlated with pathological changes in seal uteri. *Ambio*, **5**(4), 261-263.

Helle, E., Olsson, M., and Jensen, S. (1976b) DDT and PCB levels and reproduction in ringed seal from the Bothnian Bay. *Ambio*, **5** (4), 188-189.

Helle, E. (1980) Lowered reproductive capacity in female ringed seals (*Pusa hispida*) in the Bothnian Bay, northern Baltic Sea, with special reference to uterine occlusions. *Annales Zoologici. Fennici,* **17**, 147-158.

ICES (1997) Report of the ICES Working Group on seals and small cetaceans in European Seas, April 1997 (unpublished).

ICES (1998) Report of the ICES Working Group on Marine Mammal Habitats, March 1998 (unpublished).

IWC (1999) Report of the workshop on chemical pollution and cetaceans. *Journal of Cetacean Research and Management* (special issue 1) 1-42.

Jefferson, T.A., and Curry, B.E. (1996) Acoustic methods of reducing or eliminating marine mammal-fishery interactions: do they work? *Ocean and Coastal Management,* **31**, 41-70.

Kastelein, R.A., Goodson, A.D., Lien, J., and de Haan, D. (1995) The effects of acoustic alarms on harbour porpoise (*Phocoena phocoena*) behaviour. In: *Harbour porpoises - laboratory studies to reduce bycatch.* (Ed. by P.E. Nachtigall, J. Lien, W.W.L. Au, and A.J. Read), pp. 157-68. De Spil Publ., Woerden, Netherlands.

Kenward, R.E. and Hodder, K.H. (1996) Ranges V. Analysis system for biological location data. Institute for Terrestrial Ecology. Wareham, Dorset, UK. 69pp.

Kingsley, M.C.S. (1998) Population index estimates for the St. Lawrence belugas 1973-1995. *Marine Mammal Science,* **14**(3), 508-530.

Lawrence, E. (ed) (1989) *Henderson's dictionary of biological terms.* 10th edition. Longman Scientific & Technical. London.

Lurz, P.W.W., Garson, P.J. and Wauters, L.A. (1997) Effects of temporal and spatial variation in habitat quality on red squirrel dispersal behaviour. *Animal Behaviour,* **54**, 427-35.

Lønne, O.J. and Gulliksen, B. (1989) Size, age and diet of polar cod, *Boreogadus saida* (Lepechin 1773), in ice covered waters. *Polar Biology,* **9**, 187-191.

Martin, A.R. (1990) *Whales and Dolphins.* Salamander Books Ltd. London, New York.

Martineau, D., Beland, P., Desjardins, C. and Lagace, A. (1987) Levels of organochlorine chemicals in tissues of beluga whales (*Delphinapterus leucas*) from the St. Lawrence Estuary, Quebec, Canada. *Archives of Environmental Contamination and Toxicology,* **16**, 137-147.

Martineau, D., Lagacé, A., Massé, R., Morin, M. and Béland, P. (1985) Transitional cell carcinoma of the urinary bladder in a beluga whale (*Delphapterus leucas). Canadian Veterinary Journal,* **26**, 297-302.

Martineau, D., Lair, S., De Guise, S., Lipscomb, T.P. and Béland, P. (1999) Cancer in beluga whales from the St. Lawrence estuary, Quebec, Canada: a potential biomarker of environmental contamination. *Journal of Cetacean Research and Management* (special issue 1) 249-265.

Mohr, C.O. (1947) Table of equivalent populations of North American small mammals. *American Midland. Naturalist,* **37**, 223-249.

Mortensen, P., Bergman, A., Bignert, A., Hansen, H.J., Härkönen, T. and Olsson, M. (1992) Prevalence of skull lesions in harbour seals (*Phoca vitulina*) in Swedish and Danish museum collections: 1835-1988. *Ambio,* **21**(8), 520-524.

Olsson, M., Karlsson, B. and Ahnland, E. (1994) Diseases and environmental contaminants in seals from the Baltic and the Swedish west coast. *Science of the Total Environment,* **154**, 217-227.

Pribil, S. and Picman, J. (1997) The importance of using the proper methodology and spatial scale in the study of habitat selection by birds. *Canadian Journal of Zoology,* **75**, 1835-1844.

Read, A.J. (1995) New approaches to studying the foraging ecology of small cetaceans. In: *Whales, seals, fish and man.* (Ed. by A.S. Blix, L. Walløe and Ø. Ulltang), pp. 183-191. Elsevier Science, Amsterdam.

Reeves, R.R., Hofman, R.J., Silber, G.K. and Wilkinson, D. (eds) (1996) Acoustic deterrence of harmful marine mammal-fishery interactions: proceedings of a workshop held in Seattle, Washington, 20-22 March 1996. U.S. Department of Commerce, NOAA Technical Memorandum NMFS-OPR-10.

Roen, R. and Bjørge, A. (1995) Haul-out behaviour of the Norwegian harbour seal during summer. In: *Whales, seals, fish and man.* (Ed. by A.S. Blix, L. Walløe and Ø. Ulltang), pp. 61-67. Elsevier Science, Amsterdam.

Sakshaug, E., Bjørge, A., Gulliksen, B., Loeng, H. and Mehlum, F. (1994) Structure, biomass distribution, and energetics of the pelagic ecosystem in the Barents Sea: A synopsis. *Polar Biology,* **14**, 405-411.

Sanderson, G.C. (1966) The study of mammal movements - a review. *Journal of Wildlife Management,* **30**, 215-35.

Staus, N. L. (1998) Habitat use and home range of West Indian whistling-ducks. *Journal of Wildlife Management,* **62**, 171-178.

Stede, G., and Stede, M. (1990) Orientierende Untersuchungen von Seehundschädeln auf pathologische Knochenveränderungen. In: Zoologischen und ethologische Unter-suchungen zum Robbensterben, pp. 31-53. Institut für Haustierkunde, Kiel, Germany

Stewart,   B.S. (1984) Diurnal hauling patterns of harbor seals at San Miguel Island, California. *Journal of Wildlife Management,* **48**, 1459-1461.

Stewart, B.S., Leatherwood, S., Yochem, P.K. and Heide-Jørgensen, M.P. (1989) Harbor seal tracking and telemetry by satellite. *Marine Mammal Science,* **5**, 361-375.

Syvertsen, E.E. (1991) Ice algae in the Barents Sea: types of assemblages, origin, fate and role in the ice-edge phytoplankton bloom. *Polar Research,* **10**, 277-287.

Tester, J.R. and Siniff, D.B. (1965) Aspects of animal movement and home range data obtained by telemetry. *Trans.North American Wildlife and Nature Research Conference,* **30**, 379-392.

Thompson, P.M. (1989) Seasonal changes in the distribution and composition of common seal (*Phoca vitulina*) haul-out groups. *Journal of Zoology* (London), **217**, 281-294.

Thompson, P.M. (1993) Harbour seal movement patterns. In: *Marine mammals. Advances in behavioural and population biology.* (Ed. by I.L. Boyd), pp. 225-39. Symp Zool Soc Lond, 66.

Thompson, P.M. and Miller, D. (1990) Summer foraging activity and movements of radio-tagged common seals (*Phoca vitulina* L) in the Moray Firth, Scotland. *Journal of applied Ecology,* **27**, 492-501.

Thompson, P.M., Pierce, G.J., Hislop, J.R.G., Miller, D. and Diack, J.W.S. (1991) Winter foraging by common seals (*Phoca vitulina*) in relation to food availability in inner Moray Firth, N.E. Scotland. *Journal of Animal Ecology,* **60**, 283-294.

Tynan, C.T. and DeMaster, D.P. (1997) Observations and predictions of Arctic climate change: potential effects on marine mammals. *Arctic,* **50**(4), 308-322.

Wartzok, D. and Ray, G.C. (1980) The hauling out behavior of the Pacific walrus. Rep US Marine Mammal Commission, Contract MM5ACO28. US Marine Mammal Commission, Washington, DC. 1-46.

Wartzok, D. and Ray, G.C. (1981) Sea ice determinants of walrus hauling-out behavior. 4th Biennial Conference on the Biology of Marine Mammals. San Francisco, December 1981 [Abstract]

White, G.C. and Garrott, R.A. (1990) *Analysis of wildlife radio-tracking data.* Academic Press. New York. 383pp.

Wiig, Ø., Gjertz, I., Griffith, D. and Lydersen, C. (1993) Diving patterns of an Atlantic walrus *Odobenus rosmarus rosmarus* near Svalbard. *Polar Biology,* **13**, 71-72.

Woodward, B., and Goodson, A.D. (1994) Prevention of the by-catch of cetaceans by exploiting their acoustic capability. Report for the European Commission DG XIV, special study contract PEM/93/04 (unpublished).

Zachariassen, P. (1993) Pilot whale catches in the Faroe Islands, 1709-1992. *Report of the International Whaling Commission* (special issue **14**), 67-88.

Chapter 3

Ecological Aspects of Reproduction of Marine Mammals

CHRISTINA LOCKYER
Danish Institute for Fisheries Research,Charlottenlund Slot, DK 2920 Charlottenlund, Denmark. E-mail: chl@dfu.min.dk

1.   INTRODUCTION

The cetaceans fall into two main categories: balaenopterid (baleen) whales and odontocete (toothed) whales. The former are all large whales, but the second category comprises a range from the large sperm whale *(Physeter macrocephalus)*, down to the small marine harbour porpoise *(Phocoena phocoena)*, and also a number of small freshwater river dolphins. The distinction between the two categories is important because the ecology of baleen whales incorporates a regular annual cycle of seasonal migration, often over long distances involving changes of latitude, and feeding, such as the southern blue, *(Balaenoptera musculus)*, fin (*B. physalus*), and humpback whales *(Megaptera novaeangliae),* (Lockyer and Brown, 1981). The entire life history, and in particular the reproduction, is geared to this cycle (Mackintosh, 1965), and results most often in a two-year reproductive interval. The odontocetes represent a variety of life styles, some specialised to adapt to seasonal changes, but not as rigidly as the baleen whales (Gaskin, 1982). Odontocetes may have protracted inter-birth intervals of 3-5 years *e.g.* sperm and pilot whales *(Globicephala spp.)*, but may also reproduce annually *e.g.* harbour porpoise. All cetaceans spend all their lives in water and produce a single calf each pregnancy.

The pinnipeds include otariid seals (eared seals which include fur seals, sea lions and walruses) and phocid seals (true, non-eared monachine and phocine seals). Broadly speaking, all seals tend to breed seasonally, some tightly so, with a delayed implantation and gestation that fit into one year (Riedman, 1990). The walrus is, however, an exception with a longer gestation period that results in a normal two-year reproductive cycle rather than an annual one (Boyd *et al.*, 1999; Evans and Stirling, this volume). Also the otariids have a longer nursing period compared to the phocid seals where pup-rearing success depends on a rapid and intense nursing phase. All pinnipeds depend on access to land or ice for at least part of their life cycle, *e.g.* breeding and moulting. They produce a single pup each cycle.

Sirenians include manatees (freshwater and estuarine) and dugong (marine). They, like the cetaceans, are totally aquatic. Reproduction is only diffusely seasonal, and can be delayed for many reasons, most often related to food supply and energy reserves. The reproductive interval can be very varied, ranging from 2.5 - 5yr in Florida manatees to 2.5 - ~7yr in dugong (Boyd *et al.*, 1999). Like cetaceans and seals, they produce only one calf at a time.

Sea otters, whilst able to emerge from water onto land, reproduce totally in the water, giving birth and suckling at sea. They too have a protracted reproductive season with an inter-birth interval of about one year (Siniff and Ralls, 1991), sometimes two years. Sea otters also have delayed implantation. The norm is a single pup, although twins may occur in which case one may be abandoned (Garshelis and Garshelis, 1987).

Polar bears spend most of their time in water, but need fast ice to den and give birth. Delayed implantation is a reproductive feature, as is altricial (premature) birth to twins, although 3-4 is not unusual and also singletons, after a short gestation period (I. Stirling, in Macdonald, 1984). The mother polar bear invests much time and energy in suckling the cubs, during which time she remains on the ice in the den (Lentfer, 1978).

All these animals have different evolutionary origins *e.g.* cetaceans and sirenians from ungulates, the latter with pachyderm-like origins; sea otters from the mustelid branch of the carnivores; pinnipeds ancestrally from dog-like carnivores, and polar bears from the carnivorous ursids of today (Macdonald, 1984). Thus it is not surprising that they each exhibit different reproductive strategies and ecology. The paper describes some of these reproductive features, and draws comparisons between them, emphasising the ecological aspects of reproduction.

It is important to recognise the potential sources of information on reproduction in cetaceans, seals and these other species. Many studies have been based on carcase dissections and subsequent morphological and histological investigations of samples (*e.g.* Anderson and Fedak, 1987; Harrison, 1969; Kasuya and Marsh, 1984; Laws, 1961; Lockyer, 1987a; Marsh and Kasuya, 1984; Read and Hohn, 1995).

2. REPRODUCTIVE CYCLES

Here a reproductive cycle is defined as the normal minimum time period for all stages of reproduction in the female from ovulation through conception and pregnancy to birth and lactation followed by a short rest period (if any). However, the reproductive interval (meaning the actual time between the start of one cycle and another) may be extended or even shortened because of environmental, nutritional and social circumstances. In general, the chief limiting factor in reproduction is the fecundity of the female. In many marine mammals the female tends to mature earlier than the male (Evans and Stirling, this volume; Boyd *et al.*, 1999), and this is frequently related to social factors such as male competition and dominance in schooling and herding species, as well as sexual dimorphism (Ralls *et al.*, 1977). The timing of events in the reproductive cycle has evolved so that different species can optimise the resources of their habitat and exploit seasonal changes in environmental conditions to benefit the ecology of the species and favour maximal survival of the young. Therefore, the cycle will differ slightly with each species according to that species life style and ecology, and may even vary according to the population level relative to the carrying capacity of the environment (Urian *et al.*, 1996).

2.1 Cetaceans

2.1.1 Mysticetes

The fin whale *(Balaenoptera physalus)* (Fig. 1) has a two-year reproductive cycle comprising a gestation period of about 11 months and a lactation period of about 6-7 months, followed by a period of anoestrus (Laws, 1961). The cycle starts in winter in low latitudes and warmer waters with ovulation and conception leading to pregnancy, during which the female migrates to summer feeding grounds in higher latitudes. The female returns to low latitudes to give birth in winter. The calf is weaned when the female migrates into high latitudinal waters for feeding during the summer of the second year (Lockyer, 1984a; Mackintosh, 1965). The generalised cycle is shown in Figure 2 for the southern fin whale. Such a generalised cycle is applicable to other species of large baleen whales although individual timings of conception, birth and lactation may vary. Blue, humpback, sei *(B. borealis)*, and gray whales *(Eschrichtius robustus)*, all tend to follow a two-year seasonal cycle of migration, feeding and breeding. The minke whale *(B. acutorostrata)* may reproduce annually whilst the balaenid whales (bowhead *(Balaena mysticetus)* and right whales

(Eubalaena glacialis, E. australis)) may reproduce once only every three or four years (Evans, 1987).

Figure 1.   Fin whale in the Bay of Biscay (Photo: K. Young).

There is evidence for a seasonal reproductive cycle in male baleen whales. This comes from behavioural observations on coastally breeding species like right whales that mate in winter; also from seasonal increase in testis weight and spermatogenic activity *e.g.* humpback, gray, blue and fin whales (Lockyer, 1984a). In these species, male seasonal reproductive activity coincides with the period of oestrus in the female. A protracted period of breeding for both sexes has the advantage of ensuring conception, should pregnancy fail or terminate in the first instance. Whilst baleen whales may aggregate in certain areas for breeding, they do not form schools and the females remain solitary when rearing the calf.

In fin whales, the summer period of feeding has been shown to be critical for energy storage in the form of fat, in order to support the lactation which commences in the winter following (Lockyer, 1981a). In fin whales, the seasonal cycle of migration, feeding, and subsequent fattening has a great influence on reproductive interval, ovulation rate, and fecundity, as well as (by inference) survival of the calf. The summer feeding period in the Antarctic lasts about four months, but pregnant females have been observed to be the first to arrive on the summer feeding grounds and they are amongst the last to depart, thus taking the opportunity to fatten as much as possible.

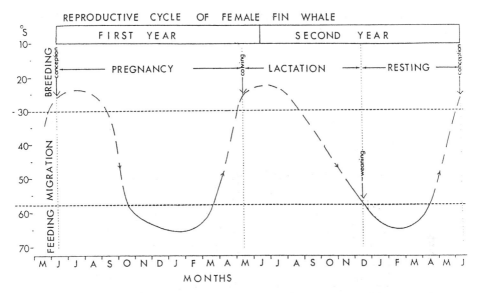

Figure 2.   Schematic two-year reproductive cycle for the southern hemisphere female fin whale (after Lockyer and Brown, 1981). The fin whale has a two-year reproductive cycle comprising a gestation period of about 11 months and a lactation period of about 6-7 months, followed by a period of anoestrus. The cycle starts in winter in low latitudes and warmer waters with ovulation and conception leading to pregnancy, during which the female migrates to summer feeding grounds in higher latitudes. The female returns to low latitudes to give birth in winter. The calf is weaned when the female migrates into high latitudinal waters for feeding during the summer of the second year.

The baleen whales are generally long-lived, with ages being recorded from growth layer groups (GLGs) in ear plugs (Lockyer, 1984b). In the balaenopterid species (Martin, 1990), ages of up to 94 yr have been recorded in the fin whale and potential longevity up to 70 yr in sei whales. The smaller minke whale may live up to 60 yr, but most of the rorquals have a long history of exploitation prior to 1985 when there was a moratorium on all commercial whaling (it had already ceased for some species), and thus their potential longevity may rarely have been attained. Nevertheless, their long life span and female ability to produce offspring into old age means that some species such as fin and minke whales could experience as many as 40 pregnancies in a lifetime if ovulations, as recorded in the form of *corpora albicantia,* are used as an indicator (Laws, 1961; Lockyer, 1987b). Recent investigations on age in balaenid whales indicate a possible longevity in excess of 100 yr (George *et al.*, 1998).

2.1.2   Odontocetes

The odontocetes comprise a broad group of species ranging greatly in body form and size, as well as reproductive strategy. The largest species is the sperm whale which exhibits extreme sexual dimorphism with the adult males being much larger than the females which are only about 65% of the adult male body length (Best, 1970; Lockyer, 1981b). Such sexual dimorphism also exists in some smaller species, but is not so marked and in some species females are larger. Virtually all species have a breeding and calving season, but unlike the baleen whales, the season is often more protracted. Many species have a gestation period of just less than a year, like the baleen whales, or the harbour porpoise, a small odontocete. Lactation is variable both within and between species, and rather than lasting a few months, can last for several years *e.g.* sperm and pilot whales. The function here can be social rather than nutritional in the latter stages of lactation, and may be linked to schooling behaviour. The reproductive interval can then be extended for three to five years, and is not necessarily associated with seasonal feeding habits or accumulation of body fat reserves. The period may be useful as a learning phase for the young in methods of co-operative feeding and foraging strategy where echo-location may be an important function (Brodie, 1969). In contrast to the baleen whales, many delphinids form large schools or at least the females are often co-operative in rearing young.

In sperm whales, synchronous oestrus in matriarchal schools may be caused by the arrival of adult bulls (Best and Butterworth, 1980), the effect of which would be advantageous in increasing the efficiency of fertilisation, minimising the period that the bull must remain with the school, and reduce the interference and disruption of the social organisation within the school, especially to mother-infant bonds. Gestation lasts 14-15 months (Best *et al.*, 1984; Whitehead *et al.*, 1989), and true lactation lasts about two years (Best, 1974), although suckling by the calf may occur up to age 7-8 yr in females and 13 yr in males (Best *et al.*, 1984). This has been interpreted as a form of social behaviour, and may not be significant nutritionally. Evidence from stomach contents suggested that solid food is taken by the calf at some stage before the end of its first year of life (Best *et al.*, 1984).

Seasonality of mating and calving has been demonstrated for all delphinids that have been studied in detail (Perrin and Reilly, 1984). However, depending on the duration of gestation and lactation, and the degree of seasonality, the reproductive interval may vary within a population because of individual circumstances. The length of gestation for small odontocetes varies from 10-16 months (Perrin and Reilly, 1984), and more recently killer whales *(Orcinus orca)* have been reported to have a gestation of 17 months (Duffield *et al.*, 1995). These times have been estimated from both observations on animals caught in fishery operations and captive live animals. Most dolphins have a gestation period of about a year or less. Long-finned pilot whales *(G. melas)* are now believed to have a gestation period closer to one year rather than the 15-16 months previously thought (Martin and Rothery, 1993).

In river dolphins, most evidence points to a gestation period of less than a year (Brownell, 1984, da Silva, 1994), and in *Inia geoffrensis*, da Silva (1994) reported that peak births occurred right after the river flood period May to July in the Amazon, although the birth season was protracted. Lactation probably lasts a year or longer in *I. geoffrensis*, and most lactating females are observed during the dry season when the river levels are low and prey become concentrated (da Silva, 1994).

Information for monodontids indicates a gestation period of about 14-15 months (Braham, 1984). However, more recently the gestation for Greenlandic belugas *(Delphinapterus leucas)*, has been estimated to be at least 330 days (Heide-Jørgensen and Teilmann, 1994), from timing of mating and birth, indicating that gestation may be less than one year, contrary to earlier findings. Calf production is about once every three years, and lactation probably lasts 20-24 months.

The smaller cetaceans vary considerably in life span according to species. Age has been determined retrospectively from growth layer groups (GLGs) in teeth (Perrin and Myrick, 1981). Bottlenose dolphins *(Tursiops spp.)* may live for more than 40 yr from individuals monitored both in the wild and captivity and from tooth GLG studies (Hohn *et al.*, 1989), and harbour porpoises *(Phocoena phocoena)* may live up to 25 yr (Lockyer, 1995a and b). The Amazonian river dolphin *(Inia geoffrensis)* may live in excess of 35 yr (da Silva, 1994), and pilot whales can live up to 60 yr (Bloch *et al.*, 1993; Kasuya and Tai, 1993). Perhaps the longest-lived odontocete is the sperm whale which may live more than 70 yr if age is calculated from tooth GLGs (Martin, 1980, and personal observation). Clearly longer-lived species can afford to invest more time in each offspring and still produce several calves during the lifetime. However, short-lived species such as harbour porpoises that generally live only till 15 yr at most (Lockyer, 1995b), must gear the reproductive cycle to enable the production of a calf annually whenever possible (Read and Hohn, 1995). The gestation period in the phocoenids is usually 10-11 months, and simultaneous pregnancy and lactation is observed (Gaskin *et al.*, 1984).

2.2   Pinnipeds

2.2.1   Phocidae

Broadly speaking, phocid (earless) seals breed seasonally, some tightly so within only a few days each year. However, the monk seals, *Monachus sp.*, representing the only tropical phocid, are an exception, with virtually year-round births. The reproductive cycle is characterised by delayed implantation (embryonic diapause) and gestation that fit into one year (Evans and Stirling, this volume; Boyd *et al.*, 1999). Implantation of the blastocyst may be delayed about four months, or longer in hooded seal, *Cystophora cristata*. Gestation itself generally lasts about 8 months from implantation. They produce a single pup (Fig. 3) each cycle, and the reproductive interval is one year. The phocid seals employ a rapid and intense nursing phase with a range from 4-60 days. In hooded and harp seals *(Phoca groenlandica)* it may last only a few days, but up to several weeks in spotted seal, *Phoca largha* and 6-7 weeks in Weddell seal *(Leptonychotes weddelli)* – see Laws (1984). During even the shortest suckling phase, the pup may double its birth weight (harp seals). There is no resting phase as such, only the period of delayed implantation. All pinnipeds depend on access to land or ice for at least part of their life cycle, *e.g.* breeding and moulting, when they usually form often large aggregations.

Figure 3.   Bearded seal pup (Photo: I. Stirling).

Photoperiod as well as nutritional status and the previous year's reproductive history appear to be important factors in determining the precise moments of ovulation and implantation in phocid seals (Boyd, 1984, 1991; Boyd *et al.*, 1999). First-time pregnancies often fail, and the younger mothers tend to give birth later in the season to a pup of lower birth weight. Birth size is generally correlated with the size (and age) of the mother, and there appears to be more investment in male pups (*e.g.* grey seals, *Halichoerus grypus* - Anderson and Fedak, 1987). Potentially, a female may give birth to many young in a lifetime, with a longevity from 20-50 years, depending on species.

2.2.2 Otariidae

The eared seals tend to breed seasonally, an exception being the Galapagos sea lion *(Arctocephalus galapagoensis)*, and like phocid seals, also exhibit delayed implantation and gestation that fit into one year (Evans and Stirling, this volume; Boyd *et al.*, 1999; Daniel, 1971, 1974). Fur seals *(Arctocephalinae)* and sea lions *(Otariinae)* come onto land for breeding and moulting, when they form large colonies. Pregnancy and birth dates may vary in Antarctic fur seals, *Arctocephalus gazella*, because of the influence of age of the mother, foetal sex, and the environmental conditions (Boyd, 1996). A single pup is produced each cycle. The otariids have a longer nursing period than the phocid seals, and lactation may last from about four months to a year or more. Lactation is characterised by intermittent suckling of the pup with intervals sometimes lasting several days while the mother is away at sea foraging. Such periodic suckling by the young has been recognised in the growth patterns in teeth of Antarctic fur seal pups, where the suckling bouts are evident as distinct laminae (Bengtson, 1988). The reproductive interval may exceed two years in some species (Riedman, 1990). Whilst maturation in the female is early, longevity is less than most phocids, at about 25 years.

2.2.3 Odobenidae

The walrus, *Odobenus rosmarus*, differs from seals in having a longer gestation period that results in a normal two-year reproductive cycle (or longer) rather than an annual one (Kenyon, 1978a). This interval tends to increase with age, so that the walrus has the lowest reproductive rate of all pinnipeds. The walrus may, however, live up to 40 years, so that several young may be produced in a lifetime after maturation at about age 6-7 years in the female (Macdonald, 1984). Gestation, including a period of about 4 months' delayed implantation, lasts about 15 months (Riedman, 1990).

Lactation lasts about six months, but the young may continue taking milk supplements for up to another year (Macdonald, 1984). The walrus forms herds of adult females and young that congregate seasonally in often large breeding colonies on land when several herds may coalesce. The bulls generally follow these herds and remain in attendance (Kenyon, 1978a). Evolutionarily, the walrus represents the more primitive reproductive cycle of the pinnipeds.

Figure 4.   Mating herd of Florida manatees in Tampa Bay (Photo: Florida Fish and Wildlife Conservation Commission) .

2.3   Sirenians

Sirenians include manatees, *Trichecus spp.* (freshwater and marine) and dugongs, *Dugong dugon* (exclusively marine). They, like the cetaceans, are totally aquatic. Reproduction is only diffusely seasonal, perhaps because of their generally tropical distribution, and can be delayed for many reasons, most often related to food supply and energy reserves (Boyd *et al.*, 1999). Mating (Fig. 4), however, appears to be independent of reproductive status, and female manatees will mate even when pregnant (Odell *et al.*, 1995). A single female dugong will attract many males, "herding", in areas used regularly for breeding. Unlike all other marine mammals, the sirenians are exclusively vegetarian browsers and dependent on primary plant productivity. The reproductive interval can be very varied, ranging from 2.5 - 5 yr in Florida manatees to 2.5 - ~7 yr in dugong (Boyd *et al.*, 1999). Like cetaceans and seals, they produce only one calf at a time. Gestation lasts about 12-14 months, and lactation is uncertain, the calves commencing eating some vegetation shortly after birth, but may last 1.5 years (Marsh, 1995). Sirenians may be long-lived at more than 50 years (dugongs may exceed 60 yr -Mitchell, 1976, 1978; Marsh, 1980), and sexual maturation in female manatees is about 5 years (Boyd *et al.*, 1999; Marmontel, 1993, 1995), whilst female dugongs tend to mature at around 10 years or even older (Mitchell, 1976, 1978; Marsh, 1980).

2.4   Mustelids

Sea Otters

Sea otters (Fig. 5), *Enhydra lutris*, whilst able to emerge from water onto land, reproduce totally in the water, giving birth and suckling at sea. The newborn is almost helpless, yet is able to remain safely at the surface floating, buoyed up by air trapped in its fur, while the mother forages (Kenyon, 1978b). Like the sirenians, sea otters have a protracted reproductive season, but with an inter-birth interval of up to two years. Sea otters have a gestation lasting nearly nine months including delayed implantation. The norm is a single pup, although twins may occur in which case one may be abandoned if the mother is sick or unable to secure adequate resources to support two pups (Garshelis and Garshelis, 1987). The lactation period is about four or more months, with the pup often remaining with the mother for extended periods after weaning. Investment in the young is thus expensive in time and energy, an important consideration in a species that is relatively short-lived (10-14 years - Garshelis, 1984; Bodkin *et al.*, 1997) with an age of sexual maturation at about six years (Garshelis, 1983). The female sea otter tends to be solitary, and is unlike most pinnipeds and some cetaceans.

2.5   Ursids

Polar Bears

Polar bears (Fig. 5), *Ursus maritimus*, spend most of their time in the pack ice or along the outer extent of the fast ice (referred to as the shear zone). Pregnant females also need appropriate fast ice or along shore habitat to den and give birth (Lentfer, 1978). Delayed implantation is a reproductive feature, and also altricial (premature) birth to twins, although singletons and 3-4 are not unusual, after a total gestation period including the long embryonic diapause, of about eight months (Macdonald, 1984). The mother polar bear invests much time and energy in suckling the cubs, during which time she remains on the ice in the den which she has excavated from snow drifts on fast coastal ice in pregnancy. This den provides insulation and protection for herself and the cubs during a relatively dormant period. The reproductive interval is usually three years or longer, although the female is able to breed again once the cubs leave her after about 28 months from birth. Females are mature at about age 4-6 years and have a longevity of up to 25 years (Macdonald, 1984).

Figure 5.   Sea otter (top) (Photo: P. Morris); Polar bear (bottom) (Photo: I. Stirling).

3. AGE AND REPRODUCTIVE PERFORMANCE

In general, it has been observed that fertility and reproductive success are depressed in the newly mature female (*e.g.* fin whales - Lockyer and Sigurjónsson, 1991; 1992; polar bears - Ramsay and Stirling, 1988; Derocher and Stirling, 1994). However, fertility is potentially high, and after reaching a peak in young animals, generally maintains a plateau throughout early and mid-life until it tends to fall in later age (Best *et al.*, 1984; Martin and Rothery, 1993; Robeck *et al.*, 1994; Lunn and Boyd, 1993; Lunn *et al.*, 1994). With advancing age, the inter-birth interval increases and also duration of lactation increases in long-finned pilot whales. Martin and Rothery (1993) propose that this may mean 1) higher survival of calves, 2) provision of milk to calves other than the mother's own, 3) increased energetic investment in later calves with advancing age of the mother. At least two of these (1 and 3) have been demonstrated in terrestrial mammals (Clutton-Brock, 1984; Clutton-Brock *et al.*, 1982), and may certainly apply to other marine mammals. Declining pregnancy rate with age has also been observed in Antarctic fur seals after age 13 years, with a peak rate at age 7-8 years, and first reproduction at age 3-4 years (Lunn *et al.*, 1994).

Reproductive senescence has not often been described in mammals, and rarely in marine mammals except for pathologies. However, it is now known that female short-finned pilot whales *(Globicephala macrorhynchus)*, may become reproductively senescent in older years (Marsh and Kasuya, 1986). Perhaps one of the most interesting discoveries is that ovaries in short-finned pilot whales become completely devoid of any follicles in later life (Marsh and Kasuya, 1984, 1986). However, old females with such ovaries may continue to lactate for several years, suckling not only their own previous offspring but other calves and juveniles in the pod (Marsh and Kasuya, 1984), and mating may still occur in post-reproductive females (Kasuya *et al.*, 1993). Such behaviours in senescent females have not been observed in other species. The function of the continuing suckling of young may be social rather than nutritional, and the mating one of social bonding and teaching. This therefore, provides important evidence that post-reproductive females have developed as a normal component of the population, with a continuing indirect reproductive and social function. The social organisation of pilot whales may predispose this species to certain characteristics. The pilot whale exists in a matriarchal schooling system where there is considerable support from other females, and the survival of young may be more favourable than in species where the mother is alone with her calf. In the former situation, the pressure to keep reproducing is allayed by potentially increased survival of the young, so that long-term investment in offspring is worthwhile. For the more solitary female, *e.g.*

baleen whale cow that lactates for only about half a year, ultimate survival may depend more on reproducing regularly at shorter intervals over a longer life span than the pilot whale.

Apparent reproductive senescence has also been reported in Antarctic and northern fur seals *(Callorhinus ursinus)* (York and Hartley, 1981; Trites, 1991; Lunn *et al.*, 1994), and in polar bears beyond age 16-20 years (Ramsay and Stirling, 1988; Derocher and Stirling, 1994), although 21 year-old females have been recorded as reproductively active.

4.   FOETAL GROWTH

The growth pattern of the foetus has been described in terms of formulae derived for a variety of mammalian species, mainly terrestrial (Huggett and Widdas, 1951; Frazer and Huggett, 1973, 1974; Sacher and Staffeldt, 1974). The growth pattern comprises a non-linear phase, t_o, from conception to the start of the linear phase, and which appears to be constant and related to the overall gestation period. In rorquals with a gestation period approaching a year, t_o could be predicted to be about 67-72 days (Huggett and Widdas, 1951), and Lockyer (1981a) calculated values of 73-74 days. In baleen whales it appears that an accelerated growth phase occurs in the last trimester (Laws, 1959; Lockyer, 1981a, 1984a). The exponential growth phase coincides with the arrival on the feeding grounds in the large rorquals, and thus may be significant in terms of energy budget. The growth pattern of odontocetes appears more steady (Gaskin *et al.*, 1984; Lockyer, 1981b; Martin and Rothery, 1993; Perrin and Reilly, 1984). Certainly the gestation period of some odontocete species lasts more than 12 months, whilst that of baleen whales is well within this period, and may demand accelerated growth in order to allow birth at a favourable time in relation to migration and feeding conditions.

One notable difference between cetaceans and pinnipeds, mustelids and ursids, is the apparent absence in the former of any delayed implantation of the blastocyst to a time when conditions are more favourable for ultimate survival of young. Clearly such a mechanism may be important to a marine mammal that only has a limited window of time for mating. Boyd *et al.* (1999) discuss the importance of photoperiod and also light intensity in triggering implantation in seals. Certainly, although the blastocyst may go into a temporary state of dormancy, once implanted, true gestation begins. As mentioned earlier, most marine mammals fit gestation into one calendar year, and even for those species *e.g.* walrus where delayed implantation occurs but gestation exceeds one year, the length of true gestation is still usually under a year.

Normally, one calf / pup is produced from each pregnancy in all species except the polar bear. The size of the calf relative to the mother is large in cetaceans and varies between about 29% of maternal length in large baleen whales to as much as 42-48% of maternal length in odontocete species as diverse as sperm whales and harbour porpoise. There is considerable overlap in body size between neonates and near-term foetuses in both long-finned pilot whales and harbour porpoises, for example. Once the calf is born, it must be able to swim, breath, and dive. This ability is particularly important in schooling species where staying with other animals represents protection. After birth, the lipid content of foetal muscle which has been increasing during gestation, drops very rapidly in long-finned pilot whales (Lockyer, 1993a), suggesting that this fat may be an important energy reserve which is drawn upon immediately after birth, perhaps before suckling is well established, thus playing an important role in calf survival.

In seals, the pup may weigh between 5.5-13.5% of the maternal body weight, depending on species. In polar bears, the young, of which there may be 1-4, are altricial and quite helpless at birth, and also very small (450-900g). The newborn sea otter is about 3kg at birth, and although fur-covered, is also fairly helpless (Garshelis and Garshelis, 1987). While it will float with trapped air in the fur, the mother generally carries it everywhere as it is unable to swim at this stage (Kenyon, 1978b).

5.   ENERGETICS OF REPRODUCTION

The energetic costs of reproduction in all species involve both sexes, but most of the discussion following will focus on the female because of the demands of pregnancy and lactation. The costs of lactation relative to those of pregnancy are greater, and different marine mammals have developed various strategies for coping with these energetic demands. However, most utilise the possibilities of storing fat, which supports the hypothesis of Pond (1984) that mammalian fat storage is a direct evolutionary consequence of the lactational aspects of reproduction. Nearly all marine mammals produce a milk high in fat content ranging from 15-60% wet weight fat depending on the stage of lactation (Oftedal *et al.*, 1987; Oftedal, 1997; Riedman, 1990). The species differ mainly in the speed and mode of transfer of this energy depot to the offspring, and how much is required for personal maintenance, all of which are geared for maximum compatibility with habitat and life style.

5.1   Cetaceans

5.1.1   Mysticetes

Feeding is a highly seasonal activity for most baleen whales, although food is taken all year-round when available in sufficient density. To survive long periods of poor food availability, baleen whales fatten intensively during the summer feeding period (Rice and Wolman, 1971; Lockyer, 1981a, 1984a, 1987a, c), and this is especially important for pregnant females preparing for lactation. The energetic demands of pregnancy have been shown to be considerably less than the overall demands of lactation (Lockyer, 1987a, c) in northern fin whales, not least the fact that the costs of pregnancy are spread over a longer period than lactation, so that the demands of lactation must be very intense on a daily basis. The thickest blubber has been described for pregnant females (Lockyer, 1981a for fin and blue whales; 1987a, c, for fin whales; Rice and Wolman, 1971, for gray whales). The fattening is not confined only to the blubber layer but is significant in the major locomotory muscles, viscera, and even the bone tissues (Lockyer, 1981a; 1987a, c).

Table 1.   Predicted energy costs of the reproductive cycle of the female fin whale off Iceland, and the calculated fat energy store from observations on carcases of different reproductive status (after Lockyer, 1987a).

Reproductive status	Carcase lipid and energy store in female						Energy costs of reproduction	
	Predicted data from girth and length measurements and lipid analysis		Observed data from individual weighings and lipid analysis, July-August			Pregnancy - 11 months	Lactation - 7 months
	Body weight, t (for 19.4 m length)	Calorific conversion[1] of carcase lipid store	Body length, m	Body weight, t	Calorific conversion[1] of carcase lipid store		
	Week 0 Week 13	Week 0 Week 13					
Anoestrous	av. 41.5 thro'out	77×10^6 102×10^6	19.2	39.0	79×10^6	Foetal growth to 1,750 kg @ 2,940kcal.kg[-1]	Milk production of 72 kg.day[-1] @ 3,320 kcal.kg[-1] at 90% gland efficiency
Pregnant	41.5 55.5	78×10^6 152×10^6	19.8	56.5	161×10^6		
Lactating	34.5 39.5 av. 37.0 thro'out	50×10^6 57×10^6 av.54$\times 10^6$ thro'out	20.5	32.0	53×10^6		
						$= 5.1 \times 10^6$ plus, Heat of gestation - $Q_0 = 4,400 M^{1,2}$ $= 34.3 \times 10^6$ (Brody, 1968)	$= 55.8 \times 10^6$
Average difference between Week 13 pregnant and near-end lactating = Reproductive costs	18.5 (range: 16-21)	98×10^6 (range:95-102$\times 10^6$)	24.5		107×10^6	Total: 39.4×10^6 + 55.8×10^6 $= 95.2 \times 10^6$	

[1] Lipid calorific value = 9,450 kcal.kg[-1]

Table 1 shows the estimated costs of reproduction in the female fin whale from Icelandic waters, and also indicates the actual observed fat energy storage in an average sized (19.4m) mature female. Lockyer (1981a, 1987a) has calculated that the difference in body mass in the form of fat energy storage between end of lactation and late pregnancy could be 50-75% of lactating body mass. The assumption is that much of this is available for milk production for the calf through catabolic breakdown of the fat. The consequences of not meeting these demands are discussed later in the next section.

5.1.2   Odontocetes

As mentioned earlier, despite some seasonality in breeding and calving, lactation is often a rather protracted business so that the costs of lactation, even if high overall, are not intense in the same way as in baleen whales. Also, feeding is generally not as intensively seasonal as in the baleen whales. Therefore, even when seasonal fat storage occurs, it is not as dramatic as in baleen whales. The energetic demands of reproduction are reflected in the measurements of girth and blubber thickness in adult females of different reproductive status, *e.g.* in harbour porpoise (Read, 1990), and some estimates of energy cost of reproduction have been made for this species by Yasui and Gaskin (1986). The energetic costs of pregnancy and lactation have been calculated for the sperm whale (Lockyer, 1981b), and long-finned pilot whale (Lockyer, 1993a).

A schematic diagram of reproductive energy costs in the long-finned pilot whale is depicted in Figure 6. The stored fat in the female represents most of the energy required for lactation during the first year after birth, and unlike baleen whales, feeding can take place year-round. Off the Faroe Islands, however, birth coincides with movement of animals into the area for feeding (in August) with subsequent fattening in winter (Lockyer, 1993a). This way, lactation is sustained by dietary intake through autumn and winter, and then sustained by fat stores in spring.

In Faroese pilot whales, the energy demands in fighting and being actively mobile (Bloch *et al.*, 1993; Bloch, 1994; Desportes *et al.*, 1993) may be great on the males which have stored fat energy during winter. The main breeding appears to take place mid-July, although Desportes *et al.* (1993) found two peaks of male reproductive activity. The body fat becomes rapidly depleted through spring and summer; firstly from the muscle and then from the blubber. By May, muscle lipid is negligible; by July, the weight of blubber is reduced; and by August, the lipid content of the blubber is less (Lockyer, 1993a).

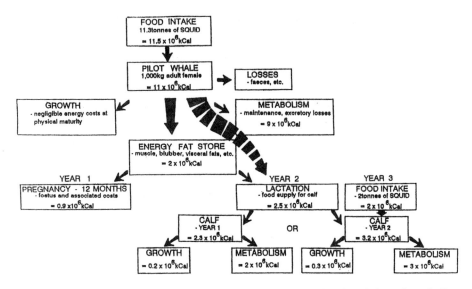

Figure 6.   Schematic diagram of reproductive energy costs in a female long-finned pilot whale of average adult size (1,000kg) off the Faroe Islands. The scheme follows energy intake and utilisation, and fat storage in year 1 used to support or supplement lactation in years 2 and 3, and the energy used for the growth and maintenance of the calf (after Lockyer, 1993a).

Investigation of harbour porpoise body condition in British waters indicated that there may be a difference in neonatal body size between the sexes, with females being larger with greater body mass and more blubber fat (Table 2), although differences, while consistent, are not significant. The adult females of reproductive size are heavier and have a greater amount of blubber fat in pregnancy compared to resting (anoestrous) females, and both categories are heavier and fatter than lactating females (Table 2). Similar findings were also found for Canadian Bay of Fundy porpoises (Read, 1990; Koopman, 1998). Clearly, the lactating female must use body fat for milk production. The neonates have relatively thin blubber which is also low in lipid content, so that the insulative properties are not efficient. The incidence of peak births in June/July (Lockyer, 1995a, b, c) means that the warmer water at this time will minimise heat losses. The greater amount of blubber fat, notably the greater thickness in female neonates, may provide extra insulation and chances of survival compared to male neonates. It is interesting to note that the age frequency distribution by sex of British strandings, indicates that while these are similar from age one year, the number of males of 0 year age class is about double that of females (Lockyer, 1995b) suggesting a higher mortality of males during the first year of life.

Table 2. Mean values ± SD for various measurements of porpoise by reproductive status (after Lockyer, 1995a, c).

Reproductive class	Length (cm)	Weight (kg)	Blubber weight (kg)	Mid-girth (cm)	Mid-lateral blubber thickness (mm)	Blubber lipid content % wet weight
CALF ≤90 cm						
Neonate (male)	76.2 ± 8.0	7.1 ± 2.0	2.7 ± 1.8	45.6 ± 5.2	9.9 ± 4.1	67.3 ± 8.6
Neonate (female)	79.3 ± 7.2	8.7 ± 4.3	4.3 ± 2.7	47.1 ± 4.1	12.8 ± 6.2	69.3 ± 11.6
ADULT FEMALE >140 cm						
Resting	156.9 ± 11.6	45.0 ± 13.0	11.9 ± 1.7	86.6 ± 6.9	13.4 ± 5.9	85.5 ± 8.1
Pregnant	151.5 ± 5.3	50.3 ± 9.1	13.6 ± 2.3	95.4 ± 9.1	16.8 ± 4.2	86.8 ± 6.3
Lactating	156.0	40.0	11.0	82.0	12.0	
Pregnant and lactating	150.0	49.0	11.0	96.0	14.0	83.0

During pregnancy and lactation, cetaceans may vary dietary habits. Bernard and Hohn (1989) reported that, regardless of year and season, there were differences in feeding habits between pregnant and lactating spotted dolphins, *Stenella attenuata*. The pregnant females had a diet of predominantly squid, similar to the main population, whilst lactating females took mainly flying fish. This is an example of a switch from low calorie diet to one of higher calorific intake, and this can be directly related to reproductive status.

5.2 Pinnipeds

In many species of pinnipeds, the breeding season is very short, and in phocid seals, the lactational period is also very short. The pup suckles intensively and grows at a steady rate of about 11-12.5% initial postnatal body weight per day in grey seals (Anderson and Fedak, 1987). The milk of seals is very high in fat content, generally 30-60% wet weight of lipid, depending on stage of lactation (Riedman, 1990). During the same period, the mother loses about 2.2% of her initial post-partum body weight per day in milk to the pup and in metabolism and other losses. Male reproductively active grey seals who are occupied fending off other male intruders during the breeding season lose about 0.85% initial body weight per day throughout the period. During a month without feeding, this may amount to a loss of about 25% of initial body mass. These costs are shown in Table 3 for the grey seal.

Table 3.   Mean rates of weight and energy changes in breeding grey seals (after Anderson and Fedak, 1987).

Age class	Initial weight (kg)	Daily weight change (kg)	[1]Energy density of daily weight change (Mjoules)	Total energy lost / gained daily (Mjoules)
Suckling pups	14	+ 1.7	26.8	+ 60.2
Adult lactating females	170	- 3.8	34.7	- 132.0
Adult breeding bulls	257	- 2.2	37.6	- 83.0

[1]Calculated from carcase analysis

On a comparative basis, the pup comprises about 5.5-8% of maternal body weight at birth in phocid species including grey seal, northern elephant seal *(Mirounga angustirostris)*, harp seal, and Weddell seal *(Leptonychotes weddelli)* (Anderson and Fedak, 1987). The duration of lactation varies between these species, which is reflected in the daily increase in weight of pups. In harp seal pups, weight gain is about 28% birth weight daily for 10 days (Stewart and Lavigne, 1984), whereas gain in Weddell seal pup weight is 8.3% birth weight daily for 45 days. Thus the harp seal pup nearly trebles its birth weight by weaning, and the northern elephant seal about 3.7 times. Such differing strategies are often related to habitat. For example, the hooded seal, like the harp seal, gives birth on unstable ice and therefore it is critical to give birth and complete lactation (four days and doubling the pup's birth weight) in the shortest possible time. Phocid seals generally feed very little, if at all, during lactation (see Evans and Stirling, this volume).

In the northern fur seal, the mother feeds intermittently during lactation which may last up to a year, so that nursing takes place as a series of bouts. During such a 7-day period with two suckling bouts, Costa and Gentry (1986) calculated that pup weight gain was 3.2% birth weight daily whilst the mother actually gained weight because of foraging. Phocids generally feed at deeper water levels than otariids and require more efficient insulation in the form of thicker blubber fat. Anderson and Fedak (1987) propose that the lactational strategies are different in phocid and otariid seals essentially because of their feeding strategies. The presence of a blubber layer permits the evolution of a lactational strategy that allows a separation of breeding and feeding in time and space, because it acts not just as an insulator but also as an energy reserve.

In summary (see Riedman, 1990; Evans and Stirling, this volume), phocid seals have a short but intense lactation, do not feed during lactation, produce milk of high fat content which reduces over time, have relatively large precocial pups that grow rapidly during the suckling phase, and are weaned abruptly. In contrast, otariid seals have a prolonged lactation with intermittent nursing, feed during lactation, produce milk of relatively lower fat content that remains constant over time, have small pups that grow gradually and are weaned gradually - see also Boyd *et al.* (1995) for a detailed discussion of the relationship between reproductive costs and life history implications in Antarctic fur seals. Walruses, for comparison, are more similar to otariid seals.

5.3   Sirenians

The Amazonian manatee, *Trichecus inungius*, has peak births during February to May when river levels are highest. In the West African manatee, *T. senegalensis*, peak births coincide with the start of the rainy season. However, all sirenian births tend to coincide with peak algal/plant food production, and for Florida manatees, *T. manatus*, births coincide with high water temperatures which are energetically favourable for the young (Boyd *et al.*, 1999).

5.4   Sea Otters

The newborn sea otter pup grows from about 3 kg to 15 kg after about six months when it may be independent. The reported growth rates vary and can be 90 g per day, but may only be 45-61 g per day (Garshelis and Garshelis, 1987). Atypical pup-rearing strategies have been reported by Garshelis and Garshelis (1987), where a young pup has been abandoned because the mother has become sick. In three cases observed, two mothers died but the third survived. The abandonment of the pup may be an attempt to conserve energy for recovery and invest in future opportunities to reproduce again.

5.5   Polar Bears

In polar bears, the female has developed a strategy to minimise energetic outlay during reproduction. The fact that the young are born prematurely means that little energy is used for the pregnancy itself. Birth occurs in winter, and the female excavates a maternity den under snow, and remains there for several months until the cubs are ready for weaning and are able to cope in more clement weather conditions. Watts and Hansen (1987) reported that in well-ventilated but insulated dens, a temperature around 0°C was maintained while the outside temperature ranged between -35°C to -15°C. They found that the female has the potential to reduce her metabolic rate to 50% of basal, although on average a reduction of 30% was observed. During the denning period, the female starved and used stored fat up to 45% of total body weight for herself and for lactation during which time a milk of high fat content is produced (Derocher *et al.*, 1993). This energy saving strategy is similar to that of dormancy found in other bears, and also is like conditions during hibernation.

6.   ENVIRONMENTAL EFFECTS ON REPRODUCTION

6.1   Natural Ecological Factors

Natural ecological factors include climate - sea temperature, salinity and ice cover, prey abundance - plankton, plant, fish and large prey production, habitat, and any normal factor in the environment that may be subject to change. Perturbations in the environment can affect the ecology of a species in several, often profound, ways, and these generally ultimately have an effect on production - most often directly through reproduction. In marine mammals, many biological parameters have been shown to be variable according to environmental changes that might affect food supply (Laws, 1962; 1977a,b; Lockyer, 1987a, 1990; Bengtson and Laws, 1985; Bengtson and Siniff, 1981; Kingsley, 1979). For example, Fowler (1984) reported that cetacean populations are regulated by density-dependent changes in reproduction and survival, and these are expressed most often in the form of birth and juvenile survival in connection with food resources. These important factors are not exclusive, and social and behavioural factors are also important.

6.1.1   Cetaceans

6.1.1.1   Mysticetes

Off Iceland, the variation in food availability from year to year can influence the amount of fat storage in female fin whales, so that in years of great abundance fat stores are very high, whereas in years of low abundance, the fat storage may be diminished. This can interfere with the two-year reproductive cycle, so that in good years the female might produce a calf in consecutive years, and in poor years the cycle could be prolonged to three years to enable fat reserves to build up. The way the effect operates appears most likely to be in suppression of ovulation if a certain threshold level of body weight or fat is not reached. This is a recognised strategy in terrestrial mammals (*e.g.* red deer, *Cervus elaphus* - Hamilton and Blaxter, 1980; and reindeer, *Tarandus rangifer* - Leader-Williams and Ricketts, 1982). The combined effects of sea temperature, plankton and food production on body fat condition and fecundity have been investigated in fin whales off Iceland (Lockyer, 1987a, c). There appeared to be a close correlation between these factors, some being directly and others indirectly linked to fecundity. The pattern of food abundance, body fat condition, and fecundity for female fin whales is shown in Figure 7.

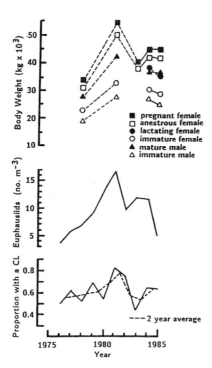

Figure 7.   Yearly variations in female fin whale body condition (weight) for different reproductive classes, whale food abundance (euphausiid density), and potential whale fecundity (proportion of females with a *corpus luteum*) off SW Iceland (after Lockyer, 1987c).

The age at sexual maturation and birth size have also been shown to vary with exploitation and external environmental factors in fin whales (Lockyer, 1990; Lockyer and Sigurjónsson, 1992). Off Iceland, it was also reported that in seasons when feeding conditions were poor, the growth of the foetus was deleteriously affected with significantly lower body weight (but not length), (Lockyer, 1990; Lockyer and Sigurjónsson, 1991, 1992). This clearly has a potential effect on calf survival and subsequent population recruitment.

There is very little field evidence for effects in other baleen whales. Perhaps the most publicised examples are related to large baleen whales where cases of age at sexual maturation change in response to exploitation and the supposed subsequent reduced intra- and inter-specific competition for food resources have been reported for fin, sei and minke whales (Lockyer, 1978). It was believed at one time that the reduction of blue, fin, sei and humpback whales in the southern oceans was anticipated to have the effect of releasing more food to the smaller competitive minke whale species and allowing earlier maturation. A similar interpretation was also given for the earlier maturation of fin and sei whales with the dramatic reduction in numbers of their species. The age at maturation has become younger in Antarctic populations (Lockyer, 1972, 1974, 1979; Masaki, 1978, 1979; Kato, 1983, 1987; Kato and Sakuramoto, 1991; Ohsumi, 1986). A similar trend was also observed in Icelandic fin whales, but was subsequently discovered to reverse (Lockyer and Sigurjónsson, 1991, 1992; Sigurjónsson, 1990; 1995), most probably because of worsening feeding conditions.

Changes in pregnancy rate were also reported for Antarctic fin and sei whales where an increasing trend was noted from the time of severe over-exploitation of southern whale stocks (Gambell, 1973; Laws, 1977a, b) and the supposed increased food availability. While this phenomenon was further investigated by Mizroch (1981a, b) and Mizroch and York (1982) for the Antarctic, and the findings were inconclusive as to whether a trend had really occurred once biases were taken into account (Lockyer, 1984a), there has been more clear-cut evidence of correlation of reproductive performance with changes in food abundance off Iceland, where the pregnancy rate for fin whale appeared closely correlated with food resource and body fat condition (Lockyer, 1986, 1987a, c; Vikingsson, 1990, 1995).

The observation that age can affect reproductive performance indicates that changes in the age frequency distribution of a population, howsoever caused, could affect population production.

6.1.1.2   Odontocetes

In the Bay of Fundy region, there is evidence for changes in growth and reproductive parameters of harbour porpoises, with animals growing faster and maturing sexually at an earlier age (Read and Gaskin, 1990). It was speculated that these changes might be mediated by increased prey availability to the porpoises, brought about by reduction in porpoise population size through incidental mortality in fisheries. There is some indirect evidence for environmental interaction with reproduction in dolphins. The effects of the 1982-1983 El Niño off Peru were experienced by dusky dolphins *(Lagenorhynchus obscurus)* in the way food abundance was diminished to the mature female population. Pregnant and lactating females deposited poorly calcified dentinal growth layer groups (GLGs) in the teeth during the period of the El Niño, indicating stress in nutrition, whereas teeth of dolphins of other reproductive classes showed no such phenomena (Manzanilla, 1989).

Similar interference in the GLG formation in teeth was observed in captive short-finned pilot whales from California. There was a link between the appearance of mineralisation interference in the teeth and reproductive events in females such as ovulation and pregnancy, suggesting dietary change as the intermediary factor (Lockyer, 1993b). Such reproductive events have now been identified in the tooth GLG patterns of known-age free-living bottlenose dolphins in the Gulf of Mexico *e.g.* sexual maturation and parturition (Hohn *et al.*, 1989). Such reproductive events were also tracked in teeth of dolphins (genus *Stenella*) and reported (Klevezal' and Myrick, 1984). While such events themselves do not affect reproductive performance, they indicate a level of stress perhaps caused partly by dietary deficiencies and hormonal changes. It may be possible that severe dietary inadequacies might lead to similar phenomena as observed in fin whales, including lowered fecundity, reduced birth weight (Lockyer, 1990), and perhaps lowered infant survival.

Clearly this is a topic that merits some investigation because of the potentially significant effects factors such as dietary change may have on subsequent recruitment to a population. The subject is addressed further below under Sirenians.

6.1.2   Pinnipeds

Evidence was given for a decline in age at sexual maturation in crabeater seals *(Lobodon carcinophagus)*, using age estimates from both teeth and ovaries, which could be related directly to changes in food availability within the southern ocean (Bengtson and Laws, 1985).

6.1.3   Sirenians

Best (1983) found evidence that Amazonian manatees can delay reproduction during prolonged dry seasons when fasting is enforced for up to seven months. A possible exacerbation of such effects may come from large-scale deforestation of the Amazon. Boyd *et al.* (1999) reported a major die-back and overgrazing of sea grass in the Torres Strait between Australia and Papua new Guinea during the mid-1970s. At this time the dugongs were found in a very emaciated body condition, coincident with few pregnancies recorded in animals fished locally at this time. However, pregnancy rates increased fourfold between 1978 and 1982, and also the proportion of sexually active males increased in this period (Marsh, 1986). Similarly, a loss of 1,000km^2 of sea grass in Hervey Bay, south Queensland in 1992 following two floods and a cyclone, was associated with a regional dugong population reduction to just over a quarter of that 21 months later (Preen and Marsh, 1995). The proportion of calves seen during surveys also dropped to a tenth (Marsh and Corkeron, 1996). Thus, environmental impacts such as this, resulting in habitat destruction, may have severe effects lasting several years.

6.1.4   Polar Bears

Fluctuations in reproductive rate have been reported in western Hudson Bay polar bears over time, correlated with changes in growth rate and body mass (Derocher and Stirling, 1998). These events have been correlated with fluctuations in prey availability caused by ice conditions and seal production in the adjacent Beaufort Sea (Kingsley, 1979).

6.2   Anthropogenic and Disease Factors

Such factors may include any foreign item introduced into the environment or disease-causing agent that could affect reproduction. Those factors include contaminants of all types *e.g.* DDT, PCB, dioxins, chlordanes, toxophenes, heavy metals, and many recently described diseases caused by viruses. There has been much interest in such factors especially in relation to reproduction, and not least the potential effects on fecundity, foetal growth, contamination of milk, and subsequent calf survival. Contaminants may also have possible teratogenic effects and cause abnormalities of reproductive organs and foetuses. Many organochlorine compounds are toxic in the way they can mimic hormones. The following briefly reviews some studies on this huge topic.

Levels of organochlorines are generally lower in adult females than males for all mammalian species. This is because of lactational transfer of compounds to the calf and placental transfer to the foetus *in utero* (Addison and Brodie, 1977; Reijnders, 1980; Subramanian *et al.*, 1987, 1988). The high transference level in milk is because of the concentration of organochlorines in lipophilic tissues such as blubber which is generally used as an energy resource for milk production. The amount of transfer is related to both actual level in the mother, but also duration of lactation, milk quality in terms of fat content, and age of the mother (Stern *et al.*, 1994).

Levels of contaminants are reportedly lower in the blubber of baleen whales than in other cetaceans, with ranges of 0.1 - 10 ppm wet weight for both DDT and PCBs (Aguilar and Borrell, 1988; Aguilar and Jover, 1982; Taruski *et al.*, 1975; Wolman and Wilson, 1970). Baleen whales are mostly feeding on euphausiids or planktonic organisms rather than fish and higher predators, and their consumption is thus closer to the start of the trophic web. This information suggests that there are likely to be few problems that might be linked to reproduction, associated with these levels of contaminants.

Death *in utero* has been reviewed by Ichihara (1962), where resorption of foetal tissue, necrosis and even calcification of the products of pregnancy were reported in fin whales. A death rate *in utero* of 0.14% was calculated by Ichihara (1962) for Antarctic fin whales, but he reported a higher rate in the region of 10° E and 40° E. Ivashin (1977) and Ivashin and Zinchenko (1982) reported nearly three times as many malformed foetuses in minke whales in Pridz Bay between longitudes 70° E and 130° E than elsewhere in the Antarctic. They speculated on the effect of some teratogenic factor, unidentified, in this region.

The diet of most odontocetes includes fish, squid, and generally higher trophic level consumers than the prey of mysticetes, so that the actual intake of contaminants is likely to be higher than that in baleen whales. In the Bay of Fundy area, harbour porpoises were reported to have relatively high PCB levels (Gaskin *et al.*, 1983), especially so in immature animals and adult males increasing with age to a maximum recorded level of 310 ppm in blubber. The adult females had relatively low levels, as anticipated from lactational mobilisation of lipids. Most other tissues had low levels of 2 ppm as expected from their low lipid content. There were no reports that indicated that the contaminants had affected reproductive health. Belugas in the St Lawrence estuary have high levels of PCBs and DDTs (>100mg per kg wet weight blubber), and Martineau *et al.* (1987) have suggested that this may be the cause of low recruitment in this population because of hormonal interference in reproduction. Belugas from this region have a high incidence of cancer and pathological conditions probably related to the immunosuppressive effects of organochlorines and the carcinogenic effects of benzopyrenes.

Organochlorine pollutants are known to suppress the immune system, but the effect these may have on cetaceans in crisis *e.g.* the striped dolphin population (*Stenella coeruleoalba*) in the Mediterranean in 1990-1991 when devastated by the morbillivirus epizootic, is uncertain (Borrell and Aguilar, 1992; Hall, 1992; Aguilar and Borrell, 1994). A similar virus, PDV (phocine distemper virus) in harbour seals *(Phoca vitulina)*, certainly caused spontaneous abortions apart from death of the adults infected (Heide-Jørgensen *et al.*, 1992). An increase in abortions and stillbirths may occur in cetaceans. Many of the events observed in this field are largely speculative. Certainly new discoveries are being made constantly. A recent example is the description of a form of genital herpes in harbour porpoise (*Phocoena phocoena*) in British waters (Ross *et al.*, 1994a) that may interfere with calf production. More seriously, another agent, a cetacean form of brucellosis, has been described in porpoise in British waters (Ross *et al.*, 1994b). This agent is known to cause abortions in cattle. It is probably safe to say that many of these "new" discoveries are not really new, but have merely failed to be recognised in previous years through ignorance or lack of interest.

Over the last 35 years, the harbour seal population in the Dutch Wadden Sea has collapsed. This has partly been because of over-exploitation, but also because of declining pup production in the last two decades, demonstrated to be mediated by toxic levels of PCBs that interfered with the female seals' fertility at the implantation phase (Reijnders, 1986). The PCBs were derived from polluted fish in the area.

In the Baltic, substantial declines in populations of ringed seal *(Phoca hispida)*, grey seals, and harbour seals have occurred since the turn of the century. Again, this is partly explained by hunting pressures, but investigations revealed that PCBs and DDTs were involved in reducing fertility (Olsson, 1978; Bergman and Olsson, 1986), so lowering the pup production. The effects of the contaminants were such as to cause spontaneous abortions in ringed seals (Olsson *et al.*, 1975), and uterine occlusions and stenoses, and tumours resulting in sterility of the females (Helle *et al.*, 1976a,b; Helle, 1980).

In March 1989, the oil tanker vessel, Exxon Valdez, was the cause of a major oil spill in the Prince William Sound and Kenai Peninsula area of Alaska. The effect was devastating on the sea otter population by oiling the fur of animals, which was subsequently licked and entered into the animals' blood. Other effects were reduction in insulation of the fur and consequent hypothermia, as well as respiratory distress, digestive disorders and anaemia. In addition, soiling of the coastal habitat required drastic environmental cleaning. During the spillage, pups were frequently abandoned, lost and separated from the mothers. Most deaths occurred within the initial three weeks. Eighteen months later the longer-term effects were assessed by comparing animals held captive with those returned to the wild (Williams, 1990). Animals both captive and free, were reproductively active and one captive female had given birth within a year of the spill. After recuperation of sick animals, the population appeared to be normal after 18 months, with representatives of all age groups feeding and behaving normally.

Recently, abnormalities of reproductive organs have been observed in female polar bears from Svalbard, north Norway. From a total of some 450 polar bears examined in the last three years, seven genetic female bears, some of which are known to have had cubs, have been observed with vestigial male organs and appear to be pseudo-hermaphrodite (Holden, 1998). A teratogenic effect of PCBs in the bears' fat, known to be 2.5 times higher level than that in Canadian waters may be a contributory cause by interfering with hormone-regulating P450 cytochromes.

The whole field of investigation of the interaction of health status of marine mammals, pollutant loads, and incidence of disease with reproductive success is one that needs to be examined carefully in the future. Contaminant loads may be significant in their effect or may be given undue importance and be misleading.

7.   CONCLUDING POINTS

Throughout this chapter, it is clear that our knowledge of reproductive matters in marine mammals are focused very much on intense studies of rather a few species, not all of which are necessarily representative of any particular group. There remain many species for which we know very little. Many marine mammals are elusive creatures, especially some of the cetaceans, and are difficult to study. However, our ability to conserve and maintain many threatened species must surely rely on better understanding of reproductive processes and the factors that influence them. Recruitment to a population depends so much on them. Much of what we know today has come to light as a result of urgent need for information after a crisis in conservation has arisen. It would be satisfying to try to anticipate future conservation needs and the information required, and so help avert potential crises.

REFERENCES

Addison, R.F. and Brodie, P.F. (1977) Organochlorine residues in maternal blubber, milk and pup blubber from grey seals (*Halichoerus grypus*) from Sable Island, Nova Scotia. *Journal of the Fisheries Research Board of Canada,* **34**, 937-941.

Aguilar, A. and Borrell, A. (1988) Age- and sex-related changes in organochlorine compound levels in fin whales (*Balaenoptera physalus*) from the Eastern North Atlantic. *Marine Enviroment Research,* **25**, 195-211.

Aguilar, A. and Borrell, A. (1994) Abnormally high polychlorinated biphenyl levels in striped dolphins (*Stenella coeruleoalba*) affected by the 1990-1992 Mediterranean epizootic. *Science of the Total Environment,* **154**, 237-247.

Aguilar, A. and Jover, L. (1982) DDT and PCB residues in the fin whale, *Balaenoptera physalus,* of the North Atlantic. *Report of the International Whaling Commission,* **32**, 299-301.

Anderson, S.S. and Fedak, M.A. (1987) The energetics of sexual success of grey seals and comparison with the costs of reproduction in other pinnipeds. *Symposium of the Zoological Society of London,* **57**, 319-341.

Bengtson, J.L. (1988) Long-term trends in the foraging patterns of female Antarctic fur seals at South Georgia. In: *Antarctic Ocean and Resources Variability.* (Ed. by D.Sahrage), pp. 286-91. Springer Verlag, Berlin and Heidelberg.

Bengtson, J.L. and Laws, R.M. (1985) Trends in crabeater seal age at maturity: an insight into Antarctic marine interactions? In: *Nutrient Cycles and Food Chains.* Proceedings of the 4th SCAR Symposium on Antarctic Biology (Ed. by W.R.Siegfried, P.R.Condy and R.M.Laws), pp. 669-675. Springer-Verlag, Berlin.

Bengtson, J.L. and Siniff, D. (1981) Reproductive aspects of female crabeater seals *(Lobodon carcinophagus)* along the Antarctic Peninsula. *Canadian Journal of Zoology,* **59**, 92-102.

Bergman, A. and Olsson, M. (1986) Pathology of Baltic grey seal and ringed seal females with special reference to adrenocortical hyperplasia: is environmental pollution the cause of a widely distributed disease syndrome? *Finnish Game Research,* **44**, 47-62.

Bernard, H.J. and Hohn, A.A. (1989) Differences in feeding habits between pregnant and lactating spotted dolphins *(Stenella attenuata). Journal of Mammalogy,* **70**, 211-15.

Best, P.B. (1970) The sperm whale *(Physeter catodon)* off the west coast of South Africa. 5. Age, growth and mortality. *Investigational Report of the Division of Sea Fisheries, South Africa,* **79**, 1-27.

Best, P.B. (1974) The biology of the sperm whale as it relates to stock management. In: *The whale problem: a status report.* (Ed. by W.E. Schevill). pp. 257-293. Harvard University Press, Cambridge, Mass.

Best, P.B. and Butterworth, D.S. (1980) Timing of oestrus within sperm whale schools. *Report of the International Whaling Commission* (special issue **2**), 137-140.

Best, P.B., Canham, P.A.S. and Macleod, N. (1984) Patterns of reproduction in sperm whales, *Physeter macrocephalus. Report of the International Whaling Commission* (special issue **6**), 51-79.

Best, R.C. (1983) Apparent dry-season fasting in Amazonian manatees (Mammalia: Sirenia). *Biotropica,* **15**, 61-64.

Bloch, D. (1994) Intermale competition in schools of long-finned pilot whales as indicated by abundance of fighting marks. ms submitted as part of a doctoral thesis, *Pilot whales in the N. Atlantic. Age, growth and social structure in Faroese grinds of the long-finned pilot whale, Globicephala melas.* Lund University, Sweden.

Bloch, D., Lockyer, C. and Zachariassen, M. (1993) Age and growth parameters of the long-finned pilot whale off the Faroe Islands. *Report of the International Whaling Commission* (special issue **14**), 163-208.

Bodkin, J.L., Ames, J.A., Kameson, R.J., Johnson, A.M. and Matson, G.M. (1997) Estimating age of sea otters with cementum layers in the first premolar. *Journal of Wildlife Management,* **61**, 967-973.

Borrell, A. and Aguilar, A. (1992) Pollution by PCBs in striped dolphins affected by the western Mediterranean epizootic. In: *The Mediterranean striped dolphin die-off.* (Ed. by X. Pastor and M. Simmonds), pp. 121-27. Proceedings of the Mediterranean striped dolphin mortality international workshop, 4-5 Nov 1991, Greenpeace, 190pp.

Boyd, I. L. (1984) The relationship between body condition and the timing of implantation in pregnant grey seals *(Halichoerus grypus). Journal of Zoology* (London), **203**, 113-123.

Boyd, I. L. (1991) Environmental and physiological factors controlling the reproductive cycles of pinnipeds. *Canadian Journal of Zoology,* **69**, 1135-1148.

Boyd, I. L. (1996) Individual variation in the duration of pregnancy and birth date in Antarctic fur seals: the role of environment, age, and sex of fetus. *Journal of Mammalogy,* **77**, 124-133.

Boyd, I. L., Croxall, J.P., Lunn, H.J. and Reid, K. (1995) Population demography of Antarctic fur seals: the costs of reproduction and implications for life-histories. *Journal of Animal Ecology,* **64**, 505-518.

Boyd, I.L., Lockyer, C.H. and Marsh, H. (1999) Reproduction in marine mammals, Chapter 8. In: *Marine Mammals, Volume 1.* (Ed. by J.E Reynolds, III and S.A.Rommel), pp. 218-286. Smithsonian Institution Press, Washington D.C.

Braham, H. (1984) Review of reproduction in the white whale, *Delphinapterus leucas,* narwhal, *Monodon monoceros,* and Irrawaddy dolphin, *Orcaella brevirostris,* with comments on stock assessment. *Report of the International Whaling Commission* (special issue **6**), 81-89.

Brodie, P.F. (1969) Duration of lactation in Cetacea: an indicator of required learning? *American Midland Naturalist,* **82**, 312-314.

Brody, S. (1968) *Bioenergetics and growth.* Hafner Publishing Co., New York. 1023pp.

Brownell, R.L.Jr (1984) Review of reproduction in Platanistid dolphins. *Report of the International Whaling Commission* (special issue **6**), 149-158.

Clutton-Brock, T.H. (1984) Reproductive effort and terminal investment in iteroparous animals. *American Naturalist,* **123**, 212-229.

Clutton-Brock, T.H., Guinness, F.E. and Albon, S.D. (1982) *Red deer: behaviour and ecology of two sexes.* University of Chicago Press, Chicago, 378pp.

Costa, D.P. and Gentry, R.L. (1986) Free ranging and reproductive energetics of the Northern fur seal. In: *Fur seals: maternal strategies on land and at sea* (Ed. by R.L. Gentry and G.L.Kooyman), pp. 79-101. Princeton University Press, Princeton.

Da Silva, V.M.F. (1994) *Aspects of the biology of the Amazonian dolphins genus Inia and Sotalia fluviatalis.* Ph.D. Thesis, University of Cambridge. 327pp.

Daniel, J. C., Jr (1971) Growth of the pre-implantation embryo of the northern fur seal and its correlation with changes in uterine protein. *Developmental Biology,* **26**, 316-322.

Daniel, J. C., Jr (1974) Circulating levels of estradiol-17ß during early pregnancy in the Alaska fur seal showing an estrogen surge preceding implantation. *Journal of Reproduction and Fertility,* **37**, 425-428.

Derocher, A.E. and Stirling, I. (1994) Age-specific reproductive performance of female polar bears *(Ursus maritmus). Journal of Zoology* (London), **234**, 527-536.

Derocher, A.E. and Stirling, I. (1998) Geographic variation in growth of polar bears *(Ursus maritimus). Journal of Zoology* (London), **245**, 65-72.

Derocher, A. E., D. Andriashek and Arnould, J. P. Y. (1993) Aspects of milk composition and lactation in polar bears. *Canadian Journal of Zoology,* **71**, 561-567.

Desportes, G., Saboureau, M. and Lacroix, A. (1993) Reproductive maturity and seasonality of male long-finned pilot whales, off the Faroe Islands. *Report of the International Whaling Commission* (special issue **14**), 234-262.

Duffield, D.A., Odell, D.A., McBain, J.F. and Andrews, B. (1995) Killer whale (*Orcinus orca*) reproduction at sea world. *Zoo Biology,* **14**, 417-430.

Evans, P.G.H. (1987) *The natural history of whales and dolphins.* Christopher Helm, London. 343pp.

Fowler, C.W. (1984) Density dependence in cetacean populations. *Report of the International Whaling Commission* (special issue **6**), 373-379.

Frazer, J.F.D and Huggett, A. St G. (1973) Specific foetal growth rates of cetaceans. *Journal of Zoology* (London), **169**, 111-126.

Frazer, J.F.D and Huggett, A. St G. (1974) Species variations in the foetal growth rates of eutherian mammals. *Journal of Zoology* (London), **174**, 481-509.

Gambell, R. (1973) Some effect of exploitation on reproduction in whales. *Journal of Reproduction and Fertility, Supplement,* **19**, 533-553.

Garshelis, D.L. (1983) *Ecology of sea otters in Prince William Sound, Alaska.* Ph.D. Thesis, University of Minnesota, Minneapolis. 321pp.

Garshelis, D.L. (1984) Age estimation of living sea otters. *Journal of Wildlife Management,* **48**, 456-463.

Garshelis, D.L. and J.A.Garshelis (1987) Atypical pup rearing strategies by sea otters. *Marine Mammal Science,* **3**, 263-270.

Gaskin, D.E. (1982) *The ecology of whales and dolphins.* Heinemann, London. 459pp.

Gaskin, D.E., Frank, R. and Holdrinet, M. (1983) Polychlorinated biphenyls in harbour porpoises *Phocoena phocoena* (L.) from the Bay of Fundy, Canada and adjacent waters, with some information on chlordane and hexachlorobenzene levels. *Archives of Environmental Contaminant Toxicology,* **12**, 211-219.

Gaskin, D.E., Smith, G.J.D., Watson, A.P., Yasui, W.Y. and Yurick, D.B. (1984) Reproduction in porpoises (*Phocoenidae*): implications for management. *Report of the International Whaling Commission* (special issue **6**), 135-148.

George, J.C., Bada, J., Zeh, J., Scott, L., Brown, S.E. and O'Hara, T. (1998) Preliminary age estimates of bowhead whales via aspartic acid racemisation. Paper SC/50/AS10, IWC Scientific Committee, April 1998, Oman.

Hall, A.J. (1992) Disease causation and the striped dolphin mortality. In: *The Mediterranean striped dolphin die-off.* (Ed. by X.Pastor and M. Simmonds), pp. 111-118. Proceedings of the Mediterranean striped dolphins mortality international workshop, 4-5 November 1991, Greenpeace, 190pp.

Hamilton, W.J. and Blaxter, K.L. (1980) Reproduction in farmed red deer. 1. Hind and stag fertility. *Journal of Agricultural Science, Cambridge,* **95**, 261-273.

Harrison, R.J. (1969) Reproduction and reproductive organs. In: *The biology of marine mammals* (Ed. by H.T. Andersen), pp. 253-348. Academic Press, New York and London, 511pp.

Heide-Jørgensen, M.P., Härkönen, T. and Åberg, P. (1992) Long-term effects of epizootic in harbour seals in the Kattegat-Skagerrak and adjacent areas. *Ambio,* **21**, 511-516.

Heide-Jørgensen, M.P. and Teilmann, J. (1994) Growth, reproduction, age structure and feeding habits of white whales (*Delphinapterus leucas*) in West Greenland waters. *Meddelelser om Grønland, Bioscience*, **39**, 195-212.

Helle, E., Olsson, M. and Jensen, S. (1976a) DDT and PCB levels and reproduction in ringed seal from the Bothnian Bay. *Ambio*, **5**, 188-189.

Helle, E., Olsson, M. and Jensen, S. (1976b) PCB levels correlated with pathological changes in seal uteri. *Ambio*, **5**, 261-263.

Helle, E. (1980) Lowered reproductive capacity in female ringed seals *(Pusa hispida)* in the Bothnian Bay, northern Baltic Sea, with special reference to uterine occlusions. *Annales Zoologici Fennici,* **17**, 147-158.

Hohn, A.A., Scott, M.D., Wells, R.S., Sweeney, J.C. and Irvine, A.B. (1989) Growth layers in teeth from known-age, free-ranging bottlenose dolphins. *Marine Mammal Science,* **5**, 315-342.

Holden, C. (Ed.) (1998) Polar bears and PCBs. *Science*, **280**, 2053.

Huggett, A.St G and W.F.Widdas (1951) The relationship between mammalian foetal weight and conception age. *Journal of Physiology*, **114**, 306-317.

Ichihara, T. (1962) Prenatal dead foetus of baleen whales. *Scientific Reports of the Whales Research Institute, Tokyo*, **16**, 47-60.

Ivashin, M.V. (1977) Some abnormalities in embryogenesis of minke whales, *Balaenoptera acutorostrata (Cetacean, Balaenoptera)* of the Indian Ocean area of the Antarctic. *Zoologicheskii Zhurnal*, **41**, 1736-1739.

Ivashin,M.V. and Zinchenko, V.L. (1982) Occurrences of pathological development of minke embryos (*Balaenoptera acutorostrata*) of the Southern Hemisphere. Document SC/34/Mi29 submitted to the Scientific Committee of the International Whaling Commission, Cambridge, June 1982.

Kasuya, T. and Marsh, H. (1984) Life history and reproductive biology of the short-finned pilot whale, *Globicephala macrorhynchus*, off the Pacific coast of Japan. *Report of the International Whaling Commission* (special issue **6**), 259-310.

Kasuya, T. and Tai, S. (1993) Life history of short-finned pilot whale stocks off Japan and a description of the fishery. *Report of the International Whaling Commission* (special issue **14**), 439-473.

Kasuya, T., Marsh, H. and Amino, A. (1993) Non-reproductive mating in short-finned pilot whales. *Report of the International Whaling Commission* (special issue **14**), 425-437.

Kato, H. (1983) Some considerations on the decline in age at sexual maturity of the Antarctic minke whale. *Report of the International Whaling Commission*, **33**, 393-9.

Kato, H. (1987) Density dependent changes in growth parameters of the southern minke whale. *Scientific Reports of the Whales Research Institute, Tokyo,* **38**, 47-73.

Kato, H. and Sakuramoto, K. (1991) Age at sexual maturity of southern minke whales: a review and some additional analyses. *Report of the International Whaling Commission*, **41**, 331-337.

Kenyon, K.W. (1978a) Walrus. In: *Marine mammals* (Ed. by D. Haley), pp. 178-183. Pacific Search Press.

Kenyon, K.W. (1978b) Sea Otter. In: *Marine mammals* (Ed. by D. Haley), pp. 226-235. Pacific Search Press.

Kingsley, M.C.S. (1979) Fitting the von Bertalanffy growth equations to polar bear age-weight data. *Canadian Journal of Zoology*, **57**, 1020-1025.

Klevezal', G.A. and Myrick, A.C. (1984) Marks in tooth dentine of female dolphins (genus *Stenella*) as indicators of parturition. *Journal of Mammalogy*, **65**, 103-110.

Koopman, H.N. (1998) Topographical distribution of the blubber of harbour porpoises *(Phocoena phocoena). Journal of Mammalogy*, **79**, 260-270.

Laws, R.M. (1959) Foetal growth rates of whales with special reference to the fin whale, *Balaenopetra physalus* (L.) *"Discovery" Report*, **29**, 281-308.

Laws, R.M. (1961) Southern fin whales. *"Discovery" Report*, 31, 327-486.

Laws, R.M. (1962) Some effects of whaling on the southern stocks of baleen whales In: *The exploitation of natural animal population* (Ed. by L.D.LeCren and M.W.Holdgate), pp. 137-158. Blackwell Scientific Publications, Oxford.

Laws, R.M. (1977a) The significance of vertebrates in the Antarctic marine ecosystem. In: *Adaptations within Antarctic ecosystems* (Ed. by G.A.Llano), pp. 411-438. 3rd Symposium on Antarctic Ecology, Smithsonian Institution, Washington, 1252pp.

Laws, R.M. (1977b) Seals and whales of the southern ocean. *Philosophical Transactions of the Royal Society, London, Series* B, **279**, 81-96.

Laws, R.M. (1984) Seals. In: *Antarctic Ecology, Vol.2*, (Ed. by R.M.Laws), pp. 621-715. Academic Press, London. 771pp.

Leader-Williams, N. and C.Ricketts (1982) Seasonal and sexual patterns of growth and condition of reindeer introduced into South Georgia. *Oikos*, **38**, 27-39.

Lentfer, J.W. (1978) Polar bear, In: *Marine mammals* (Ed. by D. Haley), pp. 218-225. Pacific Search Press.

Lockyer, C. (1972) The age at sexual maturity of the southern fin whale (*Balaenoptera physalus*) using annual layer counts in the ear plug. *Journal du Conseil International d'Exploration de Mer*, **34**, 276-294.

Lockyer, C. (1974) Investigation of the ear plug of the southern sei whale, *Balaenoptera borealis*, as a valid means of determining age. *Journal du Conseil International d'Exploration de Mer*, **36**, 71-81.

Lockyer, C. (1978) A theoretical approach to the balance between growth and food consumption in fin and sei whales, with special reference to the female reproductive cycle. *Report of the International Whaling Commission*, **28**, 243-250.

Lockyer, C. (1979) Changes in a growth parameter associated with exploitation of southern fin and sei whales. *Report of the International Whaling Commission*, **29**, 191-196.

Lockyer, C. (1981a) Growth and energy budgets of large baleen whales from the Southern Hemisphere. *FAO Fisheries Series (5), Mammals in the Seas*, **3**, 379-487.

Lockyer, C. (1981b) Estimates of growth and energy budget for the sperm whale. *FAO Fisheries Series (5), Mammals in the Seas*, **3**, 489-504.

Lockyer, C. (1984a) Review of baleen whale (*Mysticeti*) reproduction and implications for management. *Report of the International Whaling Commission* (special issue **6**), 27-50.

Lockyer, C. (1984b) Age determination by means of the earplug in baleen whales. *Report of the International Whaling Commission*, **34**, 692-696, and refs, pp. 683-684.

Lockyer, C. (1986) Body fat condition in northeast Atlantic fin whales *Balaenoptera physalus*, and its relationship with reproduction and food resource. *Canadian Journal of Fisheries and Aquatic Science*, **43**, 142-147.

Lockyer, C. (1987a) The relationship between body fat, food resource and reproductive energy costs in north Atlantic fin whales (*Balaenoptera physalus*). *Symposium of the Zoological Society of London*, **57**, 343-361.

Lockyer, C. (1987b) Observations of the ovary of the southern minke whale. *Scientific Reports of the Whales Research Institute, Tokyo*, **38**, 75-89.

Lockyer, C. (1987c) Evaluation of the role of fat reserves in relation to the ecology of North Atlantic fin and sei whales. In: *Approaches to marine mammal energetics* (Ed. by A.C.

Huntley, D.P. Costa, G.A.J. Worthy and M.A. Castellini), pp. 183-203. Society for Marine Mammalogy, Special Publication no 1, Lawrence, Kansas.

Lockyer, C. (1990) The importance of biological parameters in population assessments with special reference to fin whales from the N.E. Atlantic. *North Atlantic Studies,* **2**, 22-31.

Lockyer, C. (1993a) Seasonal changes in body fat condition of northeast Atlantic pilot whales, and their biological significance. *Report of the International Whaling Commission* (special issue **14**), 323-350.

Lockyer, C. (1993b) A report on patterns of deposition of dentine and cement in teeth of pilot whales, genus *Globicephala. Report of the International Whaling Commission* (special issue **14**), 137-161.

Lockyer, C. (1995a) Aspects of the biology of the harbour porpoise, *Phocoena phocoena,* from British waters. In: *Whales, seals, fish and men* (Ed. by A.S. Blix, L.Walløe and Ø.Ulltang), pp. 443-457. Elsevier Science, Amsterdam.

Lockyer, C. (1995b) Investigation of aspects of the life history of the harbour porpoise, *Phocoena phocoena,* in British waters. *Report of the International Whaling Commission* (special issue **15**), 189-197.

Lockyer, C. (1995c) Aspects of the morphology, body fat condition and biology of the harbour porpoise, *Phocoena phocoena,* in British waters. *Report of the International Whaling Commission* (special issue **15**), 200-209.

Lockyer, C.H. and Brown, S.G. (1981) The migration of whales. In: *Animal migration* (Ed. by D.J.Aidley), pp. 105-137. Society for Experimental Biology - seminar series 13. Cambridge University Press, Cambridge.

Lockyer, C. and Sigurjónsson, J. (1991) The Icelandic fin whale (*Balaenoptera physalus*): biological parameters and their trends over time. Document Sc/F91/F8 submitted to the Scientific Committee of the International Whaling Commission, Reykjavik 1991.

Lockyer, C. and Sigurjónsson, J. (1992) The Icelandic fin whale (*Balaenoptera physalus*): biological parameters and their trends over time. (Summary). *Report of the International Whaling Commission,***42**, 617-618.

Lunn, N. J. and Boyd, I. L. (1993) Effects of maternal age and condition on parturition and the pre-natal period of Antarctic fur seals. *Journal of Zoology,* **229**, 55-67.

Lunn, N. J., I. L. Boyd and Croxall, J. P. (1994) Reproductive performance of female Antarctic fur seals: the influence of age, breeding experience, environmental variation and individual quality. *Journal of Animal Ecology,* **63**, 827-840.

Macdonald, D. (1984) *The Encyclopaedia of Mammals: 1.* George Allen and Unwin, 447pp.

Mackintosh, N.A. (1965) *The stocks of whales.* Buckland Foundation, 232pp.

Manzanilla, S.R. (1989) The 1982-1983 El Niño event recorded in dentinal growth layers in teeth of Peruvian dusky dolphins (*Lagenorhynchus obscurus*). *Canadian Journal of Zoology,* **67**, 2120-2125.

Marmontel, M. (1993) *Age determination and population biology of the Florida manatee, Trichechus manatus latirostris.* Ph.D. Thesis, University of Florida, Gainesville. 408pp.

Marmontel, M. (1995) Age and reproduction in female Florida manatees. In: *Population Biology of the Florida Manatee* (Ed. by T.J. O'Shea, B. B. Ackermann, and H. F. Percival), pp. 98-119. U.S. Department of the Interior, National Biological Service, Information and Technology Report 1.

Marsh, H. (1980). Age determination of the dugong *(Dugong dugon* (Müller)*)* in northern Australia and its biological implications. *Report of the International Whaling Commission* (special issue **3**), 181-201.

Marsh, H. (1986) The status of the dugong in Torres Strait. In: *Torres Strait Fisheries Seminar, Port Moresby* (Ed. by A. K. Haines, G. C. Williams and D. Coates) pp. 53-76. Australian Government Publishing Service, Canberra.

Marsh, H. (1995) The life history, pattern of breeding and population dynamics of the dugong. In: *Population Biology of the Florida Manatee* (Ed. by T.J. O'Shea, B.B. Ackermann and H.F. Percival) pp. 75-83. U.S. Department of the Interior, National Biological Service, Information and Technology Report 1.

Marsh, H. and Corkeron, P. (1996) The status of the Dugong in the Great Barrier Reef region. In: *Proceedings of State of the Great Barrier Reef Workshop* (Ed. by D. Wachenfeld and J. Oliver). Great Barrier Reef Marine Park Authority, Townsville.

Marsh, H. and Kasuya, T. (1984) Changes in the ovaries of the short-finned pilot whale, *Globicephala macrorhynchus*, with age and reproductive activity. *Report of the International Whaling Commission* (special issue **6**), 331-335.

Marsh, H. and Kasuya, T. (1986) Evidence for reproductive senescence in female cetaceans. *Report of the International Whaling Commission* (special issue **8**), 57-74.

Martin, A.R. (1980) An examination of sperm whale age and length data from the 1949-78 Icelandic catch. *Report of the International Whaling Commission,* **30**, 227-231.

Martin, A.R. (1990) *Whales and dolphins.* Salamander Books Ltd, London. 192pp.

Martin, A.R. and Rothery, P. (1993) Reproductive parameters of female long-finned pilot whales (*Globicephala melas*) around the Faroe Islands. *Report of the International Whaling Commission* (special issue **14**), 263-304.

Martineau, D., Béland, P., Desjardins, C. and Lagacé, A. (1987) Levels of organochlorine chemicals in tissues of beluga whales (*Delphinapterus leucas*) from the St Lawrence Estuary, Québec, Canada. *Archives of Environmental Contaminant Toxicology,* **16**, 137-147.

Masaki, Y. (1978) Yearly change in the biological parameters for the Antarctic sei whale. *Report of the International Whaling Commission,* **28**, 421-430.

Masaki, Y. (1979) Yearly change of the biological parameters for the Antarctic minke whale. *Report of the International Whaling Commission,* **29**, 375-396.

Mitchell, J. (1976) Age determination in the dugong, *Dugong dugon* (Müller). *Biological Conservation,* **9**, 25-28.

Mitchell, J. (1978) Incremental growth layers in the dentine of dugong incisors *(Dugong dugon* (Müller)*)* and their application to age determination. *Zoological Journal of the Linnean Society,* **62**, 317-348.

Mizroch, S.A. (1981a) Analysis of some biological parameters of the Antarctic fin whale (*Balaenoptera physalus*). *Report of the International Whaling Commission,* **31**, 425-434.

Mizroch, S.A. (1981b) Further notes on Southern Hemisphere baleen whale pregnancy rates. *Report of the International Whaling Commission,* **31**, 629-634.

Mizroch, S.A. and York, A.E. (1982) Have Southern Hemisphere baleen whales pregnancy rates increased? Document SC/34/Ba4 submitted to the Scientific Committee of the International Whaling Commission, Cambridge June 1982.

Odell, D. K., G. D. Bossart, M. T. Lowe, and Hopkins, T. D. (1995) Reproduction of the West Indian manatee in captivity. In: *Population Biology of the Florida Manatee* (Ed. by T.J. O'Shea, B.B. Ackermann, and H.F. Percival), pp. 192- 93. U.S. Department of the Interior, National Biological Service, Information and Technology Report 1.

Oftedal, O.T., Boness, D.J. and Tedman, R.A. (1987) The behavior, physiology, and anatomy of lactation in the Pinnipedia. *Current Mammalogy,* **1**, 175-245.

Oftedal, O.T. (1997) Lactation in whales and dolphins: evidence of divergence between baleen- and toothed-species. *Journal of Mammary Gland Biology and Neoplasia*, **2**, 205-230.

Ohsumi, S. (1986) Ear plug transition phase as an indicator of sexual maturity in female Antarctic minke whales. *Scientific Reports of the Whales Research Institute, Tokyo*, **37**, 17-30.

Olsson, M. (1978) PCB and reproduction among Baltic seals. Proceedings from the Symposium on the Conservation of Baltic Seals in Haikko, Finland, April 26-28, 1977. *Finnish Game Research*, **37**, 40-45.

Olsson, M., Johnels, A.D. and Vaz, R. (1975) DDT and PCB levels in seals from Swedish waters. The occurrence of aborted seal pups. Proceedings from the Symposium on the Seals in the Baltic, June 1974. Lidingô, Sweden, National Swedish Environ. Protection Board PM 591.

Perrin, W.F. and Myrick, A.C. (Eds.) (1981) Report of the workshop. Age determination of toothed whales and sirenians. *Report of the International Whaling Commission* (special issue **3**), 1-50.

Perrin, W.F. and Reilly, S.B. (1984) Reproductive parameters of dolphins and small whales of the family *Delphinidae*. *Report of the International Whaling Commission* (special issue **6**), 97-133.

Pond, C. M. (1984) Physiological and ecological importance of energy storage in the evolution of lactation: evidence for a common pattern of anatomical organisation of adipose tissue in mammals. *Symposium of the Zoological Society of London*, **51**, 1-32.

Preen, A. and Marsh, H. (1995) Response of dugongs to large-scale loss of seagrass from Hervey Bay, Queensland, Australia. *Wildlife Research*, **22**, 507-19.

Ralls, K. (1977) Sexual dimorphism in mammals: Avian models and unanswered questions. *American Naturalist*, **111**, 917-938.

Ramsay, M.A. and Stirling, I. (1988) Reproductive biology and ecology of female polar bears *(Ursus maritimus)*. *Journal of Zoology* (London), **214**, 601-634.

Read, A.J. (1990) Estimation of body condition in harbour porpoises, *Phocoena phocoena*. *Canadian Journal of Zoology*, **68**, 1962-1966.

Read, A.J. and Hohn, A.A. (1995) Life in the fast lane: the life history of harbor porpoises from the Gulf of Maine. *Marine Mammal Science*, **11**, 423-440.

Read, A.J. and Gaskin, D.E. (1990) Changes in the growth and reproduction of harbour porpoises, *Phocoena phocoena*, from the Bay of Fundy. *Canadian Journal of Fisheries and Aquatic Science*, **47**, 2158-2163.

Reijnders, P.J.H. (1980) Organochlorine and heavy metal residues in harbour seals from the Wadden Sea and their possible effects on reproduction. *Netherlands Journal of Sea Research*, **14**, 30-65.

Reijnders, P.J.H. (1986) Reproductive failure in common seals feeding on fish from polluted coastal waters. *Nature* (London), **324**, 456-457.

Rice, D.W. and Wolman, A.A. (1971) *The life history and ecology of the gray whale (Eschrichtius robustus)*. Special Publication 3. The American Society of Mammalogists. 142pp.

Riedman, M. (1990) *The pinnipeds: seals, sea lions, and walruses*. University of California Press. 439pp.

Robeck, T.R., Curry, B.E., McBain, J.F. and Kraemer, D.C. (1994) Reproductive biology of the bottlenose dolphin (*Tursiops truncatus*) and the potential application of advanced reproductive technologies. *Journal of Zoo Wildlife Medicine*, **25**, 321-336.

Ross, H.M., Reid, R.J., Howie, F.E. and Gray, E.W. (1994a) Herpes virus infection of the genital tract in harbour porpoise *Phocoena phocoena. European Research on Cetaceans*, **8**, 209.

Ross, H.M., Foster, G., Reid, R.J., Jahans, K.L. and MacMillan, A.P. (1994b) *Brucella species* infection in sea mammals. *Veterinary Record*, **134**, 359.

Sacher, G.A. and Staffeldt, E.F. (1974) Relation of gestation time to brain weight for placental mammals: implications for the theory of vertebrate growth. *American Naturalist*, **108**, 593-615.

Sigurjónsson, J. (1990) Whale stocks off Iceland - assessment and methods. *North Atlantic Studies*, **2**, 64-76.

Sigurjónsson, J. (1995) On the life history and autecology of North Atlantic rorquals. In: *Whales, seals, fish and men* (Ed. by A.S. Blix, L.Walløe and Ø.Ulltang), pp. 425-441. Elsevier Science, Amsterdam.

Sinniff, D.F. and Ralls, K. (1991) Reproduction, survival and tag loss in California sea otters. *Marine Mammal Science*, **7**, 211-229.

Stern, G.A., Muir, D.C.G., Segstro, M.D., Dietz, R. and Heide-Jørgensen, M.P. (1994) PCB's and other organochlorine contaminants in white whales (*Delphinapterus leucas*) from West Greenland: variations with age and sex. *Meddelelser om Grønland, Bioscience*, **39**, 245-259.

Stewart, R.E.A. and Lavigne, D.M. (1984) Energy transfer and female condition in nursing harp seals, *Phoca groenlandica. Holarctic Ecology*, **7**, 182-194.

Subramanian, A., Tanabe, S., Tatsukawa, R., Saito, R. and Miyazaki, N. (1987) Reduction in the testosterone levels by PCB's and DDE in Dalls porpoise of North Western North Pacific. *Marine Pollution Bulletin*, **18**, 643-646.

Subramanian, A., Tanabe, S. and Tatsukawa, R. (1988) Use of organochlorines as chemical tracers in determining some reproductive parameters in *dalli*-type Dall's porpoises *Phocoenoides dalli. Marine Environmental Research*, **25**, 161-174.

Taruski, A.G., Olney, C.E. and Winn, H.E. (1975) Chlorinated hydrocarbons in cetaceans. *Journal of the Fisheries Research Board of Canada*, **132**, 2205-2209.

Trites, A. W. (1991) Foetal growth of northern fur seals: life-history strategy and sources of variation. *Canadian Journal of Zoology*, **69**, 2608-2617.

Urian, K.W., Duffield, D.A., Read, A.J., Wells, R.S. and Shell, E.D. (1996) Seasonality of reproduction in bottlenose dolphins, *Tursiops truncatus. Journal of Mammalogy*, **77**, 394-403.

Vikingsson, G.A. (1990) Energetic studies on fin and sei whales caught off Iceland. *Report of the International Whaling Commission*, **40**, 365-373.

Vikingsson, G.A. (1995) Body condition of fin whales during summer off Iceland. In: *Whales, seals, fish and men* (Ed. by A.S. Blix, L. Walløe and Ø. Ulltang), pp. 361-369. Elsevier Science, Amsterdam.

Watts, P.D. and Hansen, S.E. (1987) Cyclic starvation as a reproductive strategy in the polar bear. *Symposium of the Zoological Society of London*, **57**, 305-318.

Williams, T.M. (1990) Evaluating the long term effects of crude oil exposure in sea otters: laboratory and field observations. Special Symposium: *The Effects of Oil on Wildlife*, 13th Annual Conference of the International Wildlife Rehabilitation Council, Herndon, VI, 13pp.

Whitehead, H., Weilgart, L. and Waters, S. (1989) Seasonality of sperm whales off the Galapagos Islands, Ecuador. *Report of the International Whaling Commission*, **39**, 207-210.

Wolman, A.A. and Wilson, A.J. (1970) Occurrence of pesticides in whales. *Pesticides Monitoring Journal*, **4**, 8-10.

Yasui, W.Y. and Gaskin, D.E. (1986) Energy budget of a small cetacean, the harbour porpoise, *Phocoena phocoena*. *Ophelia*, **25**, 183-197.

York, A. and Hartley, J.R. (1981) Pup production following harvest of female northern fur seals. *Canadian Journal of Fisheries and Aquatic Science*, **38**, 84-90.

B. Sensory Systems and Behaviour

Animals perceive their environment using a suite of senses: hearing, vision, smell, taste, touch, and, in some cases, possibly through magnetism. Marine mammals are no exception, but because they live part or all of their lives in an aquatic environment, some senses are developed over others. Water has a number of properties that influence the role that different senses may play. Sound travels better through water than air, whereas light penetrates less well through water. It is therefore not surprising to find that those marine mammals most dependent upon water for living have developed sound as the most sophisticated sense for the detection of cues from their environment, and for communication to others both in close proximity and over great distances. This is exemplified by the order Cetacea, which has also been the group where sound production and hearing have been best studied, and, in chapter 4, Jonathan Gordon and Peter Tyack explore the many facets of the use of sound by cetaceans.

Marine mammals can hear sounds over a wide range of frequencies, with the smaller odontocete cetacean species being sensitive to sounds at the highest frequencies (ultrasounds up to 100-150 kHz), and the largest mysticete cetaceans apparently to sounds at the lowest frequencies (down to 20 Hz). Although audiograms have been established for several odontocete cetaceans, as well as pinnipeds, there are scarcely any data for mysticete whales, and so we have to infer their hearing sensitivities from the normal frequency range of sounds they produce. Both mysticetes and odontocetes have soft-tissue channels for sound conduction to the ear; sirenians, the other truly aquatic group, have similar hearing structures. Pinnipeds need to be able to hear sounds both in air and underwater. Above water they use their external air-filled canal for sound reception. However, it is not known whether they have specialised mechanisms underwater for maintaining the external canal as the point for sound reception or whether they too use soft-tissue channels for this function. Phocid seals appear to be better adapted to hearing underwater than otariids (with bigger differences in peak sensitivities underwater compared with in air) and this fits in with their more aquatic lifestyle (Wartzok and Ketten, 1999).

Peak hearing sensitivity in pinnipeds is between 10 and 30 kHz (with a high frequency limit of 60 kHz in phocids, and 40 kHz in otariids). Although little data exist for sirenians, manatees appear to have a hearing range of approximately 0.1-40 kHz, with peak sensitivities in the West Indian manatee near 16 kHz, and 5-10 kHz in the Amazonian manatee, although their audiograms are very flat compared with odontocete cetaceans, and more similar to phocid seals (Richardson *et al.*, 1995).

Sea otters have ears most similar to those of land mammals. There is no information on how well they hear underwater. The same applies to polar bears.

Unlike terrestrial mammals, all marine mammals have middle ears that are strongly modified structurally to reduce the impact of rapid and extreme changes in pressure. This may possibly predispose them to some protection from injury by high intensity noise (Wartzok and Ketten, 1999).

Of the other senses, vision appears to be the one that remains most important for marine mammals, and is probably used by most species even at depth. Cetaceans apparently have the most acute underwater vision, adapting the lens of their eye to accommodate for the loss of focusing power of the air-cornea interface. Sirenians are less well adapted to underwater vision, and pinnipeds, with their dual need to see in air and water, have poor visual acuity in air except in bright light. The sea otter has good visual acuity in both media, achieved by changing the radius of curvature of the lens with the use of the ciliary and iris muscles, although like others, its visual acuity is poorer in low light. The polar bear, living primarily on land, has a similar eye anatomy to terrestrial mammals.

The sense of smell, or olfaction, has been reduced in many marine mammals. It is apparently absent in cetaceans (adult odontocetes do not possess olfactory bulbs or nerves, whilst these are greatly reduced in mysticete whales), and only rudimentary in sirenians. Pinnipeds, on the other hand, use olfaction on land particularly during activities related to breeding, as do sea otters and polar bears.

Taste is also not unduly developed amongst marine mammals, although tests have shown that different chemicals can be distinguished. This applied particularly to those with a bitter or sour taste, whereas there was a less well-developed ability to detect salinity and an apparent inability to detect sugar. Most work has been conducted on captive odontocete cetaceans, although limited studies on pinnipeds gave broadly similar results. Sirenians possess more taste buds than cetaceans but they remain less than herbivorous land mammals.

Amongst marine mammals, sirenians have the most developed sensory hairs. These cover the entire body, although particularly dense around the muzzle. The manatee is unique in being able to use its bristles or vibrissae in a prehensile manner; the bristles can actually be everted and used to grasp and manipulate objects, as when feeding. The vibrissae of pinnipeds are also extensively developed, and well supplied with nerve endings. They provide seals particularly with information about form and texture and are often used in a social context (Wartzok and Ketten, 1999). Cetaceans appear to be most sensitive around the blowhole although some baleen whales also have hairs around the mouth.

The presence of magnetite, which has been implicated in magnetic field detection, has been demonstrated in a wide variety of animal taxa including cetaceans. However, magnetite is a very common contaminant from industrial processes and to distinguish this from functional magnetite, researchers have identified those characteristics most likely to indicate its use in magnetic field detection. These include single-domain crystals, magnetite with few oxides (besides iron oxide), and location of magnetite consistently in certain parts of the body, notably associated with the cerebrum and cerebellum. However, increasing magnetite concentrations with age parallels the ossification of these structures, for which it has a known function, providing structural support by increasing the hardness of chitin in teeth. Other evidence that has been used relates to correlations between locations where geomagnetic lows intersected with the coast or islands, with live strandings of cetaceans. This was demonstrated for the UK and eastern United States, but no such correlation was found in New Zealand, possibly because that country lacks the consistent orientation of magnetic contours that the North Atlantic possesses (Klinowska, 1986; Kirschvink *et al.*, 1986; Kirschvink, 1990; Brabyn and Frew, 1994). The difficulty of demonstrating in any experimental way the mechanism by which this potential sense could be used continues to limit our ability to determine its role with confidence.

Two excellent reviews of sensory systems in marine mammals are Thomas *et al.* (1992), and Wartzok and Ketten (1999), whilst Thomas and Kastelein (1990) present evidence from laboratory and field studies specifically upon cetaceans.

The ways that animals perceive their environment mould their behaviour within it. Information about the environment is first processed by the nervous system, and, in mammals, especially by the brain. This is termed cognition. We do not have a specific chapter on this subject, but Schusterman *et al.* (1986) have a book on cognition in dolphins; Tyack (1999) gives a brief review of the role of cognition in marine mammal communication; and it is explored further in Chapter 4 of this book, by Jonathan Gordon and Peter Tyack.

Cetaceans in particular are often represented as big-brained, highly intelligent creatures displaying altruistic traits and a sophisticated vocal learning and communication system that invite comparison with humans. Given the very different evolutionary paths taken by these two taxa, and the fact that cetaceans live in an aquatic medium, and humans a terrestrial one, this comparative approach may not be very helpful. True, the sperm whale has the largest brain (7.8kg) of any mammal, though not relative to its body size (and by allometric growth, larger mammals have larger brains). If body weight is taken into account, the bottlenose dolphin (and some other small odontocetes like the white-beaked dolphin) has a brain nearly as large (c. 0.87-1.25%) as a human (1.5-2.0%). Pinnipeds show relative brain sizes (termed encephalisation quotients, or EQs) close to the overall pattern for mammals, whilst sperm whales, mysticete whales, and sirenians have relatively small brains (Worthy and Hickie, 1986). Furthermore, relative brain size is not equivalent to intelligence. A more meaningful feature is probably the degree of folding of the cerebral cortex; that part of the brain most concerned with higher level neuronal processing. In this respect, odontocetes do actually exceed humans and other animals, and various theories have been proposed to explain these findings. They include the need to rapidly process complex, high frequency sounds as used in echolocation; historical redundancy of the large brain (which has been large in odontocetes over a long period of evolution); and the associated development in sociality with long term individual associations (Wood and Evans, 1980; Herman, 1980; Ridgway and Brownson, 1984; Ridgway, 1986; Worthy and Hickie, 1986; Marino, 1998). At present, we do not know which of these theories, if any of them, is correct, although there is increasing support for the latter explanation (see, for example, summaries by Evans, 1987: 202, and Tyack, 1999: 315-318).

The study of behavioural ecology attempts to answer questions about the functions of behaviour, particularly in the context of an individual's or species' environment (Krebs and Davies, 1997). In Chapter 5, James Boran, Peter Evans and Martin Rosen examine patterns of behaviour as exhibited by cetaceans in terms of two main survival strategies - foraging and mating. They review the roles of locating resources (food and mates) and avoiding predators in the development of sociality, and they show how recent advances in genetics have enabled us to have a better idea of the mating systems of species with various ecologies. Luis Cappozzo continues this theme in Chapter 6, when he reviews new perspectives on the behavioural ecology of pinnipeds, with particular reference to the evolution of pinniped social organisation and mating behaviour.

REFERENCES

Brabyn, M. and Frew, R.V.C. (1994) New Zealand herd stranding sites do not relate to geomagnetic topography. *Marine Mammal Science*, **10**, 195-207.

Evans, P.G.H. (1987) *The Natural History of Whales and Dolphins.* Christopher Helm/Academic Press, London, and Facts on File, New York. 343pp.

Herman, L.M. (1980) Cognitive characteristics of dolphins. In: *Cetacean Behavior* (Ed. by L.M. Herman), pp. 363-429. Wiley Interscience, New York, NY.

Kirschvink, J.L. (1990) Geomagnetic sensitivity in cetaceans: An update with live stranding records in the United States. In: *Sensory Abilities of Cetaceans: Laboratory and Field Evidence* (Ed. by J.A. Thomas and R.A. Kastelein), pp. 639-649. Plenum, New York, NY.

Kirschvink, J.L., Dizon, A.E., and Westphal, J.A. (1986) Evidence from strandings for geomagetic sensitivity in cetaceans. *Journal of Experimental Biology*, **120**, 1-24.

Klinowska, M. (1986) The cetacean magnetic sense - evidence from strandings. In: *Research on Dolphins* (Ed. by M.M. Bryden and R. Harrison), pp. 401-432. Clarendon Press, Oxford.

Krebs, J.R. and Davies, N.B. (1997) *Behavioural Ecology: An Evolutionary Approach*. Blackwell Scientific Publications, Oxford.

Marino, L. (1998) A comparison of encephalization between odontocete cetaceans and anthropoid primates. *Evolutionary Anthropology*, **5**, 73-110.

Richardson, W.J., Greene, C.R. Jr., Malme, C.I. and Thomson, D.H. (1995) *Marine Mammals and Noise*. Academic Press, San Diego. 576pp.

Ridgway, S. (1986) Physiological observations of dolphin brains. In: *Dolphin Cognition and Behavior: A Comparative Approach* (Ed. by R.J. Schusterman, J.A. Thomas, and F.G. Wood, F.G.), pp. 31-59. Lawrence Erlbaum, Hillsdale, NJ.

Ridgway, S.H. and Brownson, R.H. (1984) Relative brain sizes and cortical areas of odontocetes. *Acta Zoologica Fennica*, **172**, 149-152.

Schusterman, R.J., Thomas, J.A., and Wood, F.G. (Eds.) (1986) *Dolphin Cognition and Behavior: A Comparative Approach*. Lawrence Erlbaum, Hillsdale, NJ.

Thomas, J.A. and Kastelein, R.A. (Eds.) (1990) *Sensory Abilities of Cetaceans: Laboratory and Field Evidence* . Plenum, New York, NY.

Thomas, J.A., Kastelein, R.A., and Supin, A.Y. (Eds.) (1992) *Marine Mammal Sensory Systems*. Plenum, New York, NY.

Tyack, P. (1999) Communication and Cognition. In: *Biology of Marine Mammals* (Ed. by J.E. Reynolds III and S.A. Rommel), pp. 287-323. Smithsonian Institution Press, Washington DC. 578pp.

Wartzok, D. and Ketten, D.R. (1999) Marine Mammal Sensory Systems. In: *Biology of Marine Mammals* (Ed. by J.E. Reynolds III and S.A. Rommel), pp. 117-175. Smithsonian Institution Press, Washington DC. 578pp.

Wood, F.G. and Evans, W.E. (1980) Adaptiveness and ecology of echolocation in toothed whales. In: Animal Sonar Systems (Ed. by R.-G. Busnel and J. Fish), pp. 381-425. Plenum, New York, NY.

Worthy, G.A.J. and Hickie, J.P. (1986) Relative brain size in marine mammals. *American Naturalist*, **128**, 445-459.

Chapter 4

Sound and Cetaceans

[1]JONATHAN GORDON and [2]PETER L. TYACK

[1]*Wildlife Sea Mammal Research Unit, Gatty Marine Laboratory, University of St Andrews, St Andrews, Fife KY16 8LB, Scotland, UK. E-mail: jg20@st.andrews.ac.uk;* [2]*Biology Department, Woods Hole Oceanographic Institution, Woods Hole, MA 02543, USA. E-mail: ptyack@whoi.edu*

1.   INTRODUCTION

"Of all forms of radiation known, sound travels through the sea the best. In the turbid, saline water of the sea, both light and radio waves are attenuated to a far greater degree than that from the mechanical energy known as sound."

This bold statement, with which Urick opens his classic textbook on underwater sound (Urick, 1983), encapsulates one of the most fundamentally important characteristics of the physical world in which whales and dolphins live their lives. The physics of underwater sound underpins the over-riding importance of sound in the life of cetaceans.

Cetaceans, of course, live in the sea, a medium through which light travels very poorly. Even in the clearest oceanic surface waters, effective visual range is of the order of a few tens of metres. Many whales and dolphins inhabit much murkier waters than this. In general, the numbers of cetaceans are greatest in highly productive and turbid seas, where the visible range will be in the order of a few metres or less. Consider this in relation to the physical scales at which whales and dolphins operate. Most whales can see less then a body length - they seldom can see their tails - and these large animals spend their lives moving at substantial speeds through a medium in which they have effectively no visibility. Further, many whales and dolphins dive to depths where light from the surface is severely reduced, if not effectively non-existent. There is also the problem that all animals that rely on the sunlight for their visual perception have to face: the sun only shines for about half the time. By contrast, sound propagates very well through the ocean, much better than it does through air. In fact, low frequency sound in the ocean propagates over ranges more like radio waves in air than sound in air. It is little wonder then, that the acoustic modality is the dominant one for whales and dolphins. All cetaceans have sensitive hearing and most are highly vocal. As a group, they make the widest range of sounds of any mammalian order. They use sound to explore their environment and find their way around, and at least some have evolved sophisticated echolocation systems. Sound is also used to communicate - potentially over ranges of hundreds of kilometres. In addition, a host of sounds that can be acoustic cues to the presence of food or predators, and guides for orientation or navigation can be heard in the sea. It is likely that cetaceans make full use of these through passive acoustic perception.

The sea's capacity to propagate sound so efficiently, and the central role of the acoustic modality in so many aspects of the life of cetaceans, make these animals potentially vulnerable to damage or disturbance by man-made noise in the oceans. Such "acoustic pollution" can affect cetaceans in a number of ways and, as humans introduce more sources of noise into the ocean, this is becoming a cause of increasing concern to scientific conservationists (Green *et al.,* 1994; Richardson *et al.,* 1995; Würsig and Evans, this volume).

So fundamental is the importance of sound to cetaceans that it is no exaggeration to propose that to gain any real and meaningful appreciation of the world in which these animals live it is essential to understand underwater acoustics and the role of sound in their lives.

This is clearly much too large a subject to cover in the detail it warrants in a single chapter. What we aim to do here is to give an overview of most of the important subjects. We explore the subjects of particular interest to ourselves in some depth. For other topics, we point the reader towards sources of more detailed information. Our review is mainly focused towards an appreciation of the significance of acoustics for cetaceans in the world they inhabit, rather than a detailed physical understanding of cetacean sounds and how they are produced.

2.   THE NATURE OF SOUND

It is essential for a bioacoustician to understand at least a little about the physics of underwater sound. Sound consists of a series of compressional waves moving from a vibrating source through a compressible medium (which might be a gas, such as air; a fluid, such as water; or a solid). During the passage of each sound wave, the molecules in the medium move a tiny distance away from and towards the source. As they move towards each other, the medium is compressed and the local pressure is increased; as they move away the medium is "rarefied" and the pressure decreases. The molecules return to their original locations after each wave has passed so there is no net movement. A propagating sound thus consists of waves of fluctuating pressure moving away from the source coinciding with local displacement of molecules towards and away from the source. Thus, as a sound wave passes, a stationary receiver experiences both a change in pressure and a cyclical displacement of molecules.

The mammalian ear is sensitive to the changes in pressure caused by sound waves, not the movements of particles. Most man-made sensors, such as microphones and their underwater equivalents, hydrophones, also respond to changes in pressure. Typically, hydrophones contain a piezo-electric material across which a voltage is induced when it is deformed by changing pressure.

2.1   Measuring Sound

2.1.1   Intensity

The process of transmitting sound energy through a medium involves the propagation of a pressure wave. Sound intensity is the amount of energy passing through a unit area per unit time. Intensity equals the square of the pressure integrated over the time period, divided by the specific acoustic resistance of the medium, which is the product of its density, ρ, and the speed of sound, c, in that medium.

$$Intensity = \frac{pressure^2}{\rho c} \qquad \qquad \text{Eq. 1}$$

Note then, that the relationship between sound pressure and intensity always depends on the nature of the medium, and there are great differences in this respect between air and water. Water is some 800 times denser (ρ) than air, and the speed of sound (c) is some 4.5 times greater (c = 1530 ms^{-1} in water and 340 ms^{-1} in air). Hence, the specific acoustic impedance of water is about 3,500 times as great as that of air. For sounds of equal intensity in the two media, the ratio of sound pressure in water with respect to sound pressure in air will be 59.7 (3565^{-2}).

The intensity of sound is measured in decibels. This system, named after Alexander Bell, was designed to simplify the handling of the large numbers involved in acoustic calculations, but it is unfortunately often misunderstood and can be a cause of confusion. The decibel scale for intensity is ten times the log (base 10) of the intensity of that sound divided by the intensity of a reference sound.

$$Intensity = 10 \log \frac{I}{I_{ref}} \; dB \qquad\qquad Eq.\ 2$$

Because most man-made sensors (and the human ear) are sensitive to fluctuations in pressure, pressure levels are normally measured in the first place and converted to intensity later if necessary. The decibel scale for sound pressure levels is

$$Sound\ Pressure\ Level = 20 \log \frac{P}{P_{ref}} \; dB \qquad\qquad Eq.\ 3$$

(The multiplier in this case is 20 rather than 10 because intensity is proportional to (pressure)^2 (1) and numbers are squared by multiplying their logarithm by two. Multiplication by two thus maintains proportionality between the two scales so that if one sound has twice the dB value of another for intensity, it also has twice the dB value for pressure.)

There are good practical reasons for using decibels, even though they might at first seem an unnecessarily complex system. Logarithms are used because the range of SPLs of sound encountered in everyday life, and the range to which the human ear is sensitive, are enormous, and logarithms are a convenient way of managing such large numbers. Using logarithms also allows calculations involving multiplication and divisions to be completed using addition and subtraction. Finally, the use of logarithms reflects the fact that humans perceive the loudness of sound on a logarithmic rather than a linear scale. The multiplication factor (of 10 for intensity or 20 for SPL) is purely a convenience, ensuring that, in most cases, an adequate level of precision on the decibel scale can be achieved without resorting to decimals.

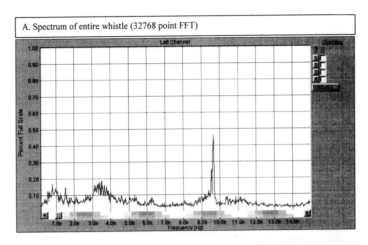

A. Spectrum of entire whistle (32768 point FFT)

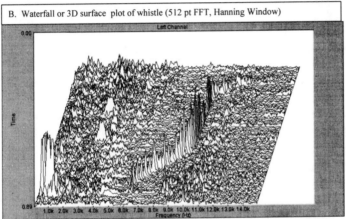

B. Waterfall or 3D surface plot of whistle (512 pt FFT, Hanning Window)

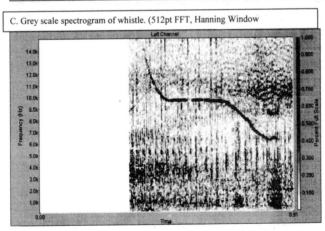

C. Grey scale spectrogram of whistle. (512pt FFT, Hanning Window

Figure 1. Three depictions of a bottlenose dolphin whistle: spectrum and spectrogram plots produced with the SpectraLAB analysis package, Sound Technology Inc.

One thing that should be immediately apparent, but which is often overlooked, is that dB measurements always refer to a ratio between a particular value of pressure (or intensity) and a reference value. Thus, a dB only has any meaning if the reference value is stated. Confusingly, a number of different reference levels have been used at different times and in different contexts. The most commonly used reference value for measurements of sound pressure levels in air is 20.4 μPa. This apparently arbitrary value was chosen because it is close to the minimum Sound Pressure Level detectable by the human ear. Underwater, a different reference value, 1 μPa, is used. The difference between these two reference levels is 20*log (20.4/1) = 26.19 dB.

Strictly, a dB value should never be quoted without stating its reference value, *e.g.* 160 dB re 1 μPa. However, not all authors are equally fastidious when it comes to including this information and making comparisons between dB levels with different reference levels has been a common source of error and confusion in the past. If you always think of dB as relating to the log of the number of times that this sound is greater than a reference sound, it should be impossible to forget the necessity of including the reference value. Chapman and Ellis (1998) provide a clear explanation of the dB scale and enlightening examples of its misuse.

Having belaboured this point, and to avoid the embarrassment of making the same mistake ourselves, let us state now that, unless specifically stated, the water standard of 1 μPa will be the reference level for all dB sound pressure levels quoted.

Remember:
- dB values refer to ratios between a measured level and a reference level.
- A dB value is meaningless unless the reference level is specified.
- The dB represents a logarithm of ratios. A doubling of sound pressure level results in a dB change of ~6, a change in sound pressure level of 100 is a change in dB of 40, 1000 is a dB change of 60, 1,000,0000 is 120 dB *etc.*

2.1.2   Frequency

If a sound is a "pure tone" the waveform of changes in pressure (and for the movements of molecules plotted against time) will be a true sine wave, and the rate at which this sound wave completes its cycles is its frequency. The units of frequency then, are cycles per second, more commonly referred to these day by the unit Hertz, named after a famous German physicist Heinrich Hertz: 1 Hz = 1 cycle per second. The wavelength of the sound is equal to the speed of sound in the medium through which it is travelling divided by its frequency. As we have seen, the speed of sound varies from medium to medium and in water, it is some 4.5 times faster than in air. Thus, for any given frequency, wavelengths are some 4.5 times greater underwater than in air - a matter of some significance for cetaceans as will be discussed later. Pure tones are rare in nature. Most sounds have a more complicated waveform and can be thought of as being made up of more than one frequency. Many sustained vocalisations, however, have some tonal component and a waveform that, though complex, repeats itself regularly.

Any signal can be represented by a linear combination of different frequencies. This means that for any signal, one can calculate how much energy there is in each frequency band. The distribution of a sound's energy within different frequency bands is known as its spectrum. The spectrum of a sound can be represented as a graph of the energy within certain frequency bands plotted against each band's mid frequency (Fig. 1a). Clearly, the amount of energy within each band will be a function of its bandwidth. Two different bandwidths are often considered in sound analysis: 1Hz bands (yielding so-called spectrum level plots) and 1/3 octave bands (an octave being a doubling in frequency). The use of this latter bandwidth may seem unusual and requires some explanation. The choice of 1/3 octave actually stemmed originally from the ease with which analogue filters of this bandwidth could be made, but it has two other features that remain desirable today. Over much of its range, the human auditory system operates as though it has filters of roughly 1/3 octave bandwidth for distinguishing signals from background noise. In addition, the bandwidth of a 1/3 octave in Hz is proportional to frequency, which reflects the way that most auditory signals are perceived by humans and other mammals.

The spectra of many biological acoustic signals vary with time and often the way in which frequency content varies temporally is an important part of the signal's characteristics. Imagine for example a complex song of a bird, or a whale, and how little of its significant detail would be revealed by a plot of the average spectrum of a whole song. What biologists often require then is a depiction that reveals changes in spectral content with time. The way that this is usually achieved is to make spectra of short time slices and then display these stacked one after another. Such displays are called spectrograms. There are three pieces of information that a spectrogram needs to depict: time, frequency and intensity. Frequency is normally displayed on the y-axis and time on the x-axis. Intensity may be plotted on a z-axis to create a three dimensional plot. Imagine, for example, that Figure 1a was the spectrum of one short time slice. Whole sequences of these could be taken and stacked up, one after another, to create Figure 1b. Such displays are called "waterfall" displays or "3D surface plots". An alternative is to code intensity in each frequency bin of each time slice as the density of shading, or as a coded colour, to create a spectrogram display like that shown in Figure 1c.

Nowadays, most spectral analysis is carried out with computers using Fourier analysis techniques, which decompose a waveform or time series into a frequency spectrum indicating the amount of energy in each frequency band. Readily available programs, running on standard personal computers, are now able to complete spectral analysis and create spectrograms from digitised sound files; some even run in real time. A more detailed, but very accessible, introduction to this subject, which is specifically aimed at biologists, is provided in the manual for one of these programs: Canary (Charif *et al.*, 1995). The ever-increasing capacity and speed of computers, and the recent introduction of multi-media computing to the general market place, has dramatically reduced the cost of, and increased the scope for, the digital collection, editing, manipulation and analysis of sound; it is now within the reach of most researchers and students. For a website which maintains an up to date index of sound analysis software, visit the Bioacoustics web page from http://cetus.pmel.noaa.gov/Bioacoustics.html.

2.1.3 Source Levels

The decibel scale is used to measure the intensity of a sound at a receiver. The power of the source of that sound is expressed as the dB level within a certain frequency band that would be measured at a standard range (usually 1 m) from it. Recalling from the previous section that a power level should include a statement to both the reference levels and the frequency band being analysed, then a sound pressure density spectrum (giving the mean square pressure within 1 Hz bands) would have units of dB re $1\mu Pa^2/Hz$ @1 m. Usually, a sound pressure level is measured at a range greater than 1 m and the expected level at 1 m is calculated from this, based on the measured, or predicted propagation conditions (see next section). Some of the most powerful sources, such as seismic air gun arrays, large ships or a blue whale, are too large and dispersed to be considered as point sources at this scale. In these cases, SPLs equal to the source level may not actually exist, even within 1 m of the source, but the source level does provide a useful value for estimating levels at greater distances.

2.2 Propagation

As a receiver moves further from a sound source, it detects a lower sound intensity. There are several different factors that contribute to this decrease, and they can have quite different effects both for different frequencies of sound and in different environments.

2.2.1   Geometric Spreading

The most obvious and usually the most substantial loss results from geometric spreading as sound spreads away from its source. Imagine that one could see a pulse of sound spreading out from a point source in free space - it would look like an expanding sphere. At the moment that the sound pulse passed a point at a certain range from the source, that range would be the radius (r) of the sphere whose surface area would be $4\pi r^2$. When the sound passes the radius of 1 metre, the sound intensity at any point would represent the source level at 1 m. As the sound spreads further, the same total sound energy is dispersed over a larger sphere. The intensity of the sound at any point is inversely proportional to the area over which it has been diluted. Thus, intensity is proportional to range^{-2}. If this is considered in terms of the dB scale, we see that for spherical spreading.

$$TL = 10 \log \frac{I}{I_{ref}} = \frac{4 \pi r^2}{4 \pi r^2_{ref}} = 20 \log \frac{r}{r_{ref}} \qquad \text{Eq. 4}$$

With a standard reference range of 1m this gives TL = 20 log r, and propagation of this form is often referred to as either spherical, or 20 log r, spreading.

In certain situations (see later) sounds can become trapped within a duct (between the surface of the sea and the seabed for example). Sound then propagates in two, rather than three, dimensions. In such situations, the shape of a spreading sound front would resemble a squat cylinder and the area of the sound front would expand at a rate proportional to the circumference of a circle with radius r: $2\pi r$. Thus for cylindrical spreading:

$$TL = 10 \log \frac{I}{I_{ref}} = \frac{2 \pi r}{2 \pi r_{ref}} = 10 \log \frac{r}{r_{ref}} \qquad \text{Eq. 5}$$

Again, with a reference range of 1m, this reduces to 10 log r. Cylindrical spreading may also be referred to as 10 log r spreading.

2.2.2   Absorption

As sound travels through a medium, some of its energy is absorbed and dissipated as heat. Typically, the rate of absorption increases with frequency. In sea water the absorption coefficient, is approximated by:

$$a = 0.036 \ f^{1.5} \ (dB / km) \qquad \text{Eq. 6}$$

where *f* is the frequency in kHz (D. Ross, *pers. comm.*, in Richardson *et al.*, 1995).

Thus, at medium-to-low frequencies, <1kHz, absorption is of little significance (<0.04 dB per km) but at high frequencies it become more significant reaching 3.2 dB per km at 20 kHz and 36 dB per kilometre at 100 kHz. This effect significantly limits the potential for long range propagation of high frequencies through seawater.

2.2.3   Scattering

Small objects suspended in seawater can reflect, refract and diffract sound passing though it, causing increased transmission loss. The effectiveness with which such objects scatter sound will depend on their composition and on the ratio of their dimensions to the wavelength of the sound. Rigid objects scatter sound most effectively if their size is of the order of a wavelength or more; so once again, higher frequency sound will be most affected. Scattering may be most significant in waters that are highly turbid due to suspended silt or high plankton populations. Air bubbles may absorb sound energy lower in frequency than the frequency at which the wavelength equals the circumference of the bubble (Clay and Medwin, 1977). Suspended gas bubbles in surf zones, or in surface waters after stormy weather, can also increase transmission loss for higher frequency sound, resulting in a sort of "acoustic fog".

2.2.4   Reflection/Transmission

When sound in the ocean reaches a boundary with a new medium, the surface or the seabed for example, some of the sound energy will pass through into the new medium and some will be reflected. The proportion that is reflected depends on the nature of the media on each side of the boundary and the angle at which the sound wave hits it (the angle of incidence). The more perpendicular the angle, the greater the proportion of sound energy that is transmitted through the boundary. An air/water interface is a very good reflector of sound, but different bottom types vary in the extent to which they reflect sound. Thus, bottom type, its general angle of incline, its roughness, and the roughness of the sea surface (which is highly weather dependent) can all affect propagation conditions, especially over ranges of several miles and more.

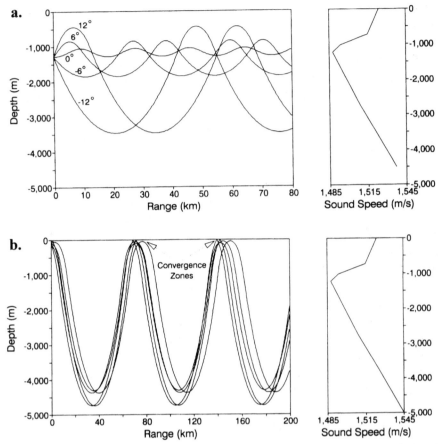

Figure 2. Ray diagrams showing typical propagation conditions for tropical and temperate deep oceans. Right hand panels show changes in speed of sound with depth. a) Ray diagram showing propagation of sound for source on axis of the deep sound channel. b) Ray diagram showing convergence zone propagation for source at shallow depth (90 m). Convergence zones at the surface are spaced approximately 80 km apart. (From Richardson *et al.*, 1995, courtesy of C.I. Malme and Academic Press, San Diego)

2.2.5 Refraction and Ducting

Just as light is bent when it passes between two media with different speeds of transmission, so sound is refracted back towards the medium with the lower transmission speed. The speed of sound in the ocean increases with both increasing temperature and with increasing pressure. One well known and dramatic consequence of this is the formation of the deep sound channel in the oceans. Figure 2a shows a typical plot of the speed of sound with depth in the open ocean. The sun heats water near the surface so that water temperature decreases, and the speed of sound also falls, as depth increases. In the deep ocean though, water temperature remains fairly constant with depth while increasing pressure causes the speed of sound to begin to increase again. Thus, there is a sound speed minimum at a depth which varies from around 1,000 m in the tropics to a few hundred metres closer to the poles. Refraction causes sound passing through these water bodies with differing transmission speeds to bend back towards the depth with minimum speed, so that the ray paths of a sound entering the channel at appropriate angles behave as though they are trapped in a duct. One significant effect of this is that such sound propagates by cylindrical (10 log r) rather than spherical (20 log r) spreading. If this sound propagates with little interaction with the sea surface or seafloor, cylindrical spreading provides the potential for propagation over very great ranges. In fact, a small explosive charge detonated in the deep sound channel can be detected right round the world (Shockley *et al.,* 1982). To benefit from this potential for long range transmission, other causes of transmission loss, such as absorption, must be minimal. Low frequency sounds, which have very low absorption coefficients (Equation 6), can potentially propagate enormous distances in the deep sound channel. For example, Payne and Webb (1971) calculated that fin whales (*Balaenoptera physalus)* producing very loud 20Hz pulse vocalisations should be able to hear each other at ranges of thousands of miles if both the vocalising and listening whale were in the sound channel. The potential for baleen whale sounds to propagate great distances in the deep sound channel has now been confirmed by the results of monitoring for whale vocalisations using military hydrophones within the sound channel (*e.g.* Gagnon and Clark, 1993 and Clark, 1994).

To make best use of the deep sound channel, an animal would have to both make its sound at that depth and also listen there. Fin whales are not known to regularly dive to the depth of the sound axis in tropical/temperate waters, or to vocalise there. Sperm whales (*Physeter macrocephalus*), on the other hand, spend much of their lives vocally active at these depths. Their higher frequency click vocalisations are not ideally suited for ultra long-range transmission. However, Hiby and Lovell (1989) calculated that sound channel propagation would allow sperm whales to hear each other out to distances of about 30 miles, which is still a very considerable range for animal communication. Figure 2b shows that some ray paths for a sound produced closer to the surface enter the sound channel and are refracted back up towards the surface. These ray paths converge in surface waters at regular intervals of about 50 km. An animal entering such a "convergence zone" would experience an increase in signal level of 10-20 dB and potentially convergence zones could allow even animals that don't dive deeply to achieve very long-range communication through the deep sound channel. (A very accessible exploration of propagation in the sound channel and the exciting potential for long range communication by whales can be found in Payne, 1995).

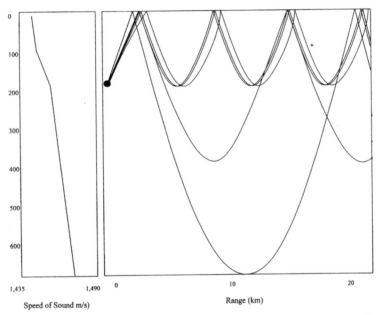

Figure 3.   Ray diagram showing deep water sound transmission in the Arctic. If the surface is covered with smooth ice or there is a smooth water surface, long range transmission of sound in a surface duct may occur. Right hand panel shows the change in speed of sound with water depth. (From Richardson *et al.*, 1995, courtesy of C.I. Malme and Academic Press, San Diego)

Figure 2a also shows that when the surface waters are warmer, sound is refracted down, away from the surface and this can serve to restrict surface propagation. Certain areas in deep water are also effectively shadowed from surface sound sources. In polar seas, surface waters are often colder than deeper waters and in such a situation the opposite effect pertains: sound is refracted back up to the surface (see Fig. 3). The dominant source of loss in these conditions depends upon how the sound interacts with the air-sea interface. When the sea surface is calm, or is covered by a uniform layer of ice, sound will be reflected efficiently by the surface and a very effective surface "duct" can result, allowing propagation in surface waters over very considerable distances.

The foregoing discussion gives some impression of how variable and complex underwater sound propagation can be (see Urick, 1983 for a more detailed treatment). With a more complete understanding of how oceanic conditions affect propagation, allied to improved computing power, acousticians are able to predict propagation conditions in particular environments with increasing precision. Such modelling may still be out of the reach of most field biologists but it is helpful to have a qualitative grasp of the sort of factors that can affect propagation. For all their sophistication, propagation models always need ground-truthing and they can sometimes give misleading results, so there is always value in recording received sound levels in the field in order to assess propagation directly.

2.3   Background Noise

The ocean is typically a noisy place. In some locations, the calls of whales can dominate certain frequency bands. In addition to the calls of marine mammals, there are sounds from other animals, the sea's surface (very weather dependent), geological processes and, increasingly, from man's activities. Figure 1, in Chapter 16, based on Wenz (1962), is one of the most frequently used illustrations in underwater acoustics. It summarises typical levels and frequencies of background noise from a variety of natural and human sources. Generally, levels of background noise are highest at lower frequencies, although this effect is less pronounced if 1/3 octave, rather than spectrum level values are plotted. At low frequencies (<100 Hz), noise is dominated by earthquakes, shipping noise and explosions; between 100 Hz and 10 kHz water noise from waves at the surface is dominant, though of course this depends on sea state. One of the most intense natural sources of local wide-band noise comes from heavy rain that can effectively obliterate everything else. At the highest frequencies, above 100 kHz, molecular noise is dominant. Intriguingly, the minimum in the sound level spectrum, at around 50-100 kHz in typical sea conditions, corresponds to the frequencies at which most dolphin and porpoise sonar operates.

Some noise, such as that from wind and waves on the sea's surface, is fairly continuous. Ships underway make noise constantly, but they move around. Other powerful noise sources, such as earthquakes, lightning, the pulses of seismic survey vessels and military sonar, are more intermittent. In addition, certain areas may be particularly noisy. Levels of medium and high frequency noise can be very high over gravel beds moving in tide rips or over rocky areas with high "snapping shrimp" populations. Busy shipping lanes are another example of consistently noisy areas. Thus, although the noise values from Wenz, give some indication of average levels to be expected in the ocean, there will be many periods in time, and patches in space, which will be either much quieter or much more noisy than this.

One significant consequence of background noise is that it can mask signals that are close to it in frequency, arrive from a similar direction, and are similar or lower in intensity. As a very rough rule-of-thumb, a noise will mask a signal if it contains the same (or greater) energy as the signal within a critical frequency band around that signal. Background noise may hamper a field biologist's ability to hear cetacean vocalisations. Equally, it will compromise a cetacean's ability to hear noises of biological significance and this can be one of the disruptive consequences of man-made noise in the oceans (see Chapter 16).

2.4   Sonar Equation

Engineers who build sonar systems have worked out equations to predict the detection range for an underwater sound (Urick, 1983). These equations are convenient for back-of-the-envelope calculations, but are only as good as their estimates of transmission loss and noise levels. A simple form of the passive sonar equation is:

$$RL\,(dB) = SL\,(dB) - TL\,(dB)$$ Eq. 7

The first two terms on the right side of the equation, source level and transmission loss, have been discussed above. Source level is the sound intensity measured in decibels with respect to a reference distance, usually 1 m. Transmission loss is the difference between the source level at the reference distance and the intensity predicted for the range of the receiver. RL is the received level of sound at the receiving hydrophone, or an animal's ear, measured in dB.

Usually, the likelihood of detecting a signal does not just depend upon the received level but also on noise, both the external noise in the environment and internal noise in the receiver. The ratio of signal to noise is critical for calculating the probability of detection. Using the logarithmic terms of the dB scale, this signal to noise ratio (SNR) is expressed as a subtraction:

$$SNR\,(dB) = RL\,(dB) - NL\,(dB)$$ Eq. 8

Many animals have strategies to minimise the consequences of masking by noise. Having better directional hearing can help if the noise and the signal are not coming from the same direction since this effectively reduces the signal to noise ratio in the direction from which the signal is arriving. We are all aware of this in normal life as the so-called "cocktail party effect". Being able to localise the signals from the single person we are attending to effectively "blocks out" noise arriving from other directions. Other ways of dealing with noise include: increasing the intensity of the signal; changing its frequency so that it overlaps less with that of the noise; introducing redundancy by repeating the signal many times or using a temporally coded signal that can be better detected in random noise.

2.5   Making Comparisons between Sound Levels in Water and in Air

Often, commentators are tempted to make comparisons between particular sound sources in water and in air, usually in an attempt to draw parallels between our own experience of the loudness of a particular source in air and a marine mammal's perception of loudness of another source underwater. (Statements along the lines of, "Such and such a sound is like a jet engine at 50 yards".) This seemingly straightforward process is fraught with difficulty and can often lead to erroneous or misleading conclusions (Chapman and Ellis, 1998). In the first place, as we have already mentioned, the standard reference levels used in calculating dBs in air and water are usually different. Typically, when the most common standards are used, this means that 26 dB should be taken from underwater dB values to compensate for differences in reference levels. However, acousticians still do not agree on whether mammals' perceptions of loudness in the two media are proportional to pressure or intensity. If the latter is the case, then further corrections are required to allow for the very different specific acoustic impedance of the two media. Even if this question is resolved, however, a fundamental problem remains: one is attempting to compare the perception of loudness for two very different mammals that may have very different acoustic sensitivities, and over different frequency ranges. A different approach to making such comparisons assumes that mammalian ears may all have about the same dynamic range, extending from the threshold of detectability to pain or damage. This approach would suggest that to compare the effects of two sounds on animals of two different species, one should subtract the received level of each sound (in dB) from the threshold level for detection of that sound for the appropriate animal. The only way one can really be sure, however, is to measure the effect of the particular sound on the appropriate animal.

3.   CETACEAN VOCALISATIONS

Cetaceans produce a wide variety of vocalisations ranging in frequency from the infrasonic (15-30 Hz moans of blue whales) to the ultrasonic (120-140 kHz) pulses of porpoises, and in duration from blue whale tones that last for 30sec to the clicks of echolocating dolphins that may last for less than a millisecond. Here we can give only a very brief overview (see Richardson *et al,* 1995 and Watkins and Wartzok, 1985 for more complete reviews, and Clark, 1990 and Edds-Walton, 1997 for detailed reviews of mysticete sounds). It should be noted that, in addition to the vocalisations considered here, which are generally produced by the passage of air through organs evolved specially for sound production, cetaceans also produce a variety of other non-vocal sounds including their blows, baleen rattle in mysticetes, tail and flipper slaps, and cavitation noise from their flukes. Some of these may be "purposeful"; sound production is at least a part of the function of the behaviour involved.   Others are "adventitious"; sound production is a functionless by-product of the activity.

Cetacean vocalisations have traditionally been divided into three main classes (Popper, 1980).

1. Tonal calls, which usually last half a second or more and have at least one narrow, emphasised frequency band, which may be constant or modulated. Examples are moans at low frequency and whistles at higher frequencies.
2. Clicks, which are short, impulsive, broadband sounds.
3. Pulsed vocalisations, which are composed of bursts of clicks repeated at such fast repetition rates that they begin to sound like broader-band tonal sounds. Pulsed sounds often sound "harsh", like "screams".

The sheer breadth and variety of cetacean vocalisations can seem bewildering. Identifying certain broad trends in vocalisation characteristics can help us to manage and understand this diversity. As in many animals, larger cetaceans are able to make lower frequency sounds. Ding *et al.* (1995) showed that the highest whistle frequency reported from the repertoires of nine different odontocete species was inversely proportional to their mean body size; while Matthews *et al.* (1999) showed a similar relationship for the mean frequency of tonal calls reported in the literature for 26 odontocete and 26 mysticete cetaceans (see Fig. 4). For example, within the mysticetes, the rorqual whales all produce quite simple pulsed calls. The largest, blue whale (*Balaenoptera musculus*), (max length 33 m) produces the lowest frequency sound, at around 15 Hz, well below the range of sensitivity of the human ear.

Only slightly smaller, the fin whales (27m maximum) produce pulses centred around 20 Hz. Bryde's whales (*Balaenoptera edeni*), (maximum length 15.5 m) produce moans between 124 and 900 Hz, while the smallest of the group, the minke whale (*Balaenoptera acutorostrata*) (10.5 m maximum) produces "thump trains" of between 100 and 2000 Hz (vocalisation values summarised from literature reviewed in Richardson *et al.*, 1995; lengths from Jefferson *et al.*, 1993).

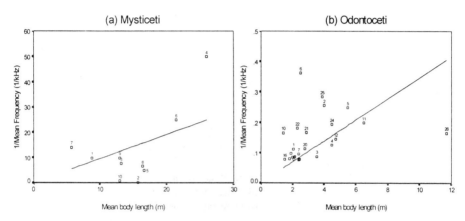

Figure 4.   Linear regressions of inverse mean call frequency on mean body length by species of a) mysticetes and b) odontocetes.   (a) 1. *B. acutorostrata*, 2. *B. borealis*, 3. *B. edeni*, 4. *B. musculus*, 5. *B. mysticetus*, 6. *B. physalus*, 7. *C. marginata*, 8. *E. australis*, 9. *E. robustus*, 10. *M. novaeangliae*, (b) 1. *D. delphis*, 2. *D. leucas*, 3. *G. griseus*, 4. *G. macrorhynchus*, 5. *G. melas*, 6. *I. geoffrensis*, 7. *L. acutus*, 8. *L. hosei*, 9. *L. obscurus*, 10. *L. vexillifer*, 11. *O. orca*, 12. *P. crassidens*, 13. *P. electra*, 14. *S. attenuata*, 15. *S. coeruleoalba*, 16. *S. fluviatilis*, 17. *S. frontalis*, 18., *S. longirostris* 19. *S. plagiodon*, 20. *T. truncatus*, 21. *M. carlhubbsi*, 22. *S. bredanensis*, 23. *S. clymene*, 24. *M. monoceros*, 25. *M. densirostris*, 26. *B. bairdii*   (From Matthews *et al.*, 1999, courtesy of *Bioacoustics*).

An object's ability to produce and effectively broadcast sound is proportional to the ratio of its size and the wavelength of that sound. Thus, if larger animals have correspondingly larger sound producing organs, then they can more effectively produce lower frequency (longer-wavelength) sounds. This provides an explanation for the correlation between the lowest frequency an animal can make and its body length. It does not explain why a similar correlation exists for the highest frequency an animal can produce (Ding *et al.*, 1995). Two explanations suggest themselves. It is difficult to make a speaker that is efficient over a broad range of frequencies. A vocal apparatus adapted for making low frequency sound might be unsuitable and inefficient when making high frequency vocalisations. In addition, an animal's hearing sensitivity is usually centred at the frequency of its vocalisations and mammalian auditory systems are only sensitive over a limited frequency range. Ketten (1997) points out that cetaceans specialised for low frequency hearing are thought not to have high frequency hearing as sensitive as cetaceans with generalised hearing, much less capabilities matching those of high frequency specialists. If a specialisation in low frequency sound production and reception were correlated with poor sensitivity at high frequencies, there might be a reduced advantage in making higher frequency sounds that conspecifics would be less able to hear.

Figure 5.   Male humpback whales sing to announce their presence so as to attract a mate and repel other males. Those songs are complex and an individual can sing non-stop for over 22 hours. (Photo: D. Glockner).

As mentioned above, whales of the genus *Balaenoptera* produce rather similar, simple and stereotyped calls. More variety exists within the other mysticetes. The bowhead (*Balaena mysticetus*), right whales (*Eubalaenidae*) and the humpback (*Megaptera novaeangliae*) all make complex calls that are remarkably similar (Clark, 1990). Bowheads sing simple songs, consisting of a few uncomplicated phrases (Clark *loc. cit.*) The humpback whale sings a long, complex and, to us, beautiful song on its breeding grounds (see Section 7.3). This trend in vocal complexity reflects similar trends in the species' social behaviour. The balaenopterid rorquals appear to have a relatively simple social organisation. They live much of their lives apart; they may come together in feeding areas but they don't appear to congregate on breeding grounds. Social organisation in the other mysticetes is more complex. Humpbacks, for example, feed co-operatively and also congregate and interact on distinct breeding grounds. Specific vocal behaviour is known to be associated with both activities (feeding: D'Vincent *et al.*, 1985; breeding: Payne *et al.*, 1983).

Odontocetes can be divided into those animals that produce both clicks and tonal whistle vocalisations and those that only produce clicks. The sperm whales (*Physeteridae* and *Kogiidae*), the porpoises (*Phocoenidae*), and the *Cephalorhynchus* dolphin subfamily appear only to produce click vocalisations. Tyack (1986) noted that these "non-whistling" species tend to be less social than other odontocetes and commented that animals producing whistles would have a greater potential for the complex signalling that might be required during social interactions. One example of such socially important signals is the signature call to allow individuals to identify themselves acoustically. Sperm whales, which are the most social of the great whales, may be an exception proving the rule. They don't produce whistles but they do, on occasions, produce clicks in stereotyped patterns called "codas," which may function as signature signals (Watkins and Schevill, 1977). The *Platanistid* river dolphins are another apparent exception to this pattern. These dolphins are not known to be particularly social and were categorised by Herman and Tavolga (1980) and Tyack (1986) as "non-whistling." However, Ding *et al.* (1995) report whistles for several platanistid species.

Narrow-band whistles are the most common tonal call in odontocetes. The frequency of whistles is often modulated rapidly and this modulation (the whistle's shape or contour) is thought to be an important carrier of information.

All odontocetes click. As was the case with tonal sounds, there is a trend for larger animals to produce clicks with lower frequencies. Watkins (1980b) noted that for five genera (*Tursiops, Pseudorca, Globicephala, Orcinus, Physeter*), their increasing size matched an increasing low frequency emphasis in their clicks. Thomas *et al.* (1988) agreed and commented that the emphasised frequencies in clicks of harbour porpoises, *Phocoena phocoena*; the boto, *Inia geoffrensis*; and Dall's porpoise, *Phocoenoides dalli* also fit the same trend. An interesting division can be drawn between those odontocetes whose clicks resemble narrow-frequency band pulses with a relatively long duration (the porpoises and the *Cephalorhynchus* dolphins) and the other odontocetes whose clicks have a wider bandwidth and, usually, a much shorter duration.

Figure 6 shows examples of each click type; a wide-band pulse from a bottlenose dolphin (*Tursiops truncatus*) (Fig. 6a) and a narrow-band pulse from a harbour porpoise (Fig. 6b). The bottlenose dolphin pulse has a sharp rise time, of the order of a few microseconds, a high peak pressure and a short duration. The peak frequency is 117 kHz and the bandwidth is 37 kHz. Peak to peak levels for *Tursiops* pulses can be greater than 220 dB re 1 µPa SPL @1 m (Au, 1993). By contrast, typical pulses from porpoises measured by Goodson and Sturtivant (1995) had durations of hundreds of microseconds, a peak frequency 147 kHz and a bandwidth of only 13kHz. Sound pressure levels for porpoise clicks of up to 170 dB re 1 µPa @1 m have been measured by Akamatsu *et al*. (1994). It is interesting to note that all of the "narrow band" species produce a pulse with a frequency of around 100-140 kHz, most of them have predominantly inshore distributions and none of them produce whistles.

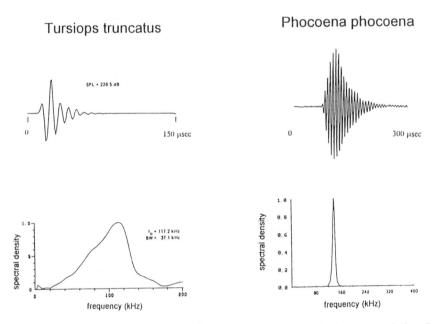

Figure 6.   Waveform and spectra of two contrasting types of odontocete echolocation clicks. (From Tyack, 1986; example waveforms from Au, 1980, and Kamminga and Weisma, 1981).

Pulling together these observations, and considering only the smaller odontocetes, we can see that a major division can be drawn between species based on their vocalisations.   Members of one group only produce narrow band clicks, don't produce tonal sounds, and are typically inshore and less social.   A second group produces broadband pulses and a variety of tonal calls, including whistles, and are generally more social and may extend into oceanic waters.

3.1   Species' Repertoires

The completeness of knowledge of the vocal repertoires of different species varies, reflecting both differences in the ease with which certain species can be encountered and recorded, and the complexity and diversity of the sounds they produce.   Vocalisations of some species are variable and many cetaceans show vocal learning (Janik and Slater, 1997); thus the simple concept of a fixed vocal repertoire must be considered questionable.

For some species that are not highly vocal and are difficult to record, even basic vocalisations have still to be fully catalogued. For example, we have very poor knowledge of the vocalisations made by sei whales (*Balaenoptera borealis),* beaked whales, and many offshore species. Species that do not occur in waters near Europe and North America have also been recorded less often than those with a distribution more closely matching that of marine mammal bioacousticians.   Because sound travels so well underwater and most hydrophone systems used to make field recordings are not directional, it can be difficult to reliably assign the recorded noises to particular animals that might be seen at the surface. It is now evident that many early reports of the vocalisations of particular cetacean species were in fact "contaminated" by vocalisations from other, unseen cetaceans. A number of field recordings with no other cetaceans in sight, all showing a good correlation between particular vocalisations and the presence of the target animals, are needed to confidently link vocalisations with a particular species.   Real time sound localisation equipment is becoming more and more available to marine mammal bioacousticians; it should ideally be used to link particular sounds with the animals under observation.

The repertoires of some species with complex vocal output, for example humpback, southern right and bowhead whales, have been well studied. Other species that produce only a few stereotyped calls, such as fin and blue whales, are inherently easier to characterise. The vocalisations of both of these groups might be considered to be reasonably completely described. Amongst the odontocetes, the physical characteristics of vocalisations of those species that only produce clicks, porpoises and sperm whales for example, are well known, although there is still much work to be done in describing and understanding the different patterns in which clicks are produced. For example, sperm whales (Whitehead and Weilgart, 1991), porpoises (Amundin, 1991) and Hector's dolphins (Dawson, 1991) all produce specific patterns of clicks in certain social settings. The repertoires of odontocetes that also produce tonal vocalisations are difficult to summarise because they are so diverse. As with the clicks of sperm whales, tonal calls may vary between individuals, groups and populations and also with motivation, emotional state, time of the day, season and age. Clearly, in these circumstances, a substantial number and diversity of recordings will need to be analysed before the entire repertoire can be described.

One fundamental problem is that of describing sounds "on paper" in the scientific literature. Using words to describe sounds is imprecise and fraught with difficulties: one person's "moan" could be another's "groan" or "grunt". Spectrograms (*e.g.* Fig. 1c) can be more helpful, but they can only be used to portray a few examples, they may be difficult to interpret and their appearance can vary greatly depending on the analysis parameters chosen (these must, of course, always be stated). Statistical measurements (minimum and maximum frequencies of calls for example) can be made from large numbers of spectrograms and are increasingly being reported, but there is no agreement on a standard set of features to describe vocalisations. Indeed, researchers asking different questions are likely to measure different parameters. A researcher interested in species identification would choose those features that most reliably distinguish species, for example, while a behavioural biologist might measure those features whose variation appeared to be of most biological significance to the individual animals of the species being studied. Early bioacousticians sometimes could convince scientific journals to publish recordings of animal sounds in papers, but this has not been accepted practice for more than a decade. One possible way forward may come from the increasingly common practice of multi-media publishing and of creating and accessing large digital databases. Papers published using an electronic format, which allowed waveform files to be linked to the text, could include examples of the relevant sounds within them. Alternatively, digital databases containing representative example vocalisations for different species could act as a useful archive and source of example vocalisations encouraging standardised approaches to description and analysis. Some initiatives to establish databases along these lines have already been completed (*e.g.* Cornell, 1996; Watkins, 1994; Felgate and Lloyd, 1997; see also review by Ranft, 1997) however, more work is required to consolidate collections and improve access to them.

4. VOCAL PRODUCTION: HOW CETACEANS MAKE SOUNDS

The mechanism by which cetaceans produce their sounds, and even the site(s) in the body at which this occurs, has been a matter of some controversy and even now the question is not fully resolved. Contrary to earlier reports (see references in Reidenberg and Laitmann, 1988), odontocetes do possess well-developed and complex larynges. The cetacean larynx has vocal folds which, though different in configuration than the vocal folds of terrestrial mammals, may vibrate to produce sound when air passes over them (Reidenberg and Laitmann, 1988). On the other hand, Norris *et al.* (1971) suggested that in odontocetes, sound is produced in the head, specifically by the release of air past the nasal plugs. Dormer (1979) proposed that both whistles and clicks are produced in the nasal plug area. Recently, Cranford *et al.* (1996) have presented compelling evidence that echolocation signals at least, are produced in the odontocete head by the passage of air past the monkey lips/dorsal bursae complex behind the melon. It remains possible, however, that tonal sounds are produced in the larynx, and the larynx remains a likely location for the production of sound in mysticetes. Even though the word "vocalisation" refers to the vocal folds, it has taken on a more general meaning. When we use "vocalisation" to refer to a cetacean sound, we want to emphasise that we are not implying that the sound is produced in the larynx as opposed to the nasal plugs. The larynx vs nasal plug debate among cetologists has obscured the need for better modelling of sound production mechanisms in general. Most modern models of sound production emphasise interactions between the sound source and the filtering effects of the vocal tract. These complexities will be particularly important and interesting for highly directional signals such as the high frequency echolocation clicks of odontocetes.

Although the passage of air is implicated in cetacean sound production, cetaceans rarely release bubbles of air while phonating; instead, the air is recycled. For a diving mammal, this is essential. A sperm whale, for example, will click nearly continuously throughout its long dives that may last an hour or more. During such dives, sperm whales may reach depths of greater than 1,000 m where the ambient pressure will be sufficient to reduce the volume of the gas in its body to 100th of its volume at the surface. Clearly, on such dives air becomes a precious commodity to be preserved and recycled to allow sound production to continue.

5.   HEARING

All cetaceans are believed to have good hearing, yet none of them have external ears. External pinnae would cause obvious hydrodynamic problems for cetaceans. As the acoustic impedance is much the same in water as in cetacean bodies, there is no requirement for specific open canals linking the outside world with the middle ear. Sound can reach the middle and inner ears of cetaceans directly through their body tissue, but there do appear to be some preferred pathways or wave-guides. In dolphins, these include a pathway through the lower jaw. In fact, the problem that cetaceans have to overcome is not how to get sound to the ear but how to isolate the ear from the rest of the body to allow binaural hearing to function properly. The bones making up the auditory bulla in cetaceans are very dense and massive compared to those of terrestrial mammals. They are not fused to the skull as are the corresponding bones in terrestrial mammals but are suspended by ligaments in a peribullar cavity and surrounded by a spongy mucosa. This helps to isolate the ear from the rest of the body and the two ears can operate independently allowing binaural hearing. Ketten (1997) provides an excellent review of the anatomy and function of cetacean ears.

The hearing sensitivities of those small odontocetes that can be kept in captivity, especially the bottlenose dolphin, have been well studied using both behavioural and electro-physiological techniques. Moore (1997) gives an up to date and comprehensive overview of work on odontocete auditory perception. Figure 7 shows audiograms for a number of cetaceans and also, for comparison, a human audiogram. Audiograms are plots of the sound level at which an animal can just detect tones at particular frequencies. Thus, they indicate how sensitive each species is to sounds of different frequencies.

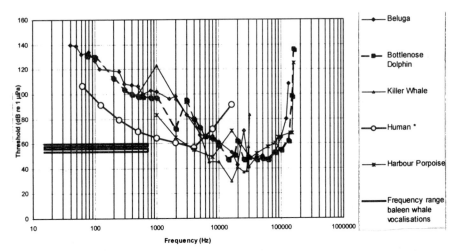

Figure 7.   Audiograms of Selected Odontocetes. Underwater audiograms: beluga (White *et al.*, 1978; Awbrey *et al.*, 1988; Johnson *et al.*, 1989, as combined by Richardson *et al.*, 1995); bottlenose dolphin (Johnson, 1967); killer whale (Hall and Johnston, 1972); harbour porpoise (Anderson, 1970). Human values are in air thresholds, from Sivian and White (1933) with 65.5 dB added to compensate for different reference levels and the impedance of water. Note: this is included to indicate the frequency range of human auditory sensitivity; no significance should be attached to actual levels. Shading indicates the frequency range over which many baleen whales might be expected to be most acoustically sensitive, based on the frequency characteristics of their vocalisations.

Most audiograms have a typical "U" shape, with best sensitivity somewhere in the middle of the auditory range. For most odontocetes, lowest threshold levels are between approximately 30 and 50 dB re 1 µPa. The frequency band over which these odontocetes operate is shifted upwards in comparison to that of humans. Humans are sensitive to sound between 20 Hz and 20 kHz approximately, while dolphins are less sensitive to lower frequency sounds (<500 Hz) but have good sensitivity up to 100 kHz or more. Dolphins also have a better ability than humans to discriminate between different frequencies and determine time intervals (Moore, 1997).

There are virtually no direct measurements of the hearing sensitivity of those larger cetaceans that cannot be kept in captivity for long periods of time. However, we might expect the sensitivity of baleen whales to be best at the frequencies at which they vocalise (<500 Hz in most cases). This assumption of a low frequency auditory range for mysticetes is also supported by studies of the comparative anatomy of cetacean ears (Ketten, 1997), and field observations of responses to a variety of noises during playback studies, (*e.g.* Dahlheim and Ljungblad, 1990; Lien *et al.*, 1990).

Clearly then, different cetaceans have different acoustic abilities and sensitivities, and operate in "acoustic worlds" that contrast with those of other cetaceans and with our own. It is vital to appreciate this when we consider the significance of acoustics in the life of a particular species, and try, as all biologists inevitably do, to "think" ourselves into their world.

6. FUNCTIONAL SIGNIFICANCE OF CETACEAN VOCALISATIONS

For the purposes of this review, we will consider two large functional divisions: the use of acoustics for orientation and localisation and the use of acoustics in communication, which we take to be the production of signals adapted to influence the behaviour of other individuals.

6.1   Passive Orientation and Location

Spend time listening with hydrophones underwater and you soon realise that the sea is full of potentially useful acoustic cues. For example, in some areas, if you are in ocean waters and you steer a directional hydrophone towards the edge of the continental shelf, you can hear it literally buzzing with life. Animals with low frequency hearing may be able to hear distant storms, or the waves breaking on distant shores. Many of the animals that are the prey of cetaceans, such as fish and other marine mammals, make sounds, and listening for these might indicate good feeding opportunities. Different bottom types can also sound quite different, the crackling of "snapping shrimp" indicating rocky bottoms, the knocking of stones in tidal currents, gravel beds, while the amount of reverberation in an environment gives clues to topography.

Of course, the calls of conspecifics can be particularly valuable sources of information, even when the animals producing these calls had no intention of providing it. The ability of sounds to propagate so well underwater will allow members of a dispersed group to keep in contact over very extensive ranges. As we have already mentioned, Payne and Webb (1971) proposed that baleen whales could keep in contact with each other over ranges of thousands of miles using the deep sound channel. Even relying solely on non-ducted propagation however, baleen whales should be able to hear each other over ranges of tens of miles, so groupings of whales can potentially keep in contact and co-ordinate their activities at this scale. Clearly then, we may have to adjust the limited perception of what constitutes a cetacean group that visual observation at the surface might give us. Indeed, Payne (1995) proposes that we should think not of whale **herds** but of **"heards"** of whales, that is, grouping of all those whales that can hear each other. Doing this involves a great leap in scale that can make even the largest oceans seem relatively small places.

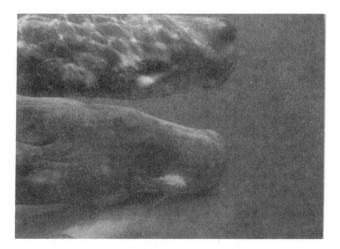

Figure 8. The exception that proves the rule? Sperm whales only make click vocalisations but have the most complex social organisation of any of the great whales (Photo: J. Gordon/IFAW).

Sperm whales provide an example of a species that almost certainly does listen for vocalisations to keep groupings together. In this species, mature females and their young live in social groupings of approximately 12-30 animals whose members remain associated over periods of many years (Best, 1979; Weilgart *et al.*, 1996). Members of these groups are typically encountered spread out over many square miles feeding. They probably disperse in this way to avoid competition, but, about once a day, they come together into larger resting and socialising groups (Gordon, 1987b). Over the years, groups may migrate over many thousands of miles. During this time, their members are rarely in visual contact yet they stay together.

The sounds that cetaceans make can also provide useful information about their behaviour to others, whether or not the sounds are produced for a communicative function. For example, in vespertilionid bats, observations by Leonard and Fenton (1984) and playback experiments by Balcombe and Fenton (1988) suggest that individuals use the echolocation calls of conspecifics as a source of information on prey availability. Miller *et al.* (1995) suggest that narwhals may also make distinctive fast series of clicks as they close on prey; these signals might broadcast that an animal has detected a prey item to neighbouring animals. While there has as yet been no definitive proof that sperm whales can echolocate, most biologists working with them do assume that some of their clicks function for echolocation. Gordon *et al.*, (1992) suggest that certain patterns of click production in sperm whales, called echolocation runs, seem to be associated with feeding. As we mentioned earlier, Hiby and Lovell (1989) calculated that sperm whales feeding at depth would be able to hear each other at ranges of around 30 miles. If they could also assess the feeding rates of other sperm whales over these ranges by attending to these echolocation runs, they would be able to move towards better foraging areas. Of course, other echolocating animals could also obtain information on feeding conditions by listening to noises associated with feeding (including those of other species) in this way, though probably not at such extreme ranges. Gordon (1987b) suggested that this predisposes echolocating animals to forage co-operatively because with this particular system of information transfer it is impossible to cheat: when an animal finds food it can only feed efficiently by making the noises which may alert others to the existence of good feeding conditions.

The use that cetaceans make of these acoustic cues has been little studied and we are probably very poorly aware of all the acoustic information that is available to cetaceans underwater. It is difficult to conceive, however, that such information does not provide vital cues, used by cetaceans when foraging for food, navigating through their environment and performing long-range migrations.

6.2   Active Orientation and Location

6.2.1   Odontocetes

All odontocetes that have been studied in sufficient detail have been found to have very impressive echolocation abilities, and it is likely that all odontocetes use sound actively to explore their environment as well as to communicate with conspecifics. The species that has been best studied is the bottlenose dolphin. Echolocation was demonstrated experimentally in this species by Norris *et al.* (1961) and the motivation behind much of the funding for this research has been to improve man-made sonar systems. Much of this research is reviewed in detail in the book by Whitlow Au (Au, 1993), while Au (1997) provides an informative comparison between the echolocation systems of bats and dolphins. The echolocation signals of the bottlenose dolphin are intense, broadband, short duration sound pulses; in contrast, most bat echolocation pulses have a narrower bandwidth, may be frequency modulated and are of a longer duration. During echolocation, a dolphin emits a sound pulse, normally in a directional beam, and listens for echoes reflected from objects in the sound's path. As mentioned above, rigid objects reflect sound efficiently if their circumference is of the order of a wavelength or more and there is an impedance mismatch. Thus, to be able to detect small objects dolphins must produce high frequency sound pulses. The peak frequency in the echolocation click shown in Figure 6a is around 100 kHz. The speed of sound in water is roughly 1,500 m/sec; thus, the wavelength of a 100 kHz pulse is around 1.5 cm and we would expect dolphins to be efficient at detecting targets with circumferences of about that size and greater.)

A number of pieces of information are potentially available in the returning echo. The time delay between the production of the click and the reception of the echo is twice the travel time to the object, so this gives its range. The horizontal bearing or azimuth of the object is available both from knowing the direction in which the beamed sound was projected and/or from determining the direction from which the echo arrived. Finally, details such as the relative spectral components of the echo and the nature of any internal reverberations can provide information on the size, direction, shape and composition of the object. An echolocating bottlenose dolphin typically emits rapid trains of clicks, often producing a click soon after the echoes from the previous one have returned, while it scans the environment with its directional click trains. An echolocating dolphin may be able to integrate information from many different echolocation pulses to build up a more detailed picture of the environment and objects within it. Controlled tests have shown the bottlenose dolphin to have impressive sonar capabilities: in tests, bottlenose dolphins were able to detect solid steel spheres with diameters of 2.54 cm and 7.64 cm at ranges of 73 and 113 m respectively (Au and Snyder 1980; Murchison 1980). Echolocating dolphins can also discriminate the composition and wall thickness of targets. For example, Evans and Powell (1967) found that dolphins could distinguish between identical plates made of copper and aluminium, and between copper plates of 0.16, 0.22 and 0.27 mm thickness.

Once bottlenose dolphins are attending to a particular target, they emit a click soon after they have received the echo from that target. Their echolocation strategy appears to depend upon processing information from the echo reflected off a target before they produce the next click. When echolocating in this mode they are said to be "locked to target". Thus, as they approach an object of interest and the travel time to it decreases, their click rate increases. Such "echolocation runs" are typical of vocal behaviour, when some bats or odontocetes are hunting (*e.g.* Miller *et al.*, 1995 for narwhals). It seems though, that not all odontocetes echolocate in this way. Turl and Penner (1989), for example, found that echolocating beluga whales produced discrete packets of clicks in bursts, resulting in click trains with interclick intervals that were less than the two way travel time to the target. Thus, echolocation mechanisms in beluga may differ from those in *Tursiops*.

Sperm whales click for most of the time that they are underwater and most biologists studying sperm whale clicks have concluded that most of these clicks serve an echolocation function, albeit often at slow click rates and therefore probably extended ranges (*e.g.* Gordon *et al.*, 1992). After a sperm whale has replenished its oxygen supply at the surface, it raises its flukes above the water to initiate a deep, near-vertical, dive. After a few minutes, the whale starts to click in a slow, monotonous manner. Listening to a hydrophone at the surface directly above a whale that has recently dived, one can often hear the echo from the seafloor arriving just before the whale's next click. (As the whale is directly below the boat, the relative timing of the echo and the click will be the same at the whale and the surface hydrophone.) In this case, it seems that the whale may be echolocating in a similar fashion as a bottlenose dolphin "locked to target", with the target being the seabed. On some occasions though, especially if the water is very deep, the whale will click two or three times as fast as the two-way travel time. In this case, the whale may either be waiting for faint echoes from a layer less deep than the seabed, or may be able to emit several pulses before hearing the first echo. If the whale pauses in its clicking, two or three bottom echoes, each the click interval apart, are often heard at the surface hydrophone after the clicks stop. When it starts clicking again, there are no echoes until after the first two or three clicks have been produced. As the whale dives deeper and gets closer to the bottom, the return time for echoes decreases and the click rate slowly increases (Fig. 9), rather like a very languid echolocation run. This pattern of timing for click and echo is like that of a bottlenose dolphin locked to target, but it all happens much more slowly. The gradual increase in click rate can usually only be seen through the first stage of the dive (and it does not happen on every dive). During the middle sequence, click rates typically fluctuate (perhaps indicating that the range to the largest reflective target is also changing for the clicking whale).

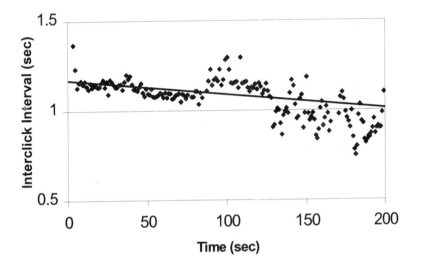

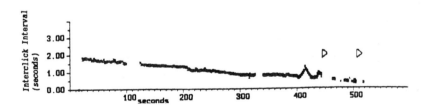

Figure 9. Sperm whale inter-click intervals. Plots of inter-click intervals against time through the initial stages of two dives. Inter-click intervals decrease (click rate increases) as whales get closer to the bottom. Triangles in lower panel indicate the occurrence of "creak" vocalisations.

Regular series of sperm whale clicks may be interrupted by silences and bursts of rapid clicks (at rates of around 20 per sec) called "creaks". Creaks sound much more like the typical echolocation runs of dolphins and may be produced when whales close in on their prey. Various pieces of evidence, presented by Gordon (1987b), suggest that this is what is happening. The normal, regular, slow clicks of sperm whales do not seem to be very directional (Watkins, 1980a but see Møhl *et al.*, 2000 for evidence of directionality in these clicks); certainly they can be heard very clearly on hydrophones at the surface directly behind a diving whale. The rapid clicks in creaks may well be beamed however. Gordon (1987b) found evidence that rapid click trains produced by sperm whales at the surface were produced directionally. It seems then that sperm whales may operate a dual-mode sonar system. A poorly-directional, longer range, lower acuity system is provided by the slow regular clicks and might provide information at greater ranges and over a broader area; and a short range, high acuity system, more akin to dolphin sonar, is represented by the "creaks". Miller *et al.* (1995) propose a similar distinction for narwhals where slow regular series of clicks may be used for orientation and initial detection of targets, with fast series analogous to the terminal buzz of some bats, produced as the animal closes on its prey, It is likely that as studies of echolocation extend beyond captive studies and into the field, and a greater range of species are investigated, more diverse echolocation systems, that have evolved to perform different tasks in a variety of environments, will be revealed (Tyack, 1997).

6.2.2   Mysticete Echolocation

The extent to which baleen whales can echolocate has long been a contentious question. At first sight, their predominantly low frequency calls, do not appear suitable for locating anything but the largest objects. (Early reports of high frequency calls from mysticete whales (*e.g.* Beamish and Mitchell, 1971; Beamish and Mitchell, 1973) have not been confirmed by others, and appear to represent sounds, perhaps from odontocetes, that were misidentified.) Although it seems unlikely that mysticetes have a high frequency, high acuity echolocation system similar to that of odontocetes, Tyack (1997) reviews suggestions that they could still receive useful information about their environments from reflections of their own sounds. For example, George *et al.* (1989) and Ellison *et al.* (1987) proposed that, by listening to the reverberation of their calls, migrating bowhead whales would be able to judge whether the ice fields ahead of them were composed of deeply keeled, multi-year ice or more easily navigable, smooth, thin ice with open water leads. The possibility that baleen whales, such as blues and fins, that make very loud low frequency sounds, could use these to locate large objects, such as the base of islands, at ranges of hundreds of miles, has been proposed by Norris (1967, 1969); Payne and Webb (1971) and Thompson *et al.* (1979). Acoustic modelling (Clark and Ellison, 1997) and tracks of phonating blue whales made using the US Navy's Sound Surveillance System (SOSUS) arrays (Gagnon and Clark, 1993; Clark and Mellinger, 1994; Clark, 1995) suggest that blue whales may indeed be able to localise islands at these ranges.

6.3   Can Cetaceans Operate their Sonars in a Bistatic Mode?

All of the examples of echolocation described in the previous two sections involve an animal listening for echoes from its own sounds. Sonar engineers call this a "monostatic" sonar, but this is not the only kind of sonar they design. For many marine applications, a "bistatic" sonar may be preferable. From our biological perspective, a bistatic sonar can be defined as a sonar in which one animal produces the sonar signal and another animal can listen for the echoes. Xitco and Roitblat (1996) demonstrated that dolphins can use their biosonar in a bistatic mode. They set up an experiment with captive dolphins in which one dolphin clicked at an artificial target and a different dolphin listened to the echoes. This second dolphin was able to identify the target based upon listening to the echoes of clicks from the first dolphin.

We do not know whether bistatic sonar plays a role in the wild, but there are a variety of circumstances where it may be beneficial. In future research, it will benefit researchers to be aware of the potential for these unusual modes of sonar (Tyack, 1997). Dolphins have such a directional sonar signal, that they are among the least likely to use sonar in a bistatic mode, unless the listening dolphin is very near the clicking dolphin (as was the case in the Xitco and Roitblat study). Baleen whales and sperm whales produce less directional sounds for which human observers can easily record echoes from targets such as the seafloor (Section 6.2.1). If we can hear the echoes, then most other whales in the area may also be able to use these sounds of other whales to learn about their environment. All monostatic sonars operate by listening for sound reflected back from a target. Another unusual feature of bistatic sonars is that they can also detect dropouts in sounds when the signal path encounters a target that absorbs sound. Tyack (1997) discusses how many targets of relevance to cetaceans may be better targets in this forward propagation mode than in the usual backscatter mode.

6.4   Do Cetaceans Use Sound to Influence the Behaviour of Prey?

Norris and Møhl (1983) presented the creative hypothesis that odontocetes may not only produce clicks to detect and track their prey, but also that the clicks may be loud enough to stun or incapacitate the prey. This hypothesis was proposed both for sperm whales and their squid prey and dolphins and their fish prey. Unfortunately, we do not have many published reports on the source levels of sperm whale clicks. Levenson (1974) estimates source levels in the 160-180 dB re 1 μPa range. Squid appear to be quite tolerant of sounds at these levels. On the other hand, Møhl *et al.* (2000) present data for source levels up to >220 dB re 1 μPa at 1 m within the beam of the click. Dolphins produce much higher peak-to-peak source levels, up to over 225 dB re 1 μPa. However, these sounds are so brief that the clicks do not actually contain as much total energy as these levels might suggest. Zagaeski (1987) found that 50% of small fish showed signs of disorientation when exposed to received levels of 236 dB re 1 μPa. None of these data make a strong case for the prey stunning hypothesis. Marten *et al.* (1988) suggest that lower frequency pulsed sounds may stun fish, but source levels have not been reported for these sounds, making further evaluation or experimentation with fish difficult.

6.5   Countermeasures

The hunting behaviour of echolocating bats has been studied much more thoroughly in the field than that of dolphins. The prey of bats (mainly flying insects) show interesting responses and countermeasures to bat sonar. Roeder (1967) discovered that certain species of moth were able to detect the high-frequency echolocation signals of bats and avoid them. It seems that high-frequency hearing evolved in these moths to allow the detection of the hunting calls of bats. (Some species of crickets and lacewings have evolved a similar ability; see Miller, 1983 for review.) Some species of moth emit high-frequency warning clicks when they hear bats approaching. Often these moths are distasteful and it is likely that these clicks are an aposematic signal to "warn" the predator (much like the warning coloration displayed by distasteful caterpillars), but it has also been suggested that, in some cases, these signals could act to "jam" the bat's sonar (*e.g.* Surlykke and Miller, 1985). We might expect similar responses and arms races to have evolved between odontocetes and their prey, but there have been relatively few detailed studies of echolocation and foraging in natural conditions (Tyack, 1997). A number of different marine mammals that are hunted by killer whales (*Orcinus orca*) seem to avoid killer whale vocalisations (Cummings and Thompson, 1971; Fish and Vania, 1971). There is some evidence that echolocating killer whales may modify their echolocation strategy depending upon the auditory sensitivities of their prey. Barrett-Leonard *et al.* (1996) reported that mammal-eating killer whales tend to produce clicks less often than do fish-eating killer whales. When mammal-eating killer whales do click, they vary the intensity, repetition rate, and spectral composition within click trains, apparently making these clicks more difficult for their acoustically sensitive prey to identify than the regular click series of fish-eating killer whales. There are also some intriguing observations that suggest that fish may be sensitive to cetacean echolocation signals. Unlike most fish, herring and some other clupeids have high frequency hearing sensitivity (Astrup and Møhl, 1993; Dunning *et al.*, 1992; Mann *et al.*, 1997; Nestler *et al.*, 1992). The anatomical basis of this high frequency sensitivity in clupeids appears to pre-date the appearance of cetaceans (Mann *et al.*, 1997). However, several clupeids respond with escape behaviour when they hear ultrasonic pulses, and this response may represent an adaptation for escaping odontocete predators (Mann *et al.*, *loc. cit.*).

7.   COMMUNICATION

Popular concepts of communication often emphasise the exchange of information between a producer and a recipient. The general perception that in human societies communication is "a good thing" can predispose us to thinking of communication in purely co-operative and benign terms. However, biologists appreciate that behaviours only evolve if they benefit the genes causing them to occur, usually through benefiting the animal initiating the communication. Thus, as Krebs and Dawkins (1984) emphasise, communication signals evolved to influence the behaviour of other animals. The animal making a signal (the actor) attempts to influence the behaviour of another animal (the receiver) in ways that will benefit the actor and may either benefit or harm the receiver. On the other side of the relationship, evolution should equip the receiver to be discriminating and only to respond to signals when it is in the long-term interests of its genes to do so. Seen from this perspective, communication is revealed not always to be a co-operative enterprise, but often to be more akin to an arms race involving mind reading and manipulation.

As we have stressed already, the physical properties of seawater favour acoustic communication in cetaceans. Visual signalling may occur in cetaceans at short ranges; the coloration patterns on many odontocetes have probably evolved to facilitate this; and chemical communication, by taste, may also be important, for example as signals of recent feeding success (Norris and Dohl, 1980) or reproductive condition. Undoubtedly though, the dominant channel for communication in cetaceans is acoustic. Dusenbery (1992) reviewed the energetic costs of transmitting information in different ways through different media and found that sound communication was generally the most efficient for long-range communication, and was 100-1,000 times more efficient through water than through air. Acoustic communication signals in cetaceans take many forms, and are adapted to perform a number of different functions.

Figure 10.   Exuberant above water displays, such as breaching (as in this sperm whale) and lobtailing, may make loud noises underwater and so serve to communicate basic information between individuals. (Photo: J. Gordon/IFAW).

7.1   Contact Calls

One of the functions that communication signals may serve is to keep animals in contact with each other. At their simplest, contact calls may contain little more information than just the caller's presence and, indirectly, location. Some calls may be specially adapted for this contact function. Efficient contact calls should be loud, so that they may travel far; easy for the intended recipient to directionalise, so that the caller can be localised; characteristic, so that they will be distinctive and will be easier to pick out from background noise; and have a duration that is at least as long as the integration time of the receiver's auditory system. They will also be easier to detect at low levels if they are produced at predictable intervals in long sequences.

Clark (1982) analysed the vocal repertoire of southern right whales (*Eubalaena australis*) and Clark (1983) identified one of these calls, the UP call, as a contact call. UP calls were most commonly recorded when a whale or group of whales was swimming towards another group. When an UP call was produced, the other group would often respond by producing an UP call. As the two groups converged, the rate of UP calling would increase until the groups joined, when it would abruptly cease. These observations led Clark (1983) to suggest that UP calls are contact calls used to broadcast both the actor's location and its motivation to join another group. The UP calls are well suited to this task. They are the loudest of the vocalisations identified by Clark (1982), and they have dominant frequencies in the 200-300 Hz band, where noise is at a minimum in the locality where Clark studied these whales.

The extremely loud, low frequency pulses made by many baleen whales (such as blue, fin and minke) in extended sequences, seem to be ideally adapted to serve as contact calls and, as we have seen, they might be audible to other baleen whales at ranges of hundreds or even thousands of miles. The work of Watkins *et al.* (1987) suggested that fin whale 20 Hz calls are reproductive displays made by males during the breeding season. However, more recent observations from the SOSUS arrays (Clark, 1994) indicate that whales call throughout the year, though rates may be lower during the summer months, and that the locations of callers move north and south with the seasons, following the whales' presumed migrations. The baleen whales that produce these calls do not form large stable groups and do not have well defined breeding grounds. It seems that breeding takes place between whales dispersed throughout the tropical oceans, in which case the need for individuals of these species to be able to keep in contact and to find each other is clear. More research is needed to evaluate the communicative functions of these calls.

7.2   Vocal Assessment

Vocalisations can be an effective means of advertising certain characteristics of a calling animal. One of the most fundamental of these is body size. As we mentioned earlier, the lowest frequency of sound that certain species can produce is inversely proportional to their body size. Call frequency has been shown to be a reliable indicator of differences in body size between different individuals of the same species too, for example: in toads (Davies and Halliday, 1978), in red deer (McCoomb, 1991) and in pig-tail macaques (Gouzoules and Gouzoules, 1990). In species where body size is a good predictor of fighting ability, as it is in deer and toads, males may assess each other from their vocalisations before deciding to escalate conflicts. Similarly, if large body size indicates that a large male would be a better mate, females might be expected to choose mates based on their body size and assess this acoustically. It is tempting to think that a process like this may have driven the evolution of the low frequency vocalisations in the rorqual whales.

Baleen whales have a sexual dimorphism that is unusual among mammals. In these species, males, which are believed to be the more vocal sex (Watkins *et al.,* 1987), are slightly smaller than females. This suggests either that there may not be a competitive advantage to male mysticetes in being larger, or perhaps this advantage does exist, but females are under even more intense selection for large size than males, perhaps because of the need to gestate and suckle an infant on fat reserves while fasting and swimming thousands of kilometres.

Sperm whales, on the other hand, are the most sexually dimorphic of cetaceans, with mature males growing to over three times the mass of mature females (Lockyer, 1981). Sperm whales make their click vocalisations at "museau de singe", at the front of their massive heads, according to Norris and Harvey (1972). Norris and Harvey (*loc. cit.)* proposed that each click reverberated inside the head between two sound reflectors, one at the very front of the head and another behind the spermaceti organ and on the front of the skull, and that this provided an explanation for the multi-pulsed structure of sperm whale clicks (Fig. 11). These authors realised that if this was the case, the interval between pulses would be twice the travel time for sound between the two sound reflectors; and that if the inter-pulse interval (IPI) could be measured, then the size of the head, and thus the size of the animal itself, could be calculated. Gordon (1991) was able to show that this was the case by comparing inter-pulse intervals and photographic estimates of body length in a sample of sperm whales.

Male sperm whales are known to fight each other: most mature males bear tooth-rake scars on their heads inflicted by their rivals. Eye-witness accounts of these conflicts attest to their fierce and bloody nature involving repeated charges, ripped flesh, and broken jaws (see review by Caldwell *et al.,* 1966). It is likely that larger males do better in these fights.

Evolutionary theory predicts that contestants should assess the fighting ability of their adversaries carefully before and during fights to avoid escalating a costly contest that they would be unlikely to win. Gordon (1993) suggested that male sperm whales might assess each other's size from the IPIs of their clicks before and during fights. Once this process was occurring, it would be advantageous for males to sound as large as possible. Sperm whales could achieve this by making their heads, and thus their IPIs, longer. In fact, in sperm whales, the head is proportionally rather longer in mature males than in females (accounting for 26% as opposed to 20% of body length) and protrudes much further forward beyond the upper jaw in males. In addition, Gordon (*loc. cit.*) observed that males had been reported to have proportionately larger heads and more prominent melons in several other species of odontocetes including, bottlenose whales, Baird's beaked whale, pilot whales, false killer whales, and spotted dolphins.

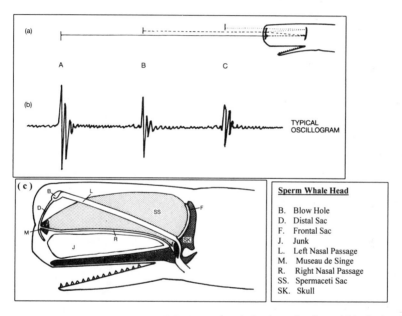

Figure 11.   Production of pulsed clicks in sperm whales by reflection within the head, as proposed by Norris and Harvey (1972). (a) depicts reflection of pulses from sound mirrors within the head and position of wavefront of three sound pulses at a moment in time; (b) shows typical changes in pressure level with time as a click passes; and (c) is a diagrammatic guide to the structure of the sperm whale head. A pulse is created at the museau de singe. Some of this sound energy passes straight out into the water (pulse A) but some is reflected from the distal sac back to the frontal sac and forward again. Some of this pulse travels out into the sea as pulse B (delayed relative to pulse A by 2x the time to travel between the distal and frontal sacs) some is reflected by the distal then frontal sacs and becomes pulse C.

It is possible, therefore, that vocal size assessment could be occurring in a number of different odontocetes. After this chapter went to press, Cranford (1999) published a paper exploring in more detail how acoustic assessment may have affected the evolution of large head size in the sperm whale.

Size is not the only important cue that might be assessed acoustically. In red deer, the rate at which a stag roared was both a good indicator of its fighting ability (Clutton-Brock and Albon, 1979) and of its attractiveness to females (McCoomb, 1991). Clutton-Brock and Albon (1979) suggested that this was because roaring was so energetically costly that only the strongest and fittest males could sustain long bouts of it. Male humpback whales on their breeding grounds sing long, complicated songs. These songs have a specific and hierarchical structure: phrases are grouped into themes, and sequences of themes make a complete song (Payne and McVay, 1971; Payne *et al.*, 1983). In any one year, all males on a particular breeding ground will sing similar songs, though songs on different breeding grounds will be completely different (Payne and Guinee, 1983). Within a breeding population, songs gradually change during the breeding season and from year to year (Payne and Payne, 1985). Individual males will sing for hours on end, hanging in mid water at a depth of about 20m. They usually return to the surface to breath at a particular point in their song, and though they continue singing, the fact that a singer has surfaced is easily detectable by a dropout of the low frequencies of song and a reduction in radiated power when the singer nears the surface. This is a consequence of the physics of sound in the sea; as the singer approaches the surface, sounds low enough in frequency to be within a fraction of a wavelength of the surface do not propagate well. It has been suggested by Chu and Harcourt (1986) that other whales, both males and females, would thus be able to assess the singer's ability to hold his breath while singing. Breath-holding ability is a valuable ability in whales and is likely to be correlated with health and fitness. Chu and Harcourt (1986) argue that some of the features of humpback song may have evolved in response to benefits accrued from signalling breath-holding ability in these whales.

There are some problems with this argument, however. Humpback whales on the same breeding ground sing songs of similar duration at any one time, and the songs can change over time with no repeated seasonal pattern (Payne *et al.*, 1983). Each individual whale is more likely to sing songs of the current average duration than of the duration they were singing a few months earlier or later (Guinee *et al.*, 1983). If humpbacks were using song to advertise their breath-holding ability, then each individual would be expected either always to sing as long as he was able, or to sing longest at that part of the breeding season when his chances of mating were highest. These predictions do not match the observations that each singer tends to match the average duration of the song on a breeding ground at any one time.

7.3 Reproductive Advertisement Displays

The songs of humpback whales are not just used for fighting assessment, but appear to be reproductive advertisement displays, with functions similar to the songs of birds. Charles Darwin (1871) argued that the evolution of these displays, directed towards females, differed enough to justify creating a new term "sexual selection" in contrast to natural selection. Since female mammals gestate the young and lactate to provide nutrition after birth, they tend to devote more resources to the young than do males. Darwin noted that in the mating systems of many animal species, females are selective in choosing a mate, and males are less selective, but tend to compete with one another for access to a female (see Chapter 5). Males often produce reproductive advertisement displays that may play a role in female choice and/or in male-male competition.

A variety of evidence supports the role of humpback song as a reproductive advertisement display. Song is heard during the winter breeding season, but seldom at other times. Efforts to determine the sex of singing humpbacks show that singers are males (Baker *et al.*, 1990; Glockner, 1983; Palsbøll *et al.*, 1992). Singing humpbacks are usually alone; they are highly motivated to join other whales, and they stop singing when they join or are joined by other whales (Tyack, 1981; 1982). Song seems to mediate male-male competition on the breeding ground. Song appears to maintain distance between singers (Tyack, 1983; Helweg *et al.*, 1992; Frankel *et al.*, 1995), and aggressive behaviour is often seen when a singer joins with a male (Tyack, 1981; 1982). Behaviour associated with sexual activity has been observed during the few interactions when a singer joins with a known female (Tyack, 1981; Medrano *et al.*, 1994), but less is known about the role that song may play in mediating a female's choice of a mate. This is easy to understand. Fights between males often involve more conspicuous behaviours than male-female interactions, and since females only rarely ovulate, their interactions with males, particularly selection of a mate, may occur much less frequently than male-male competitive interactions. However, the structure of humpback song can tell us something about its role in female choice. Displays that are produced by males and used by females to select a mate tend to be complex and costly to produce. In reviewing data on bird song, Catchpole (1982) concludes that the following features are typical of songs used to attract females: males sing continuously, do not counter-sing, and do not respond to song playback. All of these features are characteristic of humpback songs. Humpbacks sing often for hours, with little co-ordination with other singers nearby. Humpback singers showed strong approach responses to playback of the sounds of groups in which males compete for access to females; if they responded at all to song, it was a slight avoidance response (Tyack, 1983). Humpback song clearly seems to be used both in male-male competition and in female choice – it is more complex than would be expected for a signal used only for mediating interactions between males (Helweg *et al.*, 1992).

7.4   Recognising Species, Groups, and Individuals

Recognising different species, groups, and individuals so that an animal can behave appropriately to them, will often be extremely important in the social lives of cetaceans. It may be possible for animals to achieve this from noises made for other purposes, especially in the case of species recognition, but in other cases, vocalisations may be specifically adapted to be distinctive. For example, two closely related species of pilot whale have whistles that are highly distinctive; killer whales off the northwest coast of North America have group-distinctive vocal repertoires; and many dolphins have individually distinctive signature whistles.

7.4.1   Species Recognition

Calls that are adapted to aid species recognition might be expected to have evolved to prevent inter-species mating. Indeed, the need to recognise conspecifics is thought to be a major reason for divergent call characteristics in many animal groups (Harper, 1991). Selection pressure for such calls would be expected to be strongest where the ranges of similar species which have the potential to inter-breed overlap. Most cetacean species are wide ranging and inter-species mating is known to occur in cetaceans. Hybrids have been reported in the wild between fin and blue whales (Spilliaert *et al.*, 1991), and bottlenose dolphin and Risso's dolphin (Shimura *et al.*, 1986). Mechanisms to discourage inter-species mating would seem to be advantageous. Pilot whales may provide an example of the process in action. There are two species of pilot whale.   The long-finned pilot whale (*Globicephala melas*) has a cool-temperate water distribution, and the short-finned pilot whale (*G. macrorhynchus*) is found in warmer waters. The distribution of both species overlaps at about 30-40 N in the N. Atlantic and at around 35 S. The two species are very similar in appearance and are almost impossible to distinguish reliably in the field. Rendell *et al.* (1999) compared whistle vocalisations from a number of odontocetes, including both pilot whale species, in the N. Atlantic.   They found that, despite their similarity in appearance and lifestyle, the two pilot whale species had very distinctive vocalisations, with the whistles of *Globicephala melas* having a substantially lower mean frequency. They suggested that a need to prevent inter-breeding between these two very similar and spatially overlapping species, may have led to the evolution of contrasting calls in these two similar species.

7.4.2   Individual and Group Recognition

Many cetaceans are highly social animals: most live in groups and, in many species, strong, long-term associations form between specific individuals. Advantages of social living include communal care, especially of calves; defence from predators; co-operative feeding; and the formation of breeding coalitions. For members of groups to stay together they need to be able to stay in contact (we have already suggested that as they are usually out of sight of other group members, they must do this acoustically), and also to be able to recognise the identity of groups and individuals.

Cetacean species exhibit a variety of forms of social organisation, adapted to different environmental conditions and life styles (see Chapter 5). Often the form of social communication found within a species can be understood in terms of the structure of its societies, with vocalisations being distinctive at the level at which social organisation is strongest (Tyack, 1986; Tyack and Sayigh, 1997). For example; fish-eating killer whales have stable long-term groups and also distinctive, group-specific vocal dialects; bottlenose dolphins, by contrast, have dynamic groups but strong associations between individuals, and some of their vocalisations are specific to individuals.

Those killer whales in coastal waters of the Pacific Northwest that only eat fish live in some of the most stable groups of any mammal: typically, animals enter or leave groups through birth, death or very rare fissions of large groups (Bigg *et al.*, 1987). These whales have large vocal repertoires but a substantial subset of their vocalisations, discrete calls, are stereotyped, and stable over decades. Each group has a distinctive repertoire of discrete calls (Ford, 1991). It seems that each member of a group produces all of that pod's discrete calls. Different groups, especially those that often affiliate, share some of the discrete calls but none have exactly the same repertoire. Thus, it is the complete repertoire of discrete calls that is specific to a particular group and is thought to indicate group affiliation, to maintain group cohesion and to co-ordinate activities.

The social organisation in bottlenose dolphins is quite different. Inshore populations of this species, which have been well studied, do not have stable groups: theirs is a fission-fusion society in which group composition may change from minute-to-minute. Strong and stable bonds between particular individuals do exist however and some wild individual bottlenose dolphins show stable patterns of association for many years (Wells *et al.,* 1987). Bottlenose dolphins therefore, need to maintain important inter-individual bonds within a society of fluid and unstable group organisation. To do this they must be able to easily recognise other individuals with whom they share a bond, and there is a need for vocalisations that are individually distinctive and specific. Distinctive "signature whistles" were first demonstrated in captive bottlenose dolphins by Caldwell and Caldwell (1965); they showed that each individual within a captive group produced a whistle with a distinctive pattern of frequency modulation, or "contour". Early studies found that about 90% of the vocal output of isolated captive animals were made up of their "signature whistles" (reviewed in Caldwell *et al.,* 1990). More recently, Janik and Slater (1998) showed that individual captive animals kept in groups only produced their "signature whistles" when they swam into a small pool where they had no visual contact with other group members from the main pool, providing further evidence in support of their function as individually distinctive contact calls.

Figure 12.			Oceanic dolphins, such as these spotted dolphins, have a complex social organisation and produce both clicks and whistles serving a variety of functions from prey location and capture through to individual and group recognition and cohesion. (Photo: J. Gordon/IFAW)

Signature whistles are also produced by free-living bottlenose dolphins and important work on their nature and development has been carried out with wild dolphins. Sayigh *et al.* (1990) recorded whistles produced by dolphins that were restrained in a net corral so that a variety of other data could be collected from them. This allowed these researchers to record whistles of individual dolphins using hydrophones applied to restrained dolphins with suction cups. Recordings of the same whistles before and immediately after "capture" showed that the whistles produced by restrained animals were typical. These recordings from wild animals showed that signature whistles develop by two years of age at the latest (calves were not typically recorded until two years of age), and are then stable for decades. Intriguingly, about half of the male calves typically develop whistles that are very similar to those of their mother, while female calves develop a signature whistle that is distinctive (Sayigh *et al.*, 1995). This may reflect the fact that, in this species, males disperse further than females. Females and their calves which may often find themselves in the same group will need distinctive whistles; whilst there may be value for females in recognising a male with a signature whistle that is similar to that of their mothers as potential kin, to avoid inter-sibling mating (Sayigh *et al.*, 1995).

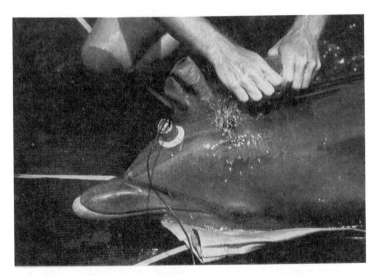

Figure 13. Photograph illustrating a method for recording whistles from a known individual wild bottlenose dolphin from the Sarasota, Florida population. This dolphin has been encircled in a net corral and lifted into a raft. A suction cup hydrophone has been attached to the melon, where it only picks up whistles from that individual. This involuntary separation from other dolphins usually stimulates the production of signature whistles that match those produced by the same individual when unrestrained in the wild. (Photo: P. Tyack)

7.5   Communication to Co-ordinate Behaviours

Cetaceans often show beautifully co-ordinated behaviour, often when out of visual range of each other. It would seem that in these cases acoustic signals must be used to co-ordinate and synchronise behaviour, but there are very few examples of specific vocalisations serving this function. So many captive dolphins show such co-ordinated behaviour that observers have concluded the behaviour must be synchronised by vocal signals, but there have been consistent surprises when hydrophones are put in the water animals are often found to be silent during these synchronous co-ordinated displays. In some cases, feeding involves co-ordinated behaviours and there are some examples of particular calls associated with co-operative feeding. D'Vincent *et al.* (1985) described specific vocalisations associated with highly co-ordinated feeding behaviour in a group of eight individual humpback whales. Each individual whale maintained a specific orientation and location in the group during co-ordinated feeding. A relatively unmodulated 500 Hz call was associated with the initiation of the co-ordinated feeding, and an upsweep was produced just before the whales simultaneously surfaced in vertical feeding lunges. Janik (2000) reported that bottlenose dolphins produced a special distinctive low frequency two-component call when feeding on salmon (*Salmo salar*) or sea trout (*Salmo trutta*) but not when feeding on smaller fish. One function of the call could have been to recruit other dolphins to help with capturing these larger fish.

7.6   Communication Between Humans and Dolphins Using Artificial Languages

Research projects in which dolphins have been taught to imitate man-made sounds or to respond to a human-imposed sign language have shown that dolphins have an impressive ability to learn new vocalisations (Richards *et al.*, 1984), and to learn to interpret strings of commands for actions and objects (Herman *et al.*, 1984). Their performance with artificial languages shows they can understand artificial syntactic structures in the sense that they understand artificial rules for ordering commands, whether these are gestural commands or sounds (Herman *et al.*, 1984; Herman, 1986). However, there are some problems with the interpretation of animal "language" studies. Linguists use a special vocabulary to describe human language – symbol, reference, syntax, grammar. Even terms such as "word" and "sentence" are linguistically loaded. The dolphin language studies use these same words to describe the sequences of commands that are used to train the dolphins. For example, in one such artificial language, the dolphin might be given a fish if it responded to the following sequence of commands PIPE HOOP FETCH by swimming to a hoop in the pool and taking it to a pipe in the pool. What was impressive about the dolphin's performance was that she was able to learn that the general rule OBJECT1 OBJECT2 ACTION meant perform the action on object2 directed towards object1. This was demonstrated by presenting the trained dolphin with the same sequence of commands, but using new objects or actions that had never been presented in the full sequence together (Herman, 1986).

Herman (1986) not only calls the individual commands "words" and the sequences of commands "sentences," but he also calls OBJECT1 in the above sequence an indirect object and OBJECT2 a direct object. This may be a useful mnemonic for us to remember the role in the sequence, but it is by no means certain that dolphins understand these strings of commands as anything really like words in a sentence in a human language, nor that they actually understand grammatical categories as humans do. The issue of whether it is appropriate to use human linguistic terms to describe the performance of highly trained animals has caused considerable controversy, even among marine mammalogists conducting animal "language" studies (Schusterman and Gisiner, 1986).

Many modern linguists argue that humans evolved a highly specialised innate ability to parse speech into the appropriate grammatical categories (*e.g.* Miller, 1990; Pinker, 1994). They point out that it takes a lab full of PhDs to train an animal to use an artificial language, but that most kids learn to speak in a much less structured environment. Almost all human children, having been exposed to a small and biased set of sentences at home, manage to develop a grammar that allows them to construct an infinite variety of sentences. In addition, kids are seldom corrected for making grammatical errors. Many linguists argue that the only way children could come up with the correct grammar, given such variable input without training, is to have an innate species-specific ability to learn grammar (Pinker, 1989).

If many linguists believe that grammatical categories are the product of a specialised mechanism for learning grammar that is species-specific to humans, then Herman *et al.* (1984) and Herman (1986) are on thin ice in suggesting that their dolphins understand words and sentences, direct and indirect objects. It is always difficult to extrapolate terms to another frame of reference for which they were not initially intended. This is particularly problematic for language, because linguists cannot agree among one another about how to think about these issues and how to interpret the data from humans, much less animals (*e.g.* Bickerton, 1990; Locke, 1993).

A major problem with animal language research is that human language involves an idiosyncratic collection of auditory processing, cognitive skills, and vocal motor abilities. It is unlikely that any other species will have exactly the same combination as our own. Animal language studies reveal fascinating performances of animals, but it simply is never clear whether the analogies with language provide insight or illusion. One reason why these studies are so impressive is that they require such a combination of skills from the dolphins. However, this complexity can render it difficult to actually determine precisely what cognitive skills the animals require to perform the task, especially if it results from years of training. As ethologists interested in these issues, we believe that it will be more promising to study from a comparative perspective using different animal groups and specific cognitive abilities that may be involved in language. We will discuss two of these abilities for which there are particularly interesting data from cetaceans: reference and vocal learning.

Reference is a critical property of human language. Words are symbols - when humans name something with a word, they understand the word to act as a symbol for the thing that has just been named. But what about dolphins trained to associate a cue with an object? Does the dolphin understand the cue to refer to the object, or has she simply learned that she will be rewarded if she performs a response when shown a stimulus? The dolphins in animal language studies can be trained to understand vocal labelling in the sense that they can use artificial sounds to label toys and other objects. Richards *et al.* (1984) conducted a study in which they were able to train a dolphin to imitate arbitrary computer generated patterns of frequency modulation. They would call the animal to station, play a model sound, and then give the dolphin a fish if it produced a close imitation of the model. They then studied labelling by pairing each kind of model with an object, say an upsweep with a frisbee and a downsweep with a ball. They then would occasionally call the animal to station, show her the ball, and then only reinforce her if she produced a downsweep. She readily learned in this way to label arbitrary manmade objects with arbitrary synthetic sounds. This study reveals fascinating abilities of dolphins to label objects, but it does not completely address the issue of symbolic reference. In order to use a signal as a symbol for an object, an animal must also be able to associate properties of the object with the symbol. For example, if you are asked what colour a banana is, the word banana allows you to remember that the object is yellow, even though there is nothing yellow about the word banana. We do not know whether dolphins can use these symbolic associations.

There is one special skill that is required for language, rare among terrestrial mammals, and for which there is overwhelming evidence among cetaceans. This is vocal learning, or the ability to modify one's vocal output based upon auditory input. One of the strongest sources of evidence for this is the kind of imitation of unusual signals studied by Richards et al. (1984) in bottlenose dolphins. The most clear cut cases have involved synthetic stimuli, but belugas (Eaton, 1974; Ridgway *et al.*, 1985) and harbour seals (Ralls *et al.*, 1985) have even been shown to create very clear imitations of human speech. It is somewhat ironic that the inability to produce speech sounds was such a roadblock in ape language studies, yet no one has used this imitative ability in dolphins as Pepperberg (1996), for example, has with accomplished avian vocal mimics. The constant changes observed in humpback song almost certainly must also result from vocal learning, with singers copying singers at any one time, and tracking the progressive changes in the song (Payne *et al.*, 1983, Payne and Payne, 1985).

Janik and Slater (1997) review the evidence for vocal learning in mammals. While this skill is relatively common among some birds, it is very rare among mammals. Vocal learning is of particular interest for humans because it is essential for our music and language. The evolutionary origin of the ability to modify vocalisations remains shrouded in mystery, since not even our closest ape relatives are capable of more than the most rudimentary vocal modifications. Marine mammals will be particularly important animals for studying the evolution of vocal learning among mammals. Vocal learning has been demonstrated for some species in the two mammalian groups known to echolocate: bats and odontocetes. This suggests that echolocation may play a role in the evolution of vocal learning. The existence of vocal learning in some seals and whales that produce reproductive advertisement displays, suggests a parallel with songbirds where sexual selection appears to have influenced the evolution of vocal learning for the production of songs. The best evidence for vocal learning among dolphins involves frequency modulated tonal sounds that are quite similar to dolphin whistles. Tyack and Sayigh (1997) review evidence that the signature whistles of dolphins develop through a process in which a young dolphin will imitate acoustic models present in its natal environment. Tyack (1986) and Janik (1997) show that adult dolphins in captivity and in the wild imitate the whistles of other adults. Smolker (1993) shows that as adult male dolphins form a coalition, they may converge on a common whistle type. Since a primary function of dolphin whistles is individual recognition, and since dolphins change their social partners throughout life, they may use vocal learning throughout their lifetime to maintain their individual-specific social relationships.

We feel that even more promising than training animals to follow artificial "languages" will be the effort to understand how animals like cetaceans use their own combination of sensory, cognitive, and motor abilities to solve the communication problems posed by their ways of life. As was true for studies of song learning in birds and echolocation in bats, once we understand what problems these systems were evolved to solve, we can delve much more intelligently into questions of what cognitive and neural mechanisms underlie the abilities.

8.   CONCLUSIONS

Behavioural ecology is similar at sea and on *terra firma*. Cetaceans use sound to solve similar problems to those solved by terrestrial mammals. However, the acoustic properties of the sea have created special opportunities for marine mammals, and cetaceans have become acoustic specialists, with unusually developed abilities of echolocation and vocal learning. Dolphins and porpoises have a specialised system for high frequency echolocation, and most whales may also use sound to explore their environment. Cetaceans respond to sounds of their predators, and also respond to the acoustic sensitivities of their prey. Reproductive advertisement displays help to structure the breeding behaviour of several whale species. Some whales produce low frequency vocalisations to communicate with conspecifics, perhaps over ranges of up to hundreds of kilometres. Cetaceans use sounds to maintain contact with critical social partners, either whole groups or individual animals, depending upon the social system typical of the species. Cetaceans depend so heavily upon sound for communication and orientation that they are vulnerable to increasing levels of noise pollution from human activities.

ACKNOWLEDGEMENTS

Jonathan Gordon was supported by the International Fund for Animal Welfare while preparing this paper. Peter Tyack was supported by ONR Grant N00014-99-130819. We are grateful to Christopher Clark and Vincent Janik for improving earlier versions with their helpful reviews and comments.

REFERENCES

Akamatsu, T., Hatakeyama, Y., Kojima, T. and Soeda, T. (1994) Echolocation rates of two harbor porpoises (*Phocoena phocoena*). *Marine Mammal Science*, **10**, 401-411.
Amundin, M. (1991) Click repetition rate patterns in communicative sounds from the harbour porpoise, *Phocoena phocoena*. pp. 91-111, *"Sound production in odontocetes with emphasis on the harbour porpoise, Phocoena phocoena"*, Ph.D. Thesis, University of Stockholm.

Andersen, S. (1970) Auditory sensitivity of the harbour porpoise (*Phocoena phocoena*). *Investigations on Cetacea*, **2**, 255-259.

Astrup, J., and Møhl, B. (1993) Detection of intense ultrasound by the cod *Gadus morhua*. *Journal of Experimental Biology*, **182**,71-80.

Au, W.W.L. (1980) Echolocation signal of the bottlenose dolphin (*Tursiops truncatus*) in open waters. In: *Animal sonar systems* (Ed. by R.G. Busnel and J.F. Fish), pp. 251-282. Plenum, New York.

Au, W.W.L. (1993) *The Sonar of Dolphins*. Springer-Verlag, New York. 277pp.

Au, W.W.L. (1997) Echolocation in dolphins with a dolphin-bat comparison. *Bioacoustics*, **8**, 137-162.

Au, W.W.L. and Snyder, K. J. (1980) Long range target detection in open waters by an echolocating Atlantic bottlenose dolphin (*Tursiops truncatus*). *Journal of the Acoustical Society of America*, **68**, 1077-1084.

Awbrey, F.T., Thomas, J.T. and Kastelein, R.A. (1988) Low-frequency underwater hearing sensitivity in belugas, *Delphinapterus leucas*. *Journal of the Acoustical Society of America*, **84**, 2273-2275.

Backus, R.H. and Schevill, W.E. (1966) *Physeter* clicks. In: *Whales, dolphins and porpoises* (Ed. by K. S. Norris), pp. 510-528. University of California Press, Berkeley.

Baker, C.S., Lambertsen, R.H., Weinrich, M.T., Calambokidis, J., Early, G., and O'Brien S.J. (1990) Molecular genetic identification of the sex of humpback whales (*Megaptera novaeangliae*). *Report of the International Whaling Commission* (special issue **13**), 105-111.

Balcombe, J.P. and Fenton, M.B. (1988) The communication role of echolocation calls in vespertilionid bats. In: *Animal Sonar: Processes and Performances* (Ed. by P. E. Nachtigall and P. W. Moore), pp. 635-628. Plenum Press, New York.

Barrett-Lennard, L.G., Ford, J.K.B. and Heise, K.A. (1996) The mixed blessing of echolocation: Differences in use by fish-eating and mammal-eating killer whales. *Animal Behaviour*, **51**, 553-565.

Beamish, P. and Mitchell, E. (1971) Ultrasonic sounds recorded in the presence of a blue whale. *Deep-Sea Research*, **25**, 469-72.

Beamish, P. and Mitchell, E. (1973) Short pulse length audio frequency sounds recorded in the presence of a minke whale. *Deep-Sea Research*, **20**, 375-386.

Best, P.B. (1979) Social organization in sperm whales, *Physeter macrocephalus*. In: *Behavior of marine animals*, vol. 3 (Ed. by H.E. Winn and B.L. Olla), pp. 227-289. Plenum Press, New York.

Bickerton, D. (1990) *Language and species*, University of Chicago Press, Chicago.

Bigg, M.A., Ellis, G. and Ford, J.K.B. (1987) *Killer whales - a study of their identification, genealogy and natural history in British Columbia and Washington State*. BC, Phantom Press, Nanaimo.

Caldwell, M.C. and Caldwell, D.K. (1965) Individualized whistle contours in bottlenosed dolphins (*Tursiops truncatus*). *Science*, **207**, 434-435.

Caldwell, D. K., Caldwell, M. C. and Rice, D. W. (1966) Behavior of the sperm whale, *Physter catodon* L. In: *Whales, Dolphins and Porpoises* (Ed. by K.S. Norris), pp. 677-717. University of California Press, Berkeley.

Caldwell, M.C., Caldwell, D.K. and Tyack, P.L. (1990) A review of the signature whistle hypothesis for the Atlantic bottlenose dolphin, *Tursiops truncatus*. In: *The bottlenose dolphin: recent progress in research*, (Ed. by S. Leatherwood and R. Reeves), pp. 199-234. Academic Press, San Diego.

Catchpole, C.K. (1982) The evolution of bird sounds in relation to mating and spacing behavior. In: *Acoustic communication in birds*, vol 1 Production, perception and design features of sounds. (Ed. by D.E. Kroodsma and E.H. Miller). Academic Press, New York.

Chapman, D.M.F. and Ellis, D.D. (1998) The elusive decibel: thoughts on sonars and marine mammals. *Canadian Acoustics*, **26**(2), 29-31.

Charif, R.A., Mitchell, S. and Clark, C.W. (1995) Canary 1.2 User's Manual. Cornell Laboratory of Ornithology, Ithaca, NY.

Chu, K. and Harcourt, P. (1986) Behavioral correlations with aberrant patterns in humpback whale songs. *Behavioral Ecology and Sociobiology*, **19**, 309-312.

Clark, C.W. (1982) The acoustic repertoire of the southern right whale, a quantitative analysis. *Animal Behaviour*, **30**, 1060-1071.

Clark C.W. (1983) Acoustic communication and behavior of the southern right whale, *Eubalaena australis*. In: *Communication and behavior of whales.* (Ed. by R. Payne), pp. 163-198. Westview Press, Boulder.

Clark, C.W. (1990) Acoustic behaviour of mysticete whales. In: *Sensory Abilities of Cetaceans* (Ed. by J. Thomas and R. Kastelein), pp. 571-583. Plenum Press, New York.

Clark, C.W. (1994) Blue deep voices: insights from the Navy's Whales '93 program. *Whalewatcher*, **28**, 6-11.

Clark, C.W. (1995) Application of US Navy underwater hydrophone arrays for scientific research on whales. Annex M, *Report of the International Whaling Commission*, **45**, 210-212.

Clark, C.W. and Mellinger, D.K. (1994) Application of Navy IUSS for whale research. *Journal of the Acoustical Society of America*, **96**, 3315, Abstract.

Clark, C.W. and Ellison, W.T. (1997) Low-frequency signaling behavior in mysticete whales. *Journal of the Acoustical Society of America*, **101**, 3163, Abstract.

Clay, C.S. and Medwin, H. (1977) *Acoustical Oceanography*. Wiley, New York.

Clutton-Brock, T.H. and Albon, S. (1979) The roaring of red deer and the evolution of honest advertising. *Behaviour*, **69**, 145-170.

Cornell University (1996) *Whales 1993*. CD prepared by Bioacoustics Research Program, Laboratory of Ornithology, Cornell University, Ithaca, New York.

Costa, D.P. (1993) The secret life of marine mammals. *Oceanography*, **6**, 120-128.

Cranford, T. W. (1999) The sperm whale's nose: sexual selection on a grand scale? *Marine Mammal Science* 15, 1133-1157.

Cranford, T.W., Amundin, M. and Norris, K.S. (1996) Functional morphology and homology in the odontocete nasal complex: Implications for sound generation. *Journal of Morphology*, **228**, 223-285.

Cummings, W.C. and Thompson, P.O. (1971) Gray whales, *Eschrichtius robustus*, avoid the underwater sounds of killer whales, *Orcinus orca. Fishery Bulletin*, **69**, 525-530.

Dahlheim, M.E. and Ljungblad, D.K. (1990) Preliminary hearing study on gray whales (*Eschrichtius robustus*) in the field. In: *Sensory abilities of cetaceans* (Ed. by J. Thomas and R. Kastelein), pp. 335-346. Plenum Press, New York.

Darwin, C. (1871) *The descent of man, and selection in relation to sex*. Murray, London.

Davies, N.B. and Halliday, T.R. (1978) Deep croaks and fighting assessment in toads. *Nature* (London), **274**, 683-685.

Davis, R.W. and Fargion, G.S. (1996) *Distribution and abundance of cetaceans in the north-central and western Gulf of Mexico*. New Orleans, Texas Institute of Oceanography and National Marine Fisheries Service, U.S . Dept. of the Interior. Minerals Mgmt. Service, Gulf of Mexico OCS Region, New Orleans, LA. 357pp.

Dawson, S. (1991) Clicks and communication: the behavioural and social contexts of Hector's dolphin vocalizations. *Ethology*, **88**, 265-276.

Ding, W., Wursig, B. and Evans, W. (1995) Comparisons of whistles among seven odontocete species. In: *Sensory systems of aquatic mammals* (Ed. by R. A. Kastelein, J. A. Thomas and P. E. Nachtigall), pp. 299-322. De Spil Publishers, Woerden.

Dormer, K. J. (1979) Mechanism of sound production and air recycling in delphinids: cineradiographic evidence. *Journal of the Acoustical Society of America*, **65**, 229-239.

Dunning, D.J., Ross, Q.E., Geoghegan, P., Reichle, J.J., Menezes, J.K., and Watson, J.K. (1992) Alewives avoid high-frequency sound. *North American Journal of Fisheries Management*, **12**, 407-416.

Dusenbery, D.B. (1992) *Sensory ecology: how organisms acquire and respond to information*. W.H.Freeman and Co., New York.

D'Vincent, C.G., Nilson R.M., and Hanna, R.E. (1985) Vocalization and coordinated feeding behavior of the humpback whale in southeastern Alaska. *Scientific Reports of the Whales Research Institute*, **36**, 41-47.

Eaton, R.L. (1979) A beluga whale imitates human speech. *Carnivore*, **2**, 22-23.

Edds-Walton, P.L. (1997) Acoustic communication signals of mysticete whales. *Bioacoustics*, **8**, 47-60.

Ellison, W.T., Clark, C.W. and Bishop, G.C. (1987) Potential use of surface reverberation by bowhead whales, *Balaena mysticetus*, in under-ice navigation: preliminary considerations. *Report of the International Whaling Commission*, **37**, 329-332.

Evans, W.E. and Powell, B.A. (1967) Discrimination of different metallic plates by an echolocating delphinid. *Proceedings of the Symposium Bionic Models of Animal Sonar*, **1**, 363-398.

Felgate, N.J. and Lloyd, L.J. (1997) The sea animal noise database system (SANDS). *European Research on Cetaceans*, **11**, 217.

Fish, J.F., and Vania, J.S. (1971) Killer whale, *Orcinus orca*, sounds repel white whales, *Delphinapterus leucas*. *Fishery Bulletin*, **69**, 531-535.

Ford, J.K.B. (1991) Vocal traditions among resident killer whales (*Orcinus orca*) in coastal waters of British Columbia (Canada). *Canadian Journal of Zoology*, **69**, 1454-1483.

Frankel, A.S., Clark, C.W., Herman, L.M. and Gabriele, C.M. (1995) Spatial distribution, habitat utilization, and social interactions of humpback whales, *Megaptera novaeangliae*, off Hawaii determined using acoustic and visual techniques. *Canadian Journal of Zoology*, **73**, 1134-1146.

Gagnon, G.J. and Clark, C.W. (1993) The use of the US Navy IUSS passive sonar to monitor the movements of blue whales. In: *Abstracts, Tenth Biennial Conference on the Biology of Marine Mammals, Galveston, Texas.*

George, J.C., Clark, C., Carroll, G.M. and Ellison, W.T. (1989) Observations on the ice-breaking and ice navigation behavior of migrating bowhead whales (*Balaena mysticetus*) near Point Barrow, Alaska, spring 1985. *Arctic*, **42**, 24-30.

Glockner, D.A. (1983) Determining the sex of humpback whales (*Megaptera novaeangliae*) in their natural environment. In: *Communication and behavior of whales.* (Ed. by R. Payne) Westview Press, Boulder.

Goodson, A.D. and Sturtivant, C.R. (1995) Sonar characteristics of the harbour porpoise, source levels and spectrum. *ICES Journal of Marine Science*, **53**(2), 465-472.

Gordon, J.C.D. (1987a) Sperm whale groups and social behaviour observed off Sri Lanka. *Report of the International Whaling Commission*, **37**, 205-217.

Gordon, J.C.D. (1987b) *The behaviour and ecology of sperm whales off Sri Lanka*. PhD Thesis, University of Cambridge.

Gordon, J.C.D. (1991) Evaluation of a method for determining the length of sperm whales (*Physeter catodon*) from their vocalizations. *Journal of Zoology* (London), **224**, 301-314.

Gordon, J. (1993) Acoustic assessment in sperm whales: a new explanation for sexual dimorphism in head size. *European Research on Cetaceans*, **7**, 85.

Gordon, J.C.D., Leaper, R., Hartley, F.G. and Chappell, O. (1992) Effects of whale watching vessels on the surface and underwater acoustic behaviour of sperm whales off Kaikoura, New Zealand. NZ Dept. *Conservation Science and Research Series*, **32**, New Zealand. 64pp.

Gouzoules, H. and Gouzoules, S. (1989) Body size effects on the acoustic structure of pigtail macaque (*Macaca nemestrina*) screams. *Ethology*, **85**, 324-334.

Green D.M., DeFerrari H.A., McFadden, D., Pearse, J.S., Popper, A.N., Richardson, W.J., Ridgway, S.H. and Tyack, P.L. (1994) *Low-frequency sound and marine mammals: current knowledge and research needs.* National Academy Press, Washington DC.

Guinee, L.N., Chu, K. and Dorsey, E.M. (1983) Changes over time in the songs of known humpback whales (*Megaptera novaeangliae*). In: *Communication and behavior of whales. AAAS Sel. Symp. 76.* (Ed. by R. Payne), pp. 59-80. Westview Press, Boulder.

Hall, J.D. and Johnson, C.S. (1972) Auditory thresholds of a killer whale *Orcinus orca* Linnaeus. *Journal of the Acoustical Society of America*, **51**, 515-517.

Harper, D.G.C. (1991) Communication. In: *Behavioural Ecology - An Evolutionary Approach* (Ed. by J.R. Krebs and N.B. Davies). Chapter 12. Blackwell Scientific Publications, Oxford.

Helweg, D.A., Frankel, A.S., Mobley Jr., J.R. and Herman, L.M. (1992) Humpback whale song: our current understanding. In: *Marine mammal sensory systems.* (Ed. by J.A. Thomas, R. Kastelein, A. and A.Y. Supin), pp. 459-483. Plenum, New York.

Herman, L.M. (1986) Cognition and language competencies of bottle-nosed dolphins. In: Dolphin cognition and behavior: a comparative approach. (Ed. by R.J. Schusterman, J.A. Thomas, and F.G. Wood), pp. 221-252. Erlbaum, Hillsdale NJ.

Herman, L.M. and Tavolga, W.N. (1980) The communication systems of cetaceans. In: *Cetacean behavior: mechanisms and functions* (Ed. by L.M. Herman), pp. 149-209. Wiley-Interscience, New York.

Herman, L.M., Richards, D.G. and Wolz, J.P. (1984) Comprehension of sentences by bottlenose dolphins. *Cognition*, **16**, 129-219.

Herman, L.M., Pack, A.A. and Wood, A.M. (1994) Bottlenose dolphins can generalize rules and develop abstract concepts. *Marine Mammal Science*, **10**, 70-80.

Hiby, A. and Lovell, P. (1989) Acoustic survey techniques for sperm whales. *International Whaling Commission Scientific Committee.* SC/41/Sp3 (unpublished).

Janik, V.M. (1997) Whistle matching in wild bottlenose dolphins. *Journal of the Acoustical Society of America*, **101**, 3136, Abstract.

Janik, J.M. (2000) Food-related bray calls in wild bottlenose dolphins (*Tursiops truncatus*). *Proceedings of the Royal Society of London B: Biological Sciences,* **267**, 923-927.

Janik, V.M., Dehnhardt, G. and Todt, D. (1994) Signature whistle variations in a bottlenose dolphin, *Tursiops truncatus. Behavioral Ecology and Sociobiology*, **35**, 243-248.

Janik, V.M., and Slater, P.J.B. (1997) Vocal learning in mammals. In: *Advances in the study of behavior, 26,* 59-99.

Janik, V.M. and. Slater, P.J.B. (1998) Context-specific use suggests that bottlenose dolphin signature whistles are cohesion calls. *Animal Behaviour*, **56**, 829-838.

Jefferson, T.A., Leatherwood, S. and Webber, M.A. (1993). *Marine Mammals of the World - FAO Identification Guide.* Rome: FAO and UNEP.

Johnson, C.S. (1967) Sound detection thresholds in marine mammals. In: *Marine bio-acoustics*, vol. 2 (Ed. by W.N. Tavolga), pp. 247-260. Pergamon. Oxford, UK.

Johnson, C.S., McManus, M.W. and Skaar, D. (1989) Masked tonal hearing thresholds in the beluga whale. *Journal of the Acoustical Society of America*, **85**, 2651-2654.

Kamminga, C. and Wiersma, H. (1981) Investigations on cetacean sonar II. Acoustical similarities and differences in odontocete sonar signals. *Aquatic Mammals*, **8**, 41-62.

Ketten, D.R. (1997) Structure and function in whale ears. *Bioacoustics*, **8**, 103-135.

Krebs, J.R. and Dawkins, R. (1984) Animal signals: mind-reading and manipulation. In: *Behavioural Ecology - An Evolutionary Approach*. Second edition (Ed. by J.R. Krebs and N.B. Davies), pp. 380-402. Blackwell Scientific Publications, Oxford.

Leonard, M.C. and Fenton, M.B. (1984) Echolocation calls of *Enderma maculata*. (Chiroptera: Vespertilionidae): Use in orientation and communication. *Journal of Mammalogy*, **65**, 122-126.

Levenson, C. (1974) Source level and bistatic target strength of the sperm whale (*Physeter catodon*) measured from an oceanographic aircraft. *Journal of the Acoustical Society of America*, **55**, 1100-1103.

Lien, J., Todd, S. and Guigne, J. (1990) Inferences about perception in large cetaceans, especially humpback whales, from incidental catches in fixed fishing gear, enhancement of nets by "Alarm" devices, and the acoustics of fishing gear. In: *Sensory abilities of cetaceans* (Ed. by J. Thomas and R. Kastelein), pp. 347-362, Plenum Press, New York.

Locke, J.L. (1993) *The child's path to spoken language*. Harvard University Press, Cambridge.

Lockyer, C. (1981) Estimates of growth and energy budget for the sperm whale, *Physeter catodon*. *FAO Fish. Ser. (5) Mammals in the Sea*. 3, 489-504.

Mann, D.A., Zhongmin, L., and Popper, A.N. (1997) A clupeid fish can detect ultrasound. *Nature* (London), **389**, 341.

Marten, K., Norris, K.S., Moore, P.W.B. and Englund, K.A. (1988) Loud impulse sounds in odontocete predation and social behavior. In: *Animal sonar: processes and performance*. (Ed. by P.E. Nachtigall and P.W.B. Moore), pp. 567-579. Plenum, New York.

Matthews, J.N., Rendell, L.E., Gordon, J.C.D. and Macdonald, D.W. (1999). A review of frequency and time parameters of cetacean tonal calls. *Bioacoustics*, **10**, 47-71.

McCoomb, K.E. (1991) Female choice for high roaring rates in deer, *Cervus elaphus*. *Animal Behaviour*, **41**, 79-88.

Medrano, L., Salinas, M., Salas, I., Ladron de Guevara, P., Aguayo, A., Jacobsen, J., and Baker, C.S. (1994) Sex identification of humpback whales, *Megaptera novaeangliae*, on the wintering grounds of the Mexican Pacific Ocean. *Canadian Journal of Zoology*, **72**, 1771-1774.

Miller, G.A. (1990) The place of language in a scientific psychology. *Psychological Science*, **1**, 7-14.

Miller, L. (1983) How insects detect and avoid bats. In: *Neuroethology and Behavioural Physiology* (Ed. by F. Huber and H. Markl), pp. 251-266. Springer Verlag, Berlin.

Miller, L.A., Pristed, J., Møhl, B. and Surlykke A. (1995) The click-sounds of narwhals (*Monodon monoceros*) in Inglefield Bay, Northwest Greenland. *Marine Mammal Science*, **11**(4), 491-502.

Møhl, B., Wahlberg, M., Madsen, P.T., and Miller, L.A. (2000) Sperm whale clicks; directionality and source level revisited. *Journal of the Acoustical Society of America*, **107**, 638-648.

Moore, P.W.B. (1997) Cetacean auditory psychophysics. *Bioacoustics*, **8**, 61-78.

Murchison, A.E. (1980) Detection range and range resolution of echolocating bottlenose porpoise (*Tursiops truncatus*). In: Animal sonar systems (Ed. by R.-G. Busnel and J.F. Fish), pp. 43-70. Plenum, New York.

Nestler, J.M., Ploskey, G.R., Pickens, J., Menezes, J. and Schilt, C. (1992) Responses of blueback herring to high-frequency sound and implications for reducing entrainment at hydropower dams. *North American Journal of Fisheries Management*, **12**, 667-83.

Norris, K.S. (1967) Some observations on the migration and orientation of marine mammals. In: *Animal orientation and navigation*. (Ed. by R.M. Storm), pp. 101-125. Oregon State University Press, Corvallis.

Norris, K.S. (1969) The echolocation of marine mammals. In: *The biology of marine mammals*. (Ed. by H.T. Andersen), pp. 391-423. Academic Press, New York.

Norris, K.S. and Harvey, G.W. (1972) A theory for the function of the spermaceti organ of the sperm whale. In: *Animal Orientation and Navigation* (Ed. by S.R. Galler, K. Schmidt-Koenig, G.J. Jacobs and R.E. Belleville), pp. 397-417, NASA Special Publication.

Norris, K.S. and Dohl, T.P. (1980) Behavior of the Hawaiian spinner dolphin, *Stenella longirostris*. *Fishery Bulletin*, **77**, 821-850.

Norris, K.S. and Møhl, B. (1983) Can odontocetes debilitate prey with sound? *American Naturalist*, **122**, 85-104.

Norris, K.S., Prescott, J.H., Asa Dorian, P.V. and Perkins, P. (1961) An experimental demonstration of echolocation behaviour in the porpoise, *Tursiops truncatus* (Montagu). *Biological Bulletin*, **120**, 163-176.

Norris, K.S., Dormer, K.J., Pegg, J. and Liese G.T. (1971) The mechanism of sound production and air recycling in porpoises: A preliminary report. In: *VII Conf. Biol. Sonar Diving Mammals, Menlo Park, CA*. Stanford Research Institute.

Palsbøll, P.J., Vader, A., Bakke, I. and El-Gewely, M.R. (1992) Determination of gender in cetaceans by the polymerase chain reaction. *Canadian Journal of Zoology*, **70**, 2166-2170.

Pavan, G., Fossati, C., Manghi, M. and Priano, M. (1998) Acoustic measure of body growth in a photo-identified sperm whale. In: *The World Marine Mammal Science Conference, abstracts*. Monaco.

Payne, R. (1995) *Among whales*. Scribner, New York, London. 431pp.

Payne, R.S. and McVay, S. (1971) Songs of humpback whales. *Science*, **173**, 587-597.

Payne, R. and Webb, D. (1971) Orientation by means of long range acoustic signalling in baleen whales. *Annals of the New York Academy of Sciences*, **188**, 110-141.

Payne, R. and Guinee, L.N. (1983) Humpback whale song as an indicator of stocks. In: *Communication and behavior of whales. AAAS Sel. Symp. 76*. (Ed. by R. Payne), pp. 371-445. Westview Press, Boulder.

Payne, K.B., Tyack, P., and Payne, R.S. (1983) Progressive changes in the songs of humpback whales. In: *Communication and behavior of whales. AAAS Sel. Symp. 76*. (Ed. by R. Payne), pp. 9-59. Westview Press, Boulder.

Payne, R. and Payne, K. (1985) Large scale changes over 19 years in songs of humpback whales in Bermuda. *Zeitschrift für Tierpsychologie*, **68**, 89-114.

Pepperberg, I.M. (1996) Effect of avian-human joint attention on allospecific vocal learning by grey parrots (*Psittacus erithacus*). *Journal of Comparative Psychology*, **110**, 286-297.

Pinker, S. (1989) *Learnability and cognition*. MIT Press, Cambridge, MA.

Pinker, S. (1994) *The language instinct*. Morrow, New York.

Popper, A.N. (1980) Sound emission and detection by delphinids. In: *Cetacean behavior: mechanisms and functions*. (Ed. by L. Herman), pp. 1-52. Wiley Interscience, New York.

Ralls, K., Fiorelli, P. and Gish, S. (1985) Vocalizations and vocal mimicry in captive harbor seals, *Phoca vitulina*. *Canadian Journal of Zoology*, **63**,1050-1056.

Ranft, R.D. (1997) Bioacoustic libraries and recordings of aquatic animals. *Marine and Freshwater Behaviour And Physiology*, **29**, 251-262

Reidenberg, J. and Laitmann, J. (1988) Existence of vocal folds in the larynx of odontoceti (toothed whales). *Anatomical Record*, **221**, 886-891.

Rendell, L.E., Matthews, J.N., Gill, A., Gordon, J.C.D. and Macdonald, D.W. (1999) Quantitative analysis of tonal calls from five odontocete species, examining interspecific and intraspecific variation. *Journal of Zoology*, **249**, 403-410.

Richards, D.G., Wolz, J.P. and Herman, L.M. (1984) Vocal mimicry of computer generated sounds and vocal labelling of objects by a bottlenosed dolphin, *Tursiops truncatus*. *Journal of Comparative Psychology*, **98**, 10-28.

Richardson, W.J., Greene, C.R. Jr., Malme, C.I. and Thomson, D.H. (1995) *Marine Mammals and Noise.*, Academic Press, San Diego, CA.

Ridgway, S.H., Carder, D.A. and M.M. Jeffries (1985) Another "talking" male white whale. *Abstracts: Sixth Biennial Conference on the Biology of Marine Mammals*, p. 67.

Roeder, K.D. (1967) Turning tendency of moths exposed to ultrasound while in stationary flight. *Journal of Insect Physiology*, **13**, 873-888.

Sayigh L.S.,Tyack P.L.,Wells R.S. and Scott M.D. (1990) Signature whistles of free-ranging bottlenose dolphins, *Tursiops truncatus*: stability and mother-offspring comparisons. *Behavioral Ecology and Sociobiology*, **26**, 247-260.

Sayigh, L.S., Tyack, P.L., Wells, R.S., Scott, M.D. and Irvine, A.B. (1995) Sex difference in signature whistle production of free-ranging bottlenose dolphins, *Tursiops truncatus*. *Behavioral Ecology and Sociobiology*, **36**, 171-177.

Schusterman, R.J. and Gisiner, R. (1986) Animal language research: marine mammals reenter the controversy. In: *Intelligence and evolutionary biology.* (Ed. by H.J. Jerison. and I. Jerison), pp. 319-350. Springer Verlag, Berlin.

Shimura, E., Numachi, K.I., Sezaki, K., Hirosaki, Y., Watabe, S. and Hashimoto, K. (1986) Biochemical evidence of hybrid formation between two species of dolphin *Tursiops truncatus* and *Grampus griseus*. *Bulletin of the Japanese Society of Scientific Fisheries*, **52**, 725-730.

Shockley R.C., Northrop J., Hansen P.G., Hartdegen C. (1982) SOFAR propagation paths from Australia to Bermuda: comparisons of signal speed algorithms and experiments. *Journal of the Acoustical Society of America*, **71**, 51-60.

Sivian, L.J. and White, S.D. (1933) On minimum audible sound fields. *Journal of the Acoustical Society of America*, **4**, 288-321.

Smolker, R.A. (1993) *Acoustic communication in bottlenose dolphins.* Ph.D. thesis, University of Michigan, Ann Arbor.

Spilliaert, R., Vikingsson, G., Arnason, U., Palsdottir, A., Sigurjonsson, J. and Arnason, A. (1991) Species hybridization between a female blue whale (*Balaenoptera musculus*) and a male fin whale (*Balaenoptera physalus*): Molecular and morphological documentation. *Journal of Heredity*, **82**, 269-274.

Surlykke, A. and Miller, L.A. (1985) The influence of arctiid moth clicks on bat echolocation: Jamming or warning? *Journal of Comparative Physiology A Sensory Neural and Behavioral Physiology*, **156**, 831-844.

Thomas, J.T., Chun, N., Au, W.W.L. and Pugh, K. (1988) Detection abilities and signal characteristics of echolocating false killer whale (*Pseudorca crassidens*). In: *Animal Sonar: Processes and Performance* (Ed. by P. Natchigall and P.W.B. Moore). Plenum Press, New York and London.

Thompson T.J., Winn H.E., Perkins P.J. (1979) Mysticete sounds. In: *Behavior of marine animals.* (Ed by H.E. Winn and B.L. Olla). vol 3, pp. 403-431. Plenum, New York.

Turl, C.W. and Penner, R.H. (1989) Differences in echolocation click patterns of the beluga *(Delphinapterus leucas)* and the bottlenosed dolphin *(Tursiops truncatus). Journal of the Acoustical Society of America,* **86** (2), 497-502.

Tyack, P. (1981) Interactions between singing Hawaiian humpback whales and conspecifics nearby. *Behavioral Ecology and. Sociobiology,* **8,** 105-116.

Tyack, P. (1982) *Humpback whales respond to sounds of their neighbors.* Ph.D. thesis, Rockefeller University, New York.

Tyack, P. (1983) Differential response of humpback whales to playbacks of song or social sounds. *Behavioral Ecology and Sociobiology,* **13,** 49-55.

Tyack, P. (1986) Population biology, social behaviour and communication in whales and dolphins. *Trends in Ecology and Evolution,* **1,** 144-150.

Tyack, P. (1997) Studying how cetaceans use sound to explore their environment. *Perspectives in Ethology,* **12,** 251-297.

Tyack, P.L. and Sayigh, L.S. (1997) Vocal learning in cetaceans. In: *Social Influences on Vocal Development* (Ed. by C. Snowdon and M. Hausberger), pp. 208-233. Cambridge University Press, Cambridge.

Urick, R.J. (1983) *Principles of underwater sound.* 3rd edition. McGraw-Hill, New York.

Urick, R.J. (1986) *Ambient noise in the sea.* Peninsula Publishing, Los Altos, CA.

Watkins, W.A. (1980a) Acoustics and the behavior of sperm whales. In: *Animal Sonar Systems* (Ed. by R.-G. Busnel and J.F. Fish), pp. 283-290. Plenum Press, New York and London.

Watkins, W.A. (1980b) Click Sounds from Animals at Sea. In: *Animal Sonar Systems* (Ed. by R.-G. Busnel and J.F. Fish), pp. 291-297. Plenum Press, New York.

Watkins, W.A. (1994) The WHOI database. In: *Underwater Animal Sounds Seminar 31 January - 1 February, Farnborough, Hants.* Underwater Systems Business Sector Defence Research Agency.

Watkins, W.A. and Schevill, W.E. (1977) Sperm whale codas. *Journal of the Acoustical Society of America,* **62,** 1485-1490.

Watkins, W.A., Tyack, P., Moore, K.E. and Bird, J.E. (1987) The 20-Hz signals of finback whales *(Balaenoptera physalus). Journal of the Acoustical Society of America,* **82,** 1901-1912.

Watkins, W.A. and Wartzok, D. (1985) Sensory biophysics of marine mammals. *Marine Mammal Science,* **1,** 219-260.

Weilgart, L., Whitehead, H. and Payne, K. (1996) A colossal convergence. *American Scientist,* **84,** 278-287.

Wells R.S., Scott M.D. and Irvine A.B. (1987) The social structure of free-ranging bottlenose dolphins. *Current Mammalogy,* **1,** 247-305.

Wenz, G.M. (1962) Acoustic ambient noise in the ocean: Spectra and sources. *Journal of the Acoustical Society of America,* **34,** 1936-1956.

White, M.J., Jr., Norris, J.C., Ljungblad, J., Baron, K. and di Sciara, G. (1978) Auditory thresholds of the two beluga whales *(Delphinapterus leucas).* HSWRI Tech. Rep. 78-109. Report from Hubbs/Sea World Res. Inst., San Diego, CA for US Naval Ocean Systems Cent,. San Diego, CA. 35pp.

Whitehead, H., and Weilgart, L. (1991) Patterns of visually observable behaviour and vocalisations in groups of female sperm whales. *Behaviour,* **118,** 275-296.

Xitco, M.J. Jr. and Roitblat, H.L. (1996) Object recognition through eavesdropping: passive echolocation in bottlenose dolphins. *Animal Learning and Behavior,* **24,** 355-365.

Zagaeski, M. (1987) Some observations on the prey stunning hypothesis. *Marine Mammal Science,* **3**(4), 275-283.

Chapter 5

Behavioural Ecology of Cetaceans

[1,2]JAMES R. BORAN, [1,3]PETER G.H. EVANS and [1,2]MARTIN J. ROSEN
[1]*Sea Watch Foundation, 11 Jersey Road, Oxford, OX4 4RT, UK;* [2]*Whalesense, 47 Parkhill Avenue, Manchester, M8 4GZ, UK;* [3]*Department of Zoology, University of Oxford, South Parks Road, Oxford, OX1 3PS, UK. E-mail (JRB):* jim.boran@umist.ac.uk

1. INTRODUCTION

The study of behavioural ecology examines the ways in which a species maximises its survival (and ultimately, its reproductive success) through optimal strategies of behaviour. It asks questions about the <u>functions</u> of behaviours, rather than describing behavioural mechanics, which is the domain of classical ethology (Krebs and Davies, 1991). In general, it recognises the close linking between a species' environment and its behaviour.

Most, if not all, of the basic tenets of mammalian behavioural ecology have been developed from studies of terrestrial mammals. For example, the relationship between the distribution of resources and an animal's mating system (Emlen and Oring, 1977) was based on studies where the fixed distribution of resources, such as fruits on a tree, could be easily quantified.

This chapter will examine relevant research on the behaviour of whales, dolphins and porpoises, order *Cetacea*, collectively known as cetaceans, in the light of some principles of behavioural ecology. Behavioural ecology can be examined in two broad categories: foraging strategies and mating strategies. These two strategies are ultimately closely intertwined. Evolutionary theory predicts that females will focus their efforts on carefully choosing a successful mate and then work to ensure the survival of an offspring in which they have a large investment. Males, on the other hand, will compete with each other to get a mate, but have a low investment in each mating; they are best served by mating as many times as possible

(Darwin, 1859; Bateman, 1948; Fisher, 1958). It has even been suggested that females are the more "ecological" sex, since they distribute themselves in relation to the distribution of resources to maximise the food they can provide for their young, while males tend to distribute themselves around the females in order to maximise their number of matings (Wrangham, 1980; Gaulin and Sailer, 1985). Both males and females attempt to maximise their reproductive success and inclusive fitness so that their genes are represented in future generations (Hamilton, 1964; Dawkins, 1976).

These basic principles can provide a framework for a review of the behavioural ecology of the order *Cetacea*. The understanding of cetacean behavioural ecology is still relatively limited, in part because this requires knowledge of two traditionally distinct disciplines: mammalogy and marine biology.

Cetaceans must first be understood in the light of their terrestrial mammalian ancestry which gives them physiological and anatomical constraints (*e.g.* the need to surface to breathe air, the need to nurse their young, *etc.*). On the other hand, cetaceans are unlike other mammals (and in some ways more like fish) in that they have completely adapted to life in a complex, three-dimensional aquatic environment with apparently few barriers to movement, where resource distribution is often poorly understood (at least by researchers) and where signals are more often vocal rather than visual.

Detailed studies of cetacean behaviour have been difficult because many species live far from land and because behavioural observations have previously been limited to surface activities, which may represent only a small proportion of a cetacean's life. The few underwater observation studies have primarily been conducted in warm, clear tropical waters, and have thus only been applicable for a few species. However, new technologies such as satellite tracking, depth-of-dive tags, and physiological sensors are improving our ability to collect information on a wider variety of species in a wider variety of habitats.

We are beginning to learn many fascinating examples of cetacean behavioural ecology. By applying the principles of behavioural ecology from studies of terrestrial mammals to our understanding of cetaceans we can gain further insight. In general, cetaceans offer exciting new insights into the ways in which habitat affects mammalian behavioural ecology.

2.   SELECTIVE PRESSURES ON CETACEANS

Before examining specific examples of cetacean behavioural ecology, it is helpful to consider the selective pressures which have shaped cetacean evolution. Cetaceans are mammals which have completely adapted to an aquatic lifestyle. Some species are linked to shorelines and shallow water, while others have colonised the deepest ocean. Most species live in marine waters, and the few species that still live in freshwater rivers (*e.g.* the river dolphins) appear to have re-colonised fresh water as an evolutionary development. However, all cetaceans share a number of selective pressures which are common to the aquatic environment.

Figure 1. The massive body size of the blue whale could only have evolved in the buoyant marine environment. (Photo: A. Aguilar)

First, simply by living in an aquatic three-dimensional environment, they have developed specialisations which for mammals are paralleled only in the bats (Norris and Schilt, 1988). The evolution of body structure has different limits for animals living in the denser medium of water than it does for those evolving on land. Life in water, where bodies are buoyant, has reduced the effects of gravity on cetacean body design (Economos, 1983). This has freed cetaceans from the structural limits on body size. Limbs are not needed for support, and so cetaceans have lost their pelvis through evolution. On the other hand, streamlining of body design for hydrodynamics has placed much greater restriction on the development of any external structures (Fish and Hui, 1991). Vision has also become less useful in certain aquatic environments (*e.g.* almost blind river dolphins, *Platanista* sp. in the turbid waters of South American rivers), but acoustic senses (the development of echolocation used by the toothed whales) have replaced this.

Second, although aquatic habitats vary greatly in physiographic complexity, ranging from complex riverine habitats to more open coastlines and islands and finally to relatively featureless pelagic habitats, the structure of the marine habitat in general is more open than are terrestrial habitats. The ocean places fewer limits to dispersal and thus there are fewer geographical boundaries which serve as isolating barriers within which species can evolve independently, as they do in terrestrial habitats on offshore islands (Darwin, 1859; MacArthur and Wilson, 1967). Of course, there may be other, less obvious boundaries in the ocean. Water temperature gradients appear to be one common barrier to dispersal, for example, in the differing distributions of the temperate long-finned pilot whale and the tropical short-finned pilot whale (van Bree *et al.*, 1978), or in the open ocean habitats of dolphins in the eastern tropical Pacific (Au and Perryman, 1985). Another example is a tidal boundary which killer whales use to divide up their habitat between two adjacent social groups (Felleman *et al.*, 1991). The more open aquatic habitat also dampens large seasonal variability and keeps most changes gradual. Some exceptions are fluctuating ice cover in the polar regions which can restrict entire seas due to the lack of breathing holes in the ice, and seasonal floods in rivers which increase habitat and disperse prey for river dolphins.

Third, the openness of the oceanic habitat also allows freer dispersal of resources, resulting in a patchy distribution. Cetaceans usually have a highly mobile lifestyle to locate these food patches.

3.   TRENDS IN CETACEAN ECOLOGY

The eighty or so species of cetaceans currently known have specialised to exploit a wide array of trophic niches. There are a number of basic organising principles or trends which have been proposed to explain cetacean diversity in ecological adaptation. The most obvious source of variation in cetaceans is body size: they range from the 30 m long, 150,000 kg blue whale, the largest animal ever evolved, to the 1.1 m, 40 kg Hector's dolphin.

Much of the general variability in cetacean anatomy, life history patterns, and behaviour has been related to the four main types of food selected by a species: whether small (often planktonic) invertebrates, fish, cephalopods (primarily squid), or other marine mammals (Slijper, 1979; Gaskin, 1982; Evans, 1987).

Planktonic invertebrates are often unpredictably dispersed in patchy clumps, are of intermediate caloric value (more than squid, but less than fish), and are generally more abundant in polar and temperate waters than in tropical waters. They are the primary food of many of the baleen whales, and long migrations appear to be based in part on locating patchy concentrations. Different types of prey selection are suggested by the different feeding methods of baleen whales - skimming, gulping or bottom feeding (Nemoto, 1959; Gaskin, 1982). The slow-moving skim feeding for plankton is exemplified by right and bowhead whales (family *Balaenidae*), where an area is slowly searched with the mouth open (Mayo and Marx, 1990). Gulpers, such as the streamlined baleen whales (family *Balaenopteridae*), pursue larger quantities of planktonic prey as well as small fish. The bottom feeding gray whale specialises on filtering benthic invertebrates out of the sea bed (Nerini, 1984). In general, species which primarily feed on plankton occur either solitarily or in small groups because they cannot find sufficient planktonic resources in a patch to support the energy demands of a large group of these large-bodied whales in their habits, whether schooling or non-schooling, migratory or sedentary. In general, the higher caloric value of fish allows fish-feeding cetaceans to occasionally aggregate into larger groups and occasionally develop co-operative feeding methods: for fish-feeding baleen whales this is exemplified by the humpback whale. Fish are also fed upon by most of the smaller, social dolphins, but relatively solitary harbour porpoise and river dolphins are exceptions to this trend. Fish feeding also seems to affect a number of life history patterns in odontocetes, with shorter gestations, shorter lactation periods and lower juvenile growth rates being found in fish-feeding odontocetes compared to those that feed on squid (see Evans and Stirling, this volume).

Squid (and other cephalopods) are a widespread, abundant prey with a low caloric value. They show similar variability in habits as do fish, but they are likely to be more difficult to catch because of the greater difficulty that echolocating odontocetes may have in detecting their soft bodies with sonar. Most of the larger species of odontocetes specialise on squid feeding: for example, the sperm whale, the various beaked whale species, and the pilot whales. All share a well-developed melon, probably to "fine-tune" their sonar, and possess reduced dentition (it is hypothesised that they feed primarily by suction).

Only killer whales and (occasionally) false killer and pygmy killer whales feed on marine mammals (Perryman and Foster, 1980; Jefferson *et al.*, 1991). The foraging specialisations of marine mammal-feeding killer whales are described below.

Another trend which has been proposed for odontocetes has been the effects of habitat on social organisation (Würsig, 1979; Wells *et al.*, 1980; Tyack, 1986; Heimlich-Boran, 1993). Some of these trends may be related to the prey selection mentioned above. Group size appears to vary over three different habitat types: riverine, coastal and offshore. River-dwelling species are most commonly seen in small groups of two to ten; coastal species, such as *Tursiops*, travel in intermediate sized groups of up to 50 animals; and pelagic dolphins, like *Stenella*, can be found in groups of hundreds or even thousands (Wells *et al.*, 1980).

Finally, it is likely that predation has a role in affecting variability in cetacean ecology (Norris and Dohl, 1980; Norris and Schilt, 1988), but this is not as well understood as the previously discussed effects of food type and habitat structure. Studies of terrestrial mammals have shown that predation rates decrease with increasing prey group size because prey in a large group have a better chance of detecting, avoiding, deterring or confusing the predator (Bertram 1978). Killer whales and sharks are known to prey on cetaceans, but exact mortality rates due to predation are generally not known, let alone the ways in which predation mortality might vary with group size. Anecdotal evidence suggesting that dolphins are aware of predators is provided by dusky dolphins 'hiding' in bays when killer whales are nearby (Würsig and Würsig, 1980) and observations of spinner dolphin schools, under stress, grouping up with younger animals in the centre and larger adult males on the perimeter (Pryor and Schallenberger, 1991). Young cetaceans are likely to be the most susceptible to predation (Norris and Dohl, 1980), and this vigilance suggests that cetaceans could form groups in order to reduce predation.

These are just a few examples of broad trends which have been proposed to explain cetacean variability. We will now review cetacean behavioural ecology through examples of strategies for optimal foraging and mating, for some of the well-studied species.

4.   THE BEHAVIOURAL ECOLOGY OF CETACEAN FEEDING

The study of the behavioural ecology of feeding examines the interaction between a species' behaviour and its habitat, within the constraints of its evolutionary history. The study of cetacean behavioural ecology requires characterisations of preferred prey distribution, preferred habitat features (*e.g.* bathymetry, temperature, currents, *etc.*), and the distribution of predators, competitors, and potential mates.

It is often difficult to categorise all of these variables in cetacean field studies. In terrestrial situations, the classic foraging studies in the wild have carefully monitored the steps of prey selection, the energetics of capture, and the final caloric intake. This is often backed up with experimental studies on captive animals to quantify specific aspects of foraging. For cetaceans, these sorts of studies have been problematic, because of the difficulties of observing underwater prey and the limited opportunities for captive research. Stomach contents from dead animals can document to some extent which species are preyed upon (Fitch and Brownell, 1965; Bowen and Siniff, 1999), but the ways the prey are caught must be inferred. Many observational studies of cetacean foraging have been based on high levels of inference that feeding was actually occurring. The behavioural categorisation of cetacean behaviour often includes a category termed "feeding", which on closer inspection, simply records close proximity of the study subject and prey, and not actual prey ingestion. Additional methods have used "non-directional" behaviour, on the implication that the animals are pursuing prey during these times. These shortcomings are often unavoidable and we can only advise care in interpreting the results of such studies.

The following are some examples of the ways in which cetaceans adapt their behaviour to local prey distribution and habitat characteristics. These foraging strategies have developed in conjunction with the predation avoidance mechanisms of the prey and thus represent a long history of co-evolution (Endler, 1991), but we are only beginning to understand the full basis of these relationships.

4.1 Resource Partitioning in Balaenopterid Whales

The rorquals of the genus *Balaenoptera* include five species of baleen whale built on a very similar body plan, differing primarily in body size (Ridgway and Harrison, 1985). They range from the 27 m blue whale (*Balaenoptera musculus*), the 22 m fin whale (*B. physalus*), the 16 m sei whale (*B. borealis*) and the 14.5 m Bryde's whale (*B. edeni*) down to the 10 m minke whale (*B. acutorostrata*). Comparative studies where more than one species have been observed feeding in the same area provide interesting insights into their behavioural ecology.

One such study was conducted in the Gulf of California off Mexico (Tershy *et al.*, 1990; Tershy, 1992; see Fig. 2). Four species were observed here: primarily Bryde's and fin, in order of decreasing abundance, but also a few rare sightings of minke and blue whales. Bryde's and fin whales appeared to divide up the habitat by temperature, with Bryde's whale occurrence being greater during the warmer summer and autumn months (sea surface temperature 20-28°C) and fin whale occurrence greater during the colder winter and spring months (temperature 15-22°C).

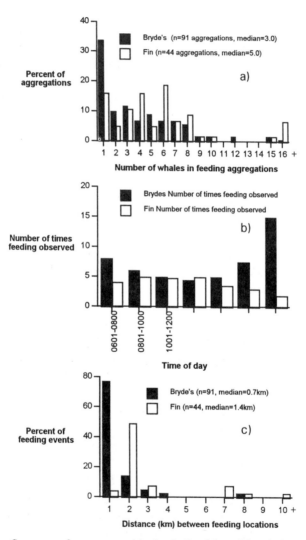

Figure 2. Summary of resource partitioning in Bryde's and fin whales: a) Bryde's occur in smaller feeding aggregations than fin whales, b) Bryde's show more marked variability in morning and evening feeding, and c) Bryde's move less between feeding locations. (Adapted from Tershy, 1992).

Bryde's and fin whales also divided the habitat by prey selection: Bryde's fed primarily on fish and fin whales fed only on planktonic euphausiids. Bryde's whales fed for a greater proportion of their time, and in smaller groups, than did fin whales and this may have been because the larger, faster, higher caloric content fish prey of Bryde's whales occurred in smaller, more predictable patches than did the euphausiid fin whale prey (Tershy, 1992). This relationship between the large fin whale and smaller Bryde's whale corresponded to the proposed relationship for different species of terrestrial antelope, in which smaller species were feeding on small, predictable, high quality food patches, while larger species were able to feed on larger patches of less predictable, lower quality food because of their lower metabolic rates and longer retention times (Jarman, 1974; Tershy, 1992). Although blue whale and minke whale were less common in this ecosystem, their preferred prey (euphausiids for the large blue whale and fish for the much smaller minke whale) fit them into this general pattern.

Another ecosystem with multiple species of baleen whales has been described in Witless Bay, Newfoundland (Piatt *et al.*, 1989). Humpback, minke and fin whales (in descending order of abundance) co-occur here, and seasonal variations in their numbers have been related to the variable abundance of their preferred prey, capelin (*Mallotus villosus*). In this case, the whales apparently did not divide up the habitat by differential prey selection, but by habitat use. Fin whales were primarily found in offshore (>70 m) areas, minke whales in nearshore (<30 m) areas and humpback whales occurred in mid-water (30-70 m) areas (Piatt *et al.*, 1989). Capelin appeared to occur in sufficient abundance to support these species and it was estimated that the large breeding population of seabirds in the region took more capelin in a day than did all of the whales in an entire summer season (Piatt *et al.*, 1989). However, this example may not be generally applicable; all three species co-occur in a small area of the Gulf of Maine where there does not appear to be a consistent pattern of resource partitioning amongst these closely-related species.

4.2 Variability in Humpback Whale Feeding Group Size

Humpback whales vary their group size depending on the types of prey. Humpbacks are the one of the most social of the baleen whales (Würsig, 1988). The species has developed elaborate feeding mechanisms which rely upon a flexible diet. Humpbacks feed primarily on small, mid-water fish and euphausiid crustaceans, resources which are patchily distributed and mobile, and which concentrate in areas of high productivity, such as coastal upwellings and frontal zones. The whales clearly have the ability to locate these high density patches of prey.

Figure 3.   Although humpback whales occasionally travel in large feeding groups, lone animals are much more common. (Photo: S. Kraus)

Optimal foraging group sizes vary depending on the size of prey concentrations and range from lone animals to a maximum recorded group size of 18 (Clapham, 1993). Humpbacks exhibit a form of fission-fusion society (as defined by Kummer, 1971), adapting their group size to the amount of available prey. In some parts of the Gulf of Maine, large groups (of six or more) were quite rare (1% of all groups), but 81% of them were associated with surface feeding groups (Clapham, 1993). This variability in feeding group size in the region has been related to the large degree of variability in group size of the favoured prey - small fish such as the American sand lance (*Ammodytes americanus*) (Whitehead, 1983). Off Alaska, where consistently large schools of herring are the favoured prey, some humpback group sizes are large and have a stable composition (Perry *et al.*, 1990). In contrast, when feeding on krill, the most common group size is small and group composition is variable (Perry *et al.*, 1990). Another factor which has been suggested to affect cetacean group size is predation (Norris and Dohl, 1980), on the principle that predation encourages prey to occur in large groups for protection and dilution of danger. For example, low humpback whale group sizes in the Gulf of Maine have been related to the virtual absence of killer whales in the area (Clapham, 1993).

4.3 Gray Whale Foraging Strategies

The gray whale, *Eschrichtius robustus*, is the only baleen whale that specialises on benthic invertebrate prey (Pike, 1962). The species may also feed in the water column or at the surface on a variety of other prey, but its primary habitats are the rich benthic communities found in and above the soft-bottomed sediment of the Bering and Chukchi seas (Nerini, 1984; Moore *et al.*, 1986; Kim and Oliver, 1989). The whales excavate their prey by sucking sediment into their mouth and then filtering out the mud. This prey is sessile and thus represents a stable food resource to which the whales can repeatedly return.

In the wake of feeding on benthic prey, gray whales leave extensive pits on the sea bed (Nerini, 1984). These pits are approximately one by two metres in area and 0.5 m deep. The steep sides open up new habitat for colonisation by additional benthic invertebrates. In this sense, gray whale feeding can increase the prey productivity in an area in a similar manner to a farmer tilling soil (Nerini and Oliver, 1983). It has been estimated that gray whales may turn over from 9-27% of the northern Bering Sea benthos each year (Nerini, 1984). However, in some cases, the whales may harvest all prey in an area and destroy future feeding prospects (Darling *et al.*, 1998).

This foraging strategy is a unique example of how a cetacean has developed behaviours to exploit a normally inaccessible resource. However, gray whales have still maintained a variety of other foraging strategies which appear to be more appropriate to other parts of their range. Since they undertake one of the longest migrations (Swartz, 1986), they must adapt to a variety of habitats. While the Bering and Chukchi seas are the prime areas for benthic feeding (and which gray whales must share with benthic feeding walrus), prey in the water column appear to be more important further south (Kim and Oliver, 1989). In a 26-yr study off Vancouver Island, British Columbia (Darling *et al.*, 1998), gray whales fed on a shifting variety of either benthic or pelagic prey in specific areas or during any given year, but over longer time periods were very consistent in their patterns of habitat use. This demonstrates that these areas between the extremes of the migration are important feeding habitats. In the vicinity of the Baja California breeding lagoons there seems to be very few suitable habitats for benthic feeding (Oliver *et al.*, 1982) and any feeding probably occurs on planktonic prey (Norris *et al.*, 1983).

In summary, gray whale foraging strategies are unique in having specialised feeding on benthic invertebrates, but have also incorporated a large degree of flexibility to feed on other prey (*e.g.* pelagic) when necessary. In fact, this flexibility, including predictable benthic, inshore resources, has been suggested to be a strong contribution as to why gray whales have exhibited such a strong recovery from population loss due to whaling (Swartz, 1986).

4.4   Minke Whale Foraging Strategies

Minke whales in the inland waters of Washington, USA have developed preferred habitats in which they also used preferred foraging specialisations (Dorsey, 1983; Hoelzel *et al.*, 1989). This differed from foraging studies of most terrestrial mammals where foraging specialisations are usually shared by all members of a species. The study identified 23 individuals on the basis of distinctive scars and pigment patterns over a five-year period. There were two types of foraging observed: lunge feeding and "bird association" feeding. Although primarily distinguished by the presence or absence of feeding seabirds (principally gulls and auks), the two feeding methods were also distinguished by respiration intervals. During lunge feeding, animals had longer pre-feeding dives and shorter feeding respiration intervals than did those feeding in association with bird aggregations.

The conclusion was that "bird association" feeding whales were exploiting ephemeral concentrations of fish presumably already corralled by other predators. The whales surfaced under these feeding flocks, scattering the birds and eating the fish. Alternatively, lunge feeding appeared to involve the whales actively pursuing and collecting the fish schools against the air/water surface interface. The longer pre-feeding dives indicated that there was a greater energy expenditure for the whales to collect these fish. However, observed feeding rates were higher for lunge feeders than for "bird association" feeders.

Each foraging method was specific to two sites within a 45 km area: 96% of all lunge foraging occurred in the northern area, while 91% of all bird-association foraging was in the southern part of the area. In one part of the northern area, where both types of foraging were recorded, the five individuals observed feeding more than once still significantly preferred a specific strategy (Hoelzel *et al.*, 1989). Thus, the foraging method was more determined by individual preference than it was by the environmental characteristics of the particular area. The conclusion was that individual minke whales learned different behaviours to exploit two types of patchily distributed prey: ephemeral groups of fish concentrated at the surface by birds, and more predictable fish in sub-surface schools which required active collection (Hoelzel *et al.*, 1989).

Figure 4. When engaged in lunge-feeding, minke whales may also be surrounded by flocks of seabirds like these kittiwakes, eager to exploit a ready source of food at the surface. (Photo: P.G.H. Evans)

4.5 Prey Selection in Killer Whales

Killer whales (*Orcinus orca*) are specialised for feeding on large prey, including pinnipeds and other cetaceans. Large body size (they are the largest member of the dolphin family), large jaws and large teeth help them to immobilise large prey, while large fins and flippers give them stability for high speed pursuit (Eschricht, 1866). Killer whale fossils date back relatively early in delphinid evolution, suggesting they developed large body size early on in their evolution (Heyning and Dahlheim, 1988). Some related species (*e.g.* the false killer whale, *Pseudorca crassidens*) have also been observed killing marine mammal prey (Perryman and Foster, 1980), but these appear to be rare incidents.

Killer whales have been studied in a wide range of habitats. Some populations appear to specialise in marine mammal predation (*e.g.* Argentina, Crozet Islands, Antarctica), while other populations specialise on fish (*e.g.* Norway, for at least part of the year). Off British Columbia,

Canada and Washington, USA, two distinct populations co-exist in the same region: one specialising on fish, and the other on marine mammals, such as pinnipeds (Felleman *et al.*, 1991). These populations have been studied since 1973, and all individuals have been identified by marks and scars on the dorsal fin and back. Some of these have been seen in the region since the 1960s (Bigg and Wolman, 1975; Ford *et al.*, 1994) and have had many years of experience in using this area.

The two forms differ in a wide variety of behavioural aspects of distribution, seasonal occurrence, acoustic dialects, and prey choice (Ford and Fisher, 1983; J.R. Heimlich-Boran, 1988; Ford, 1989; Morton, 1990; Felleman *et al.*, 1991; Baird *et al.*, 1992), as well as genetic aspects of pigmentation patterns, dorsal fin morphology and mitochondrial DNA (Duffield, 1986; Baird and Stacey, 1988; Hoelzel and Dover, 1990; Hoelzel, 1991a; Hoelzel *et al.*, 1998), indicating that they probably represent distinct races of killer whales. Genetic studies have shown the two groups to be discrete, with the last mixing probably occurring over one million years ago (Hoelzel and Dover, 1990). This degree of reproductive isolation of sympatric populations is rare, especially when the primary isolating mechanisms appear to be behavioural (Hoelzel and Dover, 1990).

The different ways in which these two groups utilise the habitat provides an interesting look at cetacean behavioural ecology. The fish-eating groups occur in the region with much greater regularity, and have been termed "residents". They travel in pods of 10-15 animals and sometimes group into "super-pods" of over 60 individuals. They primarily feed on salmon (*Onchorhynchus* sp.) that migrate through the region en route to natal spawning rivers. The salmon utilise areas of high flood current to assist their travel, moving into open channels. During ebb tides the salmon wait in protected areas out of the ebb flow. The whales appear to feed during the flood tides, travelling with the current and approaching the salmon from behind (Felleman *et al.*, 1991). They use co-ordinated behaviours of travelling in a broad flank and splashing at the water surface, probably to herd the fish (J.R. Heimlich-Boran, 1988). Favoured feeding areas are those where the whales can drive the salmon towards the shoreline or underwater seamounts, using the geographical features as barriers against which to herd the fish. Once the fish are concentrated, feeding appears to take place relatively individually (J.R. Heimlich-Boran, 1988; Hoelzel, 1993).

In contrast, the marine mammal-feeding killer whales, termed "transients" live in small pods of 2-5 and rarely form multi-pod groups. They mainly feed on the harbour seal, *Phoca vitulina*, which hauls out onto exposed rocks at low tide. These haul-out areas are primarily located out of the main current areas and thus the transient whales occur in more protected waters than do residents. The whales appear to hunt by stealth, approaching seal areas without vocalising. There is some co-operation in catching the seal, and the prey may also be shared (Baird and Dill, 1995).

Figure 5.   Killer whales live in stable groups, but they adjust their immediate travelling group size to the availability of prey. (Photo: J.R. Boran)

The effects of prey selection on group size are clearly shown by the differences between residents and transients. The variable feeding success rates of different sized groups has been used in studies of a number of terrestrial predators in order to show the most efficient sized groups (see, for example, Caraco and Wolf, 1975; Lamprecht, 1978). This has also been done for transient killer whales hunting harbour seals (Baird and Dill, 1996). The whales had the highest observed energy intake when in groups of three, which was also the most common group size observed, supporting the energy maximisation hypothesis (Baird and Dill, 1996). In contrast, resident whale foraging group size during the summer salmon season was considerably larger than that of transients, and presumably reflected the need for widespread searching for migratory fish schools (J.R. Heimlich-Boran, 1988; Felleman *et al.*, 1991; Hoelzel, 1993). In fact, the large, multi-pod groups of unrelated individuals which are supported by the abundant seasonal resource of migratory salmon may provide opportunities for outbreeding from their related pod members (Heimlich-Boran, 1993). Resident group size becomes smaller during the winter, with summer multi-pod associations breaking up into core, family groups, because salmon are not available and feeding is often dependent on resident non-schooling fish species.

This situation of two sympatrically occurring mammals of the same species utilising different ecological niches is unique. The "residents" and "transients" may even interact ecologically in a complex way that we do not yet fully understand. For example, the seals, which are the prey of the transients, appear to share a diet of smaller fish (*e.g.* herring) with salmon, the prey of the residents. The net result could be that as residents eat more salmon, the seals have reduced competition for small fish and thus prosper, in turn providing more prey for the transients (Baird *et al.*, 1992). Of course, even this complex scenario may be too simple to fully explain the system. Any indirect mutual benefit could have been a selective force leading to the evolution of the two forms developing distinct foraging specialisations, and resulting in an Evolutionary Stable State due to disruptive selection against generalists (Baird *et al.*, 1992). It is likely that the two forms are incipient species (Baird *et al.*, 1992).

Figure 6.   On a Patagonian shoreline, a killer whale intentionally strands to capture a sea lion. (Photo: B. Würsig)

This specialisation upon specific prey suggests that social learning could also play a role, where young learn to favour particular prey from the example set by their peers. The best example of this is the intentional stranding behaviour of killer whales feeding on pinnipeds on a beach. This has been observed in Argentina (Lopez and Lopez, 1985; Hoelzel, 1991b) and on the Crozet Islands in the southern Indian Ocean (Guinet, 1991; Guinet and Bouvier, 1995). It is a highly risky behaviour which takes practice. Adults have been observed encouraging young animals onto the beach and helping them off when they get stranded too high up the beach. Adults also release stunned or debilitated prey in the vicinity of younger animals to help them practice their prey handling. All of the features are suggestive of teaching (Caro and Hauser, 1992), but since all animals receive teaching, there is no teaching-free control group to compare the hunting skills of taught and non-taught individuals (Boran and Heimlich, 1999). Recent genetic studies of sperm whales suggest that they may also culturally learn both vocalisations and defence tactics from their mothers, since groups of whales with similar calls and tail scars also had similar patterns of mitochondrial DNA independent of geographical area (Whitehead and Weilgart, 1998).

4.6   Effects of Habitat on Dolphin Behaviour: Species-Specific or Environmentally Determined?

Comparative analyses of the behavioural ecology of two species of dolphin (the dusky dolphin, *Lagenorhynchus obscurus*, and the Hawaiian spinner dolphin, *Stenella longirostris*) living in two different kinds of habitats have shown how habitat can affect a species' behaviour (Würsig *et al.*, 1989; Würsig *et al.*, 1991). In fact, it appears that two different species living in similar habitats may behave more similarly than the same species living in two different habitats.

Dusky dolphins have been studied in two locations: off Golfo San José, Argentina in the south Atlantic (Würsig and Würsig, 1980) and off Kaikoura, New Zealand in the south Pacific (Würsig, 1989; Würsig *et al.*, 1991). In Argentina, the dolphins were primarily found in a relatively shallow and low relief bay in depths less than 100 m, foraging for schools of southern anchovy (*Engraulis anchovita*). The majority of feeding occurred during the day, when anchovy tend to school, and dolphins generally rested close to shore at night, probably to avoid killer whale predation. During the day, stable group units of usually less than ten dolphins were observed searching the bay, with up to 30 of these groups spread from 1-8 km apart. Once food was located, other groups apparently located prey through acoustic and visual cues such as loud splashing and associated flocks of feeding birds, and the several groups converged to herd and capture prey co-operatively. After feeding, the dolphins remained in large schools and engaged in a high activity level of socialising. Although photo-identification studies have been limited, the fission and fusion of dusky dolphin groups appears to result in continual changes in group membership.

The widespread social interactions provided by a fluid social structure probably enhance the bonding mechanisms required for the maintenance of a large number of co-operating school members (Norris and Dohl, 1980; Würsig and Würsig, 1980; Würsig, 1986; Norris and Schilt, 1988).

Off New Zealand, dusky dolphins live in a very different habitat and feed over deep water in a nearshore canyon on prey associated with the deep scattering layer, which migrate closer to the surface at night (Würsig *et al.*, 1989; Würsig *et al.*, 1991). The dolphins exhibit strong diurnal behaviour, staying close to land during the day and moving out over the canyon to feed at night. This is the opposite diurnal pattern of foraging to that found in Argentina.

The primary prey of New Zealand dolphins are mesopelagic fish and squid which are usually fairly dispersed and do not seem to require co-operative herding for capture. Subsequently, dolphin groups are large and spread out over several square kilometres, but with relatively stable subgroups which all travel together in the larger unit. There is nothing like the patterns of fission and fusion found in the Argentine dolphins. This is likely to be because the prey of the New Zealand dolphins are more dispersed and foraging is done more independently when compared with the large-scale, co-operative behaviour observed in the Argentine dusky dolphins (Würsig *et al.*, 1989; Würsig *et al.*, 1991).

The habitat of the spinner dolphin, *Stenella longirostris*, studied along the west coast of Hawaii, contains elements of both the Argentine and New Zealand habitats, and their behavioural ecology shares similarities with both groups of dusky dolphin (Würsig *et al.*, 1989; Würsig *et al.*, 1991; Norris *et al.*, 1994). Spinner dolphins feed offshore in deep water on prey associated with the deep scattering layer, and have similar diurnal behaviour patterns to the New Zealand dusky dolphins. However, during the day they are observed in resting groups in shallow, low relief bays along the Hawaiian coast which are similar to (although considerably less extensive than) Golfo San José, Argentina. During the course of a two-year study, 224 individual spinner dolphins were identified. Thirty-six were seen ten or more times throughout the study, indicating a certain degree of residency. However, even for the repeatedly sighted dolphins, only small numbers were regularly seen together. The conclusions were that spinner dolphins lived in continually changing groups except for a few core associations (Norris *et al.*, 1994). This is similar to the fission-fusion pattern of social structure observed in Argentine dusky dolphins (Würsig *et al.*, 1989; Würsig *et al.*, 1991).

Figure 7.   Spinner dolphins appear to have very fluid group memberships, frequently changing from one group to another. (Photo: B. Würsig)

Spinner dolphins use the shallow inshore and deep offshore parts of their habitat very differently during the day and the night. Small groups of dolphins spend most of the day resting and socialising, and probably also avoiding shark predation (Norris *et al.*, 1994). Group size remains relatively constant in the bays, suggesting that there may be an optimal number of resting dolphins for each bay. At dusk, the animals grow more active, eventually forming "rallying" groups which coalesce and move offshore (Norris *et al.*, 1994). Offshore groups at night are large and widely dispersed (as determined from radio tracking) and were considered to be feeding assemblages of groups from each of the bays and along the coast. These are similar in structure to the foraging groups of New Zealand dusky dolphins.

However, there are differences between these ecosystems. For example, the Hawaiian deep scattering layer sinks to depths of over twice that of New Zealand (500 m vs.100-250 m, respectively), making prey much less available during the day to the Hawaiian spinner dolphins. This makes their dependence on night time feeding much stronger. Also, although the extent of shark predation in New Zealand is unknown, it is thought to be an important factor to the Hawaiian spinner dolphins. Given the limits to the numbers of dolphins which are able to rest in the bays, Hawaiian spinners may even be practising a form of competitive exclusion for access to optimal sites for predator avoidance and resting in the bays.

In conclusion, it appears that one species (the dusky dolphin) varies its behaviour to match the differing environments it inhabits to such a degree that in one case it ends up behaving more similarly to an entirely different species (the spinner dolphin) which lives in a structurally similar habitat. This remains a simplified analysis which contains many exceptions (Würsig *et al.*, 1991), but it will help to define the structure of future research in comparative behavioural ecology.

4.7   Porpoise Foraging Ecology

Porpoises (of the family *Phocoenidae*) are small, primarily coastal, species (the Dall's porpoise *Phocoenoides dalli* is a notable exception, being largely oceanic). Unlike the social delphinids, they seldom use co-operative strategies for concentrating or obtaining prey, at least not in the same manner. The harbour porpoise (*Phocoena phocoena*) is one of the best studied of phocoenids. It feeds upon small, schooling clupeoid and gadid fish as well as a wide variety of other demersal species (Read, 1999). When foraging, porpoises often zig-zag through an area, making long dives to take fish near the sea bed. Although several individuals may aggregate in a small area, they appear to forage independently. They commonly make use of currents, feeding in or on the edge of tide rips and orienting towards the incoming flow (Evans and Borges, 1995). Satellite telemetry has greatly enhanced our understanding of foraging movements and diving behaviour of porpoises in the Gulf of Maine and Bay of Fundy, north-east United States (Westgate and Read, 1998). It has revealed a high degree of individual variation in movement patterns, although the daily distances travelled may be similar (averaging from 13.9 to 28.1 km in four out of five tagged animals, and 58.5 km in the remaining individual) (Read and Westgate, 1997). By monitoring the movements of individuals for long periods, it was shown that home ranges consist of large areas measured in tens of thousands of km^2. Although summer feeding in the region is primarily upon herring, which undergoes a diurnal vertical migration, the porpoises showed no difference in diving behaviour between night and day, at both times regularly diving to depths of 50-100 m (and sometimes to more than

150 m). Depending upon the undersea topography, in some areas they undertake shallower dives.

4.8   Habitat Use by River Dolphins

River dolphins are highly specialised to their riverine habitat: their flexible necks with unfused cervical vertebrae allow greater head movement, and their long, projecting beaks aid the capture of evading fish in a confined environment. All river dolphins have eyes oriented upwards to view objects at or near the surface (Zhou, 1989), while two sub-species (*Platanista gangetica gangetica* and *P. g. minor*) are almost completely blind because vision is useless in their highly turbid environment (Herald *et al.*, 1969). This is in contrast to other cetaceans who have quite well-developed eyesight and whose stereo vision is directed downwards (Madsen and Herman, 1980; Nachtigall, 1986).

River dolphins present a unique "natural experiment" to see the effects on the behaviour of a cetacean living in a more confined and seasonally variable habitat than the ocean. The seasonal element usually occurs as a dry season and a wet season, with the wet season caused by floods from high elevation snow melts or monsoon rains and representing an expansion of available habitat for predators and prey alike. The dry season is also characterised by increased turbidity which reduces visibility. During the wet season, the greater water input reduces turbidity and makes prey more susceptible to predation. River dolphins must adjust their behaviour to these seasonal variations.

There are six species of cetaceans which inhabit rivers. The most highly adapted are three species of "true" river dolphins (super-family *Platanistoideae*): the baiji or Yangtze River dolphin (*Lipotes vexillifer*), the boto or Amazon River dolphin (*Inia geoffrensis*), and the South Asian river dolphin divided into two sub-species, the Indus susu (*Platanista g. minor*), and the Ganges susu (*Platanista g. gangetica*). There are also two species of dolphins (family *Delphinidae*) which contain populations that live far up rivers: the tucuxi (*Sotalia fluviatilis*) and the Irrawaddy dolphin (*Orcaella brevirostris*), as well as one species of porpoise (family *Phocoenidae*): the finless porpoise (*Neophocaena phocaenoides*). All of these species are thought to have evolved in the sea and recolonised fresh water. The specialisations of the true river dolphins make them some of the most highly evolved cetaceans (Kellogg, 1928; Gaskin, 1982).

The baiji is one of the best studied species (Chen *et al.*, 1984; Chen and Hua, 1989; Hua *et al.*, 1989; Zhou, 1989; Zhou and Li, 1989; Chen and Liu, 1992). It lives in the middle and lower sections of the Chiangjiang (formerly Yangtze) River along a stretch of 1,600 km (one-quarter of the river's total length). The species has been mentioned in early Chinese texts of 2,000 years ago (Zhou and Li, 1989). The original population size is unknown, but current estimates are in the order of only 200-300 animals (Chen and Liu, 1992).

Baiji are usually observed in groups of 3-4, although they may range up to 16-17 animals (Chen and Hua, 1989; Zhou and Li, 1989). Solitary baiji are rare, and pairs (usually mother and calf) are sometimes seen. These small groups or pairs could be sub-divisions of larger social groupings. They are often dispersed along the river, separated by about 4-5 km, and have been observed to join up to make larger groups.

Baiji utilise particular areas of the river for specific behaviours. In quieter parts of the river, mother/calf pairs have been observed with two main particular behaviours: the mother 'carrying' the calf with the calf's head resting behind the mother's dorsal fin or with the mother supporting the calf with its flipper. Feeding usually occurs in the vicinity of sand bars, where tributaries enter the river and currents are stronger (Hua *et al.*, 1989; Chen and Liu, 1992). These areas are also known to be important to fish stocks. The baiji are usually seen swimming in an inverted 'V' formation with large adults in front, mothers and calves in the middle, and juveniles at the rear. However, when they are searching for prey, they appear to advance in a single line, almost abreast (Hua *et al.*, 1989). When approaching schools of fish, their formations become concave, with the larger animals at the rear. As the fish school is enclosed, the dolphins form a nearly complete circle, thus preventing the majority of fish from escaping. Echolocation occurs throughout these co-operative feeding bouts and undoubtedly gives the baiji a distinct advantage in the turbid river (Wang *et al.*, 1989; Xiao and Jing, 1989).

Baiji foraging ecology is highly variable by season. During the dry season (the winter), the baiji is limited to the deep water pool areas where fish prey are concentrated as well. In the wet season, the baiji and the fish spread out to move into previously dry areas.

Another species which lives in an environment of extreme seasonal variability due to flooding is the boto or Amazon River dolphin, *Inia geoffrensis* (Best and da Silva, 1989a, b). Botos usually occur in groups of 2-8 individuals (Trujillo, 1994; Henningsen *et al.*, 1996) and often feed at the confluence of two water channels where the current is strong, the water is deep and fish congregate. During the high water season dolphins swim into inundated forest (Igapó habitat) and newly-formed shallow lakes covering grassland. Here they pursue a great diversity of fish species, most of which are harvesting the crop of fruit and seeds which remain out of reach of aquatic organisms for much of the year. The boto's flexibility, and ability to swim backwards using extraordinarily rotatable pectoral fins, allows it to forage among branches and root systems where fish aggregate. The environment of this dolphin changes dramatically as the waters recede. Some fish species move to the main river channels or migrate away from the area altogether, but those that remain are more concentrated and can be trapped against the bank more readily. Botos have adapted their reproductive cycle to take advantage of this, and the peak of births occurs as water levels fall.

Figure 8.   When the rainy season arrives, the Amazon River dolphin or boto exploits a greater diversity of prey when it migrates from rivers into newly-formed lakes. (Photo: F. Trujillo)

Although research on river dolphins is still at an early stage, their consideration is important in a review of this nature because the environment they inhabit is so different from that of other dolphins. As research progresses, especially using modern research techniques involving individual identification (Yuanyu *et al.*, 1990; Trujillo, 1994; da Silva and Martin, 2000), the ability to track individuals using radio telemetry (Martin and da Silva, 1998) and genetic analyses, we can expect that our understanding of river dolphin behavioural ecology will improve substantially in the near future.

5.   THE BEHAVIOURAL ECOLOGY OF CETACEAN MATING SYSTEMS

The ways in which an animal utilises its habitat and collects sufficient resources to survive and grow is just one element in a species' behavioural ecology. The ultimate goal is to mate and raise offspring. Cetaceans have developed a variety of mating systems, each adapted to the specialisations of the species' distribution and behaviour.

5.1   Review of Mammalian Mating Systems

Four generalised types of mating system have been described, usually based on the number of mates for any given breeding period and the duration of the mating bond: 1) monogamy, where one male and one female maintain an exclusive mating bond, 2) polygyny, where one male mates with multiple females, maintains a prolonged mating bond (for at least one breeding season) with that group of females, and excludes other males from access, 3) polyandry, where one female mates with multiple males per breeding season, and 4) polygynandry (similar to the less precise term: promiscuity), where some males as well as females have multiple mates within a breeding season (Orians, 1969; Wilson, 1975; Emlen and Oring, 1977; Stirling, 1983; Clutton-Brock, 1989; Davies, 1992; Andersson, 1994). Mating systems will eventually be better defined by genetic methods which can document "the exact mean and variance of the number of the number of males that contribute genes to the offspring of the average female, and vice versa" (Andersson, 1994: 144).

Monogamy, although common in birds, occurs in only about 5% of mammalian species (Kleiman, 1977): primarily in canids (Moehlman, 1986) but also in primates (Rutberg, 1983). Monogamy appears to be favoured where male paternal care is required in some way for the survival of offspring (Clutton-Brock, 1989). Polyandry is also rare in mammals, as predicted by Orians (1969) on the basis of the relatively greater investment that females have in the production of young. The remaining two mating systems: polygyny and polygynandry, are the two most common forms of polygamy in mammals, and likely to be so in cetaceans. Since males obtain multiple matings in both of these systems, they are primarily differentiated by the number of matings that females obtain. While this definition still overlooks the more accurate measure of mating success (and not simply number of matings since many matings are not successful - see, for example, Stern and Smith, 1984), these simplistic terms can help to identify the most likely candidates for cetacean mating systems.

5.2   Baleen Whale Mating Systems

The mating systems of baleen whales are not well understood. Mating is often difficult to observe: for example, in dozens of studies over thousands of hours, there have been only rare observations of humpbacks copulating (Clapham, 1996; Darling, *pers. comm.*). However, enough is known about the reproductive biology and behaviour of a number of species to allow hypotheses to be formulated. In general, since many of the large baleen whales tend to fast during the breeding season, they are in many ways freed from the ecological constraints on mating which resource distribution has been shown to have (Clapham, 1996). For example, males cannot attract females to food-rich territories to maintain exclusive mating access as is found in many birds and in some mammals (Clutton-Brock, 1989). However, habitat availability may have an influence, for example in the mating lagoons of the gray whales and ice-restricted movements of bowhead whales, and will undoubtedly be important in determining where mating occurs.

An overall view of the mating systems of baleen whales has been described through an analysis of the potential for sperm competition (Brownell and Ralls, 1986). Sperm competition is a male mating strategy which is employed as an alternative strategy to direct male-male competition in the struggle to obtain a successful mating (Parker, 1984; Smith, 1984). Rather than maintaining mating access to a receptive female through the competitive exclusion of other males, a male attempts to maximise the amount of sperm he delivers in each mating through developing relatively large testes for his body size. Then, if the female has mated with a different male before or after, the larger volume of sperm will dilute that of his competitors and will increase his probability of successfully fertilising an egg. Another aspect of this strategy is to develop a longer penis to deliver the sperm farther up the vaginal tract, thus making it less likely to spill out as well as potentially delivering it closer to the cervix. Analyses of mammalian species have shown that males of species with larger testes and longer penises tend to live in societies where females mate promiscuously with multiple males, and polygynandry is the predominant mating system (Harcourt *et al.*, 1981; Harvey and Harcourt, 1984; Kenagy and Trombulak, 1986). In contrast, males of species in which a group of females mate exclusively with a single male, have smaller testes, and a greater degree of body size sexual dimorphism, implying that male-male competition plays a stronger role and that polygyny is the predominant mating system.

A comparison of testes and penis size for baleen whales shows that right whales, bowhead whales (family *Balaenidae*) and gray whales (family

Eschrichtidae) have significantly larger testes. Right and bowhead whales also have the longest penises of all baleen whales (13-14% of body length), while those of other baleen whales are shorter (11% or less of body length: Brownell and Ralls, 1986). This suggests that sperm competition is an important factor in the mating systems of right and bowhead whales, and, to a slightly lesser extent, gray whale.

There are a number of observations of mating bowhead and right whales which support the hypothesis that sperm competition is the primary strategy of their mating system. Mating groups of southern right whales are usually composed of one female surrounded by multiple males (Payne, 1972; Payne, 1976; Cummings, 1985). More than one male may surface with an erect penis and attempt to copulate. More than one male may achieve intromission. The female often appears to be avoiding the males' advances by turning upside-down and lifting her genital region out of the water. There may be some jostling between males and occasional scraping of each other with their callosities (patches of hard, roughened skin on the head), but no violent battles have been observed. There are similar (although fewer) observations of such mating groups in bowhead whales (Würsig *et al.*, 1985, 1993), where no aggressive interactions between males have been observed.

Gray whale mating groups have also been observed to contain multiple males "chasing" a lone female (Swartz, 1986). A maximum of 18 males were observed in one mating group. However, there have never been observations of males competing aggressively with one another (Swartz, 1986). This fits the findings of gray whales having testes larger than that predicted for their body size (Brownell and Ralls, 1986)

A number of elements of the mating system of the humpback whale are quite well understood, despite the paucity of observations of actual copulation (Clapham, 1996). Male humpback whales have smaller testes and penises than other baleen whales and thus are not thought to practice sperm competition. In fact, male-male competition appears to play a greater role than it does for any other cetacean: male humpbacks have been observed to have violent aggressive encounters (Tyack, 1981; Tyack and Whitehead, 1983; Baker and Herman, 1984; Clapham, 1996). They compete primarily in attempts to monopolise females: groups are commonly observed where one female is pursued by two or more males. The "principal escort", the male closest to the female, will vigorously defend his position. These interactions sometimes result in scars; in fact, principal escorts are commonly more scarred than the other males in the group (Chu and Nieukirk, 1988). This indicates the degree to which males compete with one another for the position of principal escort, which should lead them to greater opportunities for (and possibly greater success in) mating.

Females are widely and unpredictably distributed on the breeding grounds, occurring either alone or in small, unstable groups (Mobley and Herman, 1985). This means that males cannot defend either resource-based territories or groups of females, and they therefore compete for single rather than multiple females (Clapham, 1996). Possibly to help achieve this, male humpbacks also produce a song (Payne and McVay, 1971). Two functions have been proposed for humpback whale song: 1) that it is a reproductive advertisement display to attract potential mates, and has evolved through female choice (Winn and Winn, 1978; Tyack, 1981); and 2) that it is a male secondary sexual characteristic which acts as a ritualised form of male-male competition, and is used to sort out male dominance relations (Darling, 1983, 1998). The female choice hypothesis has been the most prevalent, having been proposed to play a selective role in the population-specific, annual modifications of songs (Guinee *et al.*, 1983; Payne *et al.*, 1983). It has also been hypothesised that males may use the song as an "honest" advertisement of fitness in terms of breath-holding capacity: surfacing to take a breath is detectable by characteristic patterns in the song, and females could use this as an indication of stamina and physical condition (Chu and Harcourt, 1986). The male-male competition hypothesis developed from observations of non-singing males joining singers to interact, and sometimes even taking over the singing role (Darling, 1983, 1998). Further study is needed to ascertain which hypothesis may be more generally applicable.

It has been suggested that the humpback whale mating system is similar to a form of "lek", where males aggregate at a fixed site and display within small territories and the females visit for breeding (Emlen and Oring, 1977). However, since there is no evidence for the existence of male territories nor of any spatial structuring within singing areas, the term "floating lek" has been proposed to incorporate the feature of movement of displaying males (Clapham, 1996).

5.3   Roving Male Strategies in the Sperm Whale Mating System

Female sperm whales live in closely bonded matrilineal groups which occasionally come together to form temporary aggregations (Best, 1979; Whitehead and Arnbom, 1987; Whitehead *et al.*, 1991). Females and their young offspring generally remain in tropical and warm temperate waters. Male sperm whales are more wide-ranging, seasonally entering higher latitudes. Males have developed a strategy to maximise their number of matings. Rather than becoming resident with a group of females, thus attempting to monopolise them and developing a form of harem polygyny, large males temporarily join aggregations and travel between groups of females. They spend periods of weeks or months within an aggregation, though generally spending only a few hours with each female group. In theory, males should choose to rove when the travel time between the dispersed female groups is less than the typical oestrus period of the female (Whitehead, 1990). Since oestrus usually lasts at least a few days in most mammals, and the travel time is usually in the order of hours, male roving between female groups should be an efficient strategy. This roving strategy has also been seen in bottlenose dolphin societies (discussed below).

5.4   Stable Societies of Killer Whales and Pilot Whales

The relative stability of associations between individuals in a society can provide insight into social structure and mating system (Wrangham, 1983). Long-term studies of resident killer whale social associations in the inland marine waters off Washington State, USA and British Columbia, Canada since 1973 have revealed an incredibly stable and structured society (Balcomb *et al.*, 1982; S.L. Heimlich-Boran, 1986, 1988; Bigg *et al.*, 1990; Hoelzel, 1993, 1998; Ford *et al.*, 1994). The residents live in two adjacent communities (Bigg *et al.*, 1990) which further divide into four acoustic clans (Ford, 1991).

The primary social unit of resident whales is the pod (Bigg *et al.*, 1990). A pod was defined as "the largest cohesive group of individuals within a community that travelled together ... for at least 50% of the time" (Bigg *et al.*, 1990: p. 388). Pods ranged in size from 4 to 66 animals (mean = 17, n = 19) and occasionally split into smaller "subpods" (1-3 per pod) and "intra-pod groups" (range: 2-9 individuals). The pods were found to consist of overlapping generations of females and their offspring, with genealogies determined by degree of association scaled relative to the high degree of association between known mother-offspring pairs (S.L. Heimlich-Boran, 1986, 1988; Bigg *et al.*, 1990).

Some females were classed as "post-reproductive" (Marsh and Kasuya, 1986; S.L. Heimlich-Boran, 1986, 1988), because they had no calves for at least the last ten years of the study (Olesiuk *et al.*, 1990). It was possible that some of the post-reproductive females may have been infertile young females, but the documented occurrence of such females was rare (all but one of the females who matured during the study gave birth).

Associations continued between mothers and known male and female offspring into adulthood. Changes in associations between mothers and their offspring were found to vary with the age and sex of the offspring (Fig. 9; S.L. Heimlich-Boran, 1986, 1988; Bigg *et al.*, 1990).

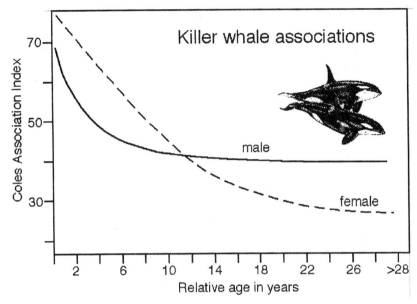

Figure 9. Changes in association rates between male and female killer whale calves and their mothers. Males maintain a higher association with their mothers into adulthood. (Adapted from Olesiuk *et al.*, 1990)

Young female calves had slightly stronger associations with their mother than did male calves, indicating a greater degree of independence for male calves. However, by the age of ten, male calves began to have higher average associations with their mothers than female calves. As young females matured and gave birth to their own calves, their association with their mothers continued to drop until the young female's early 20's, when the association index levelled out at around 25%. In contrast, associations between mothers and their sons levelled out at around ten years of age at an association index of around 40%. Thus, adult sons maintained stronger association with their mothers than did females, probably because the young females spent more time with their own calves (Bigg *et al.*, 1990). Another side of this strong male-mother relationship was shown indicated by a weakening of the bond between a female and the other members of her pod when her son matured. As females reached post-reproductive age, they often had strong associations with actively reproducing females, who were probably their daughters or younger sisters (S.L. Heimlich-Boran, 1986; 1988). Often, these older females also spent time with immature animals (possibly their grandchildren, nieces or nephews), as did some mature males, suggesting a form of allo-parental care (Haenel, 1986; Heimlich-Boran, 1986).

The resulting picture of killer whale social organisation is of a multi-level society with a high degree of stability on all levels. The long-term consistencies of intra-pod groups, subpods, pods, clans and communities all contribute to this stability. This has been verified in genetic analyses by the relatively high degrees of inbreeding (Hoelzel, 1991a, 1998). The lack of male dispersal suggests that males must either benefit from group membership, perhaps in terms of improved resource acquisition or, alternatively, they contribute something (*e.g.* allo-paternal care of young) and are allowed to remain in the groups by the matriarchs (S.L. Heimlich-Boran, 1986, 1988; Heimlich-Boran, 1993).

The mating system of the killer whale is still not completely understood. Given the high levels of relatedness within communities, successful inbreeding avoidance could only be achieved through mating between individuals from different communities, not just between different pods within the same community (Hoelzel *et al.*, 1998). Further genetic paternity studies also need to be done. It was only through genetic studies that the relationship between Alaska and British Columbia whales could be determined (Hoelzel *et al.*, 1998). Observational studies would never have predicted it.

Pilot whales also appear to live in stable societies, as evidenced by their propensity to strand in large groups (Sergeant, 1982; Klinowska, 1986) and the apparent ease with which they are caught in drive fisheries (Gibson-Lonsdale, 1990). This stability in social structure cannot be directly compared with that of killer whales because there has been no equivalent long-term study of pilot whales. However, biological research on carcasses from whaling activities, and some short-term field studies, have shown that there are a number of similarities between the social systems of the two species.

There are two species of pilot whale: the long-finned pilot whale, *Globicephala melas*, and the short-finned pilot whale, *G. macrorhynchus*. Although the two species differ in some morphological and life history traits (Kasuya *et al.*, 1988), there are general similarities in social structure, indicated by similar age and sex composition (Kasuya and Marsh, 1984) and shared indications of genetic segregation between adjacent groups (Andersen, 1988; Wada, 1988; Andersen, 1993). Genetic information of long-finned pilot whales has shown all group members to be related, including adult males and adult females within the same pod (Amos *et al.*, 1991, 1993). In all studies, pods have been primarily composed of mixed age and sex. The few rare observations of all male groups indicate only limited segregation of the sexes (Sergeant, 1962; Kasuya and Marsh, 1984; Desportes *et al.*, 1992). There is a high degree of differential mortality between the sexes: female longevity of short-finned pilot whales was over

20 years more than male longevity, resulting in a female-biased adult sex ratio (Kasuya and Marsh, 1984). Females ceased ovulating by the age of 40 and lived for an average of 23 years in an extended post-reproductive period (Kasuya and Marsh, 1984; Marsh and Kasuya, 1984, 1986, 1991). Some of these females continued to nurse their last calf for up to 15 years. This could be a form of "terminal investment" in which a female invests heavily in the survival of her last calf (Clutton-Brock, 1984).

Genetic evidence relating to the mating system came from paternity studies of long-finned pilot whales which found that males were not the fathers of the offspring in the pods (Amos *et al.*, 1991, 1993). It was also found that only one or a few related males were fathering the young, indicating a certain level of male variance in reproductive success and thus suggesting polygyny. However, it was also found that the male genetic contribution to cohorts changed from year to year and thus females were mating with multiple males (Amos *et al.*, 1991, 1993), implying a certain degree of polygynandry.

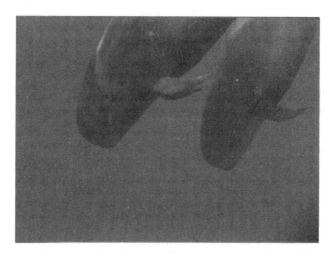

Figure 10.   The short-finned pilot whale lives in stable groups where mature males and females tend to associate only when separate pods come together. (Photo: M. Scheer)

Field studies of a free-ranging population of short-finned pilot whales were conducted for two years off Tenerife in the Canary Islands, and over 500 individuals were identified (Heimlich-Boran, 1993). Analyses of associations indicated that male-female interactions were most common when "linked" pods (separate pods with regular patterns of association) travelled together. Although mating was never observed, these associations could indicate a possible time when out-group breeding could occur. The season when members of more than one pod were observed together corresponded to the supposed peak season of conception (Heimlich-Boran, 1993). The same has been observed in long-finned pilot whales in the North-east Atlantic with temporary aggregations of very large groups (numbering 500-1,000 animals) accompanied by much social and sexual activity, during the peak season of conception (Evans, 1992; Evans, *pers. obs.*).

The ecological basis of a mating system in which related individuals of both sexes remain together throughout their entire lives and in which mating only occurs between more distantly related groups is not well understood (Heimlich-Boran, 1993; Connor *et al.*, 1998). Its apparent occurrence in killer whales in the north-east Pacific is supported by the presence of a seasonal food resource (salmon) which is sufficiently abundant to sustain large numbers of whale groups during the mating season (Felleman *et al.*, 1991). Aggregations of pilot whale groups in the Canary Islands must also be sustained by abundant resources of squid (Heimlich-Boran, 1993).

On the other hand, the primary selective pressure must be the benefit which both sexes obtain by living in a group of closely-related kin. The detection, aggregation and capture of schooling prey such as salmon and squid likely requires a large degree of group co-operation (Norris and Dohl, 1980; Würsig, 1986; Norris and Schilt, 1988). The benefits of living with kin have been well described (Hamilton, 1963; Maynard Smith, 1964; Bertram, 1976; Trivers, 1985), and, in particular, living with multiple generations of kin can allow the build-up of cultural knowledge (Bonner, 1980; Nishida, 1987; Connor, 1998; Whitehead, 1998) which can improve foraging through memories of previous feeding successes (Würsig, 1986). The stable groups of killer whales and pilot whales represent a society where both males and females can benefit from this kin selection (unlike most mammals where one sex or the other emigrates from its natal group: Greenwood, 1980; Moore and Ali, 1983), but avoid inbreeding by mating with more distantly related individuals from adjacent groups.

5.5   Fluid Dolphin Societies

In contrast to the stable societies of killer whales and pilot whales, the smaller dolphins appear to live in very fluid societies, where associations are regularly changing. We have already presented examples of fluid associations in spinner dolphins and dusky dolphins being the result of the need to interact with a wide range of unrelated individuals in the search and capture of dispersed, pelagic prey. A lesser degree of group fluidity also appears to exist in some coastal populations of bottlenose dolphins.

The most complete studies of the social organisation of bottlenose dolphins have been conducted since 1970 along 160 km of the west coast of Florida (Wells, 1991; Wells *et al.*, 1987; Scott *et al.*, 1990), but observations of habituated bottlenose dolphins off western Australia (Connor and Smolker, 1985; Connor *et al.*, 1992) have also contributed some well-documented short-term observations.

Off the west coast of Florida, a mosaic of overlapping communities of dolphins have been identified (Wells *et al.*, 1987). Data from the central community, termed the Sarasota community, have generated the bulk of the conclusions on bottlenose dolphin social organisation. This community comprised around 100 dolphins (Wells and Scott, 1990), with a female-biased adult sex ratio of 2.4 females per male (Wells *et al.*, 1987). There was a high degree of age and sex segregation (only 31% of groups were of mixed age and sex) exhibited in different patterns of habitat use (Wells *et al.*, 1987). Four "bands" of females (adjacent groups with discrete membership) were identified within the community and occasionally mixed with each other. The most common group composition was of dolphins of similar age, sex and reproductive status, resulting in the formation of female/calf groups, juvenile groups, and all-male groups (Wells *et al.*, 1987). Highly related females (indicated by a sharing of a specific marker chromosome: Duffield and Wells, 1991) demonstrated high degrees of association. This shows that, although daily swimming associates are more often correlated with reproductive condition than genetic relationships, the pool of potential associates is maintained along genetic lines, with several matrilines remaining in recurrent association over generations (Wells *et al.*, 1987).

Within the Sarasota community there were two different patterns of male associations with adult females: the resident male and the roving male patterns, distinguished by the relative duration of the association. The resident male pattern involved a single adult male repeatedly associating with a female or group of females within the range of the female activities, and only occasionally leaving to visit other female bands. Roving males, on the other hand, tended to be pairs, or occasionally trios, of males which moved between female groups and extended their movements beyond the range of the Sarasota community females (Wells *et al.*, 1987). On a few occasions, these groups of roving males were observed to interact aggressively with adult males from adjacent communities, resulting in bloodied scars (Wells, 1991). These male groups may be similar to the male coalitions observed actively herding receptive female *Tursiops* off western Australia (Connor and Smolker, 1985; Connor *et al.*, 1992). These male alliances may be competing to maintain access to receptive females (Connor *et al.*, 1992).

Figure 11.   Bottlenose dolphins are frequently covered in tooth scars from other dolphins, perhaps from aggressive encounters. (Photo: J.R. Boran)

In general, the role of aggression in cetaceans has not been well appreciated, and many cases are emerging that indicate not only intra-specific aggression (such as these *Tursiops* observations and the violent male-male contests between humpbacks mentioned earlier) but also aggression (which can lead to death) between species (*e.g.* bottlenose dolphins attacking spotted dolphins (Herzing, 1996) and harbour porpoises (Ross and Wilson, 1996). However, we do not yet know to what degree the occurrence of male-male aggression as an indicator of a polygynous mating system (as is traditionally assumed for terrestrial mammals) applies for cetaceans.

There are still many unanswered questions as to the nature of the bottlenose dolphin mating system with regards to the reproductive success of males. Do males mate with females in their own community or does mating take place between males and females from different communities? Males from adjacent communities were observed associating with the Sarasota females during the breeding season. Sarasota community males were absent from the range of Sarasota females for up to several months, during which time they probably came into contact with dolphins from the adjacent communities (Wells *et al.*, 1987). However, genetic evidence has not been conclusive. Preliminary paternity exclusion studies using DNA fingerprinting have occasionally allowed some community males to be excluded as possible fathers, but in other cases this was not possible (Duffield and Wells, 1991). Current research suggests that 70% of Sarasota calves have been fathered by males within their own community, but this is still being examined (Wells, *pers. comm.*).

Another element of the bottlenose dolphin mating system that is still unclear is the number of males who are contributing genes to any given females' offspring, and thus the degree of polygyny or polygynandry that occurs. Since there are only short-term interactions between males and females, indicated by the high degree of age/sex segregation within communities and the lack of permanent exchange of roving males between communities, females are highly likely to be mating with more than one male.

In conclusion, we hypothesise that the behavioural ecology of dolphin mating systems is based on their preference for relatively long-term residence in a coastal habitat. Such habitats are characterised by patchy but predictable clumps of prey which replenish locally and where many individuals can alternate between dispersed, core groups to locate variable-sized patches of food, and come together into larger aggregations to feed on larger patches of food, socialise and mate. These have been termed fission-fusion societies (Kummer, 1968), and are also characteristic of baboon and chimpanzee societies (Kummer, 1971). Pelagic dolphins, like *Stenella* spp., may associate in very large, fluid groups because of the wider dispersion of the prey which may be more ephemeral in its location, while coastal dolphins, like *Tursiops*, occupy more discrete areas in stable community bands with limited mixing between females, and only occasional mating between males and females from different communities.

6. CONCLUSIONS

The behavioural ecology of cetaceans, like that of terrestrial mammals, is based on the distribution of resources and on strategies of maximising food acquisition and mating success. In some ways, a less complete understanding of the distribution of resources in the oceans has limited the final conclusions which can be made about cetaceans, but it is clear that the same principles of terrestrial behavioural ecology can still be applied. Cetology has much to learn from studies of terrestrial mammals and we believe that the comparative approach needs to be used in a wider variety of cetacean studies. Technological advances are also bringing us more detailed information about the lives of individual cetaceans in the open ocean. Genetic techniques will also provide a greater understanding of mating systems (*i.e.* the exact genome representation in new generations), which will help to quantify the true benefits of the various strategies. There are still many discoveries to be made.

ACKNOWLEDGEMENTS

We would like to thank Professor Toni Raga (and the Universidad Menendez Pelayo) for hosting the conference in Valencia at which this paper was originally presented. We gratefully acknowledge the constructive comments of Phil Clapham, Jim Darling, Nick Davies, Sara Heimlich, Randy Wells and Bernd Würsig.

REFERENCES

Amos, B., Barrett, J. and Dover, G.A. (1991) Breeding behaviour of pilot whales revealed by DNA fingerprinting. *Heredity*, **67**, 49-55.

Amos, W., Schlötterer, C. and Tautz, D. (1993) Social structure of pilot whales revealed by analytical DNA profiling. *Science*, **260**, 670-672.

Andersen, L.W. (1988) Electrophoretic differentiation among local populations of the long-finned pilot whale, *Globicephala melaena*, at the Faroe Islands. *Canadian Journal of Zoology*, **66**, 1884-1892.

Andersen, L.W. (1993) Further studies on the population structure of the long-finned pilot whale, *Globicephala melas*, off the Faroe Islands. *Report of the International Whaling Commission* (special issue **14**), 219-231.

Anderssen, M. (1994) *Sexual Selection*. Princeton University Press, Princeton. 599pp.

Au, D.W.K. and Perryman, W.L. (1985) Dolphin habitats in the eastern tropical Pacific. *Fishery Bulletin*, **83**, 623-643.

Baird, R.W., Abrams, P.A. and Dill, L.M. (1992) Possible indirect interactions between transient and resident killer whales: implications for the evolution of foraging specialization in the genus *Orcinus. Oecologia*, **89**, 125-132.

Baird, R.W. and Dill, L.M. (1995) Occurrence and behaviour of transient killer whales: seasonal and pod-specific variability, foraging behaviour, and prey handling. *Canadian Journal of Zoology*, **73**, 1300-1311.

Baird, R.W. and Dill, L.M. (1996) Ecological and social determinants of group size in transient killer whales. *Behavioral Ecology*, **7**, 408-416.

Baird, R.W. and Stacey, P.J. (1988) Variation in saddle patch pigmentation in populations of killer whales (*Orcinus orca*) from British Columbia, Washington, and Vancouver Island. *Canadian Journal of Zoology*, **66**, 2582-2585.

Baker, C.S. and Herman, L.M. (1984) Aggressive behavior between humpback whales (*Megaptera novaeangliae*) on the Hawaiian wintering grounds. *Canadian Journal of Zoology*, **62**, 1922-1937.

Balcomb, K.C., Boran, J.R. and Heimlich, S.L. (1982) Killer whales in Greater Puget Sound. *Report of the International Whaling Commission*, **32**, 681-685.

Best, P.B. (1979) Social organization in sperm whales, *Physeter macrocephalus*. In: *Behavior of Marine Animals, Vol. 3: Cetaceans* (Ed. by H.E. Winn and B.L. Olla), pp. 227-289. Plenum Press, New York.

Best, R.C. and da Silva, V.M.F. (1984) Preliminary analysis of reproductive parameters of the bouto, *Inia geoffrensis*, and the tucuxi, *Sotalia fluviatilis*, in the Amazon river system. *Report of the International Whaling Commission* (special issue **6**), 361-369.

Best, R.C. and da Silva, V.M.F. (1989a) Biology, status and conservation of Inia geoffrensis in the Amazon and Orinoco River basins. In: *Biology and Conservation of the River Dolphins* (Ed. by W.F. Perrin, R.L. Brownell Jr., K. Zhou and J. Liu), pp. 23-34. Occasional Papers of the IUCN Species Survival Commission, No. 3, Gland, Switzerland.

Best, R.C. and da Silva, V.M.F. (1989b) Amazon river dolphin, boto, *Inia geoffrensis* (de Blainville, 1817). In: *Handbook of Marine Mammals, Vol. 4, River Dolphins and the Larger Toothed Whales* (Ed. by R.J. Harrison and S.H. Ridgway), pp. 1-23. Academic Press, London.

Bigg, M.A., Olesiuk, P.F., Ellis, G.M., Ford, J.K.B. and Balcomb, K.C. (1990) Social organization and genealogy of resident killer whales (*Orcinus orca*) in the coastal waters of British Columbia and Washington State. *Report of the International Whaling Commission* (special issue **12**), 383-405.

Bigg, M.A. and Wolman, A. (1975) Live-capture of killer whales (*Orcinus orca*) in Washington and British Columbia. *Journal of the Fisheries Research Board of Canada*, **32**, 1056-1065.

Bonner, J.T. (1980) *The Evolution of Culture in Animals*. Princeton, New Jersey: Princeton University Press.

Boran, J.R. and Heimlich, S.L. (1999) Social learning in cetaceans: hunting, hearing and hierarchies. In: *Mammalian Social Learning: Comparative and Ecological Perspectives* (Ed. by H. Box and K. Gibson), pp. 282-307. Cambridge University Press, Cambridge. 424pp.

Bowen, W.D. and Siniff, D.B. (1999) Distribution, Population Biology, and Feeding Ecology of Marine Mammals. In: *Biology of Marine Mammals*. (Ed. by J.E. Reynolds III and S.A. Rommel), pp. 423-484. Smithsonian Institution Press, Washington DC.

Bree, P.J.H.v., Best, P.B. and Ross, G.J.B. (1978) Occurrence of the two species of pilot whales (genus *Globicephala*) on the coast of South Africa. *Mammalia*, **42**, 323-328.

Brownell, R.L.J. and Ralls, K. 1986. Potential for sperm competition in baleen whales. *Report of the International Whaling Commission* (special issue **8**), 97-112.

Bygott, J.D., Bertram, B.C.R. and Hanby, J.P. (1979) Male lions in large coalitions gain reproductive advantage. *Nature* (London), **282**, 839-841.

Caraco, T. and Wolf, L.L. (1975) Ecological determinants of group sizes of foraging lions. *American Naturalist*, **109**, 343-352.

Caro, T.M. and Hauser, M.D. (1992) Is there teaching in non-human animals? *Quarterly Review of Biology*, **67**, 151-174.

Chen, P. and Hua, Y. (1989) Distribution, population size and protection of *Lipotes vexillifer*. In: *Biology and Conservation of the River Dolphins* (Ed. by W.F. Perrin, R.L. Brownell Jr., K. Zhou and J. Liu), pp. 81-85. Occasional Papers of the IUCN Species Survival Commission, No. 3, Gland, Switzerland.

Chen, P. and Liu, R. (1992) *Baiji, A Rare Treasure*. Tokyo: Enoshima Aquarium (incl. 50pp. English translation). 92pp.

Chen, P., Liu, R. and Lin, K. (1984) Reproduction and the reproductive system in the baiji, *Lipotes vexillifer*. *Report of the International Whaling Commission* (special issue **6**), 445-450.

Chu, K. and Harcourt, P. (1986) Behavioral correlations with aberrant patterns in humpback whale songs. *Behavioral Ecology and Sociobiology*, **19**, 309-312.

Chu, K. and Nieukirk, S. (1988) Dorsal fin scars as indicators of age, sex, and social status in humpback whales (*Megaptera novaeangliae*). *Canadian Journal of Zoology*, **66**, 416-420.

Clapham, P.J. (1993) Social organization of humpback whales on a North Atlantic feeding ground. *Symposium of the Zoological Society of London*, **66**, 131-145.

Clapham, P.J. (1996) The social and reproductive biology of humpback whales: an ecological perspective. *Mammal Review*, **26**, 27-49.

Clutton-Brock, T.H. (1984) Reproductive effort and terminal investment in iteroparous animals. *American Naturalist*, **123**, 212-229.

Clutton-Brock, T.H. (1989) Mammalian mating systems. *Proceedings of the Royal Society of London B*, **236**, 339-372.

Connor, R.C., Mann, J., Tyack, P. and Whitehead, H. (1998) Social evolution in toothed whales. *Trends in Ecology and Evolution*, **13**, 228-231.

Connor, R.C. and Smolker, R.A. (1985) Habituated dolphins (*Tursiops* sp.) in Western Australia. *Journal of Mammalogy*, **66**, 398-400.

Connor, R.C., Smolker, R.A. and Richards, A.F. (1992) Dolphin alliances and coalitions. In: *Coalitions and Alliances in Humans and Other Animals* (Ed. by A.H. Harcourt and F.B.M. De Waal), pp. 415-443. Oxford University Press, Oxford.

Cummings, W.C. (1985) Right whales - *Eubalaena glacialis* (Müller, 1776) and *Eubalaena australis* (Desmoulins, 1822). In: *Handbook of Marine Mammals, Vol. 3 The Sirenians and Baleen Whales* (Ed. by S.H. Ridgway and R. Harrison), pp. 275-304. Academic Press, London.

Darling, J.D. (1983) *Migrations, Abundance and Behavior of Hawaiian Humpback Whales, Megaptera novaeangliae* (Borowski). Ph.D. Thesis, University of California, Santa Cruz.

Darling, J.D. (1998) Evidence for the function of the humpback whale song. *Abstract to the World Marine Mammal Conference*, 21-24 January, 1998, Monaco.

Darling, J.D., Keogh, K.E. and Steeves, T.E. (1998) Gray whale (*Eschrichtius robustus*) habitat utilization and prey species off Vancouver Island, B.C. *Marine Mammal Science*, **14**, 692-720.

Darwin, C. (1859) *On the Origin of Species by Means of Natural Selection*. John Murray, London. 386pp.

Davies, N.B. (1992) *Dunnock Behaviour and Social Evolution*. Oxford University Press, Oxford. 272pp.

Dawkins, R. (1976) *The Selfish Gene*. Oxford University Press, Oxford. 224pp.

Desportes, G., Andersen, L.W., Aspholm, P.E., Bloch, D. and Mouritsen, R. (1992) A note about a male-only pilot whale school observed in Faroe Islands. *Fróðskaparrit*, **40**, 31-37.

Dorsey, E.M. (1983) Exclusive adjoining ranges in individually identified minke whales (*Balaenoptera acutorostrata*) in Washington State. *Canadian Journal of Zoology*, **61**, 174-181.

Duffield, D.A. (1986) *Orcinus orca*: taxonomy, evolution, cytogenetics and population structure. In: *Behavioral Biology of Killer Whales* (Ed. by B. Kirkevold and J.S. Lockard), pp. 19-33. A.R. Liss, New York. 457pp.

Duffield, D.A. and Wells, R.S. (1991) The combined application of chromosome, protein and molecular data for the investigation of social unit structure and dynamics in *Tursiops truncatus*. *Report of the International Whaling Commission* (special issue **13**), 155-169.

Economos, A.C. (1983) Elastic and/or geometric similarity in mammalian design. *Journal of Theoretical Biology,* **103**, 167-172.

Emlen, S.T. and Oring, L.W. (1977) Ecology, sexual selection, and the evolution of mating systems. *Science,* **197**, 215-223.

Endler, J.A. (1991) Interactions between predators and prey. In: *Behavioural Ecology - An Evolutionary Approach. 3rd edition.* (Ed. by J.R. Krebs and N.B. Davies), pp. 169-196. Blackwell Scientific Publications, Oxford.

Eschricht, D.F. (1866) On the species of the genus *Orca* inhabiting the northern seas. In: *Recent Memoirs on the Cetacea* (Ed. by W.H. Flower), pp. 153-188. Ray Society, London.

Evans, P.G.H. (1987) *The Natural History of Whales and Dolphins.* Academic Press, London and Facts on File Publications, New York. 343pp.

Evans, P.G.H. (1992) *Status Review of Cetaceans in British and Irish Waters.* Report to UK Department of the Environment. Sea Watch Foundation, Oxford. 99pp.

Evans, P.G.H. and Borges, L. (1995) Ecological studies of harbour porpoise in Shetland *European Research on Cetaceans,* **9**, 173-178.

Felleman, F.L., Heimlich-Boran, J.R. and Osborne, R.W. (1991) Feeding ecology of the killer whale (*Orcinus orca*). In: *Dolphin Societies* (Ed. by K. Pryor and K.S. Norris), pp. 113-147. University of California Press, Berkeley. 397pp.

Fish, F.E. and Hui, C.A. (1991) Dolphin swimming - a review. *Mammal Review,* **21**, 181-195.

Fisher, R.A. (1958) *The Genetical Theory of Natural Selection.* 2nd Edition. Dover Publications, New York.

Fitch, J.E. and Brownell, R.L. (1965) Fish otoliths in cetacean stomachs and their importance in interpreting feeding habits. *Journal of the Fisheries Research Board of Canada,* **25**, 2561-2574.

Ford, J.K.B. (1989) Acoustic behavior of resident killer whales (*Orcinus orca*) off Vancouver Island, British Columbia. *Canadian Journal of Zoology,* **67**, 727-745.

Ford, J.K.B. (1991) Vocal traditions among resident killer whales (*Orcinus orca*) in coastal waters of British Columbia. *Canadian Journal of Zoology,* **69**, 1454-1483.

Ford, J.K.B., Ellis, G.M. and Balcomb, K.C. (1994) *Killer Whales.* UBC Press, Vancouver, B.C. 102pp.

Ford, J.K.B. and Fisher, H.D. (1983) Group-specific dialects of killer whales (*Orcinus orca*) in British Columbia. In: *Communication and Behavior of Whales* (Ed. by R.S. Payne), pp. 129-161. Westview Press, Boulder, CO. 643pp.

Gaskin, D.E. (1982) *The Ecology of Whales and Dolphins.* London: Heinemann. 459pp.

Gaulin, S.J.C. and Sailer, L. (1985) Are females the ecological sex? *American Anthropologist,* **87**, 111-119.

Gibson-Lonsdale, J.J. (1990) Pilot whaling in the Faroe Islands - its history and present significance. *Mammal Review,* **20**, 44-52.

Greenwood, P.J. (1980) Mating systems, philopatry, and dispersal. *Animal Behaviour,* **28**, 1140-1162.

Guinee, L., Chu, K. and Dorsey, E.M. (1983) Changes over time in the songs of known individual humpback whales (*Megaptera novaeangliae*). In: *Communication and Behavior of Whales* (Ed. by R.S. Payne), pp. 59-80. Westview Press, Boulder, CO.

Guinet, C. 1991. Intentional stranding apprenticeship and social play in killer whales (*Orcinus orca*). *Canadian Journal of Zoology,* **69**, 2712-2716.

Guinet, C. and Bouvier, J. (1995) Development of intentional stranding hunting techniques in killer whale (*Orcinus orca*) calves at Crozet Archipelago. *Canadian Journal of Zoology*, **73**, 27-33.

Haenel, N.J. (1986) General notes on the behavioral ontogeny of Puget Sound killer whales and the occurrence of allomaternal behavior. In: *Behavioral Biology of Killer Whales* (Ed. by B. Kirkevold and J.S. Lockard), pp. 285-300. A.R. Liss, New York.

Hamilton, W.D. (1964) The genetical theory of social behaviour. *Journal of Theoretical Biology*, **7**, 1-52.

Harcourt, A.H., Harvey, P.H., Larson, S.G. and Short, R.V. (1981) Testis weight, body weight, and breeding system in primates. *Nature* (Lond.), **293**, 55-57.

Harvey, P.H. and Harcourt, A.H. (1984) Sperm competition, testes size, and breeding systems in primates. In: *Sperm Competition and the Evolution of Animal Mating Systems* (Ed. by R.L. Smith), pp. 589-600. Academic Press, New York.

Heimlich-Boran, J.R. (1988) Behavioral ecology of killer whales (*Orcinus orca*) in the Pacific Northwest. *Canadian Journal of Zoology*, **66**, 565-578.

Heimlich-Boran, J.R. (1993) *Social Organisation of the Short-Finned Pilot Whale, Globicephala macrorhynchus, with Special Reference to the Comparative Social Ecology of Delphinids*. Ph.D. Thesis, University of Cambridge.

Heimlich-Boran, S.L. (1986) Cohesive relationships among Puget Sound killer whales. In: *Behavioral Biology of Killer Whales* (Ed. by B. Kirkevold and J.S. Lockard), pp. 251-284. A.R. Liss, New York. 457pp.

Heimlich-Boran, S.L. (1988) *Association Patterns and Social Dynamics of Killer Whales (Orcinus orca) in Greater Puget Sound*. M.A. Thesis, Moss Landing Marine Laboratories, San Jose State University.

Henningsen, T., Knickmeier, K. and Lotter, G. (1996) Site fidelity, movement patterns and seasonal migration of dolphins in the Peruvian Amazon. *European Research on Cetaceans*, **10**, 159-162.

Herald, E.S., Brownell, R.L.J., Frye, F.L., Morris, E.J., Evans, W.E. and Scott, A.B. (1969) Blind river dolphins: first side-swimming cetacean. *Science*, **166**, 1408-1410.

Herzing, D. (1996) Vocalizations and associated underwater behaviour of free-ranging Atlantic spotted dolphins, *Stenella frontalis*, and bottlenose dolphins, *Tursiops truncatus*. *Aquatic Mammals*, **22**, 61-79.

Heyning, J.E. and Dahlheim, M.E. (1988) Orcinus orca. *Mammalian Species*, **304**, 1-9.

Hoelzel, A.R. (1991a) Analysis of regional mitochondrial DNA variation in the killer whale; implications for cetacean conservation. *Report of the International Whaling Commission* (special issue **13**), 224-233.

Hoelzel, A.R. (1991b) Killer whale predation on marine mammals at Punta Norte, Argentina; food sharing, provisioning and foraging strategy. *Behavioral Ecology and Sociobiology*, **129**, 1-8.

Hoelzel, A.R. (1993) Foraging behaviour and social group dynamics in Puget Sound killer whales. *Animal Behaviour*, **45**, 581-591.

Hoelzel, A.R., Dahlheim, M.E. and Stern, S.J. (1998) Low genetic variation amorᶾ killer whales (*Orcinus orca*) in the eastern North Pacific and genetic differentiation between foraging specialists. *Journal of Heredity*, **89**, 121-128.

Hoelzel, A.R., Dorsey, E.M. and Stern, S.J. (1989) The foraging specializations of individual minke whales. *Animal Behaviour*, **38**, 786-794.

Hoelzel, A.R. and Dover, G.A. (1990) Genetic differentiation between sympatric killer whale populations. *Heredity*, **66**, 191-195.

Hua, Y., Zhao, Q. and Zhang, G. (1989) The habitat and behavior of *Lipotes vexillifer*. In: *Biology and Conservation of the River Dolphins* (Ed. by W.F. Perrin, R.L. Brownell Jr., K. Zhou and J. Liu), pp. 92-98. Occasional Papers of the IUCN Species Survival Commission, No. 3, Gland, Switzerland.

Jarman, P.J. (1974) The social organisation of antelope in relation to their ecology. *Behaviour*, **48**, 215-267.

Jefferson, T.A., Stacey, P.J. and Baird, R.W. (1991) A review of killer whale interactions with other marine mammals: predation to coexistence. *Mammal Review*, **22**, 35-47.

Kasuya, T. and Marsh, H. (1984) Life history and reproductive biology of the short-finned pilot whale, *Globicephala macrorhynchus*, off the Pacific coast of Japan. *Report of the International Whaling Commission* (special issue **6**), 259-310.

Kasuya, T., Sergeant, D.E. and Tanaka, K. (1988) Re-examination of life history parameters of long-finned pilot whales in the Newfoundland waters. *Scientific Reports of the Whales Research Institute*, **39**, 103-119.

Kellogg, R. (1928) The history of whales - their adaptation to life in the water. *Quarterly Review of Biology*, **3**, 29-76, 74-208.

Kenagy, G.J. and Trombulak, S.C. (1986) Size and function of mammalian testes in relation to body size. *Journal of Mammalogy*, **67**, 1-22.

Kim, S.L. and Oliver, J.S. (1989) Swarming benthic crustaceans in the Bering and Chukchi seas and their relation to geographic patterns in gray whale feeding. *Canadian Journal of Zoology*, **67**, 1531-1542.

Kleiman, D.G. (1977) Monogamy in mammals. *Quarterly Review of Biology*, **52**, 39-69.

Klinowska, M. (1986) The cetacean magnetic sense-evidence from strandings. In: *Research on Dolphins* (Ed. by M.M. Bryden and R.J. Harrison), pp. 401-432. Clarendon Press, Oxford.

Krebs, J.R. and Davies, N.B. (1991) Preface. In: *Behavioural Ecology: An Evolutionary Approach* (Ed. by J.R. Krebs and N.B. Davies), pp. ix-x. Blackwell Scientific Publications, Oxford.

Kummer, H. (1968) *The Social Organization of Hamadryas Baboons*. University of Chicago Press, Chicago.

Kummer, H. (1971) *Primate Societies: Group Techniques of Ecological Adaptation*. Chicago: Aldine-Atherton.

Lamprecht, J. (1978) The relationship between food competition and foraging group size in some larger carnivores. *Zeitschrift für Tierpsychologie*, **46**, 337-343.

Lopez, J.C. and Lopez, D. (1985) Killer whales (*Orcinus orca*) of Patagonia, and their behavior of intentional stranding while hunting nearshore. *Journal of Mammalogy*, **66**, 181-183.

MacArthur, R.H. and Wilson, E.O. (1967) *The Theory of Island Biogeography*. Harvard University Press, Cambridge, MA. 203pp.

Madsen, C.J. and Herman, L.M. (1980) Social and ecological correlates of cetacean vision and visual appearance. In: *Cetacean Behavior: Mechanisms and Functions* (Ed. by L.M. Herman), pp. 101-147. J. Wiley & Sons, New York.

Marsh, H. and Kasuya, T. (1984) Ovarian changes in the short-finned pilot whale, *Globicephala macrorhynchus*. *Report of the International Whaling Commission* (special issue **6**), 311-335.

Marsh, H. and Kasuya, T. (1986) Evidence for reproductive senescence in female cetaceans. *Report of the International Whaling Commission* (special issue **8**), 57-74.

Marsh, H. and Kasuya, T. (1991) An overview of the changes in the role of a female pilot whale with age. In: *Dolphin Societies* (Ed. by K. Pryor and K.S. Norris), pp. 281-285. University of California Press, Berkeley.

Martin, A.R. and da Silva, V.F.M. (1998) Tracking aquatic vertebrates in dense tropical forest using V.H.F. telemetry. *Marine Technology Society Journal*, **32**, 82-88.

Mayo, C.A. and Marx, M.K. (1990) Surface foraging behavior of the north Atlantic right whale, *Eubalaena glacialis*, and associated zooplankton characteristics. *Canadian Journal of Zoology*, **68**, 2214-2220.

Mobley, J.R. and Herman, L.M. (1985) Transience of social affiliations among humpback whales (*Megaptera novaeangliae*) on the Hawaiian wintering grounds. *Canadian Journal of Zoology*, **63**, 762-772.

Moehlman, P.I. (1986) Ecology of cooperation in canids. In: *Ecological Aspects of Social Evolution: Birds and Mammals* (Ed. by D.I. Rubenstein and R.W. Wrangham), pp. 64-86. Princeton Univ. Press, Princeton, NJ.

Moore, J. and Ali, R. (1983). Are dispersal and inbreeding avoidance related? *Animal Behaviour*, **32**, 94-112.

Moore, S.E., Ljungblad, D.K. and van Schoik, D.R. (1986) Annual patterns of gray whale (*Eschrichtius robustus*) distribution, abundance and behavior in the northern Bering and eastern Chukchi seas, July 1980-83. *Report of the International Whaling Commission* (special issue **8**), 231-242.

Morton, A. (1990) A quantitative comparison of the behaviour of resident and transient forms of the killer whale off the central British Columbia coast. *Report of the International Whaling Commission* (special issue **12**), 245-248.

Nachtigall, P.E. (1986) Vision, audition, and chemoreception in dolphins and other marine mammals. In: *Dolphin Cognition and Behavior: A Comparative Approach* (Ed. by R.J. Schusterman, J.A. Thomas and F.G. Wood), pp. 79-113. Lawrence Erlbaum, Hillsdale, NJ.

Nemoto, T. (1959) Food of baleen whales with reference to whale movements. *Scientific Reports of the Whales Research Institute*, **16**, 149-290.

Nerini, M.K. (1984) A review of gray whale feeding ecology. In: *The Gray Whale* (Ed. by M.L. Jones, S. Swartz and S. Leatherwood), pp. 423-450. Academic Press, New York.

Nerini, M.K. and Oliver, J.S. (1983) Gray whales and the structure of the Bering Sea benthos. *Oecologia*, **59**, 224-225.

Nishida, T. (1987) Local traditions and cultural transmission. In: *Primate Societies* (Ed. by B.B. Smuts, D.L. Cheney, R.M. Seyfarth, R.W. Wrangham and T.T. Struhsaker), pp. 462-474. Chicago University Press, Chicago.

Norris, K.S. and Dohl, T.P. (1980) The structure and function of cetacean schools. In: *Cetacean Behavior: Mechanisms and Functions* (Ed. by L.M. Herman), pp. 211-261. J. Wiley & Sons, New York.

Norris, K.S. and Schilt, C.R. (1988) Cooperative societies in three-dimensional space: on the origins of aggregationsm flocks, and schools, with special reference to dolphins and fish. *Ethology and Sociobiology*, **9**, 149-179.

Norris, K.S., Villa-Ramirez, B., Nichols, G., Würsig, B. and Miller, K. (1983) Lagoon entrance and other aggregations of gray whales (*Eschrichtius robustus*). In: *Communication and Behavior of Whales* (Ed. by R.S. Payne), pp. 259-293. Westview Press, Boulder, CO.

Norris, K.S., Würsig, B., Wells, R.S. and Würsig, M. (1994) *The Hawaiian Spinner Dolphin*. University of California Press, Berkeley, CA. 408pp.

Olesiuk, P.F., Bigg, M.A. and Ellis, G.M. (1990) Life history and population dynamics of resident killer whales (*Orcinus orca*) in the coastal waters of British Columbia and Washington State. *Report of the International Whaling Commission* (special issue **12**), 209-243.

Oliver, J.S., Slattery, P.N., Silberstein, M.A. and O'Connor, E.F. (1982) A comparison of gray whale, *Eschrichtius robustus*, feeding in the Bering Sea and Baja California. *Fishery Bulletin*, **81**, 513-522.

Orians, G.H. (1969) On the evolution of mating systems in birds and mammals. *American Naturalist*, **103**, 589-603.

Osborne, R.W. (1986) A behavioral budget of Puget Sound killer whales. In: *Behavioral Biology of Killer Whales* (Ed. by B. Kirkevold and J.S. Lockard), pp. 211-249. A.R. Liss, New York.

Parker, E.D. (1984) Sperm competition and the evolution of animal mating strategies. In: *Sperm Competition and the Evolution of Animal Mating Systems* (Ed. by R.M. Smith), pp. 1-60. Academic Press, Orlando, FL.

Payne, K., Tyack, P. and Payne, R. (1983) Progressive changes in the songs of humpback whales (*Megaptera novaeangliae*): a detailed analysis of two seasons in Hawaii. In: *Communication and Behavior of Whales* (Ed. by R.S. Payne), pp. 9-57. Westview Press, Boulder, CO.

Payne, R. (1972) Swimming with Patagonia's right whales. *National Geographic*, **142**, 576-587.

Payne, R. (1976) At home with right whales. *National Geographic*, **149**, 577-589.

Payne, R. and McVay, S. (1971) Songs of humpback whales. *Science*, **173**, 585-597.

Perry, A., Baker, C.S. and Herman, L.M. (1990) Population characteristics of individually identified humpback whales in the central and eastern north Pacific. *Report of the International Whaling Commission* (special issue **12**), 307-317.

Perryman, W.L. and Foster, T.C. (1980) Preliminary report on predation by small whales, mainly the false killer whale (*Pseudorca crassidens*) on dolphin (*Stenella* spp. and *Delphinus delphis*) in the eastern tropical Pacific. *Administrative Report* LJ-80-05. Southwest Fisheries Center, La Jolla, CA.

Piatt, J.F., Methven, D.A., Burger, A.E., McLagan, R.L., Mercer, V. and Creelman, E. (1989) Baleen whales and their prey in a coastal environment. *Canadian Journal of Zoology*, **67**, 1523-1530.

Pike, G.C. (1962) Migration and feeding of the gray whale (*Eschrichtius gibbosus*). *Journal of the Fisheries Research Board of Canada*, **19**, 815-838.

Pryor, K. and Shallenberger, I.K. (1991) Social structure in spotted dolphins (*Stenella attenuata*) in the tuna purse seine fishery in the eastern tropical Pacific. In: *Dolphin Societies: Discoveries and Puzzles* (Ed. by K. Pryor and K.S. Norris), pp. 161-196. University of California Press, Berkeley.

Pusey, A.E. and Packer, C. 1987. Dispersal and philopatry. In: *Primate Societies* (Ed. by B.B. Smuts, D.L. Cheney, R.M. Seyfarth, R.W. Wrangham and T.T. Struhsaker), pp. 250-266. Chicago University Press, Chicago.

Ralls, K., Brugger, K. and Ballou, J. (1979) Inbreeding and juvenile mortality in small populations of ungulates. *Science*, **206**, 1101-1103.

Ralls, K., Brugger, K. and Glick, A. (1980) Deleterious effects of inbreeding in a herd of captive Dorcas gazelle. *International Zoo Yearbook*, **20**, 137-146.

Read, A.J. (1999) The harbour porpoise *Phocoena phocoena* (Linnaeus, 1758). In: *Handbook of Marine Mammals, Volume 6* (Ed. by S.H. Ridgway and R. Harrison), pp. 323-355. Academic Press, London.

Read, A.J. and Westgate, A.J. (1997) Monitoring the movements of harbour porpoises (*Phocoena phocoena*) with satellite telemetry. *Marine Biology*, **130**, 315-322.

Ridgway, S.H. and Harrison, R. (eds) (1980) *Handbook of Marine Mammals, Volume 3: The Sirenians and Baleen Whales*. Academic Press, London. 362pp.

Ross, H.M. and Wilson, B. (1996) Violent interactions between bottlenose dolphins and harbour porpoises. *Proceedings of the Royal Society of London*, **263**, 283-286.

Rutberg, A.T. (1983) The evolution of monogamy in primates. *Journal of Theoretical Biology*, **104**, 93-112.

Scott, M.S., Wells, R.S. and Irvine, A.B. (1990) A long-term study of bottlenose dolphins on the west coast of Florida. In: *The Bottlenose Dolphin* (Ed. by S. Leatherwood and R.R. Reeves), pp. 235-244. Academic Press, San Diego, CA.

Sergeant, D.E. (1962) The biology of the pilot or pothead whale, *Globicephala melaena* (Traill) in Newfoundland waters. *Bulletin of the Fisheries Research Board of Canada*, **132**, 1-84.

Sergeant, D.E. (1982) Mass strandings of toothed whales (*Odontoceti*) as a population phenomenon. *Scientific Reports of the Whales Research Institute*, **34**, 1-47.

da Silva, V.M.F. and Martin, A.R. (2000) A study of the boto, or Amazon river dolphin (*Inia geoffrensis*), in the Mamirauá Reserve, Brazil: operation and techniques. In: *Biology and Conservation of Freshwater Cetaceans in Asia*. (Ed. by R.R. Reeves, B.D. Smith and T. Kasuya), pp. 121-131. Occasional Papers of the IUCN Species Survival Commission, No. 23, Gland, Switzerland and Cambridge, UK.

Slijper, E.J. (1979) *Whales*. Cornell University Press, Ithaca, NY. 511pp.

Smith, R.L. (1984) *Sperm Competition and the Evolution of Animal Mating Systems*. Academic Press, Orlando, FL. 687pp.

Smuts, B.B. and Watanabe, J.M. (1990) Social relationships and ritualized greetings in adult male baboons (*Papio cynocephalus anubis*). *International Journal of Primatology*, **11**, 147-172.

Stern, B.R. and Smith, D.G. (1984) Sexual behavior and paternity in three captive groups of rhesus monkeys (*Macaca mulatta*). *Animal Behaviour*, **32**, 23-32.

Stirling, I. (1983) The evolution of mating systems in pinnipeds. *American Society of Mammalogists Special Publications*, **7**, 489-527.

Swartz, S.L. (1986) Gray whale migratory, social and breeding behaviour. *Report of the International Whaling Commission* (special issue **8**), 207-229.

Tershy, B.R. (1992) Body size, habitat use, and social behavior of *Balaenoptera* whales in the Gulf of California. *Journal of Mammalogy*, **73**, 477-483.

Tershy, B.R., Breese, D. and Strong, C.S. (1990) Abundance, seasonal distribution and population composition of balaenopterid whales in the Canal de Ballenas, Gulf of California, Mexico. *Report of the International Whaling Commission* (special issue **12**), 369-375.

Trivers, R. (1985) *Social Evolution*. Menlo Park, CA: Benjamin/Cummings. 462pp.

Trujillo, F. (1994) The use of photoidentification to study the Amazon river dolphin, *Inia geoffrensis*, in the Columbian Amazon. *Marine Mammal Science*, **10**, 348-353.

Tyack, P. (1981) Interactions between singing Hawaiian humpback whales and conspecifics nearby. *Behavioral Ecology and Sociobiology*, **8**, 105-116.

Tyack, P. (1986) Population biology, social behavior and communication in whales and dolphins. *Trends in Ecology and Evolution*, **1**, 144-150.

Tyack, P. and Whitehead, H. (1983) Male competition in large groups of wintering humpback whales. *Behaviour*, **83**, 132-145.

Wada, S. (1988) Genetic differentiation between two forms of short-finned pilot whales off the Pacific coast of Japan. *Scientific Reports of the Whales Research Institute*, **39**, 91-101.

Wang, D., Lu, W. and Wang, Z. (1989) A preliminary study of the acoustic behavior of the baiji, *Lipotes vexillifer*. In: *Biology and Conservation of the River Dolphins* (Ed. by W.F. Perrin, R.L. Brownell Jr., K. Zhou and J. Liu), pp. 137-140. Occasional Papers of the IUCN Species Survival Commission, No. 3, Gland, Switzerland.

Wells, R.S. (1991) The role of long-term study in understanding the social structure of a bottlenose dolphin community. In: *Dolphin Societies* (Ed. by K. Pryor and K.S. Norris), pp. 199-225. University of California Press, Berkeley.

Wells, R.S., Irvine, A.B. and Scott, M.D. (1980) The social ecology of inshore odontocetes. In: *Cetacean Behavior: Mechanisms and Functions* (Ed. by L.M. Herman), pp. 263-317. J. Wiley & Sons, New York.

Wells, R.S. and Scott, M.D. (1990) Estimating bottlenose dolphin population parameters from individual identification and capture-release techniques. *Report of the International Whaling Commission* (special issue **12**), 407-415.

Wells, R.S., Scott, M.D. and Irvine, A.B. (1987) The social structure of free-ranging bottlenose dolphins. In: *Current Mammalogy* (Ed. by H.H. Genoway), pp. 247-305. Plenum Press, New York.

Westgate, A.J. and Read, A.J. (1998) Applications of new technology to the conservation of porpoises. *MTS Journal*, **32**, 70-81.

Whitehead, H. (1983) Structure and stability of humpback whale groups off Newfoundland. *Canadian Journal of Zoology*, **61**, 1391-1397.

Whitehead, H. (1990) Rules for roving males. *Journal of Theoretical Biology*, **145**, 355-368.

Whitehead, H. and Arnbom, T. (1987) Social organization of sperm whales off the Galapagos Islands, February-April 1985. *Canadian Journal of Zoology*, **65**, 913-19.

Whitehead, H. (1998) Cultural selection and genetic diversity in matrilineal whales. *Science*, **282**, 1708-1711.

Whitehead, H., Waters, S. and Lyrholm, T. (1991) Social organization in female sperm whales and their offspring: constant companions and casual acquaintances. *Behavioural Ecology and Sociobiology*, **29**, 385-389.

Wilson, E.O. (1975) *Sociobiology: The New Synthesis*. Harvard University Press, Cambridge, MA. 697pp.

Winn, H.E. and Winn, L.K. (1978) The song of the humpback whale, *Megaptera novaeangliae*, in the West Indies. *Marine Biology*, **47**, 97-114.

Wrangham, R.W. (1980) An ecological model of female-bonded primate groups. *Behaviour*, **75**, 262-300.

Wrangham, R.W. (1983) Social relationships in comparative perspective. In: *Primate Social Relationships* (Ed. by R.A. Hinde), pp. 325-334. Sinauer, Sunderland, MA.

Würsig, B. (1979) Dolphins. *Scientific American*, **240**, 136-148.

Würsig, B. (1986) Delphinid foraging strategies. In: *Dolphin Cognition and Behavior: a Comparative Approach* (Ed. by R.J. Schusterman, J.A. Thomas and F.G. Wood), pp. 347-359. Erlbaum Associates, Hillsdale, NJ.

Würsig, B. (1988) The behavior of baleen whales. *Scientific American*, **258**, 102-107.

Würsig, B. (1989) Cetaceans. *Science*, **244**, 1550-1557.

Würsig, B., Cipriano, F. and Würsig, M. (1991) Dolphin movement patterns: information from radio and theodolite tracking studies. In: *Dolphin Societies* (Ed. by K. Pryor and K.S. Norris), pp. 79-111. University of California Press, Berkeley, CA.

Würsig, B., Dorsey, E.M., Fraker, M.A., Payne, R.S. and Richardson, W.J. (1985) Behavior of bowhead whales, *Balaena mysticetus*, summering in the Beaufort Sea: a description. *Fishery Bulletin*, **83**, 357-377.

Würsig, B. and Würsig, M. (1980) Behavior and ecology of the dusky dolphin, *Lagenorhynchus obscurus*, in the south Atlantic. *Fishery Bulletin*, **77**, 871-890.

Würsig, B., Guerrero, J. and Silber, G.K. (1993) Social and sexual behavior of bowhead whales in fall in the western Arctic: a re-examination of seasonal trends. *Marine Mammal Science*, **9**, 103-110.

Würsig, B., Würsig, M. and Cipriano, F. (1989) Dolphins in different worlds. *Oceanus*, **32**, 71-75.

Xiao, Y. and Jing, R. (1989) Underwater acoustic signals of the baiji, *Lipotes vexillifer*. In: *Biology and Conservation of the River Dolphins* (Ed. by W.F. Perrin, R.L. Brownell Jr., K. Zhou and J. Liu), pp. 129-136. Occasional Papers of the IUCN Species Survival Commission, No. 3, Gland, Switzerland.

Yuanyu, H., Xiafeng, Z., Zhuo, W. and Xiaogiang, W. (1990) A note on the feasibility of using photo-identification techniques to study the baiji (*Lipotes vexillifer*). *Report of the International Whaling Commission* (special issue **12**), 439-440.

Zhou, K. (1989) Review of studies of structure and function of the baiji, *Lipotes vexillifer*. In: *Biology and Conservation of the River Dolphins* (Ed. by W.F. Perrin, R.L. Brownell Jr., K. Zhou and J. Liu), pp. 99-113. Occasional Papers of the IUCN Species Survival Commission, No. 3, Gland, Switzerland.

Zhou, K. and Li, Y. (1989) Status and aspects of the ecology and behavior of the baiji, *Lipotes vexillifer*, in the lower Yangtze River. In: *Biology and Conservation of the River Dolphins* (Ed. by W.F. Perrin, R.L. Brownell Jr., K. Zhou and J. Liu), pp. 86-91. Occasional Papers of the IUCN Species Survival Commission, No. 3, Gland, Switzerland.

Chapter 6

New Perspectives on the Behavioural Ecology of Pinnipeds

HUMBERTO LUIS CAPPOZZO
Behavioural Ecology Laboratory y Estación Hidrobiológia de Puerlo Quequen, Museo Argentino de Ciencias Naturales "Bernardino Rivadavia", Av. Angel Gallardo 470, C 1405 DJR Buenos Aires, Argentina, E-mail: cappozzo@muanbe.gov.ar

1.   INTRODUCTION

Pinnipeds evolved from terrestrial carnivore ancestors and their amphibious nature has required many adaptations to life both in the water and on land or ice. In the course of their evolution they have retained certain terrestrial traits (such as giving birth on land) while having undergone many adaptations to aquatic existence (in particular, pelagic foraging). They can live for long periods of time both in the ocean and on land. Pinnipeds are classified into three living families: Otariids (fur seals and sea lions); Phocids (true seals), and Odobenids (walruses).

Pinnipeds have diverse mating systems. Many species are highly polygynous (males acquire multiple females) and sexually dimorphic (males being bigger than females). Other species are monogamous, or serially monogamous, and under this condition females and males are nearly the same size. Ecological factors are primarily responsible for the evolution of mating and social systems in pinnipeds, which have also been influenced by phylogeny (Stirling, 1983). Pinnipeds breed on land (islands or open beaches) or on ice (floating or attached to land). The environment (topography and climate conditions) plays a critical role in structuring the mating systems, the breeding behaviour, and the social organisation. Of the 33 species of pinnipeds, 20 breed on land and the remaining 13 breed on ice. Extreme polygyny is characteristic of nearly all land-breeders (18 species of the land-breeders are highly polygynous and sexually dimorphic), which breed in moderate-sized to extremely large colonies (they are gregarious). Mating occurs primarily on land as opposed to in the water in most of these species, and polygyny can take two main forms (from the male perspective): female (or harem) defence or resource defence (males compete for territories) (see below). Except for the walruses, all ice breeders are phocid seals which can mate in the water or on ice, although little is known about their reproductive behaviour.

The significance of terrestrial mating to the maintenance of organised polygyny has been discussed by several authors (Bartholomew, 1970; Stirling, 1975, 1983; Riedman, 1990). Some of them suggested that when aquatic mating occurs, males are unable to effectively maintain their harems (Stirling, 1983). Aquatic copulation, common in many ice breeding seals, only occurs in 21-30% of pinniped species, but in most polygynous pinniped species copulations occur on land. Pinnipeds that are monogamous are often serially monogamous because the male will search for another female or mother-pup pair after mating. Serial monogamy may be thought of as a sequential polygyny since one male mates with more than one female during a single breeding season (*e.g.* ringed seal, crabeater seals).

The potential for polygyny is high in sexually dimorphic pinniped species. All otariid species are polygynous, and females gather in high densities on island beaches or rocky shelves where they can find suitable birthing sites. Individual females come into oestrous a few days postpartum. Their oestrous is moderately synchronised, with most females in a given species becoming receptive within 20-35 days (in a few species this period may be more than double this length).

A model for the evolution of polygyny and sexual dimorphism in the otariids, sea lions and fur seals, must account for the extreme female gregariousness shown during the annual breeding season (Bartholomew, 1970; Boness, 1991). The theory predicts that female clumping promotes the evolution of intrasexual male competition, polygyny, and sexual dimorphism (Emlen and Oring, 1977).

2.   CURRENT VIEW OF SEXUAL SELECTION AND MATING SYSTEM

Darwin (1871) observed that male animals compete with each other to inseminate females, while females do not compete for males but are choosy in selecting a mate. Darwin argued that these behaviours helped each sex produce the largest possible number of offspring.

Sexual Selection (SS) was suggested in *the Origin of Species* (Darwin, 1859), but it was only developed as a topic in *The Descent of Man and Selection in Relation to Sex* (Darwin, 1871, 1874). Darwin put together the pieces of a puzzle by the selective forces responsible for the evolution of characters, often the sexually dimorphic characters, that seemed to hinder an organism's chances of survival and which were necessary for sexual reproduction only in the presence of competition for mates. Such characters were considered to evolve by SS (Harvey and Bradbury, 1991), and are called secondary sexual characters. Darwin excluded primary sexual characters (such as genitalia, reproductive tracts, and mammary glands) from this process. The key issue in Darwin's process was competition for mates. Two principles are involved in SS: 1) male-male competition (intra-sexual selection); 2) female choice (intersexual selection or epigamic selection).

Emlen and Oring (1977) proposed a model which explains variation in mating systems in relation to sexual selection, ecological and phylogenetic constraints. Accordingly, when conditions are such that the fitness of members of one sex can be increased by controlling access to the other sex, competition among the controlling sex increases and potential for polygamy is high. Polygamy is defined as any mating system in which some members of one sex mate with more than one member of the other within a single breeding season. Polygamy is known as polygyny when males acquire multiple females. A high potential for polygyny does not necessarily result in a polygynous mating system; both phylogenetic and ecological factors constrain the ability to capitalise on the polygamy potential (Trivers, 1972; Emlen and Oring, 1977). If substantial male parental care is required to rear offspring successfully, then polygyny is not likely to occur or it will occur at low levels (*e.g.* colonial breeding birds). Studies of mammalian species have shown that less than 5% have monogamous mating systems. The reproductive strategy of polygynous pinnipeds is for males to inseminate as many females as possible without helping to rear the offspring.

Polygyny occurs if environmental or behavioural conditions bring about the clumping of females, and males have the capacity to monopolise them. Based on the extent to which resources critical to females or females themselves are monopolisable, four major forms of polygyny have been suggested (Emlen and Oring, 1977):

a) **resource defence**: males control access to females indirectly, by monopolising critical resources (*i.e.* territories, birth sites, water, *etc.*);

b) **female (or harem) defence**: males control access to females directly, usually by virtue of female gregariousness.

c) **male dominance (lekking in mammals)**: when local female density is high, males defend small, clustered mating territories (not resources), which females visit solely for mating (The lek breeding system is not common and occurs in only 0.2% of mammalian species). Males aggregate, and females mate with one or more males.

d) **scramble competition**: solitary males search widely for receptive females which they may guard temporarily.

3.   MATING SYSTEMS, BREEDING BEHAVIOUR AND SOCIAL ORGANISATION IN PINNIPEDS

The selective factors involved in the evolution of pinniped social behaviour can be identified, and their relative importance assayed by analysis of the processes by which breeding organisation is annually re-established and maintained in pinniped rookeries. Nevertheless, some of the factors are difficult to identify: one example is site fidelity. These factors suggest that polygyny appeared early in the history of pinnipeds, probably associated their amphibious characteristics. Moreover, the fossil record shows that polygyny (as indicated by sexual dimorphism) arose early in the history of otariids (mid-Miocene). The remarkable similarity of the patterns of social structure in members of the family Otariidae and in polygynous members of the family Phocidae, such as elephant seals, suggests that the key factors which have led to the evolution of polygyny must be few in number and must also be closely related to main adaptative features which characterise the taxon Pinnipedia (Bartholomew, 1970). On the one hand, marine mammals have special adaptations like large size and subcutaneous fat, which contribute to the conservation of endogenous heat; stored fat can serve as an energy source during fasting periods. On the other hand, large mammals can go without food longer than small ones because of their low weight-relative metabolism.

Pinnipeds combine two characteristics, offshore marine feeding and terrestrial parturition, which occur together in no other mammals (Bartholomew, 1970; Stirling, 1975, 1983). These two characteristics are the most fundamental features of pinniped life history, in terms of their social behaviour. In most species, parturition is seasonal and highly synchronised. Pups are weaned at the birth site on either land or ice. Most phocids wean the pup after 6 or 7 weeks at the most, and the females of several species do not return to the water until nursing has ceased. An exception in this duration range occurs in the Mediterranean monk seal, in which the nursing period continues in most cases for over 100 days, and in some extends over 120 days (Aguilar *et al.*, unpublished). Such a long nursing period is exceptional and duplicates the maximum lactation lengths observed in other phocid species (Oftedal *et al.*, 1987; Riedman, 1990; Evans and Stirling, this volume; Lockyer, this volume). Otariids take several months to wean pups so that, after an initial intense nursing period of a week or more, females begin to make feeding trips to the sea in order to maintain their reserves for continued nursing. Some otariid species may nurse their pup for over a year. In no species does the male help feed the young. Additionally, it does not seem likely that males of any species aid or teach young when mixed sex groups are foraging. Except for walruses, ovulation and mating follow within a few days (or weeks) after parturition. Thus, all the aspects of the reproductive cycle, including intrasexual competition between males for females, occur in a relatively short time period either at the birth site or in the water immediately adjacent to it (in those species in which mating occurs in the water).

Pinnipeds provide an opportunity to study the influence of ecological variation on differences in mating systems because there is both an extreme variation in the reproductive strategies of each sex, and considerable diversity in breeding habitats and climates used by these animals (Stirling, 1983; Bonner, 1984; Boness, 1991; Le Boeuf, 1991; Boness *et al.*, 1993). Males play no role in rearing offspring and devote their entire effort during the reproductive period to the acquisition of mates. Females devote their principal effort to rearing offspring, although in some species they may also enhance fitness through mate choice (Cox and Le Boeuf, 1977; Boness *et al.*, 1982; Heath, 1989; Kuroiwa and Majluf, 1989; Cappozzo *et al.*, 1993). Sea lions, fur seals, and some seal species are best known for their extreme levels of polygyny (Emlen and Oring, 1977; Davies, 1991), and they also exhibit a wide range of levels of polygyny and mating strategies.

4.   REPRODUCTIVE BEHAVIOUR IN SOUTH AMERICAN OTARIIDS

The reproductive behaviour of individually marked southern sea lions, *Otaria flavescens*, was studied for ten years at Punta Norte rookeries in Peninsula Valdés, Argentina (Campagna, 1987; Campagna and Le Boeuf, 1988a, b; Campagna *et al.*, 1988a, b; Cappozzo *et al.*, 1991; Cappozzo, 1996). At Punta Norte rookeries, adult males defend females but not a territory. In a female-defence polygynous mating system, with males capable of forcing copulation, the female's first priority would be to survive the breeding season, and then mate with high quality males. Males should attempt to mate with as many females as possible. Nonetheless, females may also be a commodity, even for narrow-minded sexual machines. Adult males may maximise their reproductive success through the selective defence of those females that are close to oestrus (Cappozzo, 1996). Adult females could develop some kind of choice for the male with which they copulate, by changing the associated male before giving birth, or mating with more than one male during their maximum sexual receptivity.

Thermoregulatory requirements interact with rookery topography to shape mating strategies, variation in mating success, and the mating system type. At Punta Norte, the pebble substrate is homogeneous with respect to thermoregulatory advantages. Thus, sites advantageous for thermoregulation are not a limited resource that can be used to attract females. Consequently, males achieve mates through selective female defence or abducting females. Conversely, at the Puerto Piramide rookery (another sea lion reproductive area at Peninsula Valdés), variation in the quality of the substrate with respect to reducing thermal stress favours the development of a territorial system where the best territories contain water or are close to the water. Here, abduction of females or direct defence of females by males is not required since females preferentially gather in wet territories. Thus, the topography and substrate of the breeding area, coupled with thermoregulatory requirements, are driving forces that generate adaptative changes in male mating behaviour (Campagna and Le Boeuf, 1988b).

Subadult or nonterritorial males may develop alternative mating strategies: group raids (Campagna *et al.*, 1988); solitary breeding (single male with a single female or with a small isolated harem) (Campagna *et al.*, 1992); female interceptions (keeping females that leave the main breeding area on the way to and from the water) (Campagna and Le Boeuf, 1988a).

Figure 1.   Central breeding area: adult male southern sea lions defend females instead of territory. (Photo: L. Cappozzo)

All otariids are polygynous (Bartholomew 1970), and in polygynous mammals large adult size is likely to benefit males more than females (Clutton-Brock *et al.*, 1981, 1982). Among sea lions and fur seals, success in male-male competition for mates depends partially on strength, resistance, and fighting ability between contenders (Bartholomew and Hoel, 1953; Campagna and Le Boeuf, 1988). In these species, sexual selection would have favoured large, more competitive individuals (Bartholomew, 1970; Alexander *et al.*, 1979).

Sexual dimorphism of adults is striking among the otariids with males being at least three times larger in mass than adult females (Bartholomew, 1970; Alexander *et al.*, 1979; King, 1983). Under these circumstances, if the amount of parental investment partially affects adult body size, sex allocation theory predicts that mothers should invest extra resources in sons respective to daughters even at a sex ratio of 1:1 (Maynard-Smith, 1980). A male-biased investment is not predicted by traditional sex-ratio theory when the primary sex ratio of the population is 1:1 (Fisher, 1930). In accordance with Maynard Smith (1980), maternal investment in fur seals and sea lions is usually male-biased early during gestation (York, 1987), at birth (Ling and Walker, 1977; Payne, 1979; Trillmich, 1986; Ono *et al.*, 1987), and at weaning (Rand, 1956; Doidge *et al.*, 1984; Trillmich, 1986) despite a 1:1 sex ratio (Bonner, 1968; Trillmich, 1986; Kerley, 1987; Roux, 1987; Vaz-Ferreira and Ponce de Leon, 1987).

The southern sea lion, *Otaria flavescens*, is one of the largest and most dimorphic of otariids. Adult males are up to five times heavier than females (Hamilton, 1934, 1939). Differences in size between males and females have been documented also among juveniles (Hamilton, 1934, 1939), and even newborns (Cappozzo *et al.*, 1991). The sex ratio at birth for this species is 1:1 (Lewis and Ximenez 1983; Crespo 1988). But, sex ratios that slightly favour males, and differential mortality, result in an adult population that is strongly female biased. Southern sea lions born at Peninsula Valdés are sexually dimorphic in mass and length at birth. This result is consistent with those reported for juvenile individuals in other populations of the same species (Hamilton, 1934), and for newborns of other otariids (Rand, 1956; Ling and Walker, 1977; Payne, 1979; Doidge, *et al.*, 1984; Trillmich, 1986; Ono *et al.*, 1987; Costa *et al.*, 1988; see also contributing papers on fur seals in Croxall and Gentry, 1987).

The degree of dimorphism of southern sea lion pups is similar to that of other otariids. Newborn male fur seals are 9-18% heavier than females (Trillmich, 1986; Croxall and Gentry, 1987), a value similar to that found in newborn southern sea lions in Península Valdés. Assuming no sex difference in energy expenditure by foetuses, sexual dimorphism in size of newborn pups suggests that southern sea lion mothers invest more energy in sons than in daughters during gestation. Contrary to other otariids in which males increase in mass faster than females (Payne, 1979; Doidge *et al.*, 1984; Trillmich, 1986; Ono and Boness, 1996), we did not find a sex difference in growth rates in the southern sea lion. The size dimorphism present at birth in this species remains during the nursing period, suggesting that sons continue to be more costly to their mothers than daughters. We do not know if differences in size found at birth in the southern sea lions of Península Valdés remains until weaning, but there is some evidence suggesting that this may be true. Six-month old male southern sea lion pups at the Malvinas (Falkland) Islands are longer than females; the difference in length remains in eighteen-month old individuals, and is even more marked in older juveniles (Hamilton, 1934). Moreover, data for other otariids show that sexual dimorphism at birth continues at weaning (*e.g.* Doidge *et al.*, 1984; Trillmich, 1986).

Among the pinnipeds, some otariid species fit theoretical predictions about perinatal sexual dimorphism and maternal investment in polygynous species better than phocids. In all otariid species studied, females invest more in male offspring, whereas phocid females of some, but not all species, invest more in sons (Kovacs and Lavigne, 1986b). Nevertheless, some other otariid species do not fit very well the predictions about differential investment of mothers in male offspring, because weaning occurs at an early age (three months in some species) and most of the investment in size dimorphism occurs after weaning, at 3-5 years (Gentry, 1998). Conversely, in one of the most sexually dimorphic and polygynous of the pinnipeds, the southern elephant seal, *Mirounga leonina*, male pups have a similar mass at weaning than females (McCann *et al.*, 1989). In the northern elephant seal, *M. angustirostris*, mothers invest equally in sons and daughters when investment is measured in terms of the mother's future reproductive success, although body masses of the sexes differ at weaning (Le Boeuf *et al.*, 1989). Male grey seal pups are larger at birth, gain mass at a faster rate, and are weaned at a greater mass than females (Boyd and Campbell 1971; Kovacs and Lavigne, 1986a; Anderson and Fedak, 1987).

In the southern sea lion, size dimorphism of pups early in lactation is not reflected in female nursing behaviour, length of foraging trips, and periods on land attending the pup after feeding. Similar findings have been reported in the Steller sea lion, *Eumetopias jubatus*, (Higgins *et al.*, 1988). These results suggest that female behaviour does not differ by sex of the pup.

The South American fur seal, *Arctocephalus australis,* is a polygynous, highly sexually dimorphic, species. Adult males defend territories during the breeding season, and their behaviour is affected by the thermoregulatory resources available (water and shadow). We have studied this species at Isla de Lobos, Uruguay, during three consecutive reproductive seasons. The study aims to describe reproductive behaviour and male copulatory success in relation to territorial quality.

During the breeding season, territorial groups are established on the shoreline or in nearby areas. These places are provided with tidal pools or rocks giving shade. The distances between territorial males may vary according to topography.

All islands in Uruguay on which *A. australis* lives are also populated by *O. flavescens*. The competition between both species is reduced by several factors, particularly the different breeding seasons and their preference for different habitats (Vaz Ferreira and Ponce de León, 1987). As noted, the South American fur seal is a polygynous, and sexually dimorphic species. The breeding season occurs between November and December. Adult males defend territories and adult females distribute themselves on these territories once they are established by males (Vaz Ferreira and Ponce de León, 1987; Cappozzo, 1996).

Figure 2. Adult male South American fur seal giving full throat call in defence of its territory. (Photo: I. Stirling)

In order to study differences in the males' copulatory success in relation to the quality of their territories, we ranked those territories according to the presence or absence of favourable resources for thermoregulation (water and shade). Territories established in the study area remain stable during the reproductive season. Marked adult males stay in their territory more than 90% of the observed time (Vaz Ferreira, 1956; Cappozzo *et al.*, 1996). Differences found between territorial adult males, according to territory type, were correlated with the strategy of tenure. In areas without water from tidal pools, males use a strategy of "quitting-and-recovering" (QR) territories (Cappozzo, 1995). Males abandoning territories have been described for other otariid species (Gisiner, 1987; McCann, 1987; Boness, 1991). This reproductive strategy and male reproductive success was related to territorial quality. Adult females show birth site fidelity between consecutive years. From 20 adult females tagged during 1990, at least eight gave birth during 1991 at the same location as the previous year (± 2m). In the northern fur seal, female site fidelity or philopatry drives the whole system because birth sites become the resource for which males compete (see Gentry, 1998). We observed that tagged females copulated in territories other than those where they gave birth. Females move to territories with better thermoregulatory resources. Most adult females spend the hottest hours (11-14h.) forming aquatic groups of several individuals, and stay in the water at 50-100 metres from the coast. The number of females in the colony show fluctuations according to this (Cappozzo *et al.*, 1996). During the hours of highest temperatures and solar radiation, females from territories without water or shade move their pups, taking them in their mouth, to areas with tidal pools near to the sea and, after that, they return the pup to the birthplace. These daily movements are made crossing through other males' territories, but these never interfere with the displacements (Cappozzo, 1996).

In summary, males compete for territories with resources for thermoregulation. The data suggests that this characteristic is decisive for the copulatory success of a male. Territorial adult males on territories with poor resources show a reproductive strategy of "quitting-and-recovering" the same territories. Females move through territories, suggesting they may copulate with successful males in territories other than those where they give birth. These results indicate plasticity in the reproductive behaviour of the species. Successful males have high quality territories, while other males may develop strategies tending to save energy in the maintenance of territories with poor resources.

5. SOUTHERN ELEPHANT SEALS AND MEDITERRANEAN MONK SEALS: TWO DIFFERENT STRATEGIES IN PHOCIDS

Southern elephant seals, *Mirounga leonina*, have a circumpolar breeding and moulting distribution on both sides of the Antarctic Convergence. Several of the largest populations of this species have recently declined sharply in number. This species has two pelagic and two terrestrial phases in their annual cycle. The Península Valdés colony (Argentina) has distinctive ecological and demographic features (it is apparently the only colony in the world that is growing; see Campagna and Lewis, 1992). Elephant seals come ashore to breed (August ·to November) and moult (December to February). The rest of the year is spent at sea. During the breeding season, males aggressively establish a dominance hierarchy. Females are gregarious and give birth to a single pup in harems reproductively controlled by large, alpha males. About three weeks after birth, females wean their pups, mate, and return to the sea. Estimated tenure in the breeding area for dominant males (large harems of >50 females) is 57-80 days (Campagna *et al.*, 1993) and, except for small harems (2 to 5 females) and mating pairs, alpha males arrive early in the reproductive season. Most females mate inside the harems, and between 7 to 9% are mating pairs composed of one male and one female. Harems are smaller at Península Valdés than in other breeding colonies (Campagna *et al.*, 1993). This species breed in Patagonia under unusually low density conditions (Le Boeuf and Petrinovich, 1974). Female aggression, disruption of lactation and mother-pup separations are unusual events at Península Valdés. Most females wean their pup successfully around 23 days after birth; this success perhaps is associated with the particular demographic conditions in the colony.

Female southern elephant seals appear to invest equally in sons and daughters, when investment is measured in terms of mass at birth and, at the end of lactation, mass gain during nursing period, age at weaning and nursing behaviour (Campagna *et al.*, 1992b). These results are consistent with those found in male and female elephant seal pups born at South Georgia Island (McCann *et al.*, 1989). Elephant seals, being the most sexually dimorphic in size of pinnipeds at adulthood, show no differential investment and sexual dimorphism early in life. The large size of elephant seal mothers may partially explain why the species does not fit predictions based on Maynard-Smith's (1980) theory. Another possible explanation of unbiased investment by sex in elephant seals hinges on the effect of maternal investment on a pup's lifetime fitness. According to Maynard-Smith, females should invest more in male offspring if the latter's fitness would increase proportionally more than if a similar greater-than-average investment were put into female pups. If differential maternal effort does not make any significant contribution to a male's lifetime fitness, Maynard-Smith's predictions would not apply to elephant seals (Campagna *et al.*, 1992b). The available information on both elephant seal species is not conclusive about the long-term effects of maternal investment on the fitness of the offspring.

Figure 3.   Male and female southern elephant seals in Peninsula Valdez, Argentina. (Photo: J.A. Gomez)

The variation in the spatial distribution along the coastline of Península Valdés may respond to habitat preference. Females prefer sandy beaches to vegetated hillocks, pebble, cobble and rocky areas. Habitat choice may be related to thermoregulation, and as southern elephant seals are mainly subantarctic and Antarctic phocids, heat stress may be critical at the latitude of Península Valdés (approx. 42° S), where air temperature fluctuates from 18 to 25° C and solar radiation may reach 1.42 cal/cm^2/min (Campagna and Le Boeuf, 1988b). Redistribution related to habitat preference may be facilitated by a relatively weak philopatric tendency of the seals at Península Valdés compared with other colonies. Animals born and tagged at Punta Norte (P. Valdés) and adjacent areas are found moulting or reproducing as far as 120 km from their place of birth (Campagna and Lewis, 1992), but some seals are dispersed more than 1,000 km away from Patagonia, at Malvinas (Falkland) Islands (Lewis *et al.*, 1996).

The Mediterranean monk seal (*Monachus monachus*) is one of the scarcest mammalian species found worldwide. Originally restricted to the Black Sea, the Mediterranean Sea and the temperate waters of the eastern North Atlantic, it was severely depleted by sealing and other human activities (such as fishing and industrialisation), well before the end of last century. In recent decades, the overall trajectory of the various populations indicates a continuous decline in all regions still occupied by the species. Thus, many localities where monk seals were present in the 1960s are not occupied anymore, and the remaining populations are heavily fragmented, a fact that further hinders their recovery. Although there are no reliable figures for the total number that survive today, it is commonly accepted that they do not exceed 500 individuals (Reijnders *et al.*, 1993). The rarefaction of the species during the present century has obviously hampered the knowledge of its biology. The only large colony of the Mediterranean monk seal surviving today and, therefore, the only aggregation considered to still display the social structure and biological cycle typical of the species, is located in the western Sahara coast. In the last two decades, the Cabo Blanco colony of Mediterranean monk seals, the largest aggregation of the species surviving today, has had limited contact with humans because of the war in the western Sahara area (1973-1991). Fishing activity, at least on a large scale, disappeared, and the topography of the area, composed of caves at the base of high cliffs, contributed to the isolation of the species.

All phocids fast or feed very little during lactation, and maternal care revolves around a comparative brief but intensive lactation period, during which the pup is nursed several times each day (Riedman, 1990). Lactation may continue for only four days, as in the hooded seals, or may last as long as two months, as in the Baikal seal. The fasting mother lives off her own reserves in addition to nursing her pups, but in some species, as in the grey seals, *Halichoerus grypus*, during nursing the females spend the majority of their time ashore resting (Fedak and Anderson, 1982; Anderson and Harwood, 1985, Kovacs, 1987) and utilising beach sites with access to water to return to the sea between nursing bouts (Kovacs, 1987). All six species of land-breeding phocids have long lactation periods in comparison with those of seals that breed on unstable pack ice. Lactation periods range from about three to six weeks in the land breeders, which include two species of elephant seals, harbour seals, grey seals, and Hawaiian monk seals (Riedman, 1990). The sixth species, Mediterranean monk seal, has the longest lactation period among all pinniped species, lasting sixteen weeks (Aguilar *et al.*, unpublished).

Figure 4.(a). Adult female grey seals congregate on a breeding beach to suckle their young (Photo: P.G.H. Evans).

Figure 4 (b). A grey seal pup typically remains ashore for 3-4 weeks (Photo: P.G.H. Evans).

Studies on the Hawaiian monk seal, *Monachus schauinslandii*, showed that fostering behaviour (see the definition of fostering given by Riedman and Le Boeuf, 1982) is common in the species, and may be extreme in some colonies (Johnson and Johnson, 1984; Boness, 1990). This species has a long period of pupping compared to other phocids, lasting at least four months, without a distinct peak of births (Boness, 1990). Females terminate lactation abruptly by leaving the island and not returning, and the duration of lactation on this species is 36-45 days (Boness, 1990).

There is very little available information on the behavioural ecology of Mediterranean monk seals. Most papers about this species focus on censuses, abundance, and general observations on behaviour of isolated individuals in Mediterranean or north-western Atlantic waters (Boulva, 1979; Marchessaux, 1989). The mothering system and the lactation patterns of otariids and phocids, the two main suborders of the Pinnipedia, are sharply different. Preliminary results from unpublished studies indicate that in Mediterranean monk seals, these biological traits are, in most respects, substantially different from what has been traditionally proposed for the species from the fragmentary data available, and are not along the lines of typical phocid lactation patterns, including the congeneric Hawaiian monk seal, *Monachus schauinslandii*.

Figure 5.   Mediterranean monk seal in Mauritania, West Africa. (Photo: A. Aguilar)

Presence of nursing pups in the caves has been observed throughout the year, although their numbers appear to be somewhat lower from December to February (Aguilar *et al.*, unpubl.). This finding is consistent with the apparent lack of a definite seasonality in birth for this population. In pinnipeds, seasonality of reproductive events is typically associated with ecosystems with temporal heterogeneity in environmental features, mainly in the availability of food and water, or ambient temperature.

Mothers spend the first week after giving birth on land close to their newborn even during high tides, and show aggressive behaviours with any other approaching seals. The bond between mother-pup pairs appears to be well developed and recognition is established by frequent nuzzling and vocalisations. These encounters usually are initiated when the female arrives on land. She directs her head to the hauled-out seals and starts vocalising until the pup answers the call. Then, they approach and recognise each other by odour and vocal traits. Once identity has been confirmed, the mother-pup pair moves together through the group of seals until reaching an adequate location for nursing.

The most striking aspect of lactation in Mediterranean monk seals is its duration. Suckling continues in most cases for over 100 days and in some it extends for over 120 days (Aguilar *et al.*, unpubl.). Such a long nursing period is exceptional. It almost duplicates the maximum lactation lengths observed in other phocid species (for reviews, see Oftedal *et al.*, 1987; Riedman, 1990). Moreover, it indicates that previous estimates of the duration of lactation proposed for the Mediterranean monk seal were wrong. Nevertheless, monk seals are the closest to the primitive type of phocid behaviour, and are the only phocids that still inhabit the ancestral area where phocids evolved (Tethys Sea). If this is the ancestral pattern of phocids, it does not differ much from the pattern of some modern otariids.

All phocid species fast or feed very little during lactation, and maternal care involves a relatively brief but intense lactation period; the bulk of the female reproductive expenditure is in an environment in which she cannot feed (Oftedal *et al.*, 1987). Among 19 phocid species, only two feed normally during the lactation period. Conversely, the ribbon seal, harbour seal, harp seal, ringed seal and bearded seal reduce substantially their food intake (Riedman, 1990; Evans and Stirling, this volume). Of those species, only the harbour seal breeds on land, while the other species breed on pack-ice (Oftedal *et al.*, 1987).

Mediterranean monk seals show maternal attendance patterns that suggest the possibility that mothers feed at sea during the lactation period and probably have the ability to recover maternal reserves in order to allow an extended care period. Tidal cycles promote mother-pup separation, and these characteristics could affect behavioural plasticity in order to extend the nursing period. The long lactation period in the Mediterranean monk seal may be a response to the pups' lack of possibilities to suckle several times per day (their mothers are feeding at sea). Also, fostering and milk stealing appear to be common behaviours in Mediterranean monk seals, and this undoubtedly facilitates a gradual weaning process (Aguilar *et al.*, unpubl.). Overall, the results so far obtained in this study evidence that the lactation and maternal behaviour of the Mediterranean monk seal differ markedly from what has been proposed in the past from the fragmentary data available. However, further information is necessary to clarify certain aspects, like the incidence of fostering or of feeding in lactating females, to establish a definite pattern for the species.

ACKNOWLEDGEMENTS

I would like to thank Toni Raga and Peter Evans for inviting me to contribute to this book, as well as for their support and confidence. I am indebted also to Roger Gentry and Burney Le Boeuf for their kind and appropriate suggestions. Thanks also to Valeria Silvestroni for the editing work on this chapter, and Diego Golombek who helped during the early stages with his comments. I am also deeply grateful to the many persons who contributed in numerous ways to my work with pinnipeds over the past fifteen years.

REFERENCES

Aguilar, A, Gazo, M., Cappozzo, H.L., Aparicio, F., Cedenilla, M. and González, L.M. (unpublished) Lactation and mother-pup behaviour in the Mediterranean Monk Seal, *Monachus monachus*: An unusual pattern for a phocid.

Alexander, R.D., Hoogland, J.L., Howard, R.D., Noonan, K.M. and Sherman, P.W. (1979) Sexual dimorphisms and breeding systems in pinnipeds, ungulates, primates and humans. In: *Evolutionary Biology and Human Social Behaviour: An Anthropological Perspective.* (Ed. by N.A. Chagnon and W. Irons), pp. 402-435. Duxbury Press, Nort Scituate, MA.

Anderson, S.S. and Harwood, J. (1985) Time budgets and topography: how energy reserves and terrain determine the breeding behaviour of grey seals. *Animal Behaviour*, 33, 1343-1348.

Anderson, S.S. and Fedak, M.A. (1987) Grey seals, *Halichoerus grypus*, energetics: females invest more in male offspring. *Journal of Zoology* (London), 211, 667-679.

Bartholomew, G.A. (1970). A model for the evolution of pinniped polygyny. *Evolution*, 24, 546 - 559.

Bartholomew, G.A. and Hoel, P.G (1953) Reproductive behaviour in the Alaska fur seal, *Callorhinus ursinus. Journal of Mammalogy*, 34, 417-436.

Boness, D.J. (1990) Fostering behaviour in Hawaiian monk seals: is there a reproductive cost? *Behavioural Ecology and Sociobiology*, **27,** 113-122.

Boness, D.J. (1991) Determinants of mating systems in the Otariidae (Pinnipedia). In: *The behaviour of Pinnipeds* (Ed. by D. Renouf), pp. 1-44. Chapman and Hall, London & New York.

Boness D.J., Anderson S.S. and Cox, C.R. (1982) Functions of female aggressions during the pupping and mating season of grey seals, *Halichoerus grypus* (Fabricius). *Canadian Journal of Zoology*, **60,** 2270-2278.

Boness, D.J.; Bowen, W.D and Francis, J.M. (1993) Implications of DNA fingerprinting for mating systems and reproductive strategies of pinnipeds. *Symposium of Zoological Society of London*, **66,** 61-93.

Bonner, W.N. (1968) The fur seal of South Georgia. *British Antarctic Survey Scientific Report*, **56,** 1-81.

Bonner, W.N. (1984) Lactation strategies in pinnipeds: problems for a marine mammalian group. *Symposium of the Zoological Society of London,*.**51,** 253-272.

Boulva, J. (1979) Mediterranean monk seal, *FAO Fisheries Service*, **5,** 95-101.

Boyd, J.M. and Campbell, R.N. (1971) The grey seal *(Halichoerus grypus)* at North Rona, 1959 to 1968. *Journal of Zoology* (London), **164,** 469-512.

Campagna, C. (1987) *The breeding behaviour of the southern sea lion*. Ph.D. Thesis, University of California, Santa Cruz, CA. 152pp.

Campagna, C. and Le Boeuf, B.J. (1988a) Reproductive behaviour of Southern sea lions. *Behaviour*, **104,** 233-261.

Campagna, C. and Le Boeuf, B.J. (1988b) Thermoregulatory behaviour of Southern sea lions and its effect on mating strategies. *Behaviour*, **107,** 72-90.

Campagna, C., Le Boeuf, B.J. and Cappozzo, H.L. (1988a) Group raids: a mating strategy of male Southern sea lions. *Behaviour*, **105,** 224-249.

Campagna, C., Le Boeuf, B.J. and Cappozzo, H.L. (1988b) Pup abduction and infanticide in southern sea lions. *Behaviour*, **107,** 44-60.

Campagna, C., Bisioli, C., Quintana, F, Perez, F. and Vila, A. (1992a) Group breeding in sea lions: pups survive better in colonies. *Animal Behaviour*, **43,** 541-548.

Campagna, C., Le Boeuf, B.J., Lewis, M. and Bisioli, C. (1992b) Equal investment in male and female offspring in southern elephant seals. *Journal of Zoology* (London), **226,** 551-561.

Campagna, C. and Lewis, M. (1992) Growth and distribution of a southern elephant seal colony. *Marine Mammal Science*, **8,** 387-396.

Campagna, C. Lewis, M. and Baldi, R. (1993) Breeding biology of southern elephant seals in Patagonia. *Marine Mammal Science*, **9,** 34-47.

Cappozzo, H.L. (1995) Reproductive behaviour of South American Fur Seal. Eleventh Biennial Conference on the Biology of Marine Mammals. Orlando, Florida, 14-18 Dic, 1995, USA, p. 20.

Cappozzo, H.L. (1996) *Comportamiento reproductivo y selección sexual en dos especies de otáridos Sudamericanos*. Ph. D. Thesis, University of Barcelona, Spain. 146pp.

Cappozzo, H.L., Campagna, C. and Monserrat, J. (1991) Sexual dimorphism in newborn southern sea lions. *Marine Mammal Science*, **7,** 385-394.

Cappozzo, H.L., Szapkievich, V.B. and Campagna, C. (1993) Optimization of mating behaviour in South American Sea Lions. Tenth Biennial Conference on the Biology of marine Mammals, Galveston, Texas, November 1993, USA.

Cappozzo, H.L, Perez, F. and Batallés, L.M. (1996) Reproductive behaviour of South American Fur Seals in Uruguay. International Symposium and Workshop on Otariid

Reproductive Strategies and Conservation. Nat. Zool. Park, Smithsonian Institution, Washington DC, April 12-16, USA.

Clutton-Brock, T.H., Albon, S.D. and Guinness, F.E. (1981) Parental investment in male and female offspring in polyginous mammals. *Nature* (London), **289,** 487-489.

Clutton-Brock, T.H. and Albon, S.D. (1982) Parental investment in male and female offspring in mammals. In: *Current problems in sociobiology* (Ed. by King's College Sociobiology Group), pp. 223-248. Cambridge University Press, Cambridge.

Clutton-Brock, T.H. and Godfray, C. (1992) Parental investment. In: *Behavioural Ecology an evolutionary approach:.* (Ed. by J.R. Krebs and N.B. Davies), pp. 234-248. Blackwell Scientific Publications, Oxford, England.

Costa, D.P., Trillmich, F. and Croxall, J.P. (1988) Intraspecific allometry of neonatal size in the Antarctic fur seal (*Arctocephalus gazella*). *Behavioural Ecology and Sociobiology,* **22,** 361-364.

Cox, C.R. and Le Boeuf, B.J. (1977) Female incitation of male competition: a mechanism in sexual selection. *American Naturalist.,* 111 , 317-335.

Crespo, E.A. (1988) *Dinámica poblacional del lobo marino de un pelo, Otaria flavescens (Shaw, 1800), en el norte del litoral patagónico.* Ph. D. Thesis, University of Buenos Aires, Argentina. 265pp.

Croxall, J.P. and Gentry, R.L., (Eds.) (1987) Status, biology and ecology of fur seals. Proceedings of an International Workshop. *NOAA Technical Report* NMFS 51, pp 1-212.

Darwin, C. (1859) *On the Origin of Species.* Murray, London.

Darwin, C. (1871) *The Descent of Man and Selection in Relation to Sex,* First edition. Murray, London.

Darwin, C. (1874) *The Descent of Man and Selection in Relation to Sex,* Second edition. Murray, London.

Davies, N.B. (1991) Mating systems. In: *Behavioural Ecology,* 3rd. edn. (Eds. J.R. Krebs and N.B. Davies), pp. 263-299. Blackwell Scientific Publications, Oxford, England.

Doidge, D.W., Croxall, J.P. and Ricketts, C. (1984) Growth rates of Antarctic fur seal *Arctocephalus gazella* pups at South Georgia. *Journal of Zoology* (London), **203,** 87-93.

Emlen, S.Y. and Oring, L.W. (1977) Ecology, sexual selection and the evolution of mating systems. *Science,* **197,** 215-233.

Fedak, M.A. and Anderson, S.S. (1982) The energetics of lactation: accurate measurements from a large wild animal, The Grey seal *(Halichoerus grypus). Journal of Zoology* (London), **198,** 473-479.

Fisher, R.A. (1930) *The genetical theory of natural selection.* Oxford University Press. Oxford, England.

Gentry, R.L. (1998) *Behaviour and Ecology of Northern fur seal.* Princeton University Press, Princeton, NJ, USA. 376pp.

Hamilton, J.E. (1934) The southern sea lion *Otaria byronia* (de Blainville). *Discovery Reports,* Cambridge, **8,** 269-318.

Hamilton, J.E. (1939) A second report on the southern sea lion *Otaria byronia* (de Blainville). *Discovery Reports,* Cambridge, **19,** 121-164.

Harvey, P.H. and Bradbury, J.W. (1991) Sexual selection. In: *Behavioural Ecology, an evolutionary approach.* (Ed. by J.R. Krebs, and N.B. Davies), pp. 203-232. Blackwells Scientific Publications, Oxford, England.

Heath, C.B. (1989) *The Behavioural Ecology of the California Sea Lion, Zalophus californianus.* Ph. D. Thesis, University of Califonia, USA. 255pp.

Higgins, L.V., Costa, D.P., Huntley, A.C. and Le Boeuf, B.J. (1988) Behavioural and phisiological measurements of maternal investment in the Steller sea lion, *Eumetopias jubatus*. *Marine Mammal Science*, **4**, 44-58.

Johnson, B.W. and Johnson , P.A. (1984) Observations on the Hawaiian monk seal on Laysan Island from 1977 through 1980. *NOAA Technical Mem.* NMFS-SWFS-49.

Kerley, G.I.H. (1987) *Arctocephalus tropicalis* on the Prince Edward Islands. *NOAA Technical Report.* NMFS 51 (Ed. by J.P. Croxall and R.L. Gentry), 61-64.

King, J.E. (1983) *Seals of the World.* British Museum (Nat. Hist.) & Cornell University Press, New York. Second Edition. 240pp.

Krebs, J.R. and Davies, N.B. (Eds.) (1991) *Behavioural Ecology. An evolutionary approach.* Third Edition. Blackwell Scientific Publications, Oxford, England. 482pp.

Kovacs, K.M. and Lavigne, D.M. (1986a) Growth of grey seal (*Halichoerus grypus*) neonates: differential maternal investment in the sexes. *Canadian Journal of Zoology*, **64**, 1937-1943.

Kovacs, K.M. and Lavigne, D.M. (1986b) Maternal investment and neonatal growth in phocid seals. *Journal of Animal Ecology*, **55**, 1035-1051.

Kovacs, K.M. (1987) Maternal behaviour and early behavioural ontogeny of grey seals (*Halichoerus grypus*) on the Isla of May, UK.*Journal of Zoology* (London), **213**, 697-715.

Kuroiwa, M.I. and Majluf, P. (1989) Do South American fur seal males in Peru lek? Abstracts 8th. Biennial Conference Biology of Marine Mammals, 36.

Le Boeuf, B.J. (1974) Male-male competition and reproductive success in elephant seals. *American Zoologist.*, **14**, 163-176.

Le Boeuf, B.J. (1991) Pinniped mating systems on land, ice and in the water: Emphasis on the Phocidae, In: *The Behaviour of Pinnipeds.* (Ed. by D. Renouf). Chapman and Hall, London, 45-65.

Le Boeuf, B.J. and Petrinovich, L.F. (1974) Elephant seals: interspecific comparisons of vocal and reproductive behaviour. *Mammalia*, **38**, 16 - 32.

Le Boeuf, B.J., Condit, R and Reiter J. (1989) Parental investment and the secondary sex ratio in northern elephant seals. *Behavioural Ecology and Sociobiology*, **25**, 109-117.

Lewis, M. and Ximenez, I. (1983) Dinámica de la población de *Otaria flavescens* en el área de Península Valdés y zonas adyacentes (2da. parte). *Contribución Nro. 79*, Centro Nacional Patagónico, 21.

Lewis, M, Campagna, C. and Quintana, F. (1996) Site fidelity and dispersion of southern elephant seals from Patagonia. *Marine Mammals Science*, **12**, 138-146.

Ling, J.K. and Walker, G.E. (1977) Seal studies in South Australia. *The South Australia Naturalist*, **52**, 18-30.

Marchessaux, D. (1989) Recherches sur la biologie, ecologie et le status du phoque moine. *GIS Posidonie Pub.* Marsella. 280pp.

Maynard Smith, J. (1980) A new theory of sexual investment. *Behavioural Ecology and Sociobiology*, **7**, 247-251.

McCann, T.S., Fedak M.A. and Harwood, J. (1989) Parental investment in southern elephant seals, *Mirounga leonina. Behavioural Ecology and Sociobiology*, **25**, 81-87.

Oftedal, O.T.; Boness, D.J. and Tedman, R.A. (1987) The behaviour, physiology, and anatomy of lactation in the Pinnipedia. In: *Current Mammalogy*, Vol. 1. (Ed. by H. Genoways), pp. 175-245. Plenum Publishing Corporation, New York.

Ono, K.A. and Boness, D.J. (1996) Sexual dimorphism in sea lion pups: differential maternal investment, or sex-specific differences in energy allocation? *Behavioral Ecology and Sociobiology*, **38**, 31-41.

Ono, K.A.; Boness, D.J. and Oftedal, O.T. (1987) The effect of a natural environmental disturbance on patterns of maternal investment and pup behaviour in the California sea lion. *Behavioural Ecology and Sociobiology*, **21**, 166-178.

Payne, M.R. (1979) Growth in the antarctic fur seal *Arctocephalus gazella*. *Journal of Zoology* (London), **187**, 1-20.

Rand, R.W. (1956) The Cape fur seal *Arctocephalus pusillus* (Schreber). Its general characteristics and moult. *South Africa Div. Sea Fish. Invest. Report*, **21**, 1 - 52.

Reijnders, P., Brasseur, S., van der Toorn, J., van der Wolf, P., Boyd, I., Harwood, J., Lavigne, D. and Lowry, L. (1993) *Monachus monachus*. In: *Seals, Fur Seals, Sea Lions, and Walruses. Status Survey and Conservation Action Plan* IUCN/SSC, 52- 54.

Riedman, M. (1990) *The Pinnipeds. Seals, sea lions and walruses*. University of California Press, Ltd., Berkeley & Los Angeles, U.S.A. & Oxford University Press, England .

Roux, J-P. (1987) Subantarctic fur seal, *Arctocephalus tropicalis*, in French subantactic territories. In: *Status, biology, and ecology of fur seals*. (Ed. by J.P. Croxall and R.L. Gentry). Proceedings of an International Symposium and Workshop. NOAA Technical Report NMFS 51.

Stirling, I. (1975) Factors affecting the evolution of social behaviour in the Pinnipedia. *Rapp. P-V. Réun. Cons. Int. Explor. Mer.*, **169**, 205-212.

Stirling, I. (1983) The evolution of mating systems in pinnipeds. In: *Recent Advances in the study of mammalian behaviour*. (Ed. by J. Eisenberg and Kleiman), pp. 489-527. *The American Society of Mammalogists, Special Publication* No. 7.

Trillmich, F. (1986) Maternal investment and sex-allocation in the Galápagos fur seal, *Arctocephallus galapagoensis*. *Behavioural Ecology and Sociobiology*, **19**, 157-164.

Trivers, R.L. (1972) Parental investment and sexual selection. In: *Sexual selection and the descent of man 1871-1971* (Ed. by B. Campbell), pp. 136-179. Aldine, Chicago, IL.

Vaz Ferreira, R. (1956) Etología terrestre de *Arctocephalus australis* (Zimmermann) ("lobo fino") en las islas Uruguayas. *Min. Ind. Trab., Serv. Oceanog. Pesca*, **2**, 1-22.

Vaz Ferreira, R. and Ponce de León, A. (1987) South American fur seal, *Arctocephalus australis*, in Uruguay. In: *Status, biology, and ecology of fur seals*. (Ed. by J.P. Croxall and R.L. Gentry), Proceedings of an International Symposium and Workshop. NOAA Technical Report NMFS 51, 29-32.

York, A.E. (1987) Northern fur seal, *Callorhinus ursinus*, Eastern Pacific Population (Pribilof Islands, Alaska, and San Miguel Island, California). In: *Status, biology, and ecology of fur seals*. (Ed. by J.P. Croxall and R.L. Gentry), Proceedings of an International Symposium and Workshop. NOAA Technical Report NMFS 51. 9-21

C. Survey and Study Techniques

The study of marine mammals has always been constrained by their relative inaccessibility, living as they do largely in water, sometimes at depths, and often remote from human habitation. Even now, research on terrestrial ecology and behaviour dominates the major scientific journals and, largely, it is only in the last two decades that the testing of general theories has been extended to include marine mammals. For their part, marine mammalogists have been slow to embrace both theoretical and practical developments in other disciplines, and have tended to concentrate upon publishing their findings in specialised marine mammal journals. Over the last ten years, this pattern appears to have changed, due in no small measure to the significant advances that have taken place in the application of new methods for the study of marine mammals.

The investigation of an animal species often starts with trying to determine its distribution and abundance in either relative or absolute terms. In many cases, our knowledge of cetacean species in particular does not go far beyond this, and in some parts of the world we still have only scant information on these most basic aspects. The stimulus to conduct marine mammal surveys has generally been so that populations can be managed, particularly for human exploitation. Methods that were used relied heavily upon sampling already dead animals (see Allen, 1980, for a detailed review; Evans, 1987: 260-265, for a summary). But they were often imprecise, and contained a variety of potential biases that were usually difficult to correct.

From the 1970s onwards, with increasing emphasis upon protecting whale stocks from over-exploitation, attention turned increasingly to non-consumptive methods for estimating population size. These took the form either of line transect surveys by boat or plane using 'Distance' methodology (Hiby and Hammond, 1989), or the employment of mark-recapture techniques ('capturing' by photography) upon individual recognition features such as tail fluke or dorsal fin markings in cetaceans (Hammond *et al.*, 1990), or facial patterns in seals (Hiby and Lovell, 1990). In Chapter 7, Philip Hammond reviews modern methods used for the assessment of marine mammal population size and status, giving examples of their application with case studies of grey and harbour seals, humpback whale, bottlenose dolphin, and harbour porpoise.

One major handicap in surveying cetacean populations is the difficulty to sight animals particularly when sea conditions are poor. Taking advantage of the fact that some species frequently vocalise, many researchers have turned to acoustic methods for their detection. These have the advantages that they can be operational day and night, in all weather conditions, and it is easier to standardise procedures. In the past, their widespread use was restricted by cost, but now with developments in microchip technology, it is possible to gather large quantities of acoustic information and process this without recourse to a mass of specialised equipment. Jonathan Gordon and Peter Tyack have been pioneers in the development of acoustic equipment for wide use by marine mammal researchers, and in Chapter 8, they review the various techniques currently available.

Perhaps the greatest single technological innovation that has occurred in the last twenty years is the use of DNA techniques to study the genetics of animal and plant populations. This has revolutionised our understanding of various aspects of marine mammal evolutionary biology and behavioural ecology (see, for example, Hoelzel, 1991; Dizon *et al.*, 1997). A whole suite of techniques now exist ranging from nuclear amino-acid and DNA sequencing using microsatellites, through RFLP analysis, and sequencing of the D-loop of mitochondrial DNA, to the application of the class II major histocompatibility complex (MHC) loci, DRBI and DQB (Hoelzel, 1991; Dizon *et al.*, 1997; Hoelzel *et al.*, 1999; Murray *et al.*, 1999). Information obtained from the various different techniques can be used to distinguish the roles of mutation (drift), migration and selection in determining population structure: variation at MHC class II gene loci, for example, is likely maintained primarily by natural selection (Hughes *et al.*, 1994), whereas mitochondrial DNA markers reflect maternal inheritance, and variation at micosatellite loci is effectively neutral. Many of the applications of these new techniques can be seen in studies of mating systems and social structure (see examples in Chapters 1, 5 and 6), but the main area of development probably lies in a new understanding of the mechanisms for species formation. Michel Milinkovitch, Rick LeDuc, Ralph Tiedemann and Andrew Dizon address this subject in Chapter 9, examining the strengths and weaknesses of various lines of molecular evidence in defining species boundaries in cetaceans.

Other areas of marine mammal study have also seen important technical developments. The use of telemetry has revolutionised our understanding of the movement patterns of individuals of several species as well as informed us greatly on aspects of behaviour and physiology. Above the surface, very high frequency (VHF) radio-tags, satellite-linked radio-tags, and geolocation time-depth recorders (TDRs) have been widely used, and below surface, acoustic tags and TDRs have been applied (see Bowen and Siniff, 1999: 424-426 for a general review; also Martin, 1993; Read, 1995). Satellite telemetry and TDRs especially have helped us in understanding habitat use by marine mammals. Although once limited to species that could be readily recaptured for downloading of data, combination sonic/VHF tags that can remotely monitor diving behaviour and movements are now being developed (see, for example, Goodyear, 1993). Examples of findings from these techniques can be found scattered through Chapters 1-5.

The study of genetics and physiology has traditionally relied upon examination of animals after death. Now, time-depth recorders attached to free-swimming animals have become highly refined and may provide real time depth profiles, as well as physiological information on swim speeds, body temperatures and heart rates, Sampling of skin or blubber can be carried out by darting or application of a scrub pad without the need to restrain the animal. These can then be used for genetic analysis (Hoelzel, 1991), studies in toxicology (Reijnders *et al.*, 1999), and foraging ecology (Pierce *et al.*, 1993). In the case of the latter, two new approaches are currently being applied to marine mammals with some success: following on from serological methods which had a number of disadvantages, the fatty acid composition in lipids (Iverson, 1993; Iverson *et al.*, 1997), and stable isotope ratios in various tissues including skin and vibrissae (Rau *et al.*, 1992; Hobson *et al.*, 1996; Abend and Smith, 1997; Todd *et al.*, 1997; Burns *et al.*, 1998) are yielding longer-term information on dietary preferences.

Technological advances in marine mammal study continue apace. We can look forward to some exciting new developments in the coming decades. These should help bring marine mammal science further into the mainstream of biology, and meet some of the challenges needed for greater knowledge and protection of these creatures in a modern world.

REFERENCES

Abend, A.G. and Smith T.D. (1997) Differences in stable isotopes ratios of carbon and nitrogen between long-finned pilot whales (*Globicephala melas*) and their primary prey in the Western North Atlantic. *ICES Journal of Marine Science*, **54**, 500-503.

Allen, K.R. (1980) *Conservation and Management of Whales*. University of Washington Press, Seattle, Washington. 107pp.

Burns, J.M., Trumble, S.J., Castellini M.A., and Testa, J.W. (1998) The diet of Weddels seals in McMurdo Sound, Antarctica as determined from scat collections and stable isotope analysis. *Polar Biology*, **19**, 272-282.

Dizon, A. E., Chivers, S. J., and Perrin, W. F. (eds.) (1997) *Molecular genetics of marine mammals*. Special Publication Number 3, The Society for Marine Mammology. 412pp.

Evans, P.G.H. (1987) *The Natural History of Whales and Dolphins*. Christopher Helm/Academic Press, London, and Facts on File, New York. 343pp.

Goodyear, J.D. (1993) A sonic/radio tag for monitoring dive depths and underwater movements of whales. *Journal of Wildlife Management*, **57**, 503-513.

Hammond, P.S., Mizroch, S.A., and Donovan, G.P. (Eds.) (1990) Individual recognition of cetaceans. Use of photo-identification and other techniques to estimate population parameters. *Report of the International Whaling Commission* (special issue **12**), 1-440.

Hiby, A.R. and Hammond, P.S. (1989) Survey techniques for estimating abundance of cetaceans. *Report of the International Whaling Commission* (special issue **11**), 47-80.

Hiby, A.R. and Lovell, P. (1990) Computer aided matching of natural markings: A prototype system for grey seals. *Report of the International Whaling Commission* (special issue 12), 57-61.

Hobson, K.A., Schell, D.M., Renouf, D., Noseworthy, E. (1996) Stable carbon and nitrogen isotopic fractionation between diet and tissues of captive seals: implications for dietary reconstructions involving marine mammals. *Canadian Journal of Fisheries and Aquatic Sciences*, 53, 528-533.

Hoelzel, A.R. (ed.) (1991) Genetic Ecology of Whales and Dolphins. *Report of the International Whaling Commission* (special issue 13), 1-

Hoelzel, A.R., Stephens, J.C. and O'Brien, S.J. (1999) Molecular genetic diversity and evolution at the MHC DQB locus in four species of pinnipeds. *Molecular Biology and Evolution*, 16(5), 611-618.

Hughes, A.L., Hughes, M.K., Howell, C.Y. and Nei, M. (1994) Natural selection at the class II major histocompatibility complex loci of mammals. *Philosophical Transactions of the Royal Society of London B*, 345, 359-367.

Iverson, S.J. (1993) Milk secretion in marine mammals in relation to foraging. Can milk fatty acids predict diet? *Symposia of the Zoological Society of London*, 66, 263-291.

Iverson, S.J., Arnould, J.P.Y., and Boyd, I.L. (1997) Milk fatty acid signatures indicate both major and minor shifts in diet of lactating Antarctic fur seals. *Canadian Journal of Zoology*, 75, 188-197.

Murray, B.W., Michaud, R. And White, B.N. (1999) Allelic and haplotype variation of major histocompatibility complex class II DRBI and DQB loci in the St. Lawrence beluga (*Delphinapterus leucas*). *Molecular Ecology*, 8, 1127-1139.

Pierce, G.J., Boyle, P.R., Watt, J., and Grisley, M. (1993) Recent advances in diet analysis of marine mammals. *Symposia of the Zoological Society of London*, 66, 241-261.

Rau, G.H., Ainley, D.G., Bengtson, J.L., Torres, J.J., and Hopkins, T.L. (1992) 15N/ 14N and 13C/ 12C in Weddell Sea birds, seals and fish: Implications for diet and trophic structure. *Marine Ecology Progress Series*, 84, 1-8.

Read, A.J. (1995) New approaches to studying the foraging ecology of small cetaceans. In: *Whales, Seals, Fish and Man* (Ed. by A.S. Blix, L. Walløe and Ø. Ulltang), pp. 183-192. Elsevier, Amsterdam. 720pp.

Reijnders, P.J.H., Aguilar, A., and Donovan, G.P. (Eds.) (1999) Chemical Pollutants and Cetaceans. *The Journal of Cetacean Research and Management* (special issue 1), 1-273.

Todd, S., Ostrom, P., Lien, J., and Abrajano, J. (1997) Use of biopsy samples of humpback whale (*Megaptera novaengliae*) skin for stable isotope (d13C) determination. *Journal of Northwest Atlantic Fisheries Science*, 22, 71-76

Chapter 7

Assessment of Marine Mammal Population Size and Status

PHILIP S. HAMMOND

Sea Mammal Research Unit, Gatty Marine Laboratory, University of St Andrews, St Andrews, Fife KY16 8LB, Scotland, UK. E-mail: psh2@st-andrews.ac.uk

1.   INTRODUCTION

1.1   Why do We Need to Know Population Size?

"Every species of plant and animal is always absent from almost everywhere. But, a large part of the science of ecology is concerned with trying to understand what determines the abundance of species in the restricted areas where they do occur. Why are some species rare and others common? Why does a species occur at low population densities in some places and at high densities in others? What factors cause fluctuations in a species' abundance?" (Begon, Harper, and Townsend, 1996, p. 567).

One can investigate what determines animal abundance theoretically by developing population models and observing their behaviour. Indeed this has become a major tool both as a description of functional parts of an ecosystem and in the search for general rules or laws in ecology (much of Begon *et al.*, 1996, Part 2: Interactions, illustrates this well). But to test model predictions in the real world requires that abundance (or density or population size) can be determined in the field.

Marine Mammals: Biology and Conservation, edited by
Evans and Raga, Kluwer Academic/Plenum Publishers, 2002

In addition to the study of ecology as a science, *i.e.* furthering our understanding of what determines the size of populations, we may also need to estimate abundance to assess the status of species for management (this could be for harvesting *e.g.* fisheries, or for conservation of threatened species). This is often the case for marine mammals.

1.2   What do We Mean by Abundance?

Abundance, density, population size, relative abundance: these are all different ways population ecologists use to describe how many individuals there are in a population and/or area.

Note that even the term 'population' is not well-defined. For this we could use something like:

"a biological unit at the level of ecological integration where it is meaningful to speak of a birth rate, a death rate, a sex ratio, and an age-structure in describing the properties of the unit" (Cole, 1957).

Such a definition follows from the stylised equation describing population change:

Numbers (future) = Numbers (now) + births - deaths + immigration - emigration

This is perfectly acceptable in principle but, in practice, a population unit may not be able to be defined so simply. Sometimes the physical boundaries are obvious, for example fish in a lake, but often it may be very difficult, or even impossible, to determine where one population unit stops and another begins. This is especially true for marine mammals which may range over wide areas of ocean and in which 'populations' may merge into one another.

A more pragmatic definition for a population is:

"a group of organisms of the same species occupying a particular place at a particular time ... its boundaries in space and time are vague and usually fixed by the investigator, arbitrarily" (Krebs, 1972).

In many cases concerning marine mammals, there is a need to consider a population as a functional biological unit in the context of conservation and/or management.

For most purposes, it is not important whether we think in terms of population size, relative or absolute abundance, or density. But it is important to be clear that different estimation methods can give different measures. The measure could be the number of animals of a species in a defined area, or the size of a population of animals, the limits of which may or may not be completely defined by area. Density is simply the number of animals per unit area. All estimates relate to a particular time, such as a season or a year.

BOX 1. How many seals are there?

As a simple example, consider the problem of trying to estimate the number of seals in an estuary. The estuary is quite large, we might expect there to be a lot of seals in it and, as is normal when estimating abundance, we have limited resources. We decide to sample a fairly small part of the estuary by counting all the seals in it and then multiplying the sample density of seals by the total area of the estuary.

We have therefore made a number of assumptions, which include:

(i) our estimate of sample density is representative of the whole estuary;

(ii) we don't miss any seals in the sample area;

(iii) seals don't move around.

These assumptions may be violated for a number of reasons. For example:

(i) Animals are rarely distributed randomly in space, so choosing one small area to sample will not guarantee a representative sample. Animals tend to occur in clumped (non-random) distributions which means that the sampling must be random (or at least systematic) to obtain a representative sample. Ideally, samples should be replicated.

(ii) Data must be able to be collected in the way required by the estimation method. If we are likely to miss animals, our method of data collection is suspect and this will lead to biased results. In this example, if our sample area included water as well as land we would definitely miss some animals that were underwater at the time of the count.

(iii) The animals may behave in a way that violates the assumptions of the method. For example, if seals that were hauled out on land moved into the water and out of the area at the approach of the person counting them so that they were not available to be counted, the sample density will be underestimated.

Note that this example is for illustrative purposes; in practice, because of the problems described above, this is not likely to be the best method in this case (see also Example Studies).

1.3   How to Determine Abundance?

Life would be much easier if the answer to the question "how do we determine the number of animals in a population?" was "go out and count them". On very rare occasions it may be possible to conduct a census in which all individuals are counted. But this would at best be extremely tedious and in almost all situations it would be totally impracticable if not impossible. Animal numbers need to be estimated.

The reasons why censuses cannot be conducted are varied and many of them are fairly obvious. The problems posed by marine mammals illustrate this well. Access to the target species may be difficult. Most marine mammals live entirely (in the case of cetaceans and sirenians) or mostly (in the case of seals) in the sea and, when at sea, spend a large proportion of their time underwater. Scale can be a problem. Marine mammals can be present in very large numbers and can move around over very large areas. For example, rorqual whales typically inhabit whole ocean basins and may be present in tens of thousands of individuals. Remember too that resources for research are always limited; marine mammals can be particularly expensive to survey.

Estimating abundance in practice means collecting a sample of data and extrapolating from this sample to the whole population or area. In doing this, we make a number of assumptions about the representativeness of our sample, about how the estimation method works in practice and about the behaviour of individuals in the population itself; all of which may be related to each other. The validity of an estimate depends on whether the assumptions hold so that the extrapolation from sample to whole population is justified. Box 1 gives a simple example as an illustration of how the assumptions inherent in an estimation method may be violated.

1.4   Accuracy vs Precision

Whenever we have to estimate anything from a sample, there is a trade-off between accuracy (whether or not an estimate is biased) and precision (how large is the variance of the estimate). Box 2 gives some basic information about samples, accuracy and precision. Ideally, we would like an estimate that is unbiased and has a low variance. Generally speaking, the variance of an estimate will become smaller the more data are collected. But there is no point in expending lots of resources to get a very precise estimate if that estimate is seriously biased because of failure of one or more of the assumptions about the method.

In the simple example in Box 1, we assumed that all seals were counted. Suppose we miss, unknowingly, about 50% of the animals in the sample area. Then, even if we collect lots of data and end up with a very precise estimate of abundance, that estimate would be biased downwards by about 50%. If we are unaware of this, our estimate may not be very useful.

Box 2. Samples, accuracy and precision

The concept of a sample

Say we were interested in the average weight of a species of fish living in a lake. To obtain the true average of the **population,** we would need to catch and weigh every fish in the lake. This would give us the true **population mean**. Similarly, we would also be able to obtain the population variance and standard deviation. But if catching all the fish were impractical, we could catch a **sample** of fish and use their weights to calculate the **sample mean**. Similarly we could calculate the sample variance and standard deviation. In this case, we have **estimated** the mean, variance and standard deviation. Most data sets are samples. Because we only weighed a sample of fish from the population, we would not expect the value of the estimated sample mean to be exactly the same as the true population mean. However, if we took care to obtain a representative sample of the population, we would hope that the sample mean would be close to the population mean, that is it would be accurate (see below). If we took successive samples, the estimated mean values would all be different. How spread out these values are, is a measure of how precise the estimated mean is (see below).

The difference between accuracy (bias) and precision (repeatability)

The difference between precision and accuracy is often confused. **Accuracy** is how near to the true value you get when you measure something, *i.e.*, whether or not the measurements are biased. For example, if you measure a 93.7 mm plant ten times to the nearest cm and get 9 cm (90 mm) every time, your measurements are not very accurate, and the mean is biased. But if you measured it to the nearest 0.1 mm and got values between 93.6 and 93.8 mm, the measurements would be quite accurate. **Precision** is a measure of how variable or how repeatable are the measurements and estimated quantities. For example, your ten measurements to the nearest cm of 9 cm (and the mean of 9 cm) are extremely precise (there is no variability at all). But if the measurements to the nearest 0.1 mm ranged between, say, 91.2 mm and 96.2 mm, they would be quite variable and the estimated mean would be imprecise. We can calculate how precise a mean value is simply as the standard deviation divided by the square root of the number of observed values in the data. This is called the **standard error** (SE), or standard deviation of the mean. The **coefficient of variation** (of the mean), defined as SE / mean, is a useful way to portray relative precision.

But if we realised there was a problem, the solution would be to spend some time estimating the magnitude of the bias, *i.e.* the proportion of seals that were missed. Assuming fixed resources, this will take time away from counting, resulting in a smaller sample size and, therefore, a larger variance. The estimate will end up substantially less biased but also slightly less precise.

However, the best course of action depends on the size of the bias. Suppose we only missed, on average, 5% of the seals. In the real world of marine mammals at least, the precision of abundance estimates is rarely better than a coefficient of variation of about 0.15. In this case, then, we might be prepared to ignore the (relatively small) bias and expend our effort on reducing the variance. Note that all the examples discussed below achieve levels of precision of CV = 0.15 or better. This is because of careful planning, good fortune and/or because some unknown source(s) of variability have not been taken into account; a feature that is probably quite common in many studies.

This balance between the amount of bias which an investigator is prepared to accept and the level of precision of the estimated number which is necessary for the results to be useful should be an important consideration when planning a study to estimate abundance.

2. METHODS FOR ESTIMATING MARINE MAMMAL ABUNDANCE

There is quite a wide range of methods for estimating marine mammal abundance but we will focus here on three of the most frequently used. An earlier discussion of techniques for estimating the size of whale populations, that had a slightly wider scope and which uses some different examples, is given in Hammond (1987). Many other examples can be found in the annual Reports of International Whaling Commission over the last 20 years. Some of the following material is also covered in Hammond (1995).

2.1 Extrapolation of Counts

Rarely, if ever, is it possible to enumerate an entire population of animals. This is certainly true for marine mammals, which are effectively invisible for a large part of their lives. However, seals give birth and wean their pups on land or ice and also haul out during their annual moult and on other occasions, and it is possible during these periods to count the number of animals present in an area. But all the seals in a population will not be hauled out at the same time so it is necessary to extrapolate the number of animals counted to obtain an estimate of the entire population.

There are a number of ways of doing this. One is to choose a class of animal which can be readily counted (pups of the year, for example) and to use those numbers as input to an age- or stage-structured population model (Caswell, 1989) which calculates total population size. For a reliable estimate, the population model must be realistic with parameter values estimated from relevant data.

Another method of extrapolation is to count all the animals hauled out within a given period and to divide this number by an estimate of the proportion of the population that was hauled out during this time. One way to estimate this proportion is to use telemetry data; for a reliable estimate, these data must be representative of the entire population.

A major assumption for these methods is that the counts are accurate. This may be the case (as in the examples given below) but in other cases, ensuring that this assumption is not violated may be extremely difficult. Notwithstanding this, these methods are most sensitive to the quality of data used to extrapolate counts to total population size.

2.2   Mark-Recapture and Photo-Identification

Mark-recapture methods (Begon, 1979; Seber, 1982) use data on the number of animals marked and the proportion of marked animal in samples of recaptured animals to estimate population size. All mark-recapture estimates of population size are based on the simple idea that if a proportion of the whole population is marked through a first sample, then an estimate of this proportion can be obtained by observing the number of marked animals in a second sample, given various assumptions. Equating these two proportions leads to an estimate of abundance (N) known as the Peterson two-sample estimator, or sometimes the Lincoln index:

$$N = n_1 n_2 / m_2$$ Eq. 1

where n_1 is the number of animals captured, marked and released in the first sample, n_2 is the number of animals captured in the second sample, and m_2 is the number of recaptures of marked animals in the second sample.

The basic assumptions include the following: marks are unique, cannot be lost and are always reported on recovery; all animals have the same chance of being captured on any sampling occasion; and marking does not affect the subsequent catchability of an animal.

In the past, mark-recapture analyses have been used to estimate the population size of marine mammals using recoveries of artificially applied marks from harvested animals. More recently, studies have concentrated on populations of animals in which unique individual natural markings (such as pigmentation patterns or nicks on dorsal fins) can be photographed and identified. Records of re-identifications from a series of photographic samples (capture histories) can then be used as data for mark-recapture analyses to estimate population size (Hammond, 1986, 1990). Photo-identification has some advantages over the use of artificial marks but obviously cannot be used for those populations whose individuals do not possess lasting, recognisable, natural markings.

In mark-recapture analyses, the assumption most likely to be violated is that each animal must have the same probability of being captured within any one sampling occasion. It is very likely that the behaviour of individual animals will lead to heterogeneity (variability) of capture probabilities resulting in population estimates which are biased downwards (Hammond, 1986). In the extreme, some animals may never be available to be sampled and will not be included in the population estimate. There are ways to account for this analytically if closed population models can be used, but it is clearly better to minimise the problem during the process of data collection. In practice, this means designing a study that gives every animal a chance of being captured, and capturing as many animals as possible. This can be achieved most easily if the distribution of the whole population is concentrated in a limited area for a limited period of time during the year. The abundance estimate refers to the population of animals using the particular study area; care must be taken when extrapolating results to wider areas.

2.3   Sightings Surveys

Line transect methods (Burnham *et al.*, 1980; Hiby and Hammond, 1989; Buckland *et al.*, 1993) were first developed for terrestrial animals but are now widely used to estimate the abundance of cetacean populations via shipboard or aerial sightings surveys. In these sightings surveys, the study area is sampled by the survey platform searching along predetermined transects, placed so that the whole area is representatively sampled. The distance and angle to sighted animals (or groups) are measured (or estimated) allowing the calculation of perpendicular distance from the sighting to the transect line. These data are used to estimate the effective width of the strip searched so that sample density can be estimated and extrapolated to give an estimate of abundance for the whole survey area. This is necessary because the probability of detection declines with distance from the transect line; the effective strip width is an estimate of the width of a hypothetical strip (in which all animals are seen) that would yield the same number of sightings per unit of transect length.

Sightings surveys thus provide an estimate of the number of animals in a given area at a given time, not an estimate of the size of a biological population, unless the whole of that population was in the study area during the study period. The basic estimation equation can be written:

$$\hat{D} = \frac{n}{2\,\hat{w}\,L} \qquad\qquad \text{Eq. 2}$$

where D is population density (the 'hat' means that it is an estimated quantity), n is the number of animals seen (separate sighting events), w is the effective half-width of the strip searched, as estimated from the perpendicular distance data, and L is the length of the transect searched (i.e. the length of the strip). If sighting targets are groups or schools of animals, estimated density must be multiplied by an estimate of average group size. The number of animals is estimated simply by multiplying estimated density by the size of the study area.

There are a number of practical difficulties with meeting the various assumptions (see Buckland *et al.*, 1993) of line transect sampling of cetacean populations including: the design of representative and efficient surveys, obtaining accurate angle and distance measurements, estimating school size, and ensuring that animals are detected before they may move in response to the survey platform. But the most important assumption is that every animal (or school of animals if these are the sighting target) on the transect line itself is seen. That is, the probability of detecting an animal at zero perpendicular distance (i.e. on the transect line itself) is one. Clearly, this will often not be met for cetaceans, which are underwater most of the time. In practice, this probability depends on the proportion of time spent at the surface, the speed of the survey platform, and the typical detection distance.

To obtain an unbiased estimate of abundance, therefore, it is often necessary to include the probability of detection on the transect line in the estimation of abundance. The only way to determine if this is necessary is to estimate this probability. The best way to do this is through the analysis of duplicate sightings data collected from independent sighting platforms on the same vessel. However, sightings surveys cannot always collect such data and, in these cases, abundance estimates may be negatively biased by an unknown amount as a result of this. Taking the above into account, a more complete estimation equation is:

$$\hat{N} = A \frac{n\,\bar{s}}{2\hat{w}L\hat{g}(0)} \qquad\qquad \text{Eq. 3}$$

where N is animal abundance, A is the size of the study area, *s(bar)* is average group size, and $g(0)$ is the probability of detecting an animal on the transect line, estimated from duplicate sightings data. Details of these modifications to the basic methods are contained in Buckland *et al.* (1993).

3.   EXAMPLE STUDIES

The examples described below demonstrate how studies to estimate population size or abundance can take advantage of, or need to take account of, particular physical and behavioural characteristics of a species. The chosen examples are from the North Atlantic, mostly around Britain and Europe, but the methods employed and the important factors to consider are applicable everywhere.

3.1   Extrapolating Counts

Example 1. Grey Seals Around Britain

The number of grey seals (*Halichoerus grypus*) breeding around the coasts of Britain is estimated each year based on counts of pups born during autumn (Ward *et al.*, 1987). High resolution aerial photography is used to record all the pups on land at all known major breeding sites, several times during the pupping season. This gives data on the number of pups present at the site during the pupping season (see Fig. 1). A model of the birth process fitted to these data gives an estimate of total pup production (Duck *et al.*, in preparation). This model is sensitive to the timing and length of the pupping season which varies from island to island, so this is done separately for each site. The number of females in the population is estimated by feeding the annual pup production estimates into an age-structured population model (Hiby and Duck, in preparation). Total population size is obtained via a simple sex-ratio calculation.

The important factors affecting this work are: the quality of the aerial photographs; the adequacy of the model used to estimate pup production from the pup counts (including the number of counts made at each site); and the assumptions (*e.g.* about sex ratio) and adequacy of the data in the population model.

The development of a purpose-built camera system including image motion compensation (Hiby *et al.*, 1988) has resulted in excellent quality photographs. The model to estimate pup production is sensitive to the timing and length of the pupping season. These vary from site to site, and so the model is fitted to the data at each site, separately. Estimates of population size from the model are sensitive to the fecundity rate data and the sex ratio.

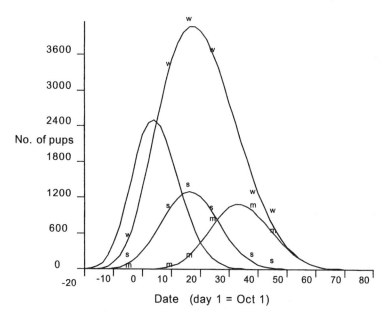

Figure 1. Curves showing the relationship between the number of whitecoated grey seal pups (w), the number of moulted pups (m) and the number of pups suckling (s) counted from aerial photographs of the island of Ceann Iar in the Monach Isles, Outer Hebrides in 1993 (Duck et al. in preparation). The maximum total pup count is 4,415 and the total pup production estimated form a model of the birth process is 5,062.

The current estimate of population size is 110,000 animals with a coefficient of variation of about 0.10 (Hiby *et al.*, 1996).

The results of annual surveys show that the population has been increasing by about 6% per annum since 1984, as indicated by the pup production estimates in Figure 2, although there is increasing evidence that the rate of growth may now be slowing down. Earlier data indicate that the population has been increasing at a similar rate since the 1960's. This has fuelled a long and continuing debate about whether or not the British grey seal population might be negatively affecting fisheries catches and, if so, whether it should be controlled in some way.

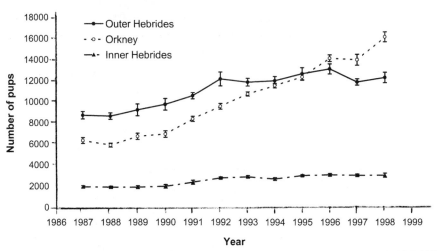

Figure 2.   Grey seal pup production at the main island groups in Scotland 1987-1998. (data from Sea Mammal Research Unit)

Example 2. Harbour Seals in the Moray Firth

The Moray Firth in North-east Scotland is home to a population of harbour seals (*Phoca vitulina*) which regularly haul out on inter-tidal sandbanks, particularly during the June/July pupping season. This makes it possible to count seals at all known sites during this period of the year with the aim of estimating population size by dividing the mean number of seals hauled out by the estimate of the proportion of time spent hauled out. This proportion was estimated from VHF telemetry data collected from 26 seals of both sexes and a range of sizes (Thompson *et al.*, 1997). Pups were not included in the estimate because they were being born throughout the counting period.

This method assumes, amongst other things, that there was no double counting of animals on haul out sites and that the telemetry data are representative of the whole population. Additional factors that had to be taken into account in this study were the timing of the counts with respect to time of year, time of day, and stage of tide.

Peak haul out counts were made during the June/July pupping season and 2 h either side of low tide; time of day did not need to be considered. This was in contrast to similar studies in Orkney where peak haul out counts were made during the moulting period and where there was no tidal relationship but time of day did have to be taken into account (Thompson and Harwood, 1990). Double counting was judged to be minimal based on the movement patterns of individuals. The telemetry data were collected over a number of years, and the data for females were biased towards pregnant females, so there is a possibility that they were not representative and that the population estimate may be biased. Obtaining representative telemetry data for any study can be difficult and is, perhaps, the major drawback of this method.

Figure 3.   VHF transmitter deployed on a harbour seal in Scotland. (Photo: Mike Fedak)

The estimate of the average number of seals hauled out in 1993 was 1,007 (SE = 32). This was extrapolated to give an estimate of population size (excluding pups) of 1,653 (SE = 93, CV = 0.056) using the estimated average proportion of low tides on which males (0.52, SE = 0.044) and females (0.70, SE = 0.036) were hauled out (Thompson *et al.*, 1997).

Harbour seals in the Moray Firth are known to forage locally and their numbers may be affected by changes in the availability of suitable prey (Thompson *et al.*, 1991). It is therefore of interest to monitor changes in abundance. If, however, the haul out behaviour of harbour seals is affected by changes in food availability, the use of telemetry data collected over a number of years to estimate abundance may be inappropriate and lead to bias (Thompson *et al.*, 1997). This highlights a general point that the most appropriate methods to study trends in abundance are not necessarily the same as methods to obtain the best estimate of the numbers of animals in a population or area at a particular time.

An alternative method of estimating the absolute abundance of harbour seals using counts and telemetry data is given in Ries *et al.* (1998).

3.2   Mark-Recapture and Photo-Identification

Example 3. Bottlenose Dolphins in the Moray Firth

Since 1989, data have been collected on the small and isolated population of bottlenose dolphins (*Tursiops truncatus*) in the Moray Firth with the aim of estimating its size (Wilson *et al.*, 1999). A similar study of bottlenose dolphins in Sarasota Bay, Florida is described by Wells and Scott (1990). In these studies, the dolphins can be identified individually by the nicks in the dorsal fin and by pigmentation patterns on the skin so there is no need to capture them bodily. They can be 'marked' simply by taking good quality photographs (Fig. 4). The left and right sides of an animal are marked differently; in the Moray Firth study, both were photographed and recorded separately. Not all animals have distinguishing natural markings so data were also collected to estimate the proportion that does. The population was sampled twice a month during the summers of 1990-92, and matches determined by eye (Wilson *et al.*, 1999).

Analysis was via a multi-sample mark-recapture model applied to the data for each summer separately. This allowed the use of identifying marks which were known to last longer than a summer but not necessarily longer than a year, thus increasing sample sizes. It also allowed heterogeneity of capture probabilities to be taken into account using a population model closed to births and deaths. The model allowing for heterogeneity gave consistently higher estimates indicating that this did need to be taken into account. The estimates in 1992, when sampling was able to cover a larger proportion of the known range, were higher indicating that there were some site preferences within the study area and that the data from this year were more representative.

Estimates of the number of animals with distinguishing marks in 1992 were 73 (SE = 12) from left side data and 80 (SE = 11) from right side data. Estimates of the proportion of animals with distinguishing marks in the population were 0.57 (SE = 0.043) for left side data and 0.61 (SE = 0.035) for right side data. Combining these results to give an average of the left and right side estimates gave a best estimate of 129 (SE = 15, CV = 0.12) (Wilson *et al.*, 1999).

Figure 4. Photographs of the dorsal fin of two bottlenose dolphins in the Moray Firth showing the nicks and other markings used for identification.(Photos: B. Wilson)

This study is a good example of the use of an extensive and detailed data set to address and take account of some of the potential problems of applying mark-recapture methods to photo-identification data. A particular point to note is that although data collection protocol was specifically designed to minimise heterogeneity of capture probabilities, it was still found to be necessary to use analytical methods to account for this. The message is that heterogeneity of capture probabilities is likely to be a factor affecting population estimates in all such studies and the implications of this need to be recognised from the outset.

Wilson *et al.* (1999) also considered the effectiveness of this method for a monitoring programme to assess trends in population size. As for the harbour seal example above, there are additional considerations when looking at changes in abundance over time, and the best method for estimating absolute abundance may not be the most appropriate method for answering questions about trends in numbers.

Example 4. Humpback Whales in the North Atlantic - Project YoNAH

At the other extreme of scale is a major multi-national study of humpback whales (*Megaptera novaeangliae*) in the North Atlantic. Humpbacks breed in winter in tropical waters, migrating annually to summer feeding areas in high latitudes. In the North Atlantic, there is known to be a high degree of fidelity of individual animals to a particular feeding area (Katona and Beard, 1990).

One of the main aims of project YoNAH (Years of the North Atlantic Humpback) was to estimate the size of the population in the whole ocean basin using mark-recapture analyses of data from photo-identification of the ventral surface of individual tail flukes (Fig. 5). Project YoNAH sampled whales in the West Indies breeding areas and in all known major feeding areas (Gulf of Maine, eastern Canada, West Greenland, Iceland and Norway) in 1992 and 1993, using standard data collection protocols so that the data would be consistent across all areas (Smith *et al.*, 1999).

Although the design of the study aimed to obtain a representative sample of data, it is clear that capture probabilities were not equal across all feeding areas because different amounts of searching and photographic effort were spent in each area. Nor were capture probabilities equal in the breeding areas because of behavioural differences related to sex and reproductive status of whales. This means that the least biased estimates of population size will result from analyses using feeding areas as one sample and breeding areas as the second sample in a simple Petersen estimate. These analyses indicate that in 1992/93 there were 10,600 (SE = 710, CV = 0.067) humpback whales in the North Atlantic (Smith *et al.*, 1999). Ongoing work is analysing previous years' data with the aim of investigating the recovery of this population from whaling.

An interesting development from this project, described by Palsbøll *et al.* (1997), is to use genetic markers identified from skin biopsies rather than photographs of natural markings as a means of recognising individuals. Advantages of genetic markers include not changing over time (as natural markings may), and providing data on the sex of sampled animals.

Figure 5.   Photograph of the ventral surface of the tail flukes of a humpback whale from the North Atlantic (Photo courtesy of S. Mayo). From a total of over 4,000 identification photographs, almost 3,000 individual whales were identified. The number of individuals identified during the different samples were as follows: 1992 breeding areas - 629; 1992 feeding areas - 787; 1993 breeding areas - 582; 1993 feeding areas - 937. Whales seen in more than one sample were used to estimate population size using simple two-sample models. (Smith *et al.*, 1999)

3.3   Sightings Surveys

Example 5. Harbour Porpoises in the North Sea and Adjacent Waters - Project SCANS

In summer 1994, a major international sightings survey was undertaken in the North Sea and adjacent waters to estimate the number of harbour porpoises (*Phocoena phocoena*) and other small cetaceans as part of project SCANS (Small Cetacean Abundance in the North Sea).

Harbour porpoises are particularly difficult to detect so it was very important to get good estimates of the proportion detected on the transect line. In addition, there were concerns that harbour porpoises might respond to approaching survey ships by moving away before they were seen. Duplicate sightings data were, therefore, collected on all ships and the observers on one platform searched farther ahead of the vessel than those on the other platform so that any responsive movement could be taken account of. New methods were developed to analyse these data (Borchers *et al.*, 1998).

The two aircraft flew in tandem (one directly behind the other) for much of the time allowing duplicate sightings data to be collected during the aerial survey. New methods were developed to analyse these data using a probabilistic approach to determining duplicates (Hiby and Lovell, 1998).

Nine ships and two aircraft surveyed the North Sea, Kattegat, Skagerrak, English Channel and Celtic Sea for about one month, covering an area of over 1 million km^2 (Hammond *et al.*, 1995). Figure 6 shows the transect lines searched over the study area and the distribution of harbour porpoise sightings.

The overall estimate of the number of harbour porpoises in the North Sea and adjacent waters was 341,000 (SE = 48,000, CV = 0.14). This estimate is very precise for a line transect sightings survey, especially considering the difficulties associated with surveying for harbour porpoises. This was achieved through careful planning and the good fortune of having excellent weather conditions for most of the survey. In addition, because of the data collection and analysis methods used, it should not be biased as result of missing animals on the transect line or responsive movement to the survey vessels.

The estimates of porpoise numbers in the North Sea and adjacent waters have been used by international organisations with responsibility for considering the impact of anthropogenic threats on small cetaceans. Although the population estimate is large, there is justifiable concern that, in some European waters at least, the incidental catch of porpoises in fishing nets is too high for the population to sustain (Tregenza *et al.*, 1997).

4.   POPULATION STATUS

Several of the examples described briefly above refer to the need to assess how numbers of animals are changing over time. It is appropriate, therefore, to consider here the topic of population status and how this can be assessed. The following is a brief introduction to this subject.

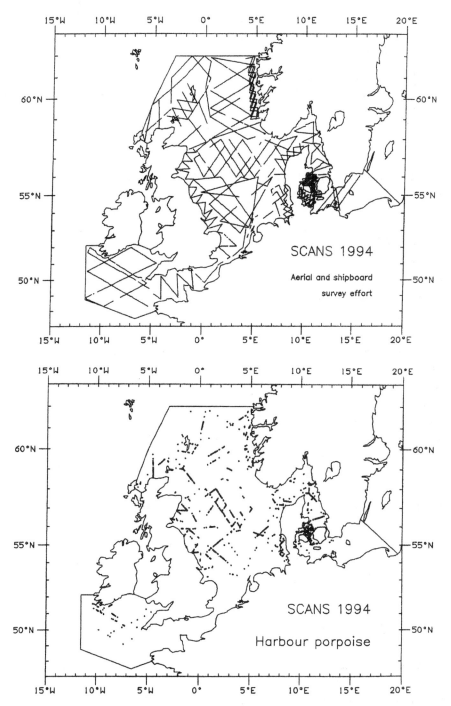

Figure 6. Maps showing the survey blocks, transect lines searched and the location of harbour porpoise sightings during project SCANS. (From Hammond *et al.*, 1995)

In common language we probably think of the status of a population as meaning whether it is 'OK' or not. But this can be measured in a number of different ways. In terms of distribution, and recalling the quote from Begon *et al.* (1996) above, status could be taken simply to mean whether or not a species is present in a particular area. An example of this is the harbour porpoise in the eastern Baltic Sea. Historical records show that porpoises used to be common in this area but are now effectively absent; their presence is indicated only by a few strandings which could be vagrants from farther west. The species in the North Atlantic, however, is widespread in the shelf waters of northern Europe and North America. Using presence or absence as a measure of status is rather crude and not particularly useful.

A more useful way to think about population status is in terms of numbers (or relative numbers) of animals. This can be done in two ways. Status could be defined as current population size relative to some time in the past. In this case, we need as a minimum an estimate of current population size and some way of determining historical abundance. This could be through a reliable estimate of population size calculated at some time in the past, but this is unlikely to be available; it is only relatively recently that we have been able to calculate reliable estimates of population size. Alternatively, if records or estimates of catches or incidental mortalities are available and there are data to estimate net reproductive rate, historical abundance could be back-calculated using a population model. An example of this kind of exercise for dolphins killed incidentally in the tuna purse seine fishery in the eastern tropical Pacific is given in Smith (1983). Note that because this method is dependent on historical directed or incidental catch data and on assumptions about net reproductive rates, which may be poorly understood, it may not be a reliable method of assessing status.

Status could also be defined as whether the current population has been increasing, decreasing, or stable in recent years. In this case, a series of abundance estimates or indices of relative abundance must be available (as in some of the examples described earlier or a population model used to infer the net reproductive rate - Barlow and Clapham, 1997). When considering such studies, it is important to be aware of the time scale that is likely to be necessary before the numbers of individuals in a population can be said to be going up or down. Power analysis (Gerrodette, 1987) is a useful tool for investigating how long a series of estimates needs to be to indicate a significant increase or decrease in abundance. Power analysis is a means of determining what the probability would be of detecting declines of various rates, given regular abundance estimates of a given precision collected over a period of years. Two examples of the use of this method for exploring how long it may take to detect a given rate of population change are provided by

Thompson *et al.* (1997) for harbour seals and Wilson *et al.* (1999) for bottlenose dolphins in the Moray Firth.

It is worth noting that concerns about status are usually linked to either conservation of populations perceived to be under threat and/or to the management of human activities such as fishing, whaling and oil exploration. Conservationists and managers would like scientists to come up with quick answers to questions about the population status of marine mammals. But the message from considering the assessment of population status is that it may require a large amount of data, many assumptions and a long period time.

REFERENCES

Barlow, J. and Clapham, P.J. (1997) A new birth-interval approach to estimating demographic parameters of humpback whales. *Ecology*, **78**, 535-546.

Begon, M. (1979) *Investigating Animal Abundance: Capture-Recapture for Biologists.* Edward Arnold, London.

Begon, M., Harper, J.L. and Townsend, C.R. (1996) *Ecology: individuals, populations and communities.* 3rd edition. Blackwell Scientific, Oxford.

Borchers, D.L., Buckland, S.T., Goedhart, P.W., Clarke, E.D. and Hedley, S.L. (1998) A Horvitz-Thompson estimator for line transect surveys. *Biometrics*, **54**, 1221-37.

Buckland, S.T., Anderson, D.R., Burnham, K.P. and Laake, J.L. (1993) *Distance sampling: Estimating Abundance of Biological Populations.* Chapman and Hall, New York.

Burnham, K.P., Anderson, D.R. and Laake, J.L. (1980) Estimation of density from line transect sampling of biological populations. *Wildlife Monographs*, **72**.

Caswell, H. (1989) *Matrix population models: construction, analysis and interpretation.* Sinauer Associates, Inc. Publishers, Sunderland, Massachusetts.

Cole, L.C. (1957) The population consequences of life history phenomena. *Quarterly Review of Biology*, **29**, 103-137.

Duck, C.D., Hiby, A.R. and Thompson, D. (in preparation). The use of aerial photography to monitor local and regional dynamics of grey seals.

Gerrodette, T. (1987) A power analysis for detecting trends. *Ecology*, **68**, 1364-1372.

Hammond, P.S. (1986) Estimating the size of naturally marked whale populations using capture-recapture techniques. *Reports of the International Whaling Commission* (Special Issue **8**), 253-282.

Hammond, P.S. (1987) Techniques for estimating the size of whale populations. In Mammal Population Studies, edited by S. Harris. *Symposia of the Zoological Society of London*, **58**, 225-245.

Hammond, P.S. (1990) Capturing whales on film - estimating cetacean population parameters from individual recognition data. *Mammal Review*, **20**, 17-22.

Hammond, P.S. (1995) Estimating the abundance of marine mammals: a North Atlantic perspective. In: *Whales, Seals, Fish and Man* (Ed. by A.S. Blix, L. Walløe and Ü. Ulltang). Developments in Marine Biology 4, 3-12. Elsevier Science B.V., Amsterdam.

Hammond, P.S., Benke, H., Berggren, P., Borchers, D., Buckland, S.T., Collet, A., Heide-Jørgensen, M.P., Heimlich-Boran, S.L., Hiby, A.R., Leopold, M.F., Øien, N. (1995) Distribution and abundance of the harbour porpoise and other small cetaceans in the North

Sea and adjacent waters. Final report to the European Commission under contract LIFE 92-2/UK/027, 240 pp. Available from the Sea Mammal Research Unit, Gatty Marine Laboratory, St Andrews, Fife KY16 8LB, UK.

Hiby, A.R. and Duck, C.D. (in preparation). Estimates of the size of the British grey seal *Halichoerus grypus* population and levels of uncertainty.

Hiby, A.R. and Hammond, P.S. (1989) Survey techniques for estimating abundance of cetaceans. *Reports of the International Whaling Commission* (Special Issue 11), 47-80.

Hiby, A.R. and Lovell, P. (1990) Computer-aided matching of natural markings: a prototype system for grey seals. *Reports of the International Whaling Commission* (Special Issue 12), 57-61.

Hiby, A.R. and Lovell, P. (1998) Using aircraft in tandem formation to estimate abundance of harbour porpoises. *Biometrics*, 54, 1280-9.

Hiby, A.R., Thompson, D. and Ward, A.J. (1988) Census of grey seals by aerial photography. *Photogrammetric Record*, 12, 589-594.

Hiby, L., Duck, C., Thompson, D., Hall, A. and Harwood, J. (1996) Seal stocks in Great Britain. *NERC news*, January 1996, 20-22.

Katona, S.K. and Beard, J.A. (1990) Population size, migrations and feeding aggregations of the humpback whale in the western North Atlantic Ocean. *Reports of the International Whaling Commission* (Special Issue 12), 295-305.

Palsbøll, P.J., Allen, J., Bérubé, M., Clapham, P.J., Feddersen, T.P., Hammond, P.S., Hudson, R.R., Jørgensen, H., Katona, S.K., Larsen, A.H., Larsen, F., Lien, J., Mattila, D.K., Sigurjönsson, J., Sears, R., Smith, T., Sponer, R., Stevick, P. and Øien, N. (1997) Genetic tagging of humpback whales. *Nature* (London), 388, 767-769.

Ries, E.H., Hiby, L.R. and Reijnders, P.J.H. (1998) Maximum likelihood population size estimation of harbour seals in the Dutch Wadden Sea based on a mark-recapture experiment. *Jounrnal of Applied Ecology*, 35, 332-339.

Seber, G.A.F. (1982) *The Estimation of Animal Abundance and Related Parameters*. Griffin, London. 2nd edition.

Smith, T.D. (1983) Changes in the size of three dolphin (*Stenella* spp.) populations in the eastern tropical Pacific. *Fishery Bulletin (U.S.)*, 81, 1-14.

Smith, T.D., Allen, J., Clapham, P.J., Hammond, P.S., Katona, S.K., Larsen, F., Lien, J., Mattila, D., Palsbøll, P.J., Sigurjónsson, J., Stevick, P.T. and Øien, N. (1999) An ocean-basin-wide mark-recapture study of the North Atlantic humpback whale (*Megaptera novaeangliae*). *Marine Mammal Science*, 15, 1-32.

Thompson, P.M. and Harwood, J. (1990) Methods for estimating the population size of common seals. *Journal of Applied Ecology*, 27, 924-938.

Thompson, P.M., Pierce, G.J., Hislop, J.R.G., Miller, D. and Diack, J.S.W. (1991) Winter foraging by common seals (*Phoca vitulina*) in relation to food availability in the inner Moray Firth. *Journal of Animal Ecology*, 60, 283-294.

Thompson, P.M., Tollit, D.J., Wood, D., Corpe, H.M., Hammond, P.S. and MacKay, A. (1997) Estimating harbour seal abundance and status in an estuarine habitat in north-east Scotland. *Journal of Applied Ecology*, 34, 43-52.

Tregenza, N.J.C., Berrow, S.D., Hammond, P.S. and Leaper, R. (1997) Harbour porpoise (*Phocoena phocoena* L.) by-catch in set gillnets in the Celtic Sea. *ICES Journal of Marine Science*, 54, 896-904.

Ward, A.J., Thompson, D. and Hiby, A.R. (1987) Census techniques for grey seal populations. In Mammal Population Studies, edited by S. Harris. *Symposia of the Zoological Society of London*, 58, 181-191.

Wells, R.S and Scott, M.D. (1990) Estimating bottlenose dolphin population parameters from individual identification and capture-release techniques. *Reports of the International Whaling Commission* (Special Issue 12), 407-415.

Wilson, B., Thompson, P.M. and Hammond, P.S. (1997) Habitat use by bottlenose dolphins: seasonal and stratified movement patterns in the Moray Firth, Scotland. *Journal of Applied Ecology*, **34**, 1365-1374.

Wilson, B., Hammond, P.S. and Thompson, P.M. (1999) Estimating size and assessing trends in a coastal bottlenose dolphin population. *Ecological Applications*, **9**, 288-300.

Chapter 8

Acoustic Techniques for Studying Cetaceans

[1]JONATHAN GORDON and [2]PETER L. TYACK
[1]*Sea Mammal Research Unit, Gatty Marine Laboratory, University of St Andrews, St Andrews, Fife KY16 8LB, Scotland, UK.. E-mail: jg20@st.andrews.ac.uk;* [2]*Biology Department, Woods Hole Oceanographic Institution, Woods Hole, MA 02543, USA. E-mail: ptyack@whoi.edu*

1. INTRODUCTION

In Chapter 4, we saw that sound propagates extremely efficiently through the sea and we explored some of the consequences of this for cetaceans. We showed that the dominant sensory modality for cetaceans is acoustic and that most cetaceans are highly vocal animals. This means that acoustic methods can often be very effective means for studying cetaceans. Studies of acoustic behaviour require a variety of specialised acoustic techniques. We will briefly review some of these here, highlighting some of the particular problems of working on cetacean acoustics and giving examples of solutions. In addition, largely because sound travels so well underwater, acoustic methods are often the most effective and efficient means of addressing questions about cetaceans which are not primarily acoustic in nature: assessing population distribution, abundance and movements for example.

It is interesting to note in passing that, when faced with the problem of finding submarines which, like whales spend most of their time submerged and are difficult to see at the surface, the world's navies have also turned to underwater acoustic systems. That they, with substantial budgets, should have made this decision should encourage us in the belief that underwater acoustic approaches will prove effective with cetaceans as well. It also has two other significant implications. In some cases, cetologists can benefit

Marine Mammals: Biology and Conservation, edited by
Evans and Raga, Kluwer Academic/Plenum Publishers, 2002

from the spin-offs of this enormous research effort, and indeed much cetacean acoustic research, especially on bio-sonar, has been funded from military sources. Less positively though, some of the active acoustic systems being developed by the Military introduce substantial new sources of noise into the ocean environment which could disrupt cetaceans' use of underwater acoustics and even damage their sensitive hearing, if not operated in an environmentally responsible manner (see Chapter 16 for review of the effects of this and other powerful noise sources).

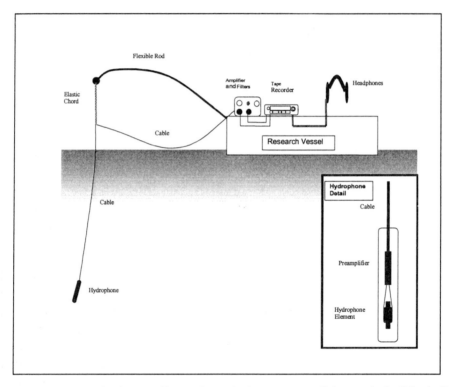

Figure 1.   A simple recording and monitoring system utilising a single "dipping" hydrophone.

2.   BASIC MONITORING AND RECORDING

In order to hear well underwater and to be able to make recordings, one normally uses a hydrophone, which is essentially an underwater microphone. The sensitive element in most modern hydrophones is a piece of piezoelectric material. Fluctuations in pressure cause corresponding fluctuations in voltage across the element. An impedance-matching preamplifier is often positioned close to the element to drive this signal up a cable and above water, where it can be monitored and amplified. (Fig. 1 illustrates a simple recording set-up). There are a huge number of different types of hydrophone systems covering an enormous price range. If research goals include a very precise physical characterisation of sounds, then expensive calibrated hydrophones and recording systems may be required. Field biologists rarely need to do this, however, and in many cases much less expensive equipment will be appropriate for biological studies. It can be very useful in the field to have a hydrophone that is directional so that phonating animals can be located and followed.

Recording equipment that also serves a consumer market is usually good value and affordable. Such equipment typically has two channels (stereo) and a frequency range that approximately matches the human auditory span (20 Hz - 20 kHz). The introduction of the rDAT format has made the collection of high quality field recordings in the audio range much more affordable. Many rDAT recorders time-stamp recordings; this is a very useful feature ensuring that underwater recordings can be reliably linked with particular encounters and data collected in other ways. Being able to monitor and record on two channels (*i.e.* in stereo) is a great advantage. Different record levels and/or filters can be set on each channel and the binaural "stereo" effect, due to differences in the time of arrival of sound at the two spaced hydrophones, can be helpful in discriminating vocalising animals. This can often allow an analyst to be confident that two calls on a recording must have come from different animals, even though there is not sufficient information to accurately locate the animal itself.

Some research generates requirements that are not covered by standard consumer products, for example if ultrasonic signals or multiple channels need to be recorded. The price of the specialised equipment required then is much higher. Watkins and Daher (1992) provide a very useful overview of techniques for recording animals underwater. Since much of the equipment in this field continues to develop rapidly, the reader is encouraged to visit relevant sites on the World Wide Web (such as the Animal Bioacoustics web site of the Acoustical Society of America, available from http://cetus.pmel.noaa.gov/Bioacoustics.html) to obtain up to date information on available equipment and software.

One persistent problem when recording sounds underwater stems from noise produced in the course of making the measurements: the noise of the research boat, and noise made as water flows past the hydrophone and its cable. Research boats used for acoustic work should be as quiet as possible and the effects of their noise can also be reduced by using long cables and floating the hydrophone away from the vessel. There are a number of tricks for reducing cable strum and hydrophone movement noise such as suspending the element in mid water on stretchy elastic, preferably from a neutrally buoyant float system with a large inherent inertia. Different systems seem to work best in different situations; it is more of an art than a science and the important thing is to persevere and to experiment. Most of this flow noise is low frequency and, if the calls of interest are predominantly higher in the spectrum, electronic filtering can be used to remove the lower frequency noise (Sayigh *et al.,* 1993). In particular, to avoid overloading the preamplifiers with low frequency noise a 1 pole high pass filter (6 dB/octave roll-off) can be introduced between the transducer and its preamplifier.

Figure 2.   Fishing for sound: bamboo poles are used to boom hydrophones out away from this small research boat as it listens for cetaceans. (Photo: J. Gordon/IFAW)

If hydrophone elements are placed in a streamlined body, then the flow noise, as it moves through the water, may be low enough to allow hydrophones to be monitored while they are being towed. For example, Leaper *et al.* (1992) placed elements in a 10 m long, 4 cm diameter, oil-filled polythene tube and were then able to monitor hydrophones effectively, at towing speeds of up to 7 knots, at frequencies above 100 Hz (*e.g.* Fig. 3). A somewhat more refined wide-band towed array has been used in the Mediterranean by Pavan and his team (see Pavan and Borsani,

1997). Miller and Tyack (1998) have added an adaptive beamformer to a multi-element towed array. Listening in a directional beam helps to reduce noise from the ship and also allows the listener to determine the direction from which a sound is coming (Thomas *et al.*, 1986).

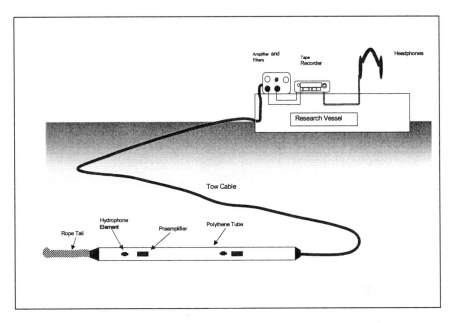

Figure 3. Schematic diagram (not to scale) showing a simple towed hydrophone system. The hydrophones used by Leaper *et al.* (1992) had a tow cable of 100 m, with hydrophone elements separated by 3 m within a 7 m polythene tube.

Real-time analysis and display of underwater sounds (for example spectrogram displays, see Chapter 4, Fig. 1) allows some features of the signal to be better appreciated, as well as providing confirmation of auditory detections and allowing the detection of signals outside the human auditory range. It is now possible to run real-time analysis programs on computers that can be readily taken into the field, and this can significantly increase the efficiency of field monitoring. (Details of available commercial and non-commercial software can be found on the Animal Bioacoustics web page from http://cetus.pmel.noaa.gov/Bioacoustics.html).

3.   STUDYING ACOUSTIC BEHAVIOUR

3.1   Determining Which Individual has Vocalised

A common requirement in studies of communication involves identifying who produces a particular signal, what context is associated with that signal, which animals receive it, and how they respond. Often this information is obvious to a terrestrial ethologist. He or she can readily determine the direction from which a sound arrived, and usually terrestrial animals behave in ways that make it obvious when they vocalise (most open their mouths for example). The situation is much less straightforward with cetaceans. A human diver cannot localise sounds underwater, and cetaceans produce few visible movements when vocalising. Difficulties in identifying which animal produced a call have been a serious obstacle to studying cetacean acoustic communication. Cetacean ethologists have been forced to create new methods to identify who is vocalising when studying communication in the wild.

3.1.1   Passive Acoustic Localisation

One set of methods compensates for our inability to determine the direction of an underwater sound by ear. These techniques use an array of hydrophones to locate sounds in order to find which animal vocalised. Greene and McLennan (1996) review techniques for acoustic localisation and tracking using arrays. One of the earliest successful attempts at this involved the deployment of four hydrophones from an oceanographic research vessel (Watkins and Schevill, 1972). Watkins and Schevill deployed the array near cetaceans and recorded all four channels of acoustic data. Back in the lab, they compared arrival times of sounds at the four hydrophones in order to calculate the location of the sound source. One problem with this technique is the long delay between observation and analysis, making it difficult to relate source location to observed behaviour. This can make it difficult to follow a vocalising animal, or to select it for focal animal observations. This hydrophone array was also non-rigid, requiring pingers on the array to calibrate the hydrophone locations. These sometimes affected the behaviour of the whales (Watkins and Schevill, 1975).

When one uses a dispersed array to locate vocalisations by measuring time delays, the method works best at locating calling animals within the array (*e.g.* Freitag and Tyack, 1993). It is possible to estimate the range to animals outside of the array if they are within several times the dimensions of the array; beyond that, one can estimate bearing to the animal, but can seldom specify range precisely. If one wants to track free-living animals (locating them in range and bearing) over several kilometres, an array with separations between hydrophones of kilometres will be required. It can be expensive and unwieldy to use kilometres of cable, and one effective solution is to use sonobuoys, which link hydrophones to radio transmitters. Clark and Ellison (1989) used this type of array deployed through sea ice to track bowhead whales, *Balaena mysticetus*, in Alaska, and Janik (2000) used a similar radio telemetry system to locate calls of bottlenose dolphins in the Moray Firth.

Improved software, and increases in the speed of computers, means that now the sort of analysis that underpinned Watkins and Schevill's techniques can be performed in real time (*e.g.* Pavan, 1992; Gillespie, 1997). Several different kinds of signal processing techniques can be used to estimate time delays for the same signal across two different channels. One widely used technique is called cross correlation. This involves correlating the waveforms received at two different hydrophones for different time lags. Peaks in the cross-correlation function often indicate time lags associated with the delay in time of arrival of a signal at the two hydrophones. This technique is described in detail for the problem of acoustic localisation in Spiesberger and Fristrup (1990). While this is an ideal method for some situations, in an environment where the sound can take several paths to the receiver, it can be difficult to identify which sounds have arrived by a direct path (Freitag and Tyack, 1993). Another technique that can be more robust, although it is less precise, is called spectrogram correlation (Clark *et al.,* 1985). For localising sounds with a sharp rise time, such as clicks, simply timing when a certain threshold is first exceeded often works well (Freitag and Tyack, 1993). Ten years ago, such localisation methods required expensive hardware and considerable technical expertise; today it can be attempted by a graduate student in biology.

There is an alternative approach to localisation that does not require a large dispersed array of hydrophones. Small rigid arrays can be used to determine the bearing to a sound source. If the actual location is required, one can use several of these arrays to triangulate source location. One of the first successful applications of this approach for marine mammal localisation was described by Clark (1980), who made a small rigid array with electronics that immediately calculated the bearing of a sound. He placed the array near shore where right whales congregated, and he was able to combine visual observations of whales with acoustic locations in the field as vocalisations were produced.

It is often useful to tow a hydrophone array from a vessel. In these settings, the optimal array is often a linear streamer of the sort described in the previous section. Signal processing techniques, such as beam-forming or cross-correlation, can be used not only to reduce noise but also to determine the direction from which a sound is coming (Miller and Tyack, 1998; Leaper *et al.,* 1992; Gillespie, 1997). With a linear array, there will be a left-right ambiguity: it is not possible to tell whether a sound arrives from the left or right hand side of the array. However, this can be easily resolved by altering course to determine whether the sound source moves further forwards or further aft. Miller and Tyack (1998) also describe how multiple towed arrays can also be used to triangulate source location, and they discuss how such arrays can be used to identify which animal is vocalising in the wild.

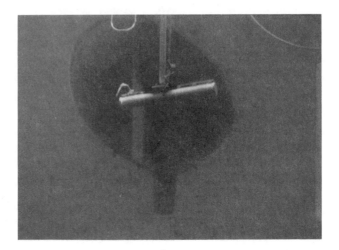

Figure 4.   A directional hydrophone mounted on the stern of a 46' sailing boat. (Photo: J. Gordon/IFAW)

3.1.2   Tagging

Another method for identifying who is vocalising involves placing on a cetacean, a tag that is sensitive to sound. It is now straightforward to make a tag incorporating a small hydrophone that can detect sounds produced by the tagged animal; however, telemetering this information can be difficult. Radio waves do not propagate well in seawater, limiting the usefulness of radio telemetry unless the animal is at the surface. As we have seen, sound travels well through the sea, but sonic telemetry can be problematic since cetaceans can hear such a broad range of frequencies and may well be disturbed by the sonic telemetry signals. Tyack (1985) developed a simple telemetry device that is useful with captive cetaceans. This device uses light for telemetry, illuminating more light emitting diodes (LEDs), the louder the sound recorded at the hydrophone. This has proved very useful for identifying which dolphin produces a sound in a captive setting. However, it is possible that the use of tags may change the behaviour of animals. For example, the dolphins in Tyack's (1985) tag study had much higher rates of whistle imitation than dolphins studied using passive acoustic localisation techniques (Janik and Slater, 1998). Advances in the power and miniaturisation of electronics have created opportunities for sound recording tags (Tyack and Recchia, 1991; Burgess *et al.*, 1998) that record sound level at the animal, and can be downloaded after removal from the animal. When such loggers are synchronised with video or human observers, visual observations can be linked retrospectively with acoustic records.

3.2   Playback Experiments

When one begins to study communication in a new species, research usually starts by observing the contexts in which signals are produced: who responds to the signal, and what kinds of responses are evoked. For example, if a sound is produced when animals are separated and when production of the call causes another animal to call back or to approach the caller, we might surmise that the vocalisation was a contact call. If a sound is produced only by males and only during the breeding season, and if females appear to use the sound to select and approach a mate, then the signal might be described as a reproductive advertisement.

However, there are problems with purely observational studies. Ethologists have developed experimental methods to test whether a specific signal actually causes a particular response or is just associated with it, and to test the specific kinds of information communicated by the signal. The dominant technique is called sound playback. This involves recording sounds used in particular interactions, usually along with observations of who makes the signal and who responds. The experimenter then plays back the signal to carefully chosen subjects in carefully chosen contexts, and observes the responses of the subject(s) when they hear the sound.

One of the earliest playback studies of cetaceans involving conspecific calls is reported by Clark and Clark (1980). Christopher Clark had used his direction finding device to discover that often when one right whale made a certain call, called an UP call, other right whales in the area would rapidly approach. This led him to think of these as contact calls. In order to test whether the calls evoked the response, he played sounds of right whales back through an underwater loudspeaker and found that most subjects rapidly approached the location of the speaker (Clark and Clark, 1980).

Playbacks need not be restricted to the sounds of conspecifics. Many cetaceans are prey of killer whales, so it was obvious to ask whether these species might detect the calls of killer whales and respond in order to try to escape before they were detected themselves. Early cetacean playback experiments verified that beluga whales and migrating gray whales both show strong avoidance and escape responses to playback of killer whale sounds (Cummings and Thompson, 1971; Fish and Vania, 1971).

The sound playback technique is capable of teasing apart complex discriminations and meanings of communication signals; such use is well established with birds and primates (*e.g.* Cheney and Seyfarth 1990). Sayigh *et al.* (1999) provide an example of using playbacks to test whether dolphins associate signature whistles with particular individual dolphins. They played back signature whistles of wild dolphins to other wild dolphins temporarily restrained in a net corral, and demonstrated that mothers and offspring respond preferentially to each others' signature whistles, even after calves become independent from their mothers.

Playback experiments have also become important methods for assessing how cetaceans respond to man-made noise. It is possible to study cetaceans in the presence of existing industrial noise sources, but unless one can control operation of the source, it is often difficult to compare responses in well-defined control and exposure conditions. Studies in which the responses of cetaceans are observed during carefully controlled sound exposure are often more powerful. For example, one sound playback study involved observation of >3,500 migrating gray whales (*Eschrichtius robustus*) and showed statistically significant responses to playback of industrial noise (Malme *et al.*, 1984; reviewed in Richardson *et al.*, 1995). Whales slowed down and started altering their course away from the sound source at ranges of 1-3 km. This resulted in an increase in the distance between the whales and the source at their closest point of approach. By comparing the source levels of the experimental source to actual industrial operations, one can predict responses of whales to industrial sources in various settings. The ranges predicted for a 50% probability of avoidance were: 1.1 km for the loudest continuous playback stimulus, a drillship; and 2.5 km for a seismic array used in oil exploration.

Conducting playbacks underwater is a far from trivial undertaking. The equipment required to project realistic levels of sound is specialised, often bulky and expensive, and this has so far restricted the use of playback techniques with cetaceans. Playbacks are also most effective after naturalistic observations allow an ethologist to predict how a particular subject should respond to the sound stimulus played back, with some other natural sound available for use as a control stimulus. Great care must be taken to ensure that the playback occurs in an appropriate context, under realistic conditions, so that the subject is presented with a naturalistic exposure to a biologically relevant signal. McGregor (1992) discusses some of the design features and controversies for ethological playback experiments.

3.3 Temporal and Spatial Scales of Cetacean Acoustic Behaviour

Cetacean acoustic behaviour can often take place at temporal and spatial scales that are challenging to a human researcher. At one extreme, the baleen whales produce infrasonic calls in extended sequences that may continue for many hours or days and could potentially affect the behaviour of other whales at ranges of tens or even hundreds of kilometres. As human observers, we are equipped with a brain that is adapted to spotting patterns, and associating behaviours, over periods of seconds to minutes, making it difficult to interpret behavioural patterns that unfold much more slowly. One solution to both detection of the low frequencies involved and the long time scale can be to play recordings back at a higher speed. There is no easy way of making observations at the spatial scales involved, but by using acoustic techniques we are at least made better aware of them. Dolphins operate at the other extreme: many of their vocalisations may be ultrasonic, and their signals are modulated and repeated with great rapidity. Playing back recordings at much reduced speed can help here, though improved perception is obtained at the cost of time, so that a few minutes of high frequency recording can take many hours to analyse at lower speed.

4. ACOUSTIC FIELD METHODS

4.1 Finding, Following, and Tracking Cetaceans Acoustically

In some cases, simple acoustic techniques can be used to find and follow cetaceans in the field so that they can be observed and studied further using other, non-acoustic methods. For example, sperm whales undertake long (45 mins or more) dives making them difficult to find and follow from the surface. They are very vocal during these dives, however; they make loud regular clicks that can be heard at ranges of 5-10 kilometres, for most of the time. Whitehead and Gordon (1986) described how fairly simple directional and non-directional hydrophone systems were used to locate sperm whales which could then be identified and measured using visual and photographic techniques. With these relatively unsophisticated techniques it was also sometimes possible to track group movements for days at a time (*e.g.* Gordon, 1987; Christal and Whitehead, 1997). Off Kaikoura, in New Zealand, commercial whale watchers have also started to use simple acoustic techniques to find sperm whales. As a result, they are more likely to be positioned close to sperm whales when they surface, allowing them to make a more leisurely approach to the whales; thus reducing whale disturbance (Gordon *et al.*, 1992).

The quite basic equipment described above simply indicates the direction from which vocalisations arrived. In order to use most of these simple hydrophones, one often must stop or slow down the vessel. More sophisticated hydrophones and arrays of hydrophones can be designed to be towed from a vessel while it is underway, facilitating the task of following and tracking animals. As discussed in Section 2, modern signal processing techniques can be used to estimate the bearing to a vocalising cetacean, using a towed linear array of hydrophones (Gillespie, 1997; Miller and Tyack, 1998). As is so often the case with underwater acoustics, advances in software, and the ever increasing power (and decreasing cost and size) of computers, greatly enhances what can be achieved affordably with such signal processing methods.

4.2   Locating Animals to Mitigate Disturbance

The acute acoustic sensitivity of cetaceans heightens concerns that they could be seriously disturbed, or even injured, by loud underwater noises (see Chapter 16 by Würsig and Evans in this volume, for example). Offshore seismic surveys, which are often conducted as part of oil exploration, use airguns to produce very intense pulses of sound every 10 seconds or so. Seismic legs may last for several hours and many legs will be completed each day working round the clock during surveys that can extend over months. Often, regulations and guidelines, designed to reduce the risk to cetaceans from these intense sounds, require operators to check that cetaceans are not within a proscribed range of the airgun before they start firing. Visual searches for cetaceans can be made before legs are started and in many cases acoustic monitoring can also be extremely helpful, especially as seismic surveys often take place in exposed offshore locations where sighting conditions are poor, and continue through the hours of darkness. Recent work in the UK has led to the development of semi-automated hydrophone systems capable of being deployed by small monitoring teams during seismic surveys. The goal of combining visual and acoustic monitoring is to test whether a relatively safe "cetacean free" area can be found ahead of survey vessels in which they can start their airguns. If too many cetaceans are detected either acoustically or visually, survey lines will be aborted. One largely automated system designed to do this, described by Chappell and Gillespie (1998), also stores a record of all automatically detected vocalisations and seismic pulses so that the operators' compliance with the regulations can be monitored.

Figure 5.   Harbour porpoises lend themselves to acoustic detection techniques since they are small and inconspicuous at the surface whilst being vocally very active. (Photo: I. Birks/SWF)

4.3   Acoustic Population Assessment

If we are to conserve cetacean populations, we need to know the following fundamental pieces of information: where they are and when, how many there are, and whether their populations are increasing or decreasing. This basic information has proved surprisingly difficult to obtain for cetaceans, largely because they are elusive creatures that can be difficult to spot, and because they inhabit a large and (for humans) inhospitable habitat. As we have stressed many times already, though, some cetaceans are more conspicuous acoustically than visually and this might lead us to expect that the detection of acoustic cues could form the basis for effective population assessment techniques.

For certain species at least, acoustic surveying techniques offer a number of advantages when compared with visual methods:

1. The acoustic range of vocalisations can be measured and predicted more precisely than is usually the case for visual range. Underwater acoustics is a sophisticated and well-developed branch of science, thanks largely to the intense interest in the subject by the military. If the source level of the vocalisations is known, the propagation conditions can be modelled and noise level measured to enable prediction of the range at which the vocalisations can be detected. In addition, it is often possible to calculate range directly, by using arrays of hydrophones (*e.g.* Greene and McLennan, 1996; Watkins, 1976; Ko *et al.*, 1986). By contrast, there is little theoretical understanding of the complex factors that affect a human observer's ability to detect cetaceans visually, and this often varies with subtle changes in physical conditions, sometimes in unexpected ways. Range to sightings is rarely directly measured during visual surveys.

2. Acoustic range is less affected by meteorological conditions than visual range. The range at which cetaceans can be spotted is curtailed rapidly by increasing sea state, and visual surveys are rarely continued at sea states above Beaufort three. Although the level of background noise increases with sea state, and this masking noise can reduce the range of detection in acoustic surveys, the effect is measurable and resultant range reductions are predictable. In practice, acoustic surveys can usually continue in higher sea states than visual surveys.

3. Acoustic range is often superior to visual range. The range at which cetaceans can be seen or heard varies from species to species and with the sophistication of the acoustic equipment being used. However, many can be detected acoustically at a greater range than they can be seen, particularly when small research vessels are used. For example, dolphins can be detected acoustically at ranges of up to 2 km and sperm whales can be reliably heard at ranges of 5-9 km using simple hydrophones (J. Gordon, unpubl. observs.) while Sparks *et al.* (1993) reported detecting sperm whales at ranges of 18 km using a towed linear array. Some of the large baleen whales can be heard with near-surface hydrophones at ranges of tens of kilometres (Clark and Fristrup, 1997).

4. Acoustic surveys are less onerous than visual surveys. Searching for whales is hard work and requires constant vigilance from experienced

observers; spotters have to be changed regularly and rested, and consequently large (expensive) field teams are required.

5. Acoustic monitoring can be conducted 24 hours a day, both day and night. Obviously, visual sightings are impossible in poor light conditions and at night. Most cetaceans continue to vocalise round the clock, although allowances may have to be made for diurnal variation in acoustic output.

6. A complete and permanent record can be made of acoustic survey cues. A high quality tape recording provides a remarkably full record of the acoustic information within its band of sensitivity, and this is then available for further analysis and can be re-analysed in the future if techniques improve.

7. There is a great potential for automation of data collection and detection. Modern digital processing techniques mean that many aspects of acoustic analysis, such as distinguishing, classifying, counting and timing vocalisations, can be performed automatically (Potter *et al.*, 1994; Gillespie, 1997; Fristrup and Watkins, 1993; Stafford *et al.*, 1994). Two distinct advantages stem from automating detection. In the first place, it further reduces the amount of human effort required to complete a survey. Secondly, and most importantly, it removes sources of human error due to inter-individual variability in ability to make detections.

8. Generally, acoustic surveys are well suited to be undertaken from small research boats or platforms of opportunity. Visual surveys typically require large vessels to provide steady elevated viewing platforms, and to accommodate large teams of spotters. Often, acoustic surveys can be conducted from very modest vessels; in fact, because smaller displacement boats are generally quieter, they may often be preferred to large vessels. Small vessels are much cheaper to charter and run than large ones, so more surveying effort can be achieved for an equivalent expenditure.

There are, however, some potential problems with an acoustic approach:

1. Our knowledge of cetacean vocal behaviour is patchy and far from complete (see Chapter 4).

2. Vocalising is not obligatory for many species. Echolocating animals may have to vocalise to feed or to orientate themselves, but rates of production of all vocalisations are potentially highly variable. By contrast, we know that all cetaceans do have to surface to breathe, even though they

are not always equally conspicuous when they do this. Current knowledge of the surfacing patterns of cetaceans is better than our understanding of vocalisation rates. In some species, vocalisation rates may vary seasonally; vocalisations that are involved with breeding provide particularly pronounced examples. We have data on inter-vocalisation intervals for very few species in very few contexts, but this situation should improve as more studies are completed.

3. Acoustic surveys may require quiet vessels; however, disturbance of subjects by noisy vessels and boat avoidance should be a concern for any type of survey.

4. Some form of hydrophone system is required which inevitably involves extra expenditure. In many cases, much can be achieved with inexpensive hydrophone equipment but for other applications, especially when one must listen at low frequencies, sophisticated and expensive gear may be required. The necessity of towing hydrophones can add to the logistic complexity of a trip, especially if one is using a long array of hydrophones to locate low frequency sounds. One disadvantage of single linear towed arrays is that there is an ambiguity as to which side of the array the sound is coming from.

5. Methods of deriving abundance from acoustic cues are still not well developed, largely because relatively little effort has so far been expended in this field. Improved methods for estimating range to the vocalising animals will make it easier to adapt existing visual techniques to acoustic data.

4.3.1   Types of Acoustic Surveys

Different types of acoustic survey of will provide different levels of information about a population. Each one requires different degrees of knowledge about the animal's vocal behaviour and different sorts of information to be determined for detections. In any survey, one is essentially making an assessment of the probability that a target animal will be detected if it is within a certain range of the survey vessel during the survey. Table 1 summarises the information required for different types of surveys.

In the most basic of surveys, detections of the vocalisations of a species could be used as an indication of its presence or absence within the survey area to reveal its geographic range and give a completely non-quantitative population assessment. For such a survey, one must at least be able to reliably identify each species from its vocalisations, and be confident that it is sufficiently vocal to be reliably detected if present in the survey area, and that not detecting a species' calls is a fair indication of its absence.

Table 1. Information required for different types of survey

Information Required for Survey	Type of Survey		
	Qualitative	Index of Abundance	Absolute Abundance estimate
Species Identity	YES	YES	YES
Effective Range at which sounds can be detected	NO	NO	YES
Variation in Range at which sounds can be detected	NO	YES or assume Constant	YES or assume Constant
"Rate" of Sound Production (this is the probability an animal will vocalise and be detected as ship passes within acoustic range)	NO	NO	YES
Variation in Rate of Sound Production (Geographic, seasonal, vessel, social.)	NO	YES or assume Constant	YES or assume Constant

If better field data can be obtained, and if the animal's vocal behaviour is sufficiently well known, measures of acoustic detection rate can provide good indicators of population density and can be used for surveys that provide an index of abundance. These might allow changes in population levels to be tracked. Quantitative surveys, capable of providing absolute abundance estimates, require the collection of the most detailed field data and a high level of knowledge of the animals' vocal behaviour.

All acoustic surveys rely on an ability to identify species from their vocalisations (though it may not be necessary to be able to identify every sound the animals might make). This emphasises the importance of establishing comprehensive databases of underwater sounds (see Chapter 6, Section 3.1) and of developing methods for systematically recognising and distinguishing between the vocalisations of different species (see for example, Fristrup and Watkins, 1994).

4.3.2   Some Examples of Acoustic Surveys

4.3.2.1   Acoustic Methods Used in Support of Visual Surveys

Thomas *et al.* (1986) used a towed hydrophone array to detect and classify cetacean vocalisations in conjunction with visual sightings surveys. They reported that the detection of acoustic cues enhanced visual detection of cetaceans, especially on the high seas. They advocated the wide use of such arrays in conjunction with sighting surveys, to provide more accurate estimates of distribution and abundance.

A major survey in the Gulf of Mexico used acoustic monitoring in conjunction with visual surveys to determine the distribution of cetaceans and relate these to oceanographic conditions (Davis and Fargion, 1996). The towed linear array they employed was mainly useful for detecting odontocetes, with sperm whales being detected at ranges of 18 km (Sparks *et al.*, 1993).

4.3.2.2   Simple Qualitative Surveys

Acoustic methods can play an important role in establishing the range and distribution of species. Useful data can be collected by using simple hydrophones to make recordings in areas where cetacean populations are poorly known to simply establish which species are present. In many cases, such recordings could be made by untrained volunteers, such as yachtsmen. This sort of survey might be thought of as an acoustic equivalent of the opportunistic visual sighting schemes which have been run in many parts of the world. In many cases, however, identifications made from tapes by experienced acousticians could be more reliable than visual observations made in the field by untrained observers.

An example from the Indian Ocean (Whitehead, 1985) illustrates the depth of information that can be gleaned from opportunistic recordings. Researchers working off the coast of Oman from a small sailing vessel in January 1982, made recordings of humpback whales singing in inshore waters at night but, because the main focus of their work was with sperm whales offshore during the day, these humpback whales were never sighted. Over several nights they were able to obtain recordings of reasonable quality and, later in the season (in February and March), they also heard humpbacks singing the same song off the coasts of India and Sri Lanka. Although the researchers never saw a humpback whale, their recordings of the whale's songs confirmed the presence of humpbacks in an area where they had not been expected. The presence of humpback song strongly suggested that these whales were breeding at that time and were therefore northern hemisphere whales, not part of a southern ocean population. Analysis also revealed that the song was quite different from songs recorded in the Pacific and Atlantic in the same year, indicating that the northern Indian Ocean whales were probably isolated from those other populations.

4.3.2.3   Standardised Monitoring

Systematic acoustic monitoring using standard equipment can provide an index of abundance for vocalising animals. The value of such indices should not be underestimated. They can reveal changes in population size and distribution, and this is usually the information that managers really need to know. They can also be useful in indicating how effort should be expended in space and time during more specialised surveys.

Recently, selected civilian scientists have been given access to data from networks of military hydrophones (the SOSUS arrays) (Fig. 6; see also "New Developments", section 5, later). Using these hydrophones, it is possible to detect and often to localise, vocalising baleen whales at ranges of hundreds or even thousands of kilometres. The thousands of locations of vocalising whales that have resulted from these programmes have been plotted to reveal general distributions and seasonal migrations of populations of blue, fin and minke whales over the large expanses of the N. Atlantic and N. Pacific covered by these hydrophone systems (Watkins *et al.*, 2000). On one occasion, the track of an individual blue whale with a very distinctive vocalisation was followed for 43 days while it migrated over 1,700 miles (Gagnon and Clark, 1993). These data have not yet been analysed to yield population abundance estimates, and this may not be feasible. However, in addition to showing offshore distribution and migration patterns over entire ocean basins (*e.g.* Clark, 1994; Clark and Charif, 1998), these data could provide a reliable index of abundance capable of showing changes in population levels over time, provided of course these hydrophone arrays are maintained as they are for several decades. This may prove a problem; the arrays were not established for this purpose; they are expensive to maintain, and many are being closed down just as their potential for marine mammal science is being demonstrated.

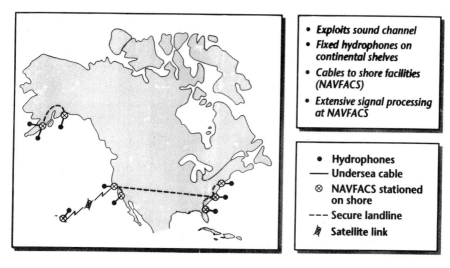

Figure 6.   The US Navy's Sound Surveillance System (SOSUS).

An attempt to obtain information on the distribution and abundance of striped dolphins in the Ligurian Sea Cetacean Sanctuary in the Mediterranean provides, by contrast, an example of relatively inexpensive and simple standardised equipment being deployed from several different platforms. Duplicate equipment, consisting of custom-built towed stereo hydrophones and DAT tape recorders, were deployed from three vessels working within the Ligurian Sea Sanctuary. The tracks of these boats were shaped according to predetermined rules. Every 15 minutes, the boat was quietened if necessary, so that a standardised 1-min recording could be made. At the same time, researchers monitored the hydrophones and scored the levels of dolphin and pilot whale whistles and clicks, and of sperm whale clicks, as well as levels of background water noise, shipping noise and self-noise. These data were recorded directly into a computer. In this area, nearly all dolphins encountered offshore were striped dolphins and species identification of dolphin vocalisations was therefore not required. Because the teams used identical equipment, deployed from similar small vessels, and they observed the same protocols, the data they collected were consistent and directly comparable. Linear regression techniques were used to model and remove the effects of covariates, such as background noise and weather conditions. These simple data provided detailed information on dolphin distributions, and relative abundance within the Ligurian Sea Sanctuary at relatively little cost, and may allow population trends to be tracked in the future (Gordon *et al.*, 2000). Methods for automating detection and measurement of whistles are being investigated by several groups (*e.g.* Fristrup and Watkins, 1994; Sturtivant and Datta, 1995; Chappell and Gillespie, 1998). Once perfected, these will further increase reliability of indices of abundance determined from such surveys.

High Frequency Click Detector

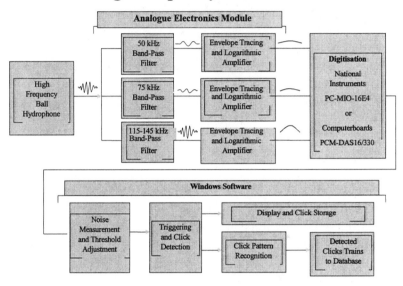

Figure 7. Schematic outline of an automated harbour porpoise detector.

An example of completely automated field detection equipment comes from recent work with harbour porpoises (Chappell *et al.*, 1996; see Fig. 7). Harbour porpoises are a notoriously difficult species to sight except in the calmest sea conditions. However, they make high-frequency vocalisations for echolocation. These are narrow-band pulses of about 140 kHz, which are quite distinct from other sources of noise in the sea (Amundin 1991; Kamminga and Wiersma 1981). In fact, the sounds are so distinctive that a relatively simple system, which monitors for sound energy in clicks in the "harbour porpoise" band and at two lower frequencies, achieves completely autonomous data collection in the field. Porpoise clicks are distinguished and logged if they contain sufficient energy in the "porpoise" band and with little energy in the other two lower frequency bands. Even though this system is analysing high frequency clicks, it only involves simple filters and a commercial analogue to digital converter board. It can run on a normal laptop computer and log days of detections on a normal hard disk.

Figure 8.   Porpoise detection equipment (Photo: J. Gordon/IFAW) .

To test the technique, six sets of equipment were deployed from vessels involved in the SCANS survey during the summer of 1994. From the perspective of the acoustic work, these vessels were essentially platforms of opportunity. The equipment was installed by inexperienced fieldworkers and ran untested. The results were encouraging. Rates of detection were similar to those obtained by large experienced visual teams. Acoustic detection rate was affected much less than the visual rate by sea state, and was superior to visual detection rate above sea state 3. The equipment is small enough to be carried by one person and can be readily deployed from a wide variety of unspecialised platforms (Chappell *et al.*, 1996). The combination of standard equipment, automated detection, and 24-hour operation by minimal survey teams, suggests that this approach could affordably provide the large quantities of abundance data from offshore areas year round that is required for the better conservation of this species. A method for deriving absolute abundance estimates with this equipment by using visual tracking methods to provide calibration data, has recently been suggested by Gordon *et al.* (1998), following an approach suggested by Buckland (1996).

4.3.2.4   Quantitative Acoustic Surveys

Acoustic methods have been used to count animals directly or to derive actual density estimates on only relatively few occasions.

The largest and most successful acoustic survey for cetaceans has been conducted off Point Barrow, Alaska, since 1984. Sonobuoys (hydrophones linked to radio transmitters) were deployed through fast-ice to create a long-baseline, near-linear array. Typical arrays consisted of 4-5 hydrophones each separated from the next nearest by roughly 800 m. Because the array was so large, the locations of vocalising whales could be calculated by measuring the relative time of arrival of sounds at the different hydrophones (Clark *et al.*, 1985a; Clark and Ellison, 1989; Zeh *et al.*, 1988). Whales were successfully located as far as 8 km offshore. Locations for sequences of vocalisations were linked to trace the likely tracks of individual whales, and these tracks were used to determine minimum numbers of whales migrating past Point Barrow. This might be considered a rather special situation in that in this area a population of whales was migrating past an ice edge, which allowed hydrophones to be deployed through the ice. In a more general case, this kind of array would need to be deployed on the seafloor by a ship or a long-baseline hydrophone array would be towed through the area to be surveyed.

Clark and Fristrup (1997) used a 1 km long, 16-element towed hydrophone array to detect and localise baleen whales during an experimental dual-mode (acoustic/visual) survey off the coast of California. A team of three observers on the survey vessel conducted a visual survey at the same time to provide comparative data. To compare acoustic and visual detection rates, Clark and Fristrup divided the time when both surveys were operational into 15-minute intervals. A comparison of the proportion of the 15-minute intervals that had reliable acoustic detections with the proportion with visual detections revealed that acoustic detection rates were six times higher than visual detections for blue whales and three times higher for fin whales.

A simple method for assessing sperm whale populations acoustically was developed by Leaper *et al.* (1992). Two identical stereo hydrophone arrays were towed on 100 m of cable behind a 14 m motor sailing vessel so that the elements formed a square array, with 3 m spacing between elements on each array. The vessel followed a pre-determined survey course at a speed of about 6 knots, 24 hours a day. One-minute recordings were made every 15 minutes. If necessary, to reduce noise, the vessel's engine was put out of gear when recordings were made. The relative times of arrival of sperm whale clicks at the different hydrophones in the array were measured from recordings using digital signal processing techniques. (The small size of the array (c. 3 m) relative to the range at which whales were detected (c. 5 km) meant that only the bearing to whales and not their locations, could be calculated.) These data were then used to calculate effective acoustic range and population densities using the "Cartwheels" technique (Hiby and

Lovell, 1989), which calculates average range to acoustic sources and their density from the patterns of change of acoustic bearings at subsequent stations separated by known distances, using a likelihood maximisation approach. During surveys in the Azores, the average range at which whales were detected was around 6 km. Acoustic detection densities are adjusted by a correction factor to allow for the proportion of time that sperm whales are not vocal to derive a population density estimate. This calibration factor was determined by measuring the vocal output of individual whales followed over long periods in the Azores.

One advantage of acoustic research, of course, is that it can be conducted at night. During some sperm whale research projects, it has proved possible to use acoustic methods to assess sperm whale distributions during the nights, leaving the daylight hours free to conduct more visually based research, such as photo-identification studies. In this way, the productivity of valuable field time can be maximised (Gordon *et al.*, 1999).

Essentially the same simple equipment has been used from the icebreaker RV. *Aurora Australis*, made available as a platform of opportunity, to survey for sperm whales in the Southern Ocean. Automation of the recording process allowed a single operator to keep the equipment running 24 hours a day over a 10-week period (Gillespie, 1997). The ease with which this was achieved, and the promising preliminary results obtained, highlight the potential for conducting very large-scale acoustic surveys for some species using platforms of opportunity.

If larger, sparse arrays of hydrophones were deployed (such as those used by Clark and Fristrup, 1997), the actual locations of vocalising sperm whales could be calculated. Sperm whale clicks are ideal signals to be localised using time of arrival techniques and Watkins and Schevill (1972) successfully tracked sperm whales using large non-rigid three-dimensional arrays over 25 years ago. Such surveys would provide higher quality data than that provided using smaller arrays as well as useful information on factors such as movements and possible ship avoidance/attraction behaviour. However, this could only be achieved using equipment that is much more complicated and expensive; and this might require a larger, possibly specialised, survey vessel.

4.3.3   Acoustic Surveys: Conclusions

Acoustic assessment techniques for cetaceans show great promise, but surprisingly few successful surveys have so far been completed. Generally, techniques for collecting, detecting, and measuring acoustic cues are well advanced. Often developments made by the military or seismic industries can be adapted for cetacean research, and cheaper faster computers and improved software are facilitating data collection and analysis. Relatively little effort has been expended on the problem of how to use these data to determine abundance parameters. On the occasions when survey mathematicians have applied themselves to this task, promising progress has been made (*e.g.* Zeh *et al.*, 1988; Hiby and Lovell, 1989). However, more attention needs to be given to this problem for the full potential of acoustic surveying to be realised.

Acoustic surveys should usually be seen as complementary to visual surveys rather than competitors or alternatives. Indeed, in many cases "dual mode" surveys, which build on the strengths of both techniques, may offer the best solution. (*e.g.* Buckland, 1996; Fristrup and Clark, 1997; Gordon *et al.*, 1998). The acoustic cues that are produced by cetaceans, and how they might be measured to assess abundance, should be considered as a matter of course in the planning of any cetacean survey. In some cases, additional costs which may be a fraction of those of ship time, could result in critical data being more than doubled.

4.4   Measuring Body Length

In the case of sperm whales at least, it is actually possible to measure an animal's body length using passive acoustic methods. Sperm whales are very sexually dimorphic and knowing body length allows mature males to be identified during surveys. Early studies of sperm whale clicks, such as those of Backus and Schevill (1966), described how sperm whales clicks often had a "burst pulse" structure, being made up of several regularly spaced pulses. Later, Norris and Harvey (1972) proposed that these pulses were formed as sound was reflected between two sound reflectors at the front and back of the sperm whale's enormous head (See Chapter 4, Fig. 9). They realised that, if this was the case, then the inter-pulse interval (IPI) would be twice the travel time between the two reflectors and that measuring the IPI would allow the length of the head of the sperm whale making the clicks, and thus its overall body length, to be determined. Others (*e.g.* Møhl *et al.*, 1976; Alder-Fenchel, 1980) explored the use of IPI measurements as a practical field technique for population assessment. By analysing the vocalisations of whales whose lengths had been measured photographically, Gordon (1991) was able to show that IPI was indeed related to head length in much the way that Norris and Harvey (1972) had suggested. Since then, Goold (1996) has investigated more reliable methods of measuring IPIs, while Pavan *et al.* (1999) were able to measure body growth in a sperm whale encountered and identified in two successive years in the Mediterranean.

5.   NEW DEVELOPMENTS

Cetacean underwater acoustics is a fast-moving field and there have been a number of exciting new developments in recent years.

For several decades, biologists and acousticians have explored the theoretical possibility that the low frequency vocalisations of baleen whales might be suitable for long-range communication (Payne and Webb 1971; Spiesberger and Fristrup 1990), but biologists have only recently been able to gather empirical data at the appropriate scale for this problem. The ability to gather these data is a windfall from the end of the cold war. One of the most closely guarded secrets of the cold war was the sophisticated system, called SOSUS, used by the U.S. Navy to track submarines. The SOSUS system linked hydrophones from all over the N. Atlantic and N. Pacific to central analysis facilities where sounds over entire ocean basins could be tracked. These sophisticated systems have recently been used to locate and track whales over long ranges, including one whale tracked for >1,700 km over 43 days (Costa 1993). The SOSUS arrays have proven capable of detecting whales at ranges of hundreds of kilometres (Stafford *et al.*, 1994, 1998), as was predicted by Payne and Webb (1971) and Spiesberger and Fristrup (1990). The SOSUS system was critical for revealing the potential of studying long range signalling in whales, but now other methods are also coming on line. Hydrophones placed on the seafloor to record seismic activity have also been used more recently to track blue and fin whales (McDonald *et al.,* 1995). Rather than having to pay for the installation of hundreds of kilometres of cable, as in the SOSUS arrays, it is now possible to deploy autonomous acoustic recorders in areas of biological interest to record for periods from days to years. This greatly increases the flexibility and usefulness to biologists of this kind of acoustic localisation.

Advances in the power and miniaturisation of electronics have also created the possibility of putting sophisticated digital acoustic recording tags directly onto marine mammals. The first such tag was developed for elephant seals, because of the ease of attachment. The elephant seal tag was able to provide calibrated measurements of received levels of ambient noise at the seal, including many natural and manmade transient sounds, from whale song and dolphin whistles to vessel noise (Fletcher *et al.,* 1996). This provides an elegant method to determine the acoustic features of sound stimuli heard by a marine mammal at sea. In addition, the acoustic sensor on the tag detects a remarkable variety of behavioural and physiological measures of use in studying responses to vocalisations produced by the tagged animal, by other animals, by playbacks or by man-made noise. The faster the seals swim, the greater the low frequency flow noise, allowing estimates of swimming speed. Rhythmic variation in flow noise appeared to correlate with swim stroke. At each surfacing, the sounds of each breath could easily be detected. When the flow noise was low, sounds of the heart beat, a phonocardiogram, proved one of the dominant signals, and heart rate could often be extracted from these acoustic records. While elephant seals had not previously been thought to vocalise underwater, this tag recorded underwater sounds similar to vocalisations heard from seals in-air (Burgess *et al.,* 1998). The acoustic recording tag thus provides an exceptional tool for studying what a marine mammal hears and how it responds to sounds, even if it may be hundreds of kilometres offshore or a kilometre below the sea surface.

6.   CONCLUSIONS

Humans are well adapted to sensing their terrestrial environment. When ethologists study the behaviour of a terrestrial species or when ecologists count their numbers, they often can rely upon their own senses, devoting most of their problem-solving abilities to analysing the data. However, students of cetaceans face major difficulties just getting to their study animals, not to mention detecting them and identifying an animal to species, age-sex class, or even to individual. These problems are particularly acute because cetaceans are so mobile, often swimming one hundred kilometres or more a day, and because many species spend the overwhelming majority of their time submerged out of sight from the surface. Cetaceans themselves appear usually to keep track of one another by listening for vocalisations, and biologists have started to attend to the same cues. We cannot rely upon our own unaided ears for locating sounds underwater, but it is getting easier and cheaper to find technological solutions to this problem. Electronics and seawater are traditionally thought to be a bad combination, but increases in the power, and decreases in the cost and size of electronics make systems. for detecting, recording, and locating marine mammal vocalisations within the purse and capabilities of most graduate students. There is little excuse for a biologist going to sea to study marine mammals not to open his or her ears to the rich acoustic world that these animals inhabit.

ACKNOWLEDGEMENTS

Jonathan Gordon was supported by the International Fund for Animal Welfare while preparing this paper. Peter Tyack was supported by ONR Grant N00014-99-130819. We are grateful to Gianni Pavan and Vincent

Janik for improving earlier versions with their helpful reviews and comments.

REFERENCES

Alder-Fenchel, H.S. (1980) Acoustically derived estimates of the size distribution of a sample of sperm whales, (*Physeter catodon*) in the western North Atlantic. *Canadian Journal of Fisheries and Aquatic Sciences*, **37**, 2358-2361.

Amundin, M. (1991) *Sound production in odontocetes with emphasis on the harbour porpoise, Phocoena phocoena*. Ph.D. Thesis, University of Stockholm, Stockholm, Sweden, 127pp.

Backus, R.H. and Schevill, W.E. (1966) *Physeter* clicks. In: *Whales, dolphins and porpoises* (Ed. by K.S. Norris), pp. 510-528. University of California Press, Berkeley.

Buckland, S.T. (1996) The potential role of acoustic surveys in estimating the abundance of cetacean populations. In: *Reports of the Cetacean Acoustic Assessment Workshop, 1996, Hobart, Tasmania*.

Burgess, W.C., Tyack, P.L., LeBoeuf, B.J. and Costa D.P. (1998) A programmable acoustic recording tag and first results from free-ranging northern elephant seals. *Deep-Sea Research*, **45**, 1327-1351

Chappell, O., Leaper, R. and Gordon, J. (1996) Development and performance of an automated harbour porpoise click detector. *Report of the International Whaling Commission*, **46**, 587-594.

Chappell, O. and Gillespie, D. (1998) *Cetacean detection software development in 1997 / 1998*. Unpubl. report to Birmingham Research and Development Limited , 9 pp.

Cheney, D.L. and Seyfarth, R.M. (1990) *How monkeys see the world*. University of Chicago Press, Chicago.

Christal, J. and Whitehead, H. (1997) Aggregations of mature male sperm whales on the Galapagos Islands breeding ground. *Marine Mammal Science*, **13**, 59-69.

Clark, C.W. (1980) A real-time direction finding device for determining the bearing to the underwater sounds of southern right whales, (*Eubalaena australis*). *Journal of the Acoustical Society of America*, **68**, 508-511.

Clark, C.W. (1994) Blue deep voices: insights from the Navy's Whales '93 program. *Whalewatcher*, **28**, 6-11.

Clark, C.W. and Clark, J.M. (1980) Sound playback experiments with Southern right whales (*Eubalaena australis*). *Science*, **207**, 663-665.

Clark, C.W., Ellison, W.T. and Beeman, K. (1985a) Progress Report on the Analysis of the Spring 1985 Acoustic Data Regarding Migrating Bowhead Whales, *Balaena mysticetus*, near Point Barrow, Alaska. *Report of the International Whaling Commission*, **36**, 587-597.

Clark, C.W., Marler, P. and Beeman, K. (1985b) Quantitative analysis of animal vocal phonology: an application to swamp sparrow song. *Ethology*, **76**, 101-115.

Clark, C.W. and Ellison, W.T. (1989) Numbers and distributions of bowhead whales, *Balaena mysticetus*, based on the 1986 acoustic study off Pt. Barrow, Alaska. *Report of the International Whaling Commission*, **39**, 297-303.

Clark, C.W. and Charif, R.A. (1998) *Acoustic monitoring of large whales to the west of Britain and Ireland using bottom-mounted hydrophone arrays, October 1996- September 1997*. Joint Nature Conservation Committee, Peterborough. 25 pp.

Clark, C.W. and Fristrup, K.M. (1997) Whales '95: a combined visual and acoustic survey for blue and fin whales of Southern California. *Report of the International Whaling Commission*, **47**, 583-599.

Costa, D.P. (1993) The secret life of marine mammals. *Oceanography,* **6**, 120-128

Cummings, W.C., and Thompson, P.O. (1971) Gray whales, *Eschrichtius robustus*, avoid the underwater sounds of killer whales, *Orcinus orca. Fishery Bulletin,* **69**, 525-530.

Davis, R.W. and Fargion, G.S. (1996) *Distribution and abundance of cetaceans in the north-central and western Gulf of Mexico.* Texas Institute of Oceanography and National Marine Fisheries Service, U.S . Dept. of the Interior. Minerals Mgmt. Service, Gulf of Mexico OCS Region, New Orleans. LA. 357pp.

Fish, J.F., and Vania, J.S. (1971) Killer whale, *Orcinus orca*, sounds repel white whales, *Delphinapterus leucas. Fishery Bulletin,* **69**, 531-535.

Fletcher, S., LeBoeuf, B.J., Costa, D.P.,Tyack, P.L., and Blackwell, S.B. (1996) Onboard acoustic recording from diving northern elephant seals. *Journal of the Acoustical Society of America,* **100(4)**, 2531-2539.

Freitag, L. E. and Tyack, P. L. (1993) Passive acoustic localization of the Atlantic bottlenose dolphin using whistles and echolocation clicks. *Journal of the Acoustical Society of America*, **93**, 2197-2205.

Fristrup, K.M. and Clark, C.W. (1997) Combining visual and acoustic survey data to enhance density estimation. *Reports of the International Whaling Commission*, **47**, 933-936.

Fristrup, K.M. and Watkins, W.A. (1994) Marine animal sound classification. *WHOI technical report.* WHOI-94-13

Gagnon, G.J. and Clark, C.W. (1993) The use of the US Navy IUSS passive sonar to monitor the movements of blue whales. In: *Abstracts, Tenth Biennial Conference on the Biology of Marine Mammals, Galveston, Texas*.

Gillespie, D. (1997) An acoustic survey for sperm whales in the Southern Ocean Sanctuary conduct from the *RV Aurora Australis. Report of the International Whaling Commission*, **47**, 897-907.

Goodson, A.D. and Sturtivant, C.R. (1995) Sonar characteristics of the harbour porpoise, source levels and spectrum. *ICES Journal of Marine Science*, **53(2)**, 465-472.

Goold, J.C. (1996) Signal processing techniques for acoustic measurement of sperm whale body lengths. *Journal of the Acoustical Society of America*, **100**, 3431-3441.

Gordon, J., Gillespie, D., Chappell, O. and Hiby, L. (1998) Potential uses of automated passive acoustic techniques to determine porpoise distribution and abundance in the Baltic Sea. In: *5th meeting of the Advisory Committee of ASCOBANS, Hel, Poland.* ASCOBANS/ADV.COM/5/DOC13.

Gordon, J., Moscrop, A., Carlson, C., Ingram, S., Leaper, R. Matthews, J. and Young, K. (1999) Distribution, movements and residency of sperm whales off Dominica, Eastern Caribbean: implications for the development and regulation of the local whale watching industry. *Report of the International Whaling Commission*, **48**, 551-557.

Gordon, J., Matthews, J.N., Panigada, S., Gannier, A., Borsani, J.F. and Notarbartolo di Sciara, G. (2000) Distribution and relative abundance of striped dolphins in the Ligurian Sea Cetacean Sanctuary: results from a collaboration using acoustic monitoring techniques. *Journal of Cetacean Research and Management*, **2**(1), 27-36.

Gordon, J.C.D. (1987) Sperm whale groups and social behaviour observed off Sri Lanka. *Report of the International Whaling Commission*, **37**, 205-217.

Gordon, J.C.D. (1991) Evaluation of a method for determining the length of sperm whales (*Physeter catodon*) from their vocalisations. *Journal of Zoology*, **224**, 301-314.

Gordon, J.C.D., Leaper, R., Hartley, F.G. and Chappell, O. (1992) Effects of whale watching vessels on the surface and underwater acoustic behaviour of sperm whales off Kaikoura, New Zealand. NZ Dep. *Conservation Science and Research Series, 32.* New Zealand. 64pp.

Greene, C.R. and McLennan, M.W. (1996) *Passive acoustic localisation and tracking of vocalising marine mammals using buoy and line arrays. Report of the Cetacean Acoustic Assessment Workshop, 1996.* Hobart, Tasmania, Biodiversity Group of Environment Australia.

Hiby, A. and Lovell, P. (1989) Acoustic survey techniques for sperm whales. *International Whaling Commission Scientific Committee.* SC/41/Sp3 presented to the International Whaling Commission Scientific Committee, 1989.

Janik, V. M. (2000) Food-related bray calls in wild bottlenose dolphins (*Tursiops truncatus*). *Proceedings of the Royal Society of London B: Biological Sciences, 267*, 923-927.

Janik, V.M. and Slater, P.J.B. (1998) Context-specific use suggests that bottlenose dolphin signature whistles are cohesion calls. *Animal Behaviour, 56*, 829-838.

Kamminga, C. and Wiersma, H. (1981) Investigations on cetacean sonar II. Acoustical similarities and differences in odontocete sonar signals. *Aquatic Mammals 8*, 41-62.

Ko, D., Zeh, J.E., Clark, C.W., Ellison, W.T., Krogman, B.D. and Sonntag, R.M. (1986) Utilization of acoustic location data in determining a minimum number of spring-migrating bowhead whales unaccounted for by the ice-based visual census. *Report of the International Whaling Commission, 36*, 325-338.

Leaper, R., Chappell, O. and Gordon, J.C.D. (1992) The development of practical techniques for surveying sperm whale populations acoustically. *Report of the International Whaling Commission, 42*, 549-560.

Malme, C.I., Miles, P.R., Clark, C.W., Tyack, P. and Bird J.E. (1984) Investigations of the potential effects of underwater noise from petroleum industry activities on migrating gray whale behavior. Phase II: January 1984 migration. Bolt Beranek and Newman Report No. 5586 submitted to Minerals Management Service, U. S. Dept. of the Interior.

McDonald, M.A., Hildebrand, J.A. and Webb, S.C. (1995) Blue and fin whales observed on a seafloor array in the Northeast Pacific. *Journal of the Acoustical Society of America, 98*, 712-721.

McGregor, P.K. (1992) *Playback and studies of animal communication.* Plenum Press, New York, 227pp.

Miller, P. and Tyack P.L. (1998) A small towed beamforming array to identify vocalizing resident killer whales (*Orcinus orca*) concurrent with focal behavioral observations. *Deep-Sea Research, 45*, 1389-1405.

Møhl, B., Larsen, E. and Amundin, M. (1981) Sperm whale size determination: Outlines of an acoustic approach. pp. 327-331, In: *Mammals in the seas*, vol. III,. Rome, United Nations Food and Agriculture Organization.

Norris, K.S. and Harvey, G.W. (1972) A theory for the function of the spermaceti organ of the sperm whale. In: *Animal Orientation and Navigation* (Ed. by S.R. Galler, K. Schmidt-Koenig, G.J. Jacobs and R.E. Belleville) NASA Special Publication, Washington D.C.

Pavan, G. (1992) A portable DSP workstation for real-time analysis of cetacean sounds in the field. *European Research on Cetaceans, 6*, 165-169.

Pavan, G. and Borsani, J.F. (1997) Bioacoustic research on cetaceans in the Mediterranean Sea. *Marine And Freshwater Behaviour and Physiology, 30*, 99-123.

Pavan, G., Fossati, C., Manghi, M. and Priano, M. (1999) Acoustic measure of body growth in a photo-identified sperm whale. *European Research on Cetaceans, 12*, 251-262.

Payne, R. and Webb, D. (1971) Orientation by means of long range acoustic signalling in baleen whales. *Annals of the New York Academy of Sciences, 188*, 110-141.

Potter, J.R., Mellinger, D.K. and Clark, C.W. (1994) Marine mammal call discrimination using artificial neural networks. *Journal of the Acoustical Society of America*, **96**, 1255-1282.

Richardson, W.J., Greene, C.R.J., Malme, C.I. and Thomson, D.H. (1995) *Marine Mammals and Noise*. Academic Press, Inc., San Diego, CA. 537pp.

Sayigh, L.S., Tyack, P.L. and Wells, R.S. (1993) Recording underwater sounds of free-ranging bottlenose dolphins while underway in a small boat. *Marine Mammal Science,* **9**, 209-213.

Sayigh, L.S., P.L. Tyack, R.S. Wells, A. Solow, M.D. Scott, A. B. Irvine. (1999) Individual recognition in wild bottlenose dolphins a field test using playback experiments. *Animal Behaviour*, **57**, 41-50.

Sparks, T.D., Norris, J.C. and Evans, W.E. (1993) Acoustically determined distributions of sperm whales in the northwestern Gulf of Mexico. In: *Abstracts, Tenth Biennial Conference on the Biology of Marine Mammals, Galveston, Texas*.

Spiesberger, J.L. and Fristrup, K.M. (1990) Passive localisation of calling animals and sensing of their acoustic environment using acoustic tomography. *The American Naturalist*, **135**, 107-153.

Stafford, K., Fox, C.G. and Mate, B.R. (1994) Acoustic detection and location of blue whales (*Balaenoptera musculus*) from SOSUS data by matched filtering. *Journal of the Acoustical Society of America*, **96**, 3250-3251.

Stafford, K.M., Fox, C.G. and Clark, D.S. (1998) Long-range acoustic detection and localization of blue whale calls in the northeast Pacific Ocean. *Journal of the Acoustical Society of America*, **104**, 3616-3625.

Sturtivant, C. and Datta, S. (1995) Techniques to isolate dolphin whistles and other tonal sounds from background noise. *Acoustics Letters*, **18(10)**, 189-193.

Thomas, J.A., Fisher, S.R., Ferm, L.M. and Holt, R.S. (1986) Acoustic detection of cetaceans using a towed array of hydrophones. *Report of the International Whaling Commission* (special issue **8**), 139-148.

Tyack, P. (1985) An optical telemetry device to identify which dolphin produces a sound. *Journal of the Acoustical Society of America*, **78**, 1892-1895.

Tyack, P.L. and Recchia, C.A. (1991) A datalogger to identify vocalizing dolphins. *Journal of the Acoustical Society of America*, **90**, 1668-1671.

Urick, R.J. (1986) *Ambient noise in the sea*. Peninsula Publishing, Los Altos, CA.

Watkins, W.A. (1976) Biological sound-source locations by computer analysis of underwater array data. *Deep-Sea Research*, **23**, 175-180.

Watkins, W.A. and Daher, M.A. (1992) Underwater sound recording of animals. *Bioacoustics*, **4**, 195-209.

Watkins, W.A., Daher, M.A., Reppucci, G.M., George, J.E., Martin, D.L., DiMarzio, N.A. and Gannon, D.P. (2000) Seasonality and distribution of whales calls in the North Pacific. *Oceanography*, **13**(1), 62-67.

Watkins, W.A. and Schevill, W.E. (1972) Sound source location by arrival-times on a non-rigid three-dimensional hydrophone array. *Deep-Sea Research*, **19**, 691-706.

Watkins, W.A. and Schevill, W.E. (1975) Sperm whales (*Physeter catodon*) react to pingers. *Deep-Sea Research*, **22**, 123-129.

Whitehead, H. (1985) Humpback whale songs from the North Indian Ocean. *Investigations on Cetacea*, **17**, 157-162.

Whitehead, H. and Gordon, J.C.D. (1986) Methods of obtaining data for assessing and modelling sperm whale populations which do not depend on catches. *Report of the International Whaling Commission* (special issue **8**), 149-165.

Zeh, J., Turet, P., Gentleman, R. and Raftery, A. (1988) Population size estimation for the bowhead whale, *Balaena mysticetus*, based on 1985 and 1986 visual and acoustic data. *Report of the International Whaling Commission*, **38**, 349-364.

Chapter 9

Applications of Molecular Data in Cetacean Taxonomy and Population Genetics with Special Emphasis on Defining Species Boundaries

[1]MICHEL C. MILINKOVITCH, [2]RICK LEDUC, [1]RALPH TIEDEMANN and [2]ANDREW DIZON

[1]*Unit of Evolutionary Genetics, Free University of Brussels (ULB, cp 300), Institute of Molecular Biology and Medicine, rue Jeener and Brachet 12, B-6041 Gosselies, Belgium E-mail: mcmilink@ulb.ac.be;* [2]*Southwest Fisheries Science Center, PO Box 271, La Jolla, CA 92109, USA*

1. INTRODUCTION

Morphological, physiological, and behavioural characters are of great interest in phylogenetic and population genetic analyses. However, the genetic basis is known for very few of these traits, and the influence of environmental factors on the observed character variance is unknown in most cases. On the other hand, molecular methods *"open the entire biological world for genetic scrutiny"* (Avise, 1994). Indeed, while the identification of the genetic bases and modes of transmission of some phenotypic traits have been possible only for humans and very few species that could rapidly and easily be crossed under controlled conditions (*e.g. Pisum sativum, Escherichia coli, Saccharomyces cerevisiae, Mus musculus, Drosophila melanogaster*), the mode of transmission of molecular characters can usually be explicitly and readily specified for any species investigated. In addition, molecular genetic techniques give access to an enormous number of characters: a typical mammalian genome contains several billion potentially informative nucleotides. Another great advantage of molecular markers is the objectivity of characters and of their character states (*i.e.* alternative conditions of a character) relative to morphological, physiological, and behavioural markers. The objectivity of defining discrete

Marine Mammals: Biology and Conservation, edited by
Evans and Raga, Kluwer Academic/Plenum Publishers, 2002

molecular character states also makes them easily repeatable by independent researchers. Finally, many molecular characters probably fit character neutrality more closely than non-molecular characters. Not only can molecular techniques provide a better understanding of character variation at the molecular level, they can also address questions in natural history and organismal evolution. This is especially relevant for cetaceans because (1) they are very mobile and often inaccessible organisms for which morphological, physiological, and behavioural characters can be exceedingly difficult to score for population studies (but see Smith *et al.*, 1999, for a large analysis of tail fluke photographs in humpback whales), and (2) their highly derived and specialised morphology reduces the utility of phenotypic data for assessing their phylogenetic position within mammals.

As with other groups, accurate taxonomy is fundamental to conservation efforts for cetaceans; the units on which conservation is based are determined partially by population structure and ultimately by species designation. Imperfect taxonomy may result, at least as much as a lack of understanding of population structure, in the loss of genetic variability (here, by unwitting extinction of a species). The rapid advances in molecular techniques of the past few decades have led to significant contributions towards improving cetacean taxonomy. At higher taxonomic levels, the increasing ease of generating useful molecular genetic data, notably DNA sequences, paralleled by theoretical advances and the development of computer programs, has stimulated reinvestigation of phylogenetic issues involving cetaceans. In some cases, results of these investigations have led to revisions of taxonomic relationships (*e.g.* Árnason *et al.*, 1992; Gatesy, 1997; Gatesy *et al.*, 1996; Hasegawa *et al.*, 1997; LeDuc *et al.*, 1999; Milinkovitch, 1995, 1997; Milinkovitch *et al.*, 1993, 1994; Rosel *et al.*, 1995).

Maybe the most dramatic example is the reappraisal of the phylogenetic origin of cetaceans within mammals. Because cetaceans experienced dramatic transformations in basically all their biological systems, phylogeneticists encountered obvious difficulties in assessing homology and polarisation of the few morphological characters that could be used for comparisons between cetaceans and other mammals. Despite these difficulties, morphological analyses have suggested that cetaceans are closely related to artiodactyls (*e.g.* Flower, 1883; Prothero *et al.*, 1988; Gingerich *et al.*, 1990; Thewissen and Hussain, 1993; Thewissen, 1994). While morphologists and molecular phylogeneticists heatedly argued over several other nodes of the mammalian phylogenetic tree, the morphological hypothesis of a close relationship between cetaceans and ungulates (and especially artiodactyls) was strongly supported by a range of molecular data: from immunological (*e.g.* Boyden and Gemeroy, 1950; Shoshani, 1986) and DNA-DNA hybridisation (Milinkovitch, 1992) studies, to analyses of mitochondrial DNA (mtDNA) and nuclear amino-acid and DNA sequences (*e.g.* Czelusniak *et al.*, 1990; Milinkovitch *et al.*, 1993). This is rather a counter-intuitive result because it makes a cow, a pig, or a camel more closely related to a dolphin or a blue whale than to a horse or a tapir. However, Graur and Higgins (1994) analysed a large data set of published amino-acid and DNA sequences and first suggested explicitly an even more provocative hypothesis: cetaceans would be nested within the artiodactyl phylogenetic tree. This hypothesis of artiodactyl paraphyly has subsequently been supported by 1) extensive cladistic analyses of DNA sequence data from multiple nuclear and mitochondrial genes (Gatesy, 1998); 2) lactalbumin DNA sequences (Milinkovitch *et al.*, 1998), and 3) phylogenetic interpretation of retropositional events that lead to the insertion of short interspersed elements (SINEs) at particular loci in the nuclear genome of various artiodactyl and cetacean ancestors (Shimamura *et al.*, 1997; Milinkovitch and Thewissen, 1997).

Recent and extensive phylogenetic re-analyses (Gatesy *et al.*, 1999) of all available data indicate that the inclusion of a monophyletic Cetacea within the phylogenetic tree of Artiodactyla has been stable to increased taxonomic sampling and to the addition of nearly nine thousand characters (of which over 1,500 are parsimony-informative) from three mtDNA genes, 12 nuclear genes, and eight SINE retroposon loci (Fig. 1). Although the concept of cetaceans as highly derived artiodactyls still raises major scepticism (*e.g.* Luckett and Hong, 1998; Heyning, 1999), it constitutes one of the best supported hypotheses of interordinal (*sensu lato*) relationship within the phylogeny of mammals.

At the intraspecific or population level, molecular data - especially on DNA sequence and microsatellite variation - have proven amenable to studies of population divergence, social structure, and mark/recapture estimates of population abundance (*e.g.* Amos *et al.*, 1993; Palsbøll *et al.*, 1997; O'Corry-Crowe *et al.*, 1997; Dizon *et al.*, 1997; Smith *et al.*, 1999). Analyses of phylogenetic relationships among, and geographic distribution of, haplotypes within a species (*phylogeographic analysis*; Avise *et al.*, 1987; Avise, 2000) can also provide insight into the history of cetacean populations both on regional and global scales (*e.g.* Tiedemann *et al.*, 1996; Rosel *et al.*, 1999; Curry *et al.*, submitted). Furthermore, classical population genetic parameters, such as Wright's F_{ST} statistic or Nei's nucleotide diversity, can be estimated from molecular data at the population level. These calculations summarise variation within and between groups. Given certain assumptions (*e.g.* mutation-drift-equilibrium, neutrality of the marker system used, constancy of dispersal rates over space and time) are fulfilled (Bossart and Prowell, 1998), they allow inference of the effective size of populations and the estimation of the direction and amount of gene flow among them (Neigel, 1996).

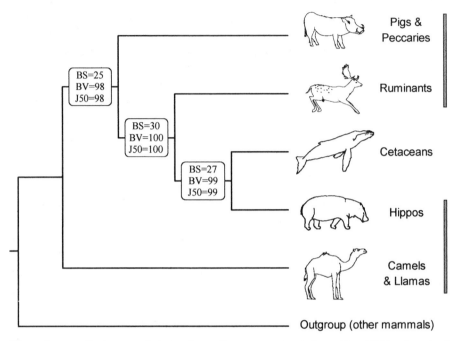

Figure 1.   Best supported topology for a concatenated matrix (8995 characters) comprised of 17 data sets from 13 taxa (Gatesy *et al.*, 1999). The 17 data sets were nuclear amino acid sequences (from five genes), mitochondrial DNA sequences (from three genes), nuclear DNA sequences (from seven genes), skeletal/dental characters, and SINE retropositional insertions. Each of the 17 data sets included at least one representative from Ruminantia (Bovidae + Cervidae + Giraffidae + Tragulidae), Camelidae, Hippopotamidae, and Suina (Suidae + Tayassuidae); only the skeletal/dental character matrix did not include cetacean representatives. Outgroup taxa were mammals, most often representatives of the order Perissodactyla. Various support indices are indicated at the nodes: BS = Bremer support (*i.e.* the number of additional character transformations necessary to collapse an internal branch; Bremer, 1994), BV = bootstrap value (informative characters were resampled [10³ replicates] with replacement from the original data set; Felsenstein, 1985), J50 = jackknife value (for 50% random deletion of characters [10³ replicates]; Penny and Hendy, 1986). Dashed bars indicate artiodactyl taxa. Modified from Figure 3 in Gatesy *et al.*, 1999, and from Figure 1 in Milinkovitch and Thewissen, 1997.

Obviously, as is the case in phylogeny inference, proper interpretation of the data often depends on restrictive assumptions that underlie the calculations. For example, interpretational errors can result from misunderstanding about particular modes of inheritance of a marker (*e.g.* the well known phenomenon of maternal *vs.* Mendelian inheritance) or from unknown differential selection coefficients among haplotypes or alleles (*i.e.* deviation from the assumption of neutrality). Extrapolation of the genetic structure found at a particular marker to the genetic structure of the population itself implicitly assumes that deviations from these assumptions are negligible.

The main focus of this chapter lies at the junction of the two aforementioned levels of research (studies of the relationships among species, and studies of intraspecific population structure), *i.e.* the study of the differentiation of species. Mayr (1969) termed this *alpha* taxonomy; it involves the number and names of species in a classification. He termed two additional levels of study: *beta* taxonomy (the interrelationships among species, genera, *etc.*) and *gamma* taxonomy (the pattern of variation within species). These levels are clearly not mutually exclusive. For example, an increase in the understanding of *gamma*-level variation may lead to changes in the number of species recognised.

2. THE CENTRAL ROLE OF *ALPHA* TAXONOMY

In practice, the number and names of cetacean species provide the framework for phylogenetic and population genetic research. If two groups of individuals show disjunction of their state distributions in one morphological character (assuming sex and age have been accounted for), their status as separate non-interbreeding entities (*i.e.* biological species, cf. below) can in principle be suggested (Mayr, 1969; the convention of strict non-overlap is meant to be conservative; strata showing some overlap could actually be good biological species). This operational principle often used in morphological analysis is based on the assumption that if disjointed state distributions are observed among parapatric/sympatric groups, it either originated in parapatry/sympatry and was produced by disruptive selection (*e.g.* Thoday and Gibson, 1962) and reinforcement through fitness reduction of the hybrids, or was initiated in allopatry by the combined effects of random drift, mutations, and differential selection within each of the reproductively isolated groups (and secondary contact occurred afterwards). In the former, disruptive selection was/is the obligated original driving force responsible for gene flow reduction, while in the latter, differential selection and/or mutations and/or genetic drift made the two groups diverge sufficiently such that prezygotic and/or postzygotic isolation mechanisms had incidentally developed, preventing the merging of the two strata after secondary contact.

Although theoretically possible, parapatric/sympatric speciation through reinforcement is thought to be very rare (but see Coyne and Orr, 1989); hence, the working assumption is that non-overlap of state distributions in morphological characters for groups in parapatry/sympatry arose from allopatric speciation followed by secondary contact. When the groups are allopatric, the existence of one or several reproductive isolation mechanisms (other than habitat isolation) is difficult to test for cetaceans, since the frequency of captive hybrids is probably a poor indicator of potential interbreeding in the wild. Moreover, natural hybrids between species such as bottlenose and Risso's dolphins (Fraser, 1940), blue and fin whales (Árnason *et al.*, 1991; Spilliaert *et al.*, 1991), and common and dusky dolphins (Reyes, 1996) have been reported, despite these species being well defined both on morphological and molecular grounds.

By no means do we claim that reduction of gene flow (with or without selection) is the only interpretation for bimodal distribution of quantitative morphological characters (see for example Scharloo, 1987), but discussing these aspects is beyond the scope of the present chapter. Nevertheless, in practice, most morphologists interpret non-overlap in a morphological character as indication of barriers to interbreeding, usually invoking differential selection as a major mechanism driving reproductively isolated populations apart. When species are thus defined, it is necessarily implied that they are biological species (*i.e.* interbreeding populations which are reproductively isolated from other such groups; Mayr, 1942, 1963). Within this logical framework, several morphological characters exhibiting similar non-overlapping distributions (*i.e.* separating the groups into identical strata) are interpreted as independent indicators of differential selection. For example, different tooth counts may arise from selection for feeding on different prey items, but differences in another cranial character may indicate feeding at different depths. Taken together, these may indicate a greater degree of differential selection than either does on its own; the case is stronger if the characters are both *functionally* and genetically unlinked.

The rationale used in these morphological analyses (with selection assumed) is therefore diametrical to that used in molecular phylogenetics and population genetics (where character neutrality is assumed). In the latter approach, the case for separating species is also strengthened when multiple and independent (*i.e. genetically*-unlinked) diagnostic characters covary. In that case, species could probably be regarded as groups of individuals showing a reticulate pattern of relationships rather than a cladogenetic hierarchy (see below). However the transition between these two patterns (reticulation and hierarchy) is not necessarily abrupt and cannot be known *a priori* (*i.e.* before analysis).

An important distinction between, on the one hand, the *alpha*- and *gamma*-level studies, and, on the other hand, the phylogenetic approach most often used at the *beta* level, is that the former usually rely on the examination of variation within and among predefined groups, while the latter infers groups of taxa through identification of the topology(ies) that optimise(s) an objective function (*i.e.* the use of an optimality criterion such as maximum parsimony or maximum likelihood). For geneticists, who focus on one level of research or the other, the species designations provide an important link between the two, often in the form of assumptions used in the analysis of data or the interpretation of results.

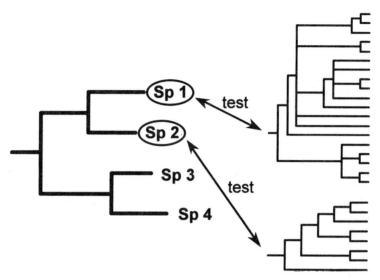

Figure 2. Individual samples included in a phylogenetic analysis most often are assumed to represent monophyletic species. This assumption is rarely tested by the inclusion of multiple representatives of the species.

For example, starting with an accepted *alpha* taxonomy, most *beta*-level studies examine the relationships among assumed monophyletic species. This assumption can be either explicitly stated, or implicit in the species names assigned to specific sequences, but is rarely tested by inclusion of multiple representatives of the species in the phylogenetic analysis. The fundamental entities of phylogenetic analyses should be individual organisms, not species (*e.g.* Vrana and Wheeler, 1992). Nevertheless, only one or a few individuals from putative biological species can, in practice, usually be included in a phylogeny because of constraints on sampling (*e.g.* lack of access to the full range of variation across the species) and/or on analysis (*i.e.* computing time increases explosively with the number of included terminal taxa when optimality-criterion methods are used). An inferred tree may corroborate species status in the phylogenetic sense (*i.e.* monophyletic) if multiple individuals that were nominally conspecific form a clade (Fig. 2). Conversely, when a particular nominal species is shown to be para- or polyphyletic in a tree, the integrity of the species is questioned (see below), exactly as in higher-taxa phylogeny where para- or polyphyly of a family, of an order, or of any other taxonomic unit is potentially detectable (*cf.* Fig. 1).

In population genetic studies, a similar common underlying assumption is that the taxa under consideration represent a single conspecific group, with possibly some amount of gene flow between predefined "clusters". Typically, when these clusters are defined by geographic criteria and are either allopatric or parapatric, they are deemed populations, and one attempts to describe the degree and pattern of population subdivision, a function of genetic drift, differential selection pressures, and gene flow. The divergence among clusters is often only expressed as an overall relative measure of differentiation, such as F_{ST}, which is used to provide an indirect estimate of overall gene flow, N_m (*e.g.* Nei, 1987). However, as outlined above, these *gamma*-level studies have the potential to detect reproductively isolated clusters (as does inference of cladogenetic hierarchy, if phylogenetic tools are used) and can hence contribute to *alpha* taxonomy, *i.e.* the definition of species.

Our point is that in both higher and lower level studies, it is too often overlooked that a vast number of species designations in the current taxonomy have not been tested in terms of actual non-interbreeding. Researchers (probably more often in molecular studies) frequently assume that populations which share the same binomial scientific name are indeed conspecific, or even that subspecific designations are well-founded. It is our contention here that these assumptions are often not fulfilled, which has serious implications for the ultimate analysis of data and interpretation of results. One must realise that groups of individuals very often received the same binomial name not because interbreeding or lack of cladogenetic hierarchy had been demonstrated (see below for suggestions on how to achieve such demonstration), but because evidence for splitting them into several reproductively-isolated groups had not accumulated at the time of species designation. Similarly, previously recognised species groups might be found to form a single interbreeding group and might then be merged into a single species.

3.   CURRENT *ALPHA* TAXONOMY IN CETACEANS

The number of extant cetacean species remains controversial. Cetacean systematics is rapidly changing for a variety of reasons, including advances in analytical techniques, application of molecular markers, and increases in the amount of material available. Therefore, revisions can be expected to continue at all levels. Mead and Brownell (1993) and Rice (1998) present recent lists of species, and Hershkovitz (1966) gives a nomenclatural review, although original research papers (*e.g.* Heyning and Perrin, 1994) often provide more complete summaries for specific groups. For the most part, the species names now in use originated in the 19th century, as did many junior synonyms. As a broad generalisation, one might say that most of these synonyms were assigned to geographic variants of already recognised species. Some species of cetaceans show extensive geographic variation in details of morphology and coloration (see Perrin, 1984). Because these variants were often only represented by one or at most a few individuals, sample sizes (in both number and geographic coverage) were inadequate to examine the ranges of variation within and between groups, so that later taxonomists were unable to determine if the variants warranted specific recognition. In recent decades, the taxonomic approach in these cases has been a conservative one, favouring the recognition of a single wide-ranging and highly variable species when the data are insufficient for designating the geographic variants as separate biological species.

The classification for any given species group cannot always be expected to accurately reflect the theoretical criterion of (non)interbreeding or even the operational criterion of morphological discontinuity but, in many cases (*e.g.* one species of bottlenose dolphin or two species of right whale), simply may reflect the need for further study. The main limitation is access to adequate sample sizes, as a series of adult animals are required for the documentation of geographic morphological variation. Such series may take decades to accumulate in museums and research institutions, unless large-scale fishery mortality accelerates the process. An excellent example of this problem was the confusion over the number of species of *Stenella* (*e.g.* Hershkovitz, 1966, recognised eight species). Using in part the large number of by-caught specimens from the eastern tropical Pacific (ETP) yellowfin tuna fishery, it was possible to delimit the five species now recognised (Perrin, 1975; Perrin *et al.*, 1981, 1987). However, few groups of cetacean species have been studied nearly as thoroughly as *Stenella* spp. with regards to geographic variation. For some examples of recent revisions, Rice (1998) recognises only one species of black right whale, but two species each of minke and Bryde's whales, and he emphasises the uncertainty regarding the number of species in genera such as *Tursiops* and *Sousa*. Even well-studied groups, such as minke whales and bottlenose dolphins, have been recently revised. For large whales or rare cetaceans such as the beaked whales (*e.g.* Reyes *et al.*, 1991), the lack of specimens and resulting taxonomic uncertainty is even more pervasive (Fig. 3). Our point here is simple: gaps in our understanding of geographic variation mean that the list of currently recognised species of cetaceans will probably undergo serious revisions. Here, molecular genetics can provide significant contributions to taxonomic understanding.

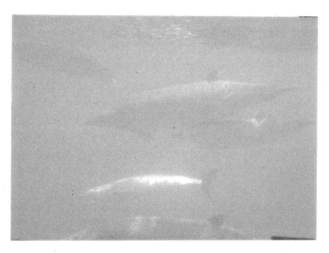

Figure 3.   Very little material is available for investigating the alpha taxonomy, phylogeny and population structure of beaked whales of the genus *Mesoplodon*. Some of the 13 morphologically very similar species of this genus are scarcely known, but the dense-beaked whale *M. densirostris*, photographed here in the Canary Islands, has been frequently observed at close quarters. (Photo: F. Ritter)

Thus, for the cetacean molecular geneticist faced with assigning taxon names to sequences or interpreting particular results, whether inferring a tree or calculating an F_{ST}, it is important to examine the assumptions inherent in the taxonomic context of his or her study. The fact that different populations are regarded as conspecific in any particular classification does not necessarily mean that the evidence favouring this status is strong. In other cases, the converse may be true, where species now regarded as separate may be lumped together with the collection of more data. Furthermore, the implications of taxonomic assumptions extend beyond issues of classification. For example, on the basis of comparing genetic diversity in different cetacean species, Whitehead (1998) recently proposed that cultural selection in some species has led to lower natural levels of mitochondrial genetic diversity. The strength of his hypothesis aside (*e.g.* Mesnick *et al.*, 1999; Schlötterer, 1999; Tiedemann and Milinkovitch, 1999; Amos 1999), this evidence may be partly compromised by taxonomic uncertainty, if some of the "species" that were compared (*e.g.* bottlenose dolphin) actually represent multiple reproductively-isolated groups (*i.e.* biological species).

4.   MOLECULAR APPROACHES TO *ALPHA*-LEVEL TAXONOMY

As mentioned above, the morphological criterion often used for separating species is a simple one – basically, disjunction of their state distributions in at least one morphological character. Using this approach, new findings are easily incorporated and old ones reinterpreted. When using molecular data, several approaches have been used for separating groups into different species: (1) lack of shared haplotypes/sequences, (2) fixed differences, (3) reciprocal monophyly, and (4) degree of divergence.

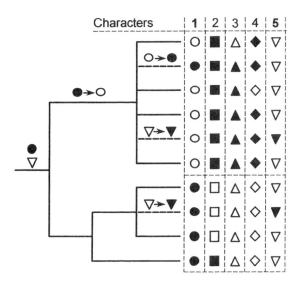

Figure 4.   Since multiple changes can occur on a single character, two reciprocally-monophyletic groups can exhibit character state distributions with no fixed differences between the two clades. Characters 1 and 5 are traced on the branches; character 1 experienced reversal while character 5 experienced convergence.

Before we discuss each of these in turn, there are some general considerations that deserve mention. First, any method is susceptible to sampling problems; if the number and/or geographic scope of the samples is too small, inferences will be suspect. Conceivably, an investigator could use single samples to represent each taxon and produce results that would trivially satisfy all of the enumerated criteria for splitting groups. Second, one should not expect the different enumerated criteria to necessarily suggest identical groupings. It is easy to imagine a data set in which two groups show reciprocal monophyly but no fixed differences (because of reversals and/or convergences; Fig. 4); if, for any reason, phylogeny inference yields an erroneous tree, the converse (groups not monophyletic but showing fixed differences) could trivially be true. Third, due to the wide geographic ranges of many cetacean species, allopatric distribution of populations is generally the rule rather than the exception. These allopatric populations may occupy different ocean basins and/or different hemispheres, or involve peripheral isolates in smaller bodies of water. In these cases, geographic barriers often preclude contemporaneous gene flow. However, genetic distinctness may evolve much more rapidly than morphological divergence. This is especially true for mtDNA markers, whose clonal maternal inheritance renders them especially sensitive to population bottlenecks and rapid genetic drift (*e.g.* Avise, 1994). For example, Pichler *et al.* (1998) found fixed differences in mitochondrial control region sequences between Hector's dolphin populations on either side of New Zealand's South Island. These populations are not thought to have diverged morphologically, but the social structure and behaviour of the species, coupled with the small population sizes, may explain their molecular distinctness over such a short geographic distance.

In other instances, where selective pressures accelerate fixation of morphological differences or large population size decelerates genetic drift, molecular genetic distinctness can lag behind the development of selective morphological differences. Similarly, if one considers the criterion of degree of divergence, some *bona fide* species can be genetically very similar (*e.g.* Meyer *et al.*, 1990; Sturmbauer and Meyer, 1992) and some morphologically-recognised conspecific populations can show high levels of molecular divergence. Examining this in a systematic manner by comparing molecular genetic distinctness with the criteria used by many morphologists represents a formidable challenge to molecular geneticists. Finally, for evaluating these enumerated criteria in a practical framework (*i.e.* for producing a useful classification), molecular geneticists should suggest conditions that are sufficient for prompting to recognition of new species.

4.1   Lack of Shared Haplotypes/Sequences

One could argue that, given well-sampled strata and a moderate amount of variation, the absence of shared haplotypes indicates that contemporaneous or recent gene flow between strata is not occurring. However, this is not a sufficient condition for species resolution because the criterion completely disregards relationships among haplotypes and variation within strata. If, for example, every individual in the entire data set each had a unique haplotype (*i.e.* none shared even within groups), the absence of shared haplotypes between groups would be rather meaningless.

Conversely, the presence of shared haplotypes among groups does not necessarily demonstrate the occurrence of gene flow. Here, the calculation of fixation indices (F_{ST}) can put the relative significance of shared *vs.* unshared haplotypes into perspective.

Additionally, it could indicate the need to infer phylogenetic relationships among haplotypes and/or collect new data. Indeed, recently diverged species may still demonstrate some shared alleles or haplotypes for slowly-evolving markers, while further investigations of markers that evolve more quickly could reveal lack of gene flow. Furthermore, and as an extreme case, one could even imagine haplotypes, from members of two different biological species, that would be identical by state but not by descent, *i.e.* they would have arisen independently in the two strata. The likelihood of such a scenario is however exceedingly low as soon as enough variable characters are scored (*i.e.* sampling scope is not restricted to the number of haplotypes). In conclusion, the lack of shared haplotypes/sequences for any given marker is neither a necessary nor a sufficient criterion for species designation.

4.2　Fixed Differences

These can be in the form of two strata having different fixed nucleotides at one or more sites in a DNA sequence or having fixed but different alleles at a given microsatellite or allozyme locus. Hey (1991) argues for the strength of this kind of evidence. When the strata represent two distinct (based on non-molecular criteria) parapatric or sympatric groups, such as the common dolphin morphotypes studied by Rosel *et al.* (1994), fixed molecular differences corroborate the existence of biological barriers to interbreeding and warrant separate species status. The likelihood of even a single fixed difference to correspond by chance to an *a priori* morphological designation is indeed exceedingly low. Other cases include inshore and offshore populations of bottlenose dolphins in the western North Atlantic/Gulf of Mexico (Curry, 1997; Curry *et al.*, submitted; Hoelzel *et al.*, 1998a) and resident and transient killer whales in the eastern North Pacific (Hoelzel *et al.*, 1998b). It should be mentioned that the morphological differences in bottlenose dolphins and killer whales are, contrary to the common dolphin morphotypes (cf. below), not diagnostic (complete disjunction of state distributions), and designation to type was based at least partly on behavioural and ecological factors (habitat choice, parasite load, and stomach contents) which probably do not have a direct genetic basis.

In all three of these cases, fixed differences between strata were nonetheless observed in mtDNA sequences. Given the maternal inheritance of the mt genome, these results are also compatible with the alternative hypothesis of strong female philopatry with male-mediated gene flow still occurring. We, however, argue that significant male-mediated gene flow should be detected, at least in some cases, by sampling the haplotypes of the migrants themselves (Tiedemann *et al.*, 2000). In addition, in the case of the bottlenose dolphins, Hoelzel *et al.* (1998a) also found a nearly fixed difference at one microsatellite locus, indicating that male-mediated gene flow is unlikely or negligible. This illustrates the usefulness of independent lines of evidence (here, mt and nuclear markers); congruence makes a stronger case for species status. In conclusion, we argue that, in parapatry/sympatry (a) fixed molecular differences which are independent (*i.e.* at genetically unlinked loci) and which split the analysed group of individuals into identical strata, or (b) a single fixed molecular difference that strictly corresponds to *a priori* morphological/physiological/ behavioural designations (through discontinuous state distribution, *cf.* above) are each a sufficient criterion for species designation. Fixed differences between allopatric populations confirm that gene flow is negligible among populations but do not reveal whether interbreeding would occur in the case of a secondary contact.

4.3   Reciprocal Monophyly

This is the only one of our enumerated criteria that relies on the estimate of a phylogeny for testing. The criterion used for the phylogenetic species concept (PSC) is identity by descent rather than interbreeding. The phylogenetic approach to species designation seeks therefore to identify each smallest group of individuals forming a clade (*i.e.* a monophyletic group) diagnosed on the basis of shared derived character states. The PSC is not restricted (while the biological species concept, BSC, is) to sexually reproducing organisms, and it favours the identification of patterns rather than processes (*e.g.* Wheeler and Nixon, 1990; Nixon and Wheeler, 1990). However, this honourable quest for objectivity neglects one indisputable fact: hierarchies of groups (the pattern) in sexually reproducing *organisms* are shaped by population-genetic parameters (the processes) such as gene flow (interbreeding), recombinations, and genetic drift. Figure 5a illustrates that a non-recombining piece of DNA from a given individual will have a strictly non-reticulated furcating history which will be the result of a specific genealogical pathway through the organismal pedigree. However, given the rules of Mendelian genetics, that piece of non-recombining DNA could have (but has not) followed any one of the huge number of other possible pathways (2^n-1; where n = number of generations in organisms with non-overlapping generations) through the same organismal pedigree.

Any other (unlinked) non-recombining autosomal piece of DNA from the same individual has a very high probability ($=1-(1/2^n)$) to follow one of these other genealogical trails. Figure 5b illustrates the matrilineal pathway of transmission, which is just one of the 2^n possible pathways for autosomal loci while it is the obligated route of transmission for mitochondrial genes (which are maternally inherited and all form a single non-recombining piece of DNA).

The important point is that, given two different genealogical pathways within a single species tree, the two corresponding gene trees may also conflict in their branching patterns, although each of them indicates the correct historical relationships with respect to the corresponding gene. For example, in the genealogy of the "locus x" (Fig. 5a), all individuals from "species B" and "species C" are grouped in a clade excluding all individuals from "species A"; in other words, all haplotypes from species B and C trace genealogically to a single ancestral haplotype that was transmitted by the female individual indicated by a black symbol at generation "g_{-14}".

Conversely, if one considers the genealogy of "locus y" (Fig. 5b), all individuals from "species A" and "species B" are grouped in a clade excluding all individuals from "species C".

These two gene trees conflict in their branching patterns (although both correctly indicate the respective genealogies of the loci) and the "locus y" gene tree happens to correspond to the species tree. This well-known problem of multiple, possibly conflicting, gene trees *versus* a single species tree (*e.g.* Pamilo and Nei, 1998; Tajima, 1983; Takahata, 1989; Avise, 1994; Maddison, 1995; Avise and Wollenberg, 1997) originates from a process called "stochastic lineage sorting" where each allelic lineage has a non-null probability to go extinct. It should be noted that (1) conflict in branching patterns between a *species* tree and a given gene tree can occur only if ancestral polymorphism is maintained for this gene through two or more speciation events (*e.g.* Avise, 1994), and (2) lineage sorting is a phenomenon relevant to any heritable character, hence, to morphological monogenic neutral characters as well.

When one tries to identify biological species (interbreeding groups), one is in fact faced with identifying a set of intermingled gene trees whose nodes and branches are shared by the members of the species; every sexually reproducing individual organism is sitting simultaneously on the branches of multiple gene trees. In that sense, we think it is appropriate to extend the "gene pool" concept (*e.g.* Dobzhansky, 1955) of a biological species to a concept of "tree pool", which incorporates phylogenetic considerations. For example, in Figure 5c, representing the superposition of the gene trees from loci x and y for five generations, the individual indicated as a black symbol at generation "g_{-15}" possessed simultaneously in

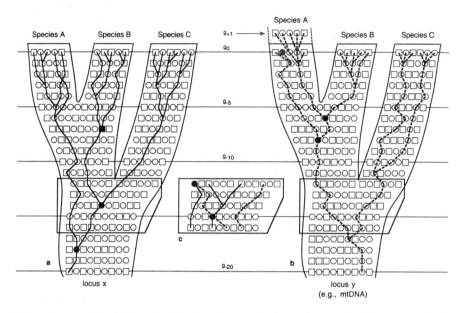

Figure 5.   Two independent haplotypic transmission pathways (*i.e.* for two genetically unlinked markers, loci x and y) through 20 generations of the same three species phylogeny ("Species A" and "Species B" form a clade to the exclusion of "Species C"). "g_0"= current generation. Circles and squares represent female and male individuals, respectively. (a) all haplotypes from "Species B" and "Species C" are direct descendants of the single ancestral haplotype that was transmitted by the female individual indicated as a black symbol at generation "g_{-14}", while the single most recent ancestor for the haplotypes from "Species A" and "Species B" is older (indicated as a black symbol at generation "g_{-18}"). Hence, with respect to that specific marker and in conflict with the species phylogeny (as well as with the "locus y" genealogy), haplotypes from "Species B" and "Species C" form a clade to the exclusion of "Species A" haplotypes. The black symbol at generation "g_{-7}" indicates the male individual in which all "Species B" haplotype genealogies coalesce. (b) matrilineal pathway (*e.g.* the pathway of mtDNA) which, here, is in agreement with the species phylogeny, *i.e.* "Species A" and "Species B" haplotypes form a clade to the exclusion of "Species C" haplotypes. However, "Species A" is paraphyletic with respect to "Species B" because some "Species A" haplotypes are cladogenetically more closely related to "Species B" haplotypes (through the individual indicated by the black symbol at generation "g_{-6}") than to other "Species A" haplotypes (through the individual indicated by the black symbol at generation "g_{-8}"). However, if the "Species A" female indicated as a black symbol at generation "g_0" does not have any offspring while one or both of the other females do (as indicated on the figure), "Species A" will become monophyletic at generation "g_{+1}". (c) superposition of "locus x" and "locus y" genealogies for five generations (g_{-16} to g_{-12}); the individuals indicated by the black symbols (at generations "g_{-15}" and "g_{-12}") carried, simultaneously for both loci, haplotypes whose descendants survived until generation "g_0".

its genome (1) a haplotype of "locus x" which descendants eventually got fixed both in "species B" and "Species C" and (2) a haplotype of "locus y" which followed in subsequent generations a different pathway than that of "locus x", and which descendants eventually got fixed both in "species A" and "Species B". If "locus y" represents mtDNA, then that individual also transmitted this same haplotype to all its offspring at generation "g_{-14}" but these alternative pathways terminated (before reaching generation "g_9") because some of them either eventually reached a male individual (which cannot transmit its mtDNA to its offspring) or a female which did not have any offspring. Because its causes are at work within each extant reproductively isolated group, the phenomenon of lineage sorting has at least two implications relevant to the present discussion.

First, the survival probability of allelic lineages is a function of population genetic parameters such as number of generations and population size. Increased *reproductive* isolation time of a population (as well as reduced population size) will increase, for any given marker, the likelihood of all its alleles to coalesce at a single allelic ancestor within that lineage (*e.g.* Nei, 1987). This means that reproductively isolated populations will tend to evolve from polyphyly (with regard to any gene tree) to paraphyly, then to reciprocal monophyly (*e.g.* Avise, 1994). For example, "Species A" is paraphyletic at generation "g_0" with respect to the gene tree of the "locus y", while it becomes monophyletic at generation "g_{+1}" because of lineage sorting (*i.e.* the female, at generation "g_0", indicated as a black symbol, did not have any offspring at generation "g_{+1}"). The important point is to consider this phenomenon in a multilocus perspective. Given long enough reproductive isolation of a population, *any* piece of non-recombining DNA will have all its alleles coalesce within that lineage and will therefore be consistent with *any other* genetically-unlinked marker in supporting the grouping of all individuals from the population in a monophyletic group. In other words, because we are considering populations of finite size, concordance among independent (genetically-unlinked) markers in revealing identical cladistic cohesion (*i.e.* the identification of shared derived characters) of such a group will inevitably increase with increased duration of its *reproductive* isolation.

We therefore agree with Avise and Wollenberg (1997) that "*reproductive barriers are important, even within a strictly phylogenetic species framework, because they generate through time increased genealogical depth and concordance across allelic pathways.*" The same authors proposed two approaches ("qualitative concordance" and "quantitative co-ancestry") which incorporate both phylogenetic and population genetic aspects to biological species identification

Second, monophyletic groups within organismal genealogies can merge as can populations; *i.e.* the permanent or temporary condition of organismal clades is similar to (*i.e.* linked to) that of barriers to interbreeding. The two points prompt us to consider (1) that the usually articulated opposition between the PSC and the BSC is greatly exaggerated, (2) that integration of multiple genetically unlinked molecular markers (or the comparisons of molecular and morphological markers) should be very much promoted in investigations on species designation.

On the practical side, using monophyly as a sufficient criterion while focusing on a single molecular marker (*i.e.* failing to recognise the "tree pool" concept) would lead to a great increase in the number of phylogenetic species, especially if the marker is fast evolving. More importantly, the species thus defined may also be unstable to the addition of data from unlinked loci (which could support a conflicting branching pattern, cf. above). However, even if two strata are reproductively isolated, one should not expect all molecular markers to necessarily demonstrate reciprocal monophyly (or fixed differences, cf. above) for these strata, because any marker might not have drifted to the point of fixed differences or reciprocal monophyly. In other words, finding multiple independent markers revealing identical cladistic cohesion is a sufficient criterion for raising the two groups to separate species status, but the lack of concordance does not *demonstrate* interbreeding. As discussed above, for the criterion of fixed differences: (1) a single molecular marker which indicates reciprocal monophyly and which corresponds to an *a priori* morphological designation is a sufficient criterion for species designation, and (2) reciprocal monophyly among allopatric populations confirm that gene flow is negligible among these populations but does not reveal whether interbreeding would occur in the case of a secondary contact.

Furthermore, given the substantial influence of population size on coalescence time, a small interbreeding group will be much more easily recognised as a phylogenetic species than a larger interbreeding group; given identical reproductive isolation time, larger populations will tend more to be paraphyletic with respect to smaller populations. For example, in the case of the inshore and offshore bottlenose dolphins (Curry *et al.*, submitted), or the long- and short- beaked common dolphins (Rosel *et al.*, 1994), one group is nested within the other in the phylogeny (cf. below). In other words, the smaller population (inshore bottlenose or long-beaked common dolphins) is more likely to become randomly fixed for derived character states than the larger, wider ranging population (due to more rapid genetic drift in the smaller population). In these cases, recognition of long-beaked common dolphins or inshore bottlenose dolphins as separate species renders the other biological species paraphyletic.

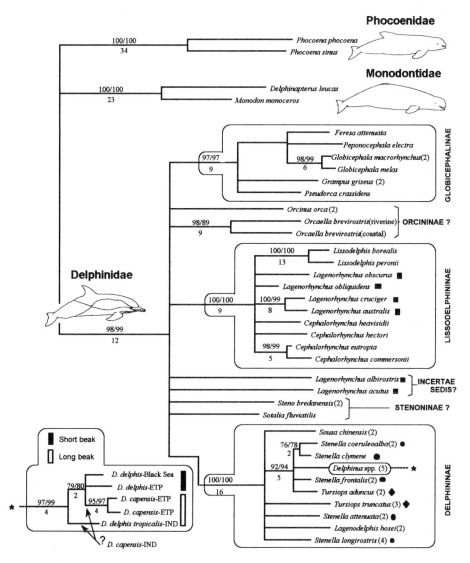

Figure 6. Seventy percent bootstrap consensus tree (unweighted parsimony analysis of mtDNA cytochrome *b* sequence data; LeDuc *et al.*, 1999) and proposed systematic revision for subfamily affiliations. When a species is represented by multiple samples, the number of included sequences is given between parentheses. The filled symbols indicate possibly polyphyletic or paraphyletic genera: *Lagenorhynchus* (squares), *Stenella* (circles), and *Tursiops* (diamonds). The asterisk points at the inferred phylogeny among the *Delphinus* specimens included in the analyses. Bootstrap values for Maximum Parsimony/Neighbour Joining analyses and Bremer support are indicated above and below the nodes, respectively. Branch lengths are those obtained under maximum parsimony. "ETP", Eastern tropical Pacific; "IND", Indian Ocean.

We are therefore left with the question whether it is legitimate to accept paraphyletic species (such as "Species A" at generation "g_0" in Fig. 5b) despite that multilocus concordance cannot be observed. We think it is indeed legitimate because interbreeding as a grouping criterion has an utilitarian rationale: mortality or removals of groups of individuals will impact biodiversity very differently depending on whether these groups are or are not separated by barriers to interbreeding. Furthermore, paraphyletic biological species with large populations will tend to remain paraphyletic for long periods for the very reason that they are large (cf. above). We do not view alpha taxonomy as a statement of absolute objectivity but as a multi-purpose tool, *e.g.* for management of biodiversity. Our suggestion is to accept paraphyletic biological species for management because, in doing so, we will avoid partitioning species into multitudinous groups which do not make cladistic sense in a multilocus framework, while we will still nearly maximise preservation of cladogenetic diversity.

In a larger and simpler context, the test of monophyly is of course dependent on the particular taxa included in the analysis. For example, the inclusion of sei whales in the studies of Wada and Numachi (1991) and Yoshida and Kato (1999) revealed that the pelagic Bryde's whale and the sei whale form a clade exclusive of the pygmy Bryde's whale (which was sister to that clade). Failure to include the sei whale in the analysis would simply have revealed a monophyletic pelagic/pygmy Bryde's clade, with the two forms being monophyletic sister groups. In some cases, the effect of the phylogenetic context may be more subtle. For example, LeDuc *et al.* (1999) proposed a systematic revision of Delphinidae based on mtDNA cytochrome *b* phylogenetic analyses (Fig. 6). If this phylogeny is correct, the *tropicalis* form of common dolphin is basal within the genus (*i.e.* sister to a clade consisting of all the other included *Delphinus* sequences; Fig. 6). Considering this form to be a race of *Delphinus capensis* (the long-beaked common dolphin, to which *tropicalis* is most similar morphologically) will have very different systematic implications than considering it to be a variant of the short-beaked common dolphin *D. delphis* or a distinct species. In this sense, the taxon designation of a specimen should not be taken lightly.

If there is any question of the taxonomic status of a population or species (and there is for many cetaceans), one risks confounding initial assumptions with possible conclusions.

The situation is in fact very similar to that encountered in higher-taxa phylogeny where families, orders, or even higher taxonomic units, are often *assumed* monophyletic (and a single or very few representatives of these units are included in the analyses). Indeed, in the example discussed above regarding the phylogenetic position of cetaceans within mammals (Fig. 1), the hypothesis of their sister group relationship with artiodactyls became a famous example of congruence between molecules and morphology (*e.g.* Novacek, 1992). However, the congruence was restricted to the hypothesis that cetaceans and artiodactyls form a clade while the congruence or incongruence between molecules and morphology regarding the resolution of relationships *within* that clade had not been assessed. Indeed, most molecular studies published before 1994 could not test the hypothesis of artiodactyl monophyly because very few representatives of artiodactyls had been used. It was implicitly *assumed* at that time (even among molecular phylogeneticists) that molecular data would ultimately support artiodactyl monophyly. Similarly, the delphinid cytochrome *b* phylogeny of LeDuc *et al.* (1999) indicates that *Lagenorhynchus*, *Stenella*, and *Tursiops* (cf. filled squares, circles, and diamonds, respectively; Fig. 6) might be polyphyletic or paraphyletic genera.

In conclusion, we propose that, in parapatry/sympatry, reciprocal monophyly of two strata or paraphyly of one stratum with respect to a monophyletic other stratum is sufficient for raising the two groups to separate biological species status, as long as phylogenetic inference involves either multiple genetically unlinked loci or an agreement between at least a single gene tree and disjunction of state distribution of at least one morphological/physiological character.

4.4 Degree of Divergence

Árnason and Gullberg (1994) and Árnason *et al.* (1993) supported species status for two forms of minke whales because the single sequences included for each form diverged more from each other than did sequences from Bryde's and sei whales. Despite other studies (with good geographic coverage and which included several individuals from each form of minke whale; Wada and Numachi, 1991; Pastene *et al.*, 1994) that supported the recognition of the Antarctic form as a distinct species, the criterion of divergence must be applied with caution.

The mean level of genetic divergence (for neutral markers) among reproductively isolated groups will be proportional to the time since isolation; however, the degree of divergence between groups is only loosely correlated with the presence/absence of isolation mechanisms. Therefore, some pairs of biological species can be very close genetically, while conspecifics can be very divergent. For example, in their delphinid cytochrome *b* phylogeny, LeDuc *et al.* (1999) found overlap between levels of inter- and intraspecific pairwise distances. For example, some species of *Cephalorhynchus* were genetically more similar to each other than were samples of *Stenella attenuata* from different populations.

In another study, Hoelzel *et al.* (1998b) concluded that the two sympatric forms of killer whales in the Northeast Pacific were conspecific because the genetic divergence between these two groups was similar to that among allopatric groups which are considered conspecific. The important point is that this perspective ignores the fixed differences seen between the two sympatric forms. Although we agree that a high degree of divergence could serve as a starting point for the investigation of additional characters (in order to search for fixed differences or phylogenetic cohesion), the minke and killer whale samples illustrate that a given degree of genetic divergence alone relative to other species or other populations is neither necessary nor sufficient for designating two populations as separate or identical species.

5.   CASE   STUDIES: DEFINING   CETACEAN SPECIES BOUNDARIES WITH GENETIC DATA

To provide a framework for the interpretation and use of genetic data, we briefly examine some chosen case studies, where molecular evidence has already contributed or might potentially contribute to the clarification of species boundaries in cetaceans.

5.1   *Delphinus*

This genus is a typical case of a wide-ranging, variable taxon. Until recently, *Delphinus delphis* was generally considered to be a single, highly variable species. Morphological variants were known and described but had not been thoroughly studied. These variants were mainly distinguishable based on their rostrum length, with similar types occurring in more than one ocean basin and different types often coexisting in certain areas. Heyning and Perrin (1994) demonstrated the existence of sympatric adult morphotypes in the eastern North Pacific that had no overlap in a number of morphological characters (Fig. 7). This was solid evidence for reproductive isolation in this region between the two forms. These species were assigned to *D. capensis* and *D. delphis* (the long- and short-beaked forms, respectively), based on the similarity of the former to a species previously described from the Cape of Good Hope, and of the latter to the common dolphin binomial first proposed by Linneaus.

The fixed mtDNA differences between the two forms in the eastern Pacific found by Rosel *et al.* (1994) provided additional evidence of their reproductive isolation. Heyning and Perrin noted the need for further study,

Figure 7.   Typical colour patterns and beak sizes in short-beaked *D. delphis* (top), long-beaked *D. capensis* (middle), and Indian Ocean *D. tropicalis* (bottom) common dolphins. (Photos reproduced with permission from Heyning & Perrin, 1994; and R. Salm)

especially in the Indian Ocean, where an additional nominal species may occur. This Indian Ocean form, referred to as the *tropicalis* form, has a colour pattern similar to *D. capensis* but a rostrum that is even longer. At present, research material from the Indian Ocean is scarce for common dolphins of any type, so no studies have yet been conducted on their levels of variation. However, Heyning and Perrin note that if no morphological intergradation is found between *D. capensis* and the *tropicalis* form from the Indian Ocean, then *tropicalis* may represent a third species of common dolphin. In the mitochondrial cytochrome *b* phylogeny of LeDuc *et al.* (1999), the *tropicalis* form was basal within the genus *Delphinus* (Fig. 6), and, hence, was not a sister group to the eastern Pacific *D. capensis*. This means that considering *tropicalis* to be conspecific with, and a geographic variant of, *D. capensis* would render that species paraphyletic. Although the sample size was far too small to be definitive, these findings invoke yet another possibility; perhaps the long-beaked and/or the short-beaked form arose independently in different ocean basins. The relationship of the Indian Ocean *D. capensis* (samples of which were not available to Rosel *et al.* or LeDuc *et al.*) to the other forms from different areas is critical to the discussion.

If further studies were to show *D. capensis* from the Indian Ocean to be closely related genetically to *tropicalis*, then the eastern Pacific *D. capensis* would be distinct from the *capensis* form found in the Indian Ocean. In other words, assuming the relationships shown in Figure 6 are correct, where eastern Pacific ("ETP") *D. capensis* and Indian Ocean ("IND") *tropicalis* are not sister taxa, then any Indian Ocean *D. capensis* added to the phylogeny cannot unite the three into a monophyletic group. Of course, as discussed above, the status of any species shown to be paraphyletic on a given single gene tree needs careful consideration, but the indication of possible independent origins for a specific morphotype has interesting biological implications. On a taxonomic note, since the type specimen of *capensis* is from the Cape of Good Hope, that name would apply to animals from that region and the eastern Pacific long-beaked form would take on the next most senior synonym, if the two were split into different species.

5.2   *Tursiops aduncus/truncatus*

Ross (1977) detected a number of morphological characters which showed no overlap between two forms off the coast of South Africa, proposing that *T. aduncus* be recognised as a separate species (Fig. 8). Later, Ross and Cockroft (1990) found intergrades between the two forms along the Australian coast and concluded that they should be considered the same species. However, the intergradation in Australia does not negate the reproductive isolation indicated for South African populations.

Figure 8. The taxonomy of bottlenose dolphins remains unresolved but the genus comprises at least two species, Top) *T. truncatus* and Bottom) *T. aduncus*, the latter being generally smaller with a longer beak, slimmer body and contrasting cape (Photos: P.G.H. Evans and B. Haase/C. Smeenk)

Molecular genetic analyses provide additional perspectives and support the initial designations proposed by Ross (1977). In a study by Smith-Goodwin (1997), ten closely related haplotypes were found in a sample set of 74 South African *aduncus* samples. In a global genetic study of the genus, Curry (1997) included seven of these haplotypes, adding 13 more from the Timor Sea, Indo-Pacific region, and the western North Pacific, plus samples (73 haplotypes) of *truncatus*-type dolphins from the Pacific, Atlantic, and Indian Oceans. She found two nearly-fixed base substitutions and a fixed insertion/deletion between *aduncus* and *truncatus*, which were also reciprocally monophyletic. Furthermore, in their cytochrome *b* phylogeny of delphinids, LeDuc *et al.* (1999) found the *aduncus* samples to be more closely related to *Delphinus* and some species of *Stenella* than they were to *Tursiops truncatus*, indicating that *aduncus* and *truncatus* were not even sister taxa (cf. Fig. 6).

These molecular results support the recognition of *Tursiops aduncus* as a distinct species. However, most recently, Hoelzel *et al.* (1998a) found some South African *aduncus*-type dolphins to have haplotypes typical of *truncatus*. The fact that they found two of their six *aduncus* samples to have haplotypes also found in *truncatus*, while none of Smith-Goodwin's 74 samples did, raises questions about the frequency of shared haplotypes in the population. The discrepancy may arise from differences in geographic sampling (Smith-Goodwin not sampling a small-scale zone of introgression) or problems with sample designation (two *truncatus* incorrectly designated as *aduncus*).

Neither of these possibilities seems very likely, given the sampling scheme of Smith-Goodwin and the reference museum specimens used by Hoelzel *et al.* In addition, fixed differences or lack of shared haplotypes are not *necessary* conditions for reproductive isolation. In the larger sense, the status of the Australian populations may be critical. If the morphological evidence of isolation off South Africa and intergradation off Australia is correct, isolation-by-distance may exist, *i.e.* *truncatus* and *aduncus* in South Africa might represent the ends of a geographic continuum where gene flow is possible via intermediate populations but not between the two sympatric ends of the continuum. However, this would have to be reconciled with the apparent paraphyly detected by LeDuc *et al.* (cf. Fig. 6). Obviously, more work is needed. Apart from Timor Sea samples, none of the molecular studies included animals from the Australian area and the morphological studies cited did not include *aduncus* specimens from the Indo-Pacific or western Pacific regions.

5.3 *Orcinus orca*

The Northeast Pacific populations of killer whales have been extensively studied with molecular methods (Hoelzel and Dover, 1991; Hoelzel *et al.*, 1998b). In the Antarctic, Berzin and Vladimirov (1983) described the morphology of another, smaller form of killer whale that is at least seasonally sympatric with the standard form. This case seems to be parallel to the situation in the Pacific, with one form being a fish specialist and the other feeding primarily on mammals, with no observations of the two schooling together. Furthermore, the differences in size, colour pattern, life history, and osteology are much more dramatic for the Antarctic forms. A molecular genetic study is desirable, not only to examine the relationship of the Antarctic forms to each other, but also to place them in a context of world-wide genealogy of killer whales.

5.4 *Orcaella brevirostris*

Although the riverine and marine forms of this species have been described as separate species in the past, they are currently regarded as one (Marsh *et al.*, 1989). In a mitochondrial cytochrome *b* phylogeny of delphinids, LeDuc *et al.* (1999) included a Laotian riverine sample and an Australian marine sample (Fig. 6). The divergence between these two sequences was far greater than that observed in any other intraspecific pair in the dataset, including those that originated from different ocean basins. In fact, this difference was greater than most intergeneric comparisons within subfamilies (cf. Fig. 6). Once more, we want to emphasise the fact that the degree of divergence between groups is only loosely correlated with the presence/absence of isolation mechanisms (cf. above). However, it does suggest that additional research is desirable in order to detect possible multiple synapomorphies or fixed differences; molecular genetic distinctness might exist between freshwater and marine forms, or perhaps between those of Southeast Asia and Australia.

5.5 *Mesoplodon* spp.

Mead (1981) published records of Hector's beaked whale (*Mesoplodon hectori*) strandings on the west coast of the United States, the first such records in the northern hemisphere. A DNA sequence from one of these stranded animals was used by Henshaw *et al.* (1997) as representative of *M. hectori*. However, when Dalebout *et al.* (1998) compared this sequence to that from a South Australian Hector's beaked whale, they found them to be more divergent than either was to some of the other species of *Mesoplodon*, raising the possibility that Hector's beaked whale is polyphyletic. Dalebout *et al.* (1998) suggested that the northern hemisphere strandings may represent an as yet undescribed species. In another situation involving beaked whales, Pitman *et al.* (1987) described sightings of an unknown beaked whale with a distinct male colour pattern in the eastern Pacific, which could represent yet another species if found to be different from the US strandings (colour pattern was not discernible on the stranded animals). In another case, Pitman *et al.* (1999) proposed that another beaked whale seen in the tropical Pacific may represent *M. (Indopacetus) pacificus*, currently known in museums from only two worn skulls. In both cases, tissue from strandings or biopsies may provide genetic links between sightings, strandings, and museum specimens sufficient to resolve the taxonomic questions.

5.6   *Balaenoptera* spp.

As discussed above, current evidence favours the recognition of a pygmy form of Bryde's whale as a separate species occurring in coastal areas of the tropical west Pacific. This species is recognised by Rice (1998) in a review of nomenclature and taxonomic history. However, there are still more questions to be answered. In a mtDNA phylogeny, Yoshida and Kato (1999) found Japanese coastal populations of Bryde's whales to be distinct from the ones farther south. Furthermore, their work indicated some discrepancy between relationships inferred by d-loop sequences and those inferred from cytochrome *b* sequences. These two genes are linked on the mitochondrial genome and should share the same phylogenetic history. The uncertainty of these results raises questions about possible multiple origins for coastal populations. There is also much uncertainty regarding the morphological similarity of the various populations; the sample sizes for the different groups are as yet far too small for any comprehensive analysis.

6.   CONCLUSIONS

Molecular approaches are potentially of great utility for distinguishing species because they bring a wealth of new markers that can be compared against previously available data (*e.g.* morphological data). In some cases, such as that of inshore and offshore bottlenose dolphins, much of the non-molecular evidence for separation can be equivocal. Differences in food habits or parasite loads may not be genetically determined characters, and the morphological differences detected can be problematic. In these cases, molecular data can reveal two or more forms to be reproductively isolated, at least long enough for fixed differences and monophyly of at least one of the groups to arise. Of course, within a population where complete stratification has not been detected with morphological, ecological, or behavioural characters, congruent fixed differences in multiple independent genetic markers could be revealed. In this case, evidence of cryptic species must be considered and may indicate the need for more detailed morphological and ecological study. We suggest the operational principle that, in parapatry and sympatry, (a) covariation between *a priori* morphological/physiological designation and a minimum of one molecular character, and/or (b) fixed molecular differences at several unlinked loci, and/or (c) multilocus reciprocal monophyly, are sufficient conditions to identify lack of interbreeding, hence for biological species recognition.

Defining biological species boundaries is probably a good and efficient starting point for taxonomy and management of biodiversity. Multilocus analysis of molecular data provides a formidable source of markers for effective identification of reproductively-isolated groups in cetaceans. One must however realise that speciation is probably always relatively gradual and that the underlying evolutionary processes and outcomes are of such complexity that their reality cannot be fully captured by a necessarily simplified binomial summary (*i.e.* alpha taxonomy). Molecular techniques also open the possibility of directly investigating the genes that influence reproductive isolation, and would therefore allow biologists to study the very finest mechanisms involved in the fascinating phenomenon of speciation.

ACKNOWLEDGEMENTS

We thank John C. Avise, Brian Bowen, Insa Cassens, Rus Hoelzel, Patrick Mardulyn and Xavier Vekemans for commenting on previous versions of the manuscript. We are particularly grateful to William Perrin for participating in valuable and informative discussions. M.C.M. is supported by grants from the National Fund for Scientific Research Belgium (FNRS), the Free University of Brussels (ULB), the Defay Fund (Belgium), the Van Buuren Fund (Belgium), and the "Communauté Française de Belgique" (ARC 98/02-223).

REFERENCES

Amos, W. (1999) Culture and Genetic Evolution in Whales (4). *Science*, **284**, 2055a.

Amos, W., Schlotterer, C., and Tautz, D. (1993) Social structure of pilot whales revealed by analytical DNA profiling. *Science*, **260**, 670-672.

Árnason, Ú, Spilliaert, R., Pálsdóttir, Á., and Árnason, A. (1991) Molecular identification of hybrids between the two largest whale species, the blue whale (*Balaenoptera musculus*) and the fin whale (*B. physalus*). *Hereditas*, **115**, 183-189.

Árnason, Ú., Grétarsdóttir, S., and Widegren, B. (1992) Mysticete (baleen whale) relationships based upon the sequence of the common cetacean DNA satellite. *Molecular Biology and Evolution*, **9**, 1018-1028.

Árnason, Ú, Gullberg, A., and Widegren, B. (1993) Cetacean mitochondrial DNA control region: sequences of all extant baleen whales and two sperm whale species. *Molecular Biology and Evolution*, **10**, 960-970.

Árnason, Ú, and Gullberg, A. (1994) Relationship of baleen whales established by cytochrome *b* gene sequence comparison. *Nature* (London), **367**, 726-728.

Avise, J. C. (1994) *Molecular markers, natural history and evolution*. Chapman & Hall, New York.

Avise, J.C. (2000) *Phylogeography*. Harvard University Press, Cambridge.

Avise, J. C., Arnold, J., Ball, R. M., Bermingham, E., Lamb, T., Neigel, J. E., Reeb, C. A., and Saunders, N. C. (1987) Intraspecific phylogeography: The mitochondrial DNA bridge between population genetics and systematics. *Annual Review in Ecology and Systematics*, **18**, 489-522.

Avise, J. C., and Wollenberg, K. (1997) Phylogenetics and the origin of species. *Proceedings of the National Academy of Sciences, USA*, **94**, 7748-7755.

Berzin, A.A. and Vladimirov, V.L. (1983) Novyi vid kosatki (Cetacea: Delphinidae) iz vod Antarktiki. *Zool. Zhurn.*, **62**, 287-295.

Bossart, J. L. and Prowell, D. P. (1998) Genetic estimates of population structure and gene flow: limitations, lessons and new directions. *Trends in Ecology and Evolution*, **13**, 202-206.

Bremer, K. (1994) Branch support and tree stability. *Cladistics*, **10**, 295-304.

Boyden, A. and Gemeroy, D. (1950) The relative position of the Cetacea among the orders of Mammalia as indicated by precipitin tests. *Zoologica*, **35**, 145-151

Coyne J. A., and Orr, H. A. (1989) Two rules of speciation. In: *Speciation and its consequences* (Ed. by D. Otte and J. A. Endler), pp. 180-207. Sinauer, Sunderland, Massachusetts.

Curry, B. E. (1997) *Phylogenetic Relationships Among Bottlenose Dolphins (genus Tursiops) in a Worldwide Context*. Ph. D. Thesis, Texas A & M University, Galveston, 138pp.

Curry, B. E., Milinkovitch, M.C., LeDuc, R. and Dizon, A.E. (in prep.) MtDNA evidence for reproductive isolation between inshore and offshore bottlenose dolphins (genus *Tursiops*) in the western North Atlantic Ocean/Gulf of Mexico. *Submitted*.

Czelusniak, J., Goodman, M. Koop, B.F., Tagle, D.A., Shoshani, J., Braunitzer, G., Kleinschmidt, T.K., De Jong, W.W., and Matsuda, G. (1990) Perspectives from Amino acid and nucleotide equences on cladistic relationships among higher taxa of Eutheria. In: *Current Mammalogy, Vol.2* (Ed. by H. H. Genoways), pp. 545-572. Plenum, New York.

Dalebout, M. L., van Helden, A., Van Waerebeek, K., and Baker, C.S. (1998) Molecular genetic identification of southern hemisphere beaked whales (Cetacea: Ziphiidae). *Molecular Ecology*, **7**, 687-694.

Dizon, A. E., Chivers, S. J., and Perrin, W. F. (eds.) (1997) *Molecular genetics of marine mammals*. Special Publication Number 3, The Society for Marine Mammology. 412pp.

Dobzhansky, Th. (1955) A review of some fundamental concepts and problems of population genetics. *Cold Spring Harbor Group Symposium of Quantitative Biology.*, **20**, 1-15.

Felsenstein, J. (1985) Confidence limits on phylogenies: an approach using the bootstrap. *Evolution*, **39**, 783-791.

Flower, W.H. (1883) On whales, present and past and their probable origin. *Proceedings of the Zoological Society, London*, **1883**, 466-513.

Fraser, F. C. (1940) Three anomalous dolphins from Blacksod Bay, Ireland. *Proceedings of the Royal Irish Academy*, **45**, 413-455.

Gatesy, J. (1997) More DNA support for a Cetacea/Hippopotamidae clade: the blood-clotting protein gene g-fibrinogen. *Molecular Biology and Evolution*, **14**, 537-543.

Gatesy, J. (1998) Molecular Evidence for the Phylogenetic Affinities of Cetacea. In: *The Emergence of Whales: Evolutionary Patterns in the Origin of Cetacea* (Ed J.G.M. Thewissen), pp. 63-111. Plenum, New York.

Gatesy, J., Hayashi, C., Cronin, M.A., and Arctander, P. (1996) Evidence from milk casein genes that cetaceans are close relatives of hippopotamid artiodactyls. *Molecular Biology and Evolution*, **13**, 954-963.

Gatesy, J., Milinkovitch, M.C., Waddell, V., and Stanhope, M. (1999) Stability of Cladistic Relationships between Cetacea and Higher-Level Artiodactyl Taxa. *Systematic Biology*, **48**, 6-20.

Gentry, A. and Hooker, J. (1988) The phylogeny of the Artiodactyla. In: *The Phylogeny and Classification of the Tetrapods*, Volume 2: Mammals (Ed. M. Benton), pp. 235-272. Clarendon Press, Oxford.

Gingerich, P.D., Smith, B.H. and Simons, E.L. (1990) Hind limbs of Eocene Basilosaurus: evidence of feet in whales. *Science*, **249**, 154-157.

Graur, D., and Higgins, D. G. (1994) Molecular evidence for the inclusion of cetaceans within the order Artiodactyla. *Molecular Biology and Evolution*, **11**, 357-364.

Hasegawa, M., Adachi, J., and Milinkovitch, M. C. (1997) Novel phylogeny of whales supported by total molecular evidence. *Journal of Molecular Evolution*, **44**, S117-S120.

Henshaw, M.D., LeDuc, R.G., Chivers, S.J., and Dizon, A.E. (1997) Identification of beaked whales (family Ziphiidae) using mtDNA sequences. *Marine Mammal Science*, **13**, 487-495.

Hershkovitz, P. (1966) Catalog of living whales. *U. S. National Museum Bulletin*, **246**, 1-259.

Hey, J. (1991) The structure of geneologies and the distribution of fixed differences between DNA sequence samples from natural populations. *Genetics*, **128**, 831-840.

Heyning, J.E. (1999) Whale Origins - Conquering the Seas. *Science*, **283**, 943.

Heyning, J.E. and Perrin, W.F. (1994) Evidence for two species of common dolphins (genus *Delphinus*) from the eastern North Pacific. *Natural History Museum Los Angeles County Contr. Sci.*, **442**, 1-35.

Hoelzel, A.R., and Dover, G.A. (1991) Genetic differentiation between sympatric killer whale populations. *Heredity*, **66**, 191-195.

Hoelzel, A.R., Potter, C.W., and Best, P.B. (1998a) Genetic differentiation between parapatric 'nearshore' and 'offshore' populations of the bottlenose dolphin. *Proceedings of the Royal Society London*, **265**, 1177-1183.

Hoelzel, A.R., Dahlheim, M., and Stern, S.J. (1998b) Low genetic variation among killer whales (*Orcinus orca*) in the eastern North Pacific and genetic differentiation between foraging specialists. *Journal of Heredity*, **89**, 121-128.

LeDuc, R.G., Perrin, W.F., and Dizon, A.E. (1999) Relationships among the delphinid cetaceans based on cytochrome *b* sequences. *Marine Mammal Science*, **15**, 619-648.

Luckett, W.P., and Hong, N. 1998. Phylogenetic relationships between the orders Artiodactyla and Cetacea: a combined assessment of morphological and molecular evidence. *Journal of Mammalian Evolution*, **5**, 127-182.

Maddison, W. (1995) Phylogenetic histories within and among species. In: *Experimental and Molecular Approaches to Plant Biosystematics* (Ed. by P.C. Hoch and A.G. Stephenson), pp. 273-287. Missouri Botanical Garden, St. Louis.

Marsh, H.R., Lloze, R., Heinsohn, G.E., and Kasuya, T. (1989) Irrawaddy dolphin *Orcaella brevirostris* (Gray, 1866). In: *Handbook of Marine Mammals, Volume 4: River Dolphins and the Larger Toothed Whales* (Ed. by S.H. Ridgway and R. Harrison), pp. 101-118. Academic Press, London.

Mayr, E. (1942) *Systematics and the Origin of Species*. Columbia University Press, New York.

Mayr, E. (1963) *Animal Species and Evolution*. Harvard University Press, Cambridge, Massachusetts.

Mayr, E. (1969) *Principles of Systematic Zoology*. McGraw Hill, New York. 428pp.

Mead, J. G. (1981) First records of *Mesoplodon hectori* (Ziphiidae) from the northern hemisphere and a description of the adult male. *Journal of Mammology*, **62**, 430-432.

Mead, J. G. and Brownell, R. L. Jr. (1993) Order Cetacea. In: *Mammal Species of the World: a Taxonomic and Geographic Reference*, 2nd ed. (Ed. by D. E. Wilson and D. M. Reeder), pp. 349-364. Smithsonian Institution Press, Washington and London.

Mesnick, S., Taylor, B., LeDuc R.G., Escorza Treviño, S., and Dizon, A.E. (1999) Culture and Genetic Evolution in Whales (1). *Science*, **284**, 2055a

Meyer, A., Kocher, T.D., Basasibwaki, P., and Wilson, A.C. (1990) Monophyletic origin of Lake Victoria cichlid fished suggested by mitochondrial DNA sequences. *Nature* (London), **347**, 550-553.

Milinkovitch, M.C. (1992) DNA-DNA hybridizations support ungulate ancestry of Cetacea. *Journal of Evolutionary Biology*, **5**, 149-160.

Milinkovitch, M.C. (1995) Molecular phylogeny of cetaceans prompts revision of morphological transformations. *Trends in Ecology and Evolution*, **10**, 328-334.

Milinkovitch, M.C. (1997) The Phylogeny of Whales: a Molecular Approach. In: *Molecular Genetics of Marine Mammals*, (Ed. by A. Dizon *et al.*), pp. 317-338. Special Publication Number 3; The Society for Marine Mammology.

Milinkovitch, M.C., Orti, G., and Meyer, A. (1993) Revised phylogeny of whales suggested by mitochondrial ribosomal DNA sequences. *Nature* (London), **361**, 346-348.

Milinkovitch, M.C., Meyer, A., and Powell, J. (1994) Phylogeny of all major groups of cetaceans based on DNA sequences from three mitochondrial genes. *Molecular Biology and Evolution*, **11**, 939-948.

Milinkovitch, M.C., and Thewissen, J.G.M. (1997) Eventoed Fingerprints on Whale Ancestry. *Nature* (London), 388, 622-624.

Milinkovitch, M. C., M. Bérubé, and P. J. Palsbøll (1998) Cetaceans Are Highly Specialized Artiodactyls. In: *The Emergence of Whales: Evolutionary Patterns in the Origin of Cetacea* (Ed. by J. G. M. Thewissen), pp. 113-131. Plenum, New York.

Nei, M. (1987) *Molecular evolutionary genetics*. Columbia University Press, New York.

Neigel, J.E. (1996) Estimation of effective population size and migration parameters from genetic data. In: *Molecular genetic approaches in conservation* (Ed. by T. B. Smith and R. K. Wayne), pp. 329-346. Oxford University Press, New York.

Nixon, K.C., and Wheeler, Q.D. (1990) An amplification of the phylogenetic species concept. *Cladistics*, **6**, 211-223.

Novacek, M.J. (1992) Mammalian phylogeny: shaking the tree. *Nature* (London), **356**, 121-125.

O'Corry-Crowe, G.M., Suydam, R.S., Rosenberg, A., Frost, K.J., and Dizon, A.E. (1997) Phylogeography, population structure and dispersal patterns of the beluga whale *Delphinapterus leucas* in the western Nearctic revealed by mitochondrial DNA. *Molecular Ecology*, **6**, 955-970.

Palsbøll, P., Allen, J., Berube, M., Clapham, P.J., Feddersen, T.P., Hammond, P.S., Hudson, R.R., Jørgensen, H., Katona, S., Larsen, A.H., Larsen, F., Lien, J., Mattila, D.K., Sigurjonsson, J., Sears, R., Smith, T., Sponer, R., Stevick, P., and Øien, N. (1997) Genetic tagging of humpback whales. *Nature* (London), **388**, 767-769.

Pamilo, P. and Nei, M. (1988) Relationship between gene trees and species trees. *Mol. Biol. Evol.* **5**, 568-583.

Pastene, L.A., Fujise, Y., and Numachi, K. 1994. Differentiation of mitochondrial DNA between ordinary and dwarf forms of southern minke whale. *Reports of the International Whaling Commission*, **44**, 277-281.

Penny, D., and Hendy, M. (1986) Estimating the reliability of evolutionary trees. *Molecular Biology and Evolution*, **3**, 403-417.

Perrin, W.F. (1975) Variation of spotted and spinner porpoise (genus *Stenella*) in the Eastern Pacific and Hawaii. *Bulletin of the Scripps Institution of Oceanography*, **21**, 1-206.

Perrin, W.F. (1984) Patterns of geographical variation in small cetaceans. *Acta Zoologica Fennica*, **172**, 137-140.

Perrin, W.F., Mitchell, E.D., Mead, J.G., Caldwell, D.K., and van Bree, P.J.H. (1981) *Stenella clymene*, a rediscovered tropical dolphin of the Atlantic. *Journal of Mammology*, **62**, 583-598.

Perrin, W.F., Mitchell, E.D., Mead, J.G., Caldwell, D.K., Caldwell, M.C., van Bree, P.J.H., and Dawbin, W.H. (1987) Revision of the spotted dolphins, *Stenella* spp. *Marine Mammal Science*, **3**, 99-170.

Pichler, F.B., Dawson, S.M., Slooten, E., and Baker, C.S. (1998) Geographic isolation of Hector's dolphin populations described by mitochondrial DNA sequences. *Conservation Biology*, **12**, 676-682.

Pitman, R.L., Aguayo L.A., Urbán R.J. (1987) Observations of an unidentified beaked whale (*Mesoplodon* sp.) in the eastern tropical Pacific. *Marine Mammal Science*, **3**, 345-352.

Pitman, R.L., Palacios, D.M., Brennen, P.L.R., Brennen, B.J., Balcomb, K.C., and Miyashita, T. (1999) Sightings and possible identity of a bottlenose whale in the tropical Indo-Pacific: *Indopacetus pacificus*? *Marine Mammal Science*, **15**, 531-549.

Prothero, D. Manning, E. and Fischer, M. (1988) The phylogeny of the ungulates. In: *The Phylogeny and Classification of the Tetrapods, Volume 2: Mammals* (Ed. by M. Benton), pp. 201-234. Clarendon Press, Oxford.

Reyes, J. C. (1996) A possible case of hybridism in wild dolphins. *Marine Mammal Science*, **12**, 301-307.

Reyes, J.C., Mead, J.M., and Van Waerebeek, K. (1991) A new Species of Beaked Whale *Mesoplodon peruvianus* sp. n. (Cetacea: Ziphiidae) from Peru. *Marine Mammal Science*, **7**, 1-24.

Rice, D.W. (1998) Marine mammals of the world: systematics and distribution. *Society Marine Mammology Special Publication*, **4**, 231pp.

Rosel, P.E., Dizon, A.E., and Heyning, J.E. (1994) Genetic analysis of sympatric morphotypes of common dolphins (genus *Delphinus*). *Marine Biology*, **119**, 159-167.

Rosel, P.E., Haygood, M.G., and Perrin, W.F. (1995) Phylogenetic relationships among the true porpoises (Cetacea: Phocoenidae). *Molecular Phylogenetic and Evolution*, **4**, 463-474.

Rosel, P.E., Tiedemann, R., and Walton, M. (1999) Genetic evidence for restricted trans-Atlantic movements of the harbour porpoise, *Phocoena phocoena*. *Marine Biology*, **133**, 583-591.

Ross, G.J.B. (1977) The taxonomy of bottlenosed dolphins *Tursiops* species in South African waters, with notes on their biology. *Annals of the Cape Provincial Museum (Natural History)*, **11**, 135-194.

Ross, G.J.B., and Cockcroft, V.G. (1990) Comments on Australian bottlenose dolphins and the taxonomic status of *Tursiops aduncus* (Ehrenberg, 1832). In: *The Bottlenose Dolphin* (Ed. by S. Leatherwood and R.R. Reeves), pp. 101-128. Academic Press, San Diego, Ca.

Scharloo, W. (1987) Constraints in Selective Response. In: *Genetic Constraints on Adaptive Evolution* (Ed. by V. Loeschcke), Springler Verlag, Berlin.

Shimamura, M., Yasue, H., Ohshima, K., Abe, H., Kato, H., Kishiro, T., Goto, M., Munechika, I., and Okada, N. (1997) Molecular evidence from retroposons that whales form a clade within even-toed ungulates. *Nature* (London), **388**, 666-670.

Schlötterer, C. (1999) Culture and Genetic Evolution in Whales (2). *Science*, **284**, 2055a

Shoshani, J. (1986) Mammalian phylogeny: comparison of morphological and molecular results. *Molecular Biology and Evolution*, **3**, 222-242.

Siddal, M. (1995) Another monophyly index: Revisiting the jackknife. *Cladistics*, 11, 33-56.

Smith, T.D., Allen, J., Clapham, P. J., Hammond, P.S., Katona, S., Larsen, F., Lien, J., Mattila, D., Palsbøll, P.J., Sigurjónsson, J., Stevick, P.T., and Øien, N. (1999) An ocean-basin-wide mark-recapture study of the North Atlantic humpback whale (*Megaptera novaeangliae*). *Marine Mammal Science*, **15**, 1-32.

Smith-Goodwin, J.A. (1997) A Molecular Genetic Assessment of the Population Structure and Variation in Two Inshore Dolphin Genera on the East Coast of South Africa. Ph. D. dissertation, Rhodes University, Grahamstown, South Africa, 248 pp.

Spilliaert, R., Víkingsson, G., Árnason, Ú., Pálsdóttir, A., Sigurjónsson, J., and Árnason, A. (1991) Species hybridization between a female blue whale (*Balaenoptera musculus*) and a male fin whale (*B. physalus*): Molecular and morphological documentation. *Journal of Heredity*, **82**, 269-274.

Sturmbauer, C. and Meyer, A. (1992) Genetic divergence, speciation and morphological stasis in a lineage of African cichlid fishes. *Nature* (London), **358**, 578-581.

Tajima, F. (1983) Evolutionary relationship of DNA sequences in finite populations. *Genetics*, **105**, 437-460.

Takahata, N. (1989) Gene Genealogy in three related populations: consistency probability between gene and population trees. *Genetics*, **122**, 957-966.

Thewissen, J.G.M (1994) Phylogenetic aspects of cetacean origins: a morphological perspective. *Journal of Mammalian Evolution*, **2**, 157-183

Thewissen, J.G.M and Hussain, S.T. (1993) Origin of underwater hearing in whales. *Nature* (London), **361**, 444-445

Thoday, J.M. and Gibson, J.B. (1962) Isolation by disruptive selection. *Nature* (London), **193**, 1164-1166.

Tiedemann, R., Harder, J., Gmeiner, C., and Haase, E. (1996) Mitochondrial DNA sequence patterns of Harbour porpoises (*Phocoena phocoena*) from the North and the Baltic Sea. *Zeitschrift für Säugetierkunde*, **61**, 104-111.

Tiedemann, R., Hardy, O., Vekemans, X., and Milinkovitch, M.C. (2000) Higher impact of female than male migration on population structure in large mammals. *Molecular Ecology*, **9**, 1159-1163.

Tiedemann, R., and Milinkovitch, M.C. (1999) Culture and Genetic Evolution in Whales (3). *Science*, **284**, 2055a.

Vrana, P. and Wheeler, W. (1992) Individual Organisms as Terminal Entities: Laying the Species Problem to Rest. *Cladistics*, **8**, 67-72.

Wada, S., and Numachi, K. (1991) Allozyme analysis of genetic differentiation among the populations and species of the *Balaenoptera. Report of the International Whaling Commission* (special issue **13**), 125-154.

Wheeler, Q.D. and Nixon, K.C. (1990) Another way of looking at the species problem: a reply to de Querioz and Donoghue. *Cladistics*, **6**, 77-81.

Whitehead, H. (1998) Cultural selection and genetic diversity in matrilineal whales. *Science*, **282**, 1708-1711.

Yoshida, H., and Kato, H. (1999) Phylogenetic relationships of Bryde's whales in the western North Pacific and adjacent waters inferred from mitochondrial DNA sequences. *Marine Mammal Science*, **15**, 1269-1286.

D. Health, Parasites and Pathogens

The first studies of marine mammal parasites and pathogens were mostly faunistic, dating back to the Linnaean period. It was not until the polar expeditions carried out in the 19th and early 20th centuries by different European and North American nations, that there was any significant advance in this type of study (Delyamure, 1955). In the late 1950s and early 1960s, attempts to maintain marine mammals in captivity stimulated investigations on health, pathogens and other biological aspects. For instance, in the First Symposium on Cetacean Research held in 1963, scientists were challenged to establish a system of information exchange on medical care and husbandry of these animals (Ridgway, 1972). Research carried out in the 1960s and 1970s greatly contributed to our current knowledge on physiology, husbandry, and environmental conditions required by sea mammals (Dierauf, 1990; Howard, 1983). In the last two decades, the focus of marine mammal health has gradually shifted towards wild populations as a result of increased public concern for environment degradation and nature conservation. In addition, the increase in the number of reports of marine mammal die-offs since the 1980s and the urge to explain these phenomena have given a decisive stimulus to the study of health, parasites and pathogens of marine mammals. Two important publications that aided a standardised approach to marine mammal post mortem analysis were those of Kuiken and Hartmann (1991) arising from a special workshop organised by the European Cetacean Society, and a more detailed field guide for strandings by Geraci and Lounsbury (1993) based upon the American experience. However, much more work is needed. For instance, not a single virus had been isolated from any marine mammal prior to 1968, and information on viral diseases was scanty until the 1980s (Lauckner, 1985a). In fact, research on microparasites in general can be considered as a fairly new addition to the field of marine mammal health.

The subtle equilibrium between physiological adaptations and environmental factors in terms of marine mammal health is analysed in Chapter 10 by Joseph R. Geraci and Valerie Lounsbury. The sea represents a harsh environment for a warm-blooded mammal. Heat loss is much faster in water than in air; food is often found only at considerable depths; and dehydration is favoured by the higher osmotic concentration of seawater. Marine mammals have developed several adaptations to cope with these physiological challenges. However, if any of these adaptations fails, the entire equilibrium is disturbed and, eventually, diseases may arise. Such failures can be triggered by natural factors (*e.g.* food shortage and pathogens) or by human activities (*e.g.* pollution and habitat degradation). Obviously, this distinction is academic, as it is often difficult to disentangle the origin of diseases in marine mammal populations. For instance, anthropogenic changes such as global warming may affect prey distribution leading to local shortages and starvation. The accumulation of pollutants in the body tissues may reduce immunity against viruses, bacteria or metazoan parasites and predispose individuals to infectious diseases. Joseph Geraci and Valerie Lounsbury argue that mass mortality events are occurring with greater frequency, probably as a consequence of degradation of coastal environments and climatic changes.

Parasites, including microparasites (viruses, bacteria and protozoa) and macroparasites (typically helminths and arthropods), certainly represent an important source of mortality and unfulfilled fecundity in several animal populations. Microparasites are typically small and possess the ability to multiply directly and rapidly within the host. Macroparasites are much larger and, generally, do not multiply directly within their hosts (Anderson, 1993). Conservation biologists are increasingly aware of the potential role of parasites in the dynamics of animal populations. In aquatic mammals, recent lethal viral epizootics have raised concern about the possibility of transmission to endangered populations, such as those of the Mediterranean monk seal (*Monachus monachus*). Parasites can undoubtedly cause severe diseases in marine mammals (Dailey, 1985; Lauckner, 1985a,b,c). Helminths are considered as one of the primary causes of single and mass strandings of toothed whales. Lungworm infections are thought to be a common cause of mortality in diving mammals. Viruses have caused epizootics killing thousands of pinnipeds and odontocetes (Kennedy, 1998). The impact of parasitism on the population dynamics of marine mammals is still poorly known because quantitative data are missing (Geraci and St. Aubin, 1987; Raga et al., 1997; Van Bressem *et al.*, 1999). However, the combination of empirical data with mathematical modelling has provided some valuable and interesting insights, as highlighted in Chapter 11 by F. Javier Aznar, Juan A. Balbuena, Mercedes Fernandez and Juan Antonio Raga.

Besides their detrimental effect on the health of their host, parasites also deserve attention because they represent an important segment of biological diversity (Raga *et al.*, in press). Practically all vertebrates harbour several parasite species, and the majority of living organisms and over 50% of all species are parasites. Parasites live in close association with their hosts which makes them valuable markers of their host's ecology and evolution. Despite potential sampling biases, parasites have proven useful to disclose relevant information for the conservation and management of marine mammals populations, particularly regarding stock identity, social structure and host movements (Balbuena *et al.*, 1995). Besides, parasites can be used as elegant markers of evolutionary events and phylogenetic relationships of marine mammal taxa. These aspects are also covered in Chapter 11.

Marine mammal die-offs have attracted widespread public and scientific attention in recent years (Geraci, 1999). These mortalities have often been interpreted as a premonitory sign of the poor health of our oceans and have raised questions about how natural and anthropogenic environmental factors may influence them as well as about their likelihood to drive populations to extinction, to be predicted or avoided. The primary causes of several recent marine mammal die-offs are mostly well understood. However, it is difficult to establish an integrated explanatory framework for these events because the intervention and relative contribution of additional, underlying, factors is obscure. In particular, the role of pollution as contributing to, or enhancing the severity of, these outbreaks remains unresolved and is still the subject of much controversy (Ross *et al.*, 1996; Simmonds and Mayer, 1997). Whatever the cause, mass mortalities might be the most important factor determining long-term population size in the absence of human exploitation (Harwood and Hall, 1990). In Chapter 12, Mariano Domingo, Seamus Kennedy and Marie Françoise Van Bressem provide valuable insights into these issues. They analyse the etiology of some recent mass mortalities, focusing on morbillivirus lethal epizootics. Besides describing case studies and pathological findings, the authors also discuss epidemiological features that can explain the emergence and transmission of virus infections.

REFERENCES

Anderson, R.M. (1993) Epidemiology. In: *Modern Parasitology* (Ed. by F.E.G. Cox), pp. 75-116. Blackwell, Oxford.

Balbuena, J.A., Aznar, F.J., Fernandez, M. and Raga, J.A. (1995) The use of parasites as indicators of social structure and stock identity of marine mammals. In: *Whales, Seals, Fish and Man* (Ed. by A.S. Blix, L. Walløe and Ø. Ulltang), pp. 133-139. Elsevier Science, Amsterdam. 720pp.

Dailey, M.D. (1985) Diseases of Mammalia: Cetacea. In: *Diseases of Marine Animals. Vol. IV, Part 2* (Ed. by O. Kinne), pp. 805-847. Biologische Anstalt Helgoland, Hamburg.

Delyamure, S.L. (1955) *Helminth fauna of marine mammals (ecology and phylogeny)*, Akademiya Nauk SSSR, Moscow. (Translated by Israel Program for Scientific Translation, Jerusalem, 1968).

Dierauf, L.A. (1990) *Handbook of Marine Mammal Medicine: Health, Diseases, and Rehabilitation.* CRC Press, Boca Raton, 735 pp.

Geraci, J.R. and Lounsbury, V.J. (1993) *Marine Mammals Ashore. A Field Guide for Strandings.* Texas A&M Sea Grant Publication, Galveston, Texas, USA. 305pp.

Geraci, J.R. (1999) Marine Mammal Die-Offs: causes, investigations, and issues. In: *Conservation and Management of Marine Mammals* (Ed. by J.R. Twiss J.R. and R.R. Reeves), pp. 367-395. Smithsonian Institution Press, Washington.

Geraci, J. R., and St. Aubin, D. J. (1987) Effects of parasites on marine mammals. *International Journal for Parasitology* **17**, 407-414.

Harwood, J. and Hall, A. (1990) Mass mortality in marine mammals: its implications for population dynamics and genetics. *Trends in Ecology and Evolution*, 5, 254-257.

Howard, E.B. (Ed) (1983) *Pathobiology of Marine Mammal Diseases*. Vol. I and II. CRC Press, Boca Raton.

Kennedy, S. (1998) Morbillivirus infections in aquatic mammals. *Journal of Comparative Pathology*, **119**, 201-225.

Kuiken, T. and Hartmann, M.G. (1991) *Cetacean Pathology: Dissection Techniques and Tissue Sampling*. ECS Newsletter No. **17**, Special Issue, 1-43.

Lauckner, G. (1985a) Diseases of Mammalia: Carnivora. In: *Diseases of marine animals, Vol. IV, Part 2* (Ed. by O. Kinne), pp. 645-682, Biologische Anstalt Helgoland, Hamburg.

Lauckner, G. (1985b) Diseases of Mammalia: Pinnipedia.. In: *Diseases of marine animals, Vol. IV, Part 2* (Ed. by O. Kinne), pp. 683-793, Biologische Anstalt Helgoland, Hamburg.

Lauckner, G. (1985c) Diseases of Mammalia: Sirenia. In: *Diseases of marine animals, Vol. IV, Part 2* (Ed. by O. Kinne), pp. 795-803. Biologische Anstalt Helgoland, Hamburg.

Raga, J.A., Aznar, F.J., Fernandez, M., and Balbuena, J.A. (in press) Parasites. In: *Encyclopedia of Marine Mammals* (Ed. by W.F. Perrin, B. Würsig & H.G.M. Thewissen.). Academic Press, San Diego.

Raga, J. A., Balbuena, J. A., Aznar, F. J., and Fernandez, M. (1997) The impact of parasites on marine mammals: a review. *Parassitologia,* 39: 293-296.

Ridgway, S.H. (Ed.) (1972) *Mammals of the Sea. Biology and Medicine*. Charles C. Thomas Publisher, Springfield.

Ross, P., De Swart, R., Addison, R., Van Loveren, H., Vos, J., and Osterhaus A.D.M.E. (1996) Contaminant-induced immunotoxicity in harbour seals: wildlife at risk? *Toxicology*, **112**,157-169.

Simmonds, M.P. and Mayer, S.J. (1997) An evaluation of environmental and other factors in some recent marine mammal mortalities in Europe: implications for conservation and management. *Environmental Review*, **5**, 89-98.

Van Bressem, M.F., Van Waerebeek, K., and Raga, J.A. (1999) A review of virus infections of cetaceans and the potential impact of morbilliviruses, poxviruses and papillomaviruses on host population dynamics. *Diseases of Aquatic Organisms*, **38**, 53-65.

Chapter 10

Marine Mammal Health: Holding the Balance in an Ever-changing Sea

[1,2]JOSEPH R. GERACI and [1]VALERIE J. LOUNSBURY
*[1]National Aquarium in Baltimore, 501 East Pratt Street, Baltimore, MD 21202-3194, USA;
[2]University of Maryland School of Medicine, Comparative Medicine Program, Baltimore,
MD 21201-1192, USA. E-mail (JRG): jgeraci@aqua.org and (VJL): vlounsbury@aqua.org*

1.   INTRODUCTION

Whales, dolphins, seals, and sea otters live in an environment that, for a mammal, may be the harshest on earth - too much salt, no oxygen to breathe, cold that drains body heat - and where food is hard to find, difficult to catch, and never guaranteed. Over millions of years, marine mammals have evolved adaptive mechanisms that allow them to survive - even thrive - under these conditions. Animals that come ashore weak or dying often reveal evidence of failed adaptations - such as electrolyte imbalance and emaciation - that can obscure or complicate other factors that may be at work. Here, we will review briefly a few of the physiological and anatomical specialisations essential for mammals in the marine environment. Then, using a young harbour seal and a sea otter as examples, we will show what happens to an animal as these adaptations gradually break down, and how disease often appears towards the end of that process. We conclude with a brief discussion of general concepts in marine mammal health, trends in the health of populations, and implications for the future.

Marine Mammals: Biology and Conservation, edited by
Evans and Raga, Kluwer Academic/Plenum Publishers, 2002

2.   ADAPTATIONS

Health is a state of being well. To be healthy, a marine mammal must be unimpaired by illness and fit enough to control body temperature, pursue prey (often at high speeds or at great depths), migrate between sometimes distant feeding areas and breeding grounds, and store reserves sufficient to withstand the demands of reproduction and seasonal fasts.

Marine mammals have generally evolved mechanisms to conserve body heat. Of these, blubber has arguably been the key to evolutionary success. This coat of fat provides cetaceans and certain pinnipeds with mechanical protection (a bumper of sorts), warmth, buoyancy, nutrients when food is scarce, and - like a camel's hump - fresh water in reserve. Otariids have thinner blubber and less body fat than phocids or the walrus and are thus less tolerant of cold, and depend to a certain extent on their pelage for insulation. This is especially true for otariid pups: young northern fur seals (*Callorhinus ursinus*), for example, do not acquire an adult coat or adequate fat until they are about three months old (Gentry *et al.*, 1986) and, in the meantime, are prone to hypothermia when they become wet. The sea otter (*Enhydra lutris*), with little blubber, relies entirely on a high metabolic rate to generate heat and dense, well-groomed fur to prevent its loss (Morrison *et al.*, 1974). Sirenians have low metabolic rates and little ability to control surface heat loss - traits that suit slow-moving animals in warm habitats but effectively restrict their range to tropical and subtropical waters (Gallivan *et al.*, 1983; Irvine, 1983).

Cetaceans and pinnipeds forage at all depths. A northern elephant seal (*Mirounga angustirostris*) may spend 90% of its time foraging at 500 to 700 m (Le Boeuf *et al.*, 1989; Stewart and DeLong, 1991), and a sperm whale (*Physeter macrocephalus*) may remain submerged for as long as two hours (Watkins *et al.*, 1985). While these may be extreme examples, many species have evolved remarkable adaptations for coping with pressure and potentially deadly nitrogen (Kanwisher and Ridgway, 1983; Kooyman, 1989). These include a flexible rib cage, airways with cartilaginous supports, and modifications of the venous system that provide mechanisms for counterbalancing pressure on the thorax, ears, and nasal sinuses. They have evolved lungs that rapidly extract oxygen, which is circulated in a large volume of haemoglobin-rich blood and channelled to organs that need it most, such as the heart and brain. One-third of the body's oxygen may be stored in muscle myoglobin, a haemoglobin-like protein that releases its oxygen during prolonged dives (Lenfant *et al.*, 1970). Deep dives may require a shift to anaerobic metabolism; such dives may be useful for escape and exploration but are costly in terms of time and energy (Kooyman *et al.*, 1980; Gentry *et al.*, 1986). For most animals, survival depends on obtaining sufficient prey within the depth and time limits imposed by aerobic diving capacity, which in turn depends on the species and the size, age, and health of the individual. Large animals, because of their relatively greater capacity to store oxygen, tend to be better divers. It is not surprising that juveniles may find it difficult to reach prey that is easily accessible to adults.

Marine mammals also have evolved features that protect them from the high osmotic concentration of the sea (Ridgway 1972; Ortiz *et al.*, 1978; Gentry, 1981; Whittow, 1987): 1) their external surfaces are impermeable to seawater; 2) their body water is highly conserved - sweat glands are either reduced or absent, and the kidneys efficiently concentrate urine; 3) they drink very little seawater and acquire most of their fresh water from food (water makes up about 70% of a fish, 80% of a squid, and over 90% of aquatic plants), and each gram of fat in the diet or drawn from blubber yields close to its weight in fresh water when it is metabolised. Unlike other mammals, pinnipeds and cetaceans under stress (those that have been studied) secrete large amounts of aldosterone (St. Aubin and Geraci, 1986; Thomson and Geraci, 1986), a hormone produced by the adrenal cortex that promotes the resorption of sodium from the kidney, thereby drawing water back into the body. Why? One explanation is that by doing so, an animal in trouble or without food can recycle its own water and avoid drinking seawater, which has dangerously high concentrations of salts. Maintaining electrolyte balance thus depends on adequate blubber, well-functioning kidneys, hormonal balance, and a healthy, intact epidermis.

3. WHEN ADAPTATIONS FAIL

These adaptations work together, such that when any one fails, the entire equilibrium is disturbed (Fig. 1). Imagine, for example, a female harbour seal pup (*Phoca vitulina*) in mid-winter that is unable to get enough food - either because she was weak when weaned, has lungworms and cannot dive deeply enough to forage, or simply because she finds herself in an area where food is not abundant. For a time, the seal will be able to make up any nutrient deficit by using blubber reserves for both energy and water. Eventually, though, if conditions do not change, her reserves will run out and the animal will die, presumably of starvation. It seems a simple enough scenario.

Now, let us review the circumstances again, this time taking into account the adaptations that make this animal an effective cold-water predator. Add to this what is known about natural history, physiology, and disease, and we will see that the story is quite complex, has numerous sub-plots, and a less certain ending. First, this young female is not the only pup on the rookery

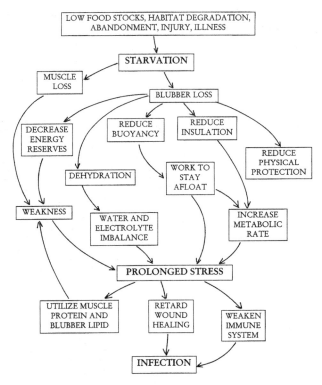

Figure 1.   For a marine mammal, starvation initiates a cascade of events that may ultimately lead to infection.

facing such problems. More than 25% of the harbour seals born in this population died before weaning and another 25% will not reach sexual maturity (Boulva and McLaren, 1979; Brodie and Beck, 1983). Some were abandoned prematurely and died quickly of starvation; others were taken by sharks and other predators, or acquired fatal bacterial infections after dragging their fresh umbilical wounds over contaminated sand and rocks. Many pups had insufficient blubber to begin life at sea and divided their time between foraging close to shore and hauling out on the rookery. A few ingested whatever was available - gravel and stones - and consequently died of an impacted stomach. Others fed in shallow waters and there consumed fish that had a higher than average probability of serving as intermediate hosts for certain potentially harmful parasites. Hauling out for extended periods created another problem, because prolonged time out of water creates favourable conditions for seal lice (*Echinophthirius horridus*) to proliferate on the animals' skin (Figs 2 and 3). Lice feed on blood and while doing so consume larvae of seal heartworm, *Dipetalonema spirocauda*.

Within the lice, the larvae transform into infective forms, which are then inoculated into the same or other seals (Geraci *et al.*, 1981). Thus, the seals resting out of water to conserve energy may have developed more serious heartworm infections (Fig. 4) - infections that would impair circulation and further force them into shallow waters where the cycle would inevitably continue.

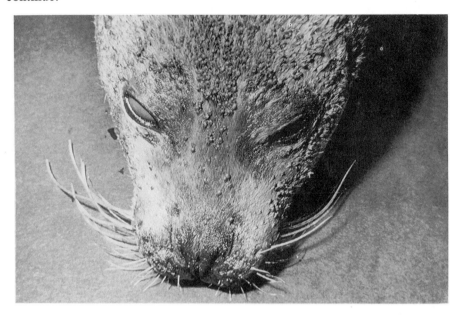

Figure 2.   Heavy infestations of seal lice are possible only on seals that spend prolonged periods out of the water. (Photo: J.R. Geraci)

As the weak seals became stressed, their cortisol levels increased, upsetting thyroid hormone balance; this in turn led to an incomplete moult that left their hair coats dull and patchy (Geraci *et al.*, 1979). This was just an outward sign of more serious illness. The high cortisol levels also led to weakening of connective tissue and muscle, and slower wound healing, and increased the likelihood of infection by the numerous bacteria and fungi that are normally present in the environment (Geraci and Medway, 1973; Thomson and Geraci, 1986; St. Aubin and Geraci, 1989). Stress also increased the demand on the adrenal cortex to produce aldosterone as a way to offset the diminishing supply of dietary water. In some of the seals, this led to exhaustion of the aldosterone-producing zone of the cortex, consequent loss of salt and water balance, and death (St. Aubin and Geraci, 1986). A few ailing seals lingered for weeks because late summer temperatures were still favourable and they could retain their body heat despite the substantial loss of blubber. In colder climates, such seals would have died sooner and with disease conditions that would have been less advanced. (For that reason, the most complete expression of illness can be found in animals in the warmest parts of the range - in those that can survive until they have exhausted their last stores of blubber fat).

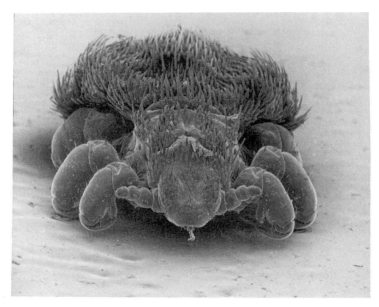

Figure 3.   The seal louse (*E. horridus*) may infect its harbour seal host with heartworm (*Dipetalonema spirocauda*). (Photo: B. Hicks)

The fate of the young female seal in this model depends on the reason it was unable to obtain food in the first place, and if and when it will be able to break away from this destructive cycle of events. Even then, as a coastal-dwelling species, the seal's future is intimately tied to the health of the nearshore environment, one that is influenced by commercial and recreational interests, industrial activities, and residential development. Contaminants consumed through prey will eventually accumulate in her tissues and will raise concerns for the animal's health and fertility (O'Shea, 1999). If she survives to reproduce, she will pass organochlorines on to her pup, a burden it could well do without at the outset of its first difficult year.

The story of an Alaskan sea otter, whose thermoregulatory mechanisms fail, would be much shorter. Imagine, for example, a young female, newly independent in late winter, inexperienced and unable to dive as deeply as the larger adults. Dwelling in an area depleted of optimum prey, the otter could be forced to spend more than 70% of her time foraging to fuel her

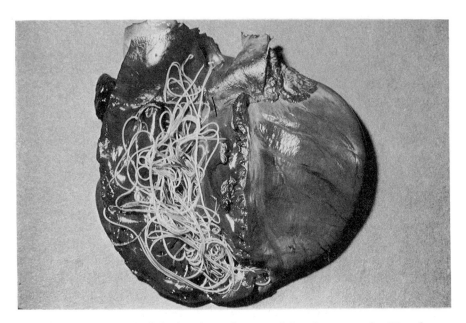

Figure 4.   Heartworm infection in seals, caused by the nematode *Dipetalonema spirocauda*, may cause pulmonary and cardiovascular complications that would inhibit the animals' ability to dive and feed. (Photo: J.R. Geraci)

high energy demands (Garshelis *et al.*, 1986). Like others in the same predicament, she might conserve some heat by going ashore, perhaps trading the risk of metabolic stress for a quicker death from terrestrial predators. By summer, if fortunate, she could be one of the 40-50% of her first-year age class to survive (Monnett and Rotterman, 1988). Imagine then that one day she surfaced briefly in a patch of fresh oil, fouling just a fraction of her body. Experience tells us that she would groom frantically and thereby spread the oil, scratch her already irritated eyes, and swallow harmful petroleum compounds (Costa and Kooyman, 1982; Osborn and Williams, 1990). Her coat would lose its insulative properties. In less than two days, the otter would probably die from the combined effects of stress, shock, and gastrointestinal damage.

4.   BASIC CONCEPTS

Constructing stories of this kind, hypothetical as they are, is easier to do for pinnipeds than for any other marine mammal group because they normally spend much of their time on land where they are easily observed, many come ashore stranded, and their biology and life history have been more intensively studied. Less is known about manatees and sea otters, and even less about cetaceans. Still, enough information exists to develop some basic concepts that apply to all groups.

- Mortality in most species is highest in the very young and the very old (Ralls *et al.*, 1980) - in the former because they are dependent, inexperienced, may be easier prey, and have yet to acquire an effective immune system; in the latter, arguably because of the cumulative effects of physiological, behavioural, and environmental stressors, past episodes of illness, and the gradual loss of immune function.
- Marine mammals host numerous macroparasites (Geraci and St. Aubin, 1987; Aznar *et al.*, this volume). Some, such as ectoparasites and gastrointestinal helminths, generally are not debilitating for healthy animals. Others are harmful enough to affect the well-being of individuals and even segments of a population. These include heartworms, some lungworms, and the hookworm *Uncinaria lucasi* in pinnipeds and, in cetaceans, the nematodes *Crassicauda* sp. (Fig. 5) in the mammary glands, cranial sinuses, and kidneys; the trematodes *Nasitrema* sp. in the cranial sinuses; and *Campula* sp. in the liver and pancreas. Certain ectoparasites provide visible clues to an animal's health. A heavy infestation of seal lice, which are transmitted and proliferate only when the seal is on land, requires a long time ashore - one sign that the animal may be ill. A fast-swimming odontocete offers barnacles little opportunity to attach, so the presence of species such as *Lepas* or *Xenobalanus* on the flukes, dorsal fin, or commissure of a dolphin's mouth suggests that the animal has been moving unusually slowly, a common sign of illness.
- Microparasites flourish in marine mammal habitats, but most are not inherently pathogenic and pose little or no threat to a healthy animal (Aznar *et al.*, this volume). Certain bacteria, such as *Leptospira* sp. (Smith *et al.*, 1974; Dierauf *et al.*, 1985) and *Mycobacteria* sp. (Forshaw and Phelps, 1991; Cousins *et al.*, 1993), are primary pathogens. Other bacteria, viruses, and fungi take hold in animals that have been weakened previously by injury, stress, starvation, or illness (Baker and Baker, 1988; Baker and McCann, 1989; Banish and Gilmartin, 1992). Such opportunistic infections may overwhelm the host and obscure the underlying cause of death. The nature and severity of infections can be influenced by the animals' age and behaviour, time of year, and environmental conditions.
- The adaptations that serve marine mammals so well in the water render them vulnerable to injury and physiological stress on land. A stranded animal's chances for survival diminish by the hour (Geraci and

Lounsbury, 1993). Sea otters and pinnipeds face the threat of hyperthermia, injury from terrestrial predators, and starvation. A cetacean has difficulty shedding heat even in cold weather, and a larger one may develop respiratory fatigue and distress as the chest cavity is compressed under its own weight. Within a few hours of stranding, some cetaceans begin to show evidence of shock or vascular collapse, which leads to poor circulation and impaired function of organs such as the skin, muscles, and liver. Compounds normally metabolised by the liver, including metabolites and hormones, can accumulate to dangerous levels. Whatever the reason for the initial stranding, the onset of shock further impairs the animal's health and may prevent its recovery, even if it is returned to sea in what appears to be good condition.

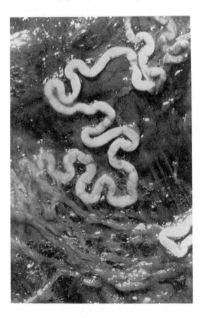

Figure 5.   *Crassicauda* sp. (nematode) larvae dwelling beneath the blubber of a cetacean will find their way to a location where the female can discharge eggs into the environment. These parasites damage bone in the cranial sinuses, destroy milk-secreting tissue of the mammary glands, and injure veins of the kidney. (Photo: D. Halliday/J.R. Geraci)

5.   TRENDS IN MARINE MAMMAL HEALTH

Over the past twenty-five years, we have gained tremendous knowledge about marine mammals, their physiology, and factors that affect their health. At the same time, we recognise that environmental changes resulting from human activities such as habitat disturbance, pollution, and over-fishing exert their own influence on individuals and populations. Determining the extent of these effects - direct or indirect - requires thoughtful analysis and serious investigation.

Judging from the past two decades alone, however, we can say with some certainty that unusual mortality events are occurring with greater frequency. Large-scale mortalities in the 1980s and 1990s have been associated with three primary factors: viruses, toxic algal blooms, and climatic anomalies (Geraci *et al.*, 1999; Domingo *et al.*, this volume). To a large extent, such die-offs have provided the impetus and the means to advance our understanding of the health of marine mammals and their environments.

Twenty years ago, only a few viruses were known to infect marine mammals, and none were known to cause large-scale mortality. Since 1980, the list has grown rapidly. Outbreaks of influenza continue to occur in harbour seals along the North Atlantic coast of the United States (Callan *et al.*, 1995), where the first marine mammal die-off demonstrated to be caused by a virus, avian influenza, occurred in the winter of 1979/80 (Geraci *et al.*, 1982). The likely source of these outbreaks is waterfowl, which serve as a reservoir worldwide for influenza viruses (Webster *et al.*, 1992). More devastating were the outbreaks of morbillivirus infection in European waters between 1987 and 1992, which killed thousands of Baikal seals (*Phoca sibirica*), nearly 17,000 harbour seals, and over 2,500 striped dolphins (*Stenella coeruleoalba*), and affected cetaceans and pinnipeds on a smaller scale in the Northwestern Atlantic (Hall, 1995; Geraci *et al.*, 1999). Morbillivirus infection is now known to be common in many species or populations, often with little or no evidence of serious illness (Bengtson *et al.*, 1991; Duignan *et al.*, 1995a,b,c,d, 1996; Van Bressem *et al.*, 1998). These events demonstrate that marine mammals are, and will continue to be, threatened by the introduction of a virus or other novel pathogen into previously unexposed populations.

Certain species of marine phytoplankton produce toxins that concentrate through the food chain and kill marine mammals along the way. Fatal poisonings have occurred in animals representing all major groups (cetaceans, pinnipeds, sea otters, and manatees) (Geraci *et al.*, 1989; O'Shea *et al.*, 1991; Bossart *et al.*, 1998; Geraci *et al.*, 1999). Toxic algal blooms seem to be intensifying and their range extending, in part as a result of human activities (Hallegraeff, 1993; Anderson, 1994). Toxic species have been introduced in ballast water to many areas of the world where they were previously unknown. In some coastal waters, the increased frequency and severity of blooms, or occurrence of toxic blooms associated with previously non-toxic species, have been attributed to 'cultural eutrophication', *i.e.* nutrient enrichment. In other cases, blooms have been precipitated by unusual meteorological and oceanographic conditions, some of which may be associated with climate change. Considering the degradation of coastal environments and predictions of global warming, these trends - and thus incidents of marine mammal poisoning - are likely to continue, if not increase.

We have seen other dramatic effects of climatic anomalies since 1980. When changes in water temperatures and nutrient availability result in disappearance of prey vital to marine mammal populations, cetaceans and some pinnipeds may wander beyond their normal range in search of food. Isolated populations, or those constrained by environmental or behavioural factors, may have no recourse. For example, the El Niño Southern Oscillation Event of 1982/83 caused widespread starvation among Galapagos Island pinnipeds and had serious impacts on other Eastern Pacific pinniped populations (Trillmich *et al.*, 1991); unusual oceanographic conditions along southwestern Africa in 1993/94 ultimately led to massive starvation among Namibia's Cape fur seal (*Arctocephalus pusillus*) population (Geraci *et al.*, 1999). Such events generally have the greatest impacts on pup production and first-year survival (Trillmich *et al.*, 1991).

6.   OTHER HEALTH RISKS

Only rarely have oil or other toxic spills killed more than a few marine mammals (Geraci and St. Aubin, 1990). Fortunately, events such as the 1989 Exxon Valdez spill in Alaska (Loughlin, 1994) - which killed several hundred harbour seals and several thousand sea otters - are uncommon. Most species are at greater risk from other, more routine types of habitat degradation. Entanglement in, or ingestion of, marine debris (Fig. 6) can result in a quick death from drowning or a lengthy battle with debilitation, infection, and starvation (Fowler, 1987; Laist, 1987). Incidental take in fishing gear continues to have serious impacts on certain species or stocks, such as harbour porpoises (*Phocoena phocoena*) in the Northwestern Atlantic (Read and Gaskin, 1988; Marine Mammal Commission, 1998).

The threat posed by chemical pollution is more insidious. Marine mammals accumulate and store organochlorines and other contaminants ingested in their prey. The consequences are not clear, and the evidence is mixed. Based on knowledge of other species and limited studies of marine mammals, these effects - for animals in highly polluted waters - could include endocrine disruption, reproductive impairment, and increased susceptibility to disease (Helle, 1980; Safe, 1984; Reijnders, 1986; Bergman *et al.*, 1992; Lahvis *et al.*, 1995; Ross *et al.*, 1996; Hall, this volume).

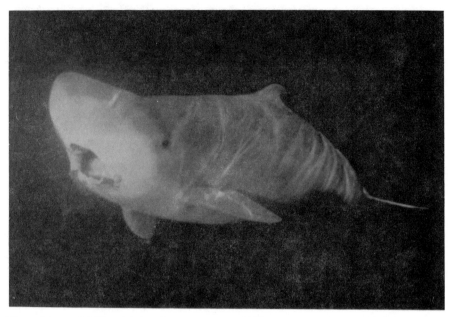

Figure 6.   Marine debris of various types injures or kills marine mammals through entanglement or ingestion. This pygmy sperm whale (*Kogia breviceps*), found stranded, was given a second chance after plastic trash was removed from its stomach. (Photo: the National Aquarium in Baltimore)

However, although exposure to anthropogenic substances was widely suspected as a contributing factor to many of the die-offs of the past decade, studies to date have failed to demonstrate a cause-and-effect relationship between contaminant burdens and mortality in any of these events (Hall *et al.*, 1992; Geraci *et al.*, 1999; O'Shea, 1999).

Marine mammals in coastal waters also face increased exposure to pathogens normally associated with humans or domestic animals. Some organisms, such as *Salmonella* sp., have been implicated in disease but also have been isolated from apparently healthy animals (Gilmartin *et al.*, 1979; Banish and Gilmartin, 1992; Baker *et al.*, 1995). Evidence of infection with bacteria representing an apparently new strain or species of *Brucella* has been found in many marine mammal species from the coasts of North America and the United Kingdom (Ewalt *et al.*, 1994; Ross *et al.*, 1994; Jepson *et al.*, 1997). In the Southern Hemisphere, mycobacteria of the complex associated with tuberculosis (*Mycobacterium bovis, M. tuberculosis*) appear to be endemic in some wild populations of sea lions and fur seals from Australia and South America (Cousins *et al.*, 1993; Woods *et al.*, 1995; Bernardelli *et al.*, 1996). Such infections, while perhaps not fatal, may cause chronic illness that leaves animals more susceptible to other pathogens, or impairs their ability to forage efficiently or their chances of producing healthy offspring.

Largely due to protective legislation, many marine mammal populations have grown during the past few decades. Concurrent with these increases have been advances in fishing technology and growing human demands on marine food resources. Thus, in some regions of the world we see more marine mammals competing for dwindling stocks of prey. Shortage of prey may result in nutritional stress that leaves large segments of a population more susceptible to disease and less able to cope with the demands of the marine environment. Some individuals or populations may switch to alternate prey and, in doing so, ingest parasites or biotoxins for which they have evolved no protection. In other cases, the new prey may be another species of marine mammal. The decline in sea otter populations in Alaska, for example, has been linked to an apparent increase in killer whale (*Orcinus orca*) predation, which in turn has been attributed to the decline in the numbers of Steller sea lions (*Eumetopias jubatus*) and harbour seals - preferred prey for some whale groups (Estes *et al.*, 1998; Hatfield *et al.*, 1998). [The factors causing the collapse of these western North Pacific pinniped populations have yet to be determined but probably include reduced abundance of prey due to intense regional fisheries operations (Estes *et al.*, 1998)]. In response to prey shortages, other marine mammals may wander far beyond their normal range and, in the process, expose themselves or populations along their routes to various health risks. The unusual harp seal (*Phoca groenlandica*) 'invasions' into Norwegian waters in the late 1980's, for example, coincided with depletion of fish stocks in the Barents Sea (Geraci *et al.*, 1999). These invasions had one, possibly two, notable results: an estimated 100,000 or more harp seals died in coastal fishing nets in 1987 alone; and, as suggested by studies subsequent to the 1988 European harbour seal die-off, the harp seals may have brought with them the morbillivirus responsible for that event (Markussen and Have, 1992).

7.   THE FUTURE

The health of a marine mammal depends on functional and behavioural adaptation to the marine environment. Beyond the basic mechanisms for diving and maintaining thermal and electrolyte balance, are many others that have evolved in response to millions of years of environmental pressures. For example, most species appear to have physiological mechanisms to cope with the high levels of metals and petroleum compounds that are common in the marine environment - levels that are often toxic to terrestrial species (Gaskin, 1982; Neff, 1990; O'Shea, 1999). Species subject to occasional catastrophic mortality, *e.g.* many pinnipeds, have reproductive strategies that allow rapid population recovery. Baleen whales and many pinnipeds have evolved complex annual cycles involving migration and prolonged periods of fasting - all dependent on ample blubber stores. Thus, although the marine environment is harsh for a mammal, the conditions and risks within that environment have been generally stable over the long term, allowing these kinds of adaptations to evolve over hundreds, thousands, even millions of generations.

What if the "ever-changing sea" is changing in ways that strain or exceed some populations' ability to adapt? Bioaccumulation of anthropogenic compounds, increased exposure to novel pathogens and biotoxins, prolonged stress associated with disturbance and habitat loss, or reduction in prey due to over-fishing are factors that can profoundly influence the health of individuals or populations. Add to this the almost unfathomable potential effects of climate change, and we can predict that unusual mortalities will continue, and that survival of some populations may be at risk (Harwood and Hall, 1990; Geraci *et al.*, 1999). We might also expect that as the factors causing ecosystem disruption become increasingly complex, determining the underlying causes of certain health conditions or die-offs may also become more difficult.

On a positive note, we are beginning to understand the complex effects of both natural phenomena and human-related activities on marine mammal health. Our ability to recognise clinical and pathological effects of disease agents and toxins is growing rapidly. Through advances in communications technology and tools such as GIS (Geographic Information Systems), we are on the threshold of developing effective international capability to investigate die-offs and other unusual events, and to utilise methods that will maximise the value of the samples and data obtained. We must continue to work toward this goal and to use the information gained to promote sound environmental policies worldwide. In the end, responsible stewardship of the oceans may be the key to the future health of marine mammals.

REFERENCES

Anderson, D.M. (1994) Red tides. *Scientific American*, **271**, 62-68.

Baker, J.R. and Baker, R. (1988) Effects of environment on grey seal (*Halichoerus grypus*) pup mortality. Studies on the Isle of May. *Journal of Zoology* (London), **216**, 529-537.

Baker, J.R. and McCann, T.S. (1989) Pathology and bacteriology of adult male Antarctic fur seals, *Arctocephalus gazella*, dying at Bird Island, South Georgia. *British Veterinary Journal*, **145**, 263-275.

Baker, J.R., Hall, A., Hiby, L., Munro, R., Robinson, I., Ross, H.M. and Watkins, J.F. (1995) Isolation of salmonellae from seals from UK waters. *Veterinary Record*, **136**, 471-472.

Banish, L.D. and Gilmartin, W.G. (1992) Pathological findings in the Hawaiian monk seal. *Journal of Wildlife Diseases*, **28**, 428-434.

Bengtson, J.L., Boveng, P., Franzén, U., Have, P., Heide-Jørgensen, M.P. and Härkönen, T.J. (1991) Antibodies to canine distemper virus in Antarctic seals. *Marine Mammal Science*, **7**, 85-87.

Bergman, A., Olsson, M. and Reiland, S. (1992) Skull-bone lesions in the Baltic grey seal (*Halichoerus grypus*). *Ambio*, **21**, 517-519.

Bernardelli, A., Bastida, R., Loureiro, J., Michelis, H., Romano, M.L., Cataldi, A. and Costa, E. (1996) Tuberculosis in sea lions and fur seals from the south-western Atlantic coast. *Revue Scientifique et Technique (France)*, **15**, 985-1005.

Bossart, G.D., Baden, D.G., Ewing, R.Y., Roberts, B. and Wright, S.D. (1998) Brevetoxicosis in manatees (*Trichechus manatus latirostris*) from the 1996 epizootic: gross, histologic and immunohistochemical features. *Toxicologic Pathology*, **26**, 276-282.

Boulva, J. and McLaren, I.A. (1979) Biology of the harbour seal, *Phoca vitulina*, in eastern Canada. *Bulletin of the Fisheries Research Board of Canada*, **200**, 1-24.

Brodie, P. and Beck, B. (1983) Predation by sharks on the grey seal (*Halichoerus grypus*) in eastern Canada. *Canadian Journal of Fisheries & Aquatic Sciences*, **40**, 267-271.

Callan, R.J., Early, G., Kida, H. and Hinshaw, V.S. (1995) The appearance of H3 influenza viruses in seals. *Journal of General Virology*, **76**, 199-203.

Costa, D.P. and Kooyman, G.L. (1982) Oxygen consumption, thermoregulation, and the effect of fur oiling and washing on the sea otter, *Enhydra lutris. Canadian Journal of Zoology*, **60**, 2761-2767.

Cousins, D.V., Williams, S.N., Reuter, R., Forshaw, D., Chadwick, D., Coughran, D., Collins, P. and Gales, N. (1993) Tuberculosis in wild seals and characterisation of the seal bacillus. *Australian Veterinary Journal*, **70**, 92-97.

Dierauf, L.A., Vandenbroek, D., Roletto, J., Koski, M., Amaya, L. and Gage, L. (1985) An epizootic of leptospirosis in California sea lions. *Journal of the American Veterinary Medical Association,* **187**, 1145-1148.

Duignan, P.J., House, C., Walsh, M.T., Campbell, T., Bossart, G.D., Duffy, N., Fernandes, P.J., Rima, B.K., Wright, S. and Geraci, J.R. (1995a) Morbillivirus infection in manatees. *Marine Mammal Science*, **11**, 441-451.

Duignan, P.J., House, C., Geraci, J.R., Duffy, N., Rima, B.K., Walsh, M.T., Early, G., St. Aubin, D.J., Sadove, S., Koopman, H. and Rinehart, H. (1995b) Morbillivirus infection in cetaceans of the western Atlantic. *Veterinary Microbiology*, **44**, 241-249.

Duignan, P.J., House, C., Geraci, J.R., Early, G., Copland, H., Walsh, M.T., Bossart, G.D., Cray, C., Sadove, S., St. Aubin, D.J. and Moore, M. (1995c) Morbillivirus infection in two species of pilot whales (*Globicephala* sp.) from the western Atlantic. *Marine Mammal Science*, **11**, 150-162.

Duignan, P.J., Saliki, J.T., St. Aubin, D.J., Early, G., Sadove, S., House, J.A., Kovacs, K. and Geraci, J.R. (1995d) Epizootiology of morbillivirus infection in North American harbor (*Phoca vitulina*) and grey seals (*Halichoerus grypus*). *Journal of Wildlife Diseases*, **31**, 491-501.

Duignan, P.J., House, C., Odell, D.K., Wells, R.S., Hansen, L.J., Walsh, M.T., St. Aubin, D.J., Rima, B.K. and Geraci, J.R. (1996) Morbillivirus infection in bottlenose dolphins: evidence for recurrent epizootics in the western Atlantic and Gulf of Mexico. *Marine Mammal Science,* **12**, 499-515.

Estes, J.A., Tinker, M.T., Williams, T.M. and Doak, D.F. (1998) Killer whale predation on sea otters linking oceanic and nearshore ecosystems. *Science*, **282**, 473-476.

Ewalt, D.R., Payeur, J.B., Martin, B.M., Cummins, D.R. and Miller, W.G. (1994) Characteristics of a *Brucella* species from a bottlenose dolphin (*Tursiops truncatus*). *Journal of Veterinary Diagnostic Investigation*, **6**, 448-452.

Forshaw, D. and Phelps, G.R. (1991) Tuberculosis in a captive colony of pinnipeds. *Journal of Wildlife Diseases*, **27**, 288-295.

Fowler, C.W. (1987) Marine debris and northern fur seals: a case study. *Marine Pollution Bulletin*, **18**, 326-335.

Gallivan, G.J., Best, R.C. and Kanwisher, J.W. (1983) Temperature regulation in the Amazonian manatee *Trichechus inunguis*. *Physiological Zoology*, **56**, 255-262.

Garshelis, D.L., Garshelis, J.A. and Kimker, A.T. (1986) Sea otter time budgets and prey relationships in Alaska. *Journal of Wildlife Management*, **50**, 637-647.

Gaskin, D.E. (1982) *The Ecology of Whales and Dolphins*. Heinemann Publishers, Exeter, NH.

Gentry, R.L. (1981) Seawater drinking in eared seals. *Journal of Comparative Biochemistry & Physiology*, **68A**, 81-86.

Gentry, R.L., Costa, D.P., Croxall, J.P., David, J.H.M., Davis, R.W., Kooyman, G.L., Majluf, P., McCann, T.S. and Trillmich, F. (1986) Synthesis and conclusions. In: *Fur Seals: Maternal Strategies on Land and at Sea*. (Ed. by R.L. Gentry and G.L. Kooyman), pp. 220-264. Princeton University Press, Princeton, NJ.

Geraci, J.R. and Lounsbury, V.J. (1993) *Marine Mammals Ashore: A Field Guide for Strandings*. Texas A&M University Sea Grant Publications, Galveston, TX. 305pp.

Geraci, J.R. and Medway, W. (1973) Simulated field blood studies in the bottle-nosed dolphin *Tursiops truncatus*. 2. Effects of stress on some hematologic and plasma chemical parameters. *Journal of Wildlife Diseases*, **9**, 29-33.

Geraci, J.R. and St. Aubin, D.J. (1987) Effects of parasites on marine mammals. *International Journal of Parasitology*, **17**, 407-414.

Geraci, J.R. and St. Aubin, D.J. (1990) Summary and conclusions. In: *Marine Mammals and Oil: Confronting the Risks* (Ed. by J.R. Geraci and D.J. St. Aubin), pp. 253-256. Academic Press, San Diego, CA.

Geraci, J.R., St. Aubin, D.J. and Smith, T.G. (1979) Influence of age, condition, sampling time and method on plasma chemical constituents in free-ranging ringed seals, *Phoca hispida*. *Journal of Fisheries Research Board of Canada*, **32**, 2559-2564.

Geraci, J.R., Fortin, J.F., St. Aubin, D.J. and Hicks, B.D. (1981) The seal louse, *Echinophthirius horridus*: an intermediate host of the seal heartworm, *Dipetalonema spirocauda* (Nematoda). *Canadian Journal of Zoology*, **59**, 1457-1459.

Geraci, J.R., St. Aubin, D.J., Barker, I.K., Webster, R.G., Hinshaw, V.S., Bean, W.J., Ruhnke, H.L., Prescott, J.H., Early, G., Baker, A.S., Madoff, S. and Schooley, R.T. (1982) Mass mortality of harbor seals: pneumonia associated with influenza A virus. *Science*, **215**, 1129-1131.

Geraci, J.R., Anderson, D.M., Timperi, R.J., St. Aubin, D.J., Early, G.A., Prescott, J.H. and Mayo, C.A. (1989) Humpback whales (*Megaptera novaeangliae*) fatally poisoned by dinoflagellate toxin. *Canadian Journal of Fisheries & Aquatic Sciences*, **46**, 1895-1898.

Geraci, J.R., Harwood, J. and Lounsbury, V.J. (1999) Marine mammal die-offs: causes, investigations, and issues. In: *Conservation and Management of Marine Mammals*, (Ed. by J.R. Twiss, Jr. and R.R. Reeves), pp. 367-395. Smithsonian Institution Press, Washington, D.C.

Gilmartin, W.G., Vainik, P.M. and Neill, V.M. (1979) Salmonellae in feral pinnipeds off the southern California coast. *Journal of Wildlife Diseases*, **15**, 511-514.

Hall, A.J. (1995) Morbilliviruses in marine mammals. *Trends in Microbiology*, **3**, 4-9.

Hall, A.J., Law, R.J., Wells, D.E., Harwood, J., Ross, H.M., Kennedy, S., Allchin, C.R., Campbell, L.A. and Pomeroy, P.P. (1992) Organochlorine levels in common seals (*Phoca vitulina*) which were victims and survivors of the 1988 phocine distemper epizootic. *Science of the Total Environment,* **115**, 145-162.

Hallegraeff, G.M. (1993) Phycological Reviews 13: A review of harmful algal blooms and their apparent global increase. *Phycological Reviews*, **32**, 79-99.

Harwood, J. and Hall, A.J. (1990) Mass mortality in marine mammals: its implications for population dynamics and genetics. *Trends in Ecology and Evolution,* **5**, 254-257.

Hatfield, B.B., Marks, D., Tinker, T.M., Nolan, K. and Peirce, J. (1998) Attacks on sea otters by killer whales. *Marine Mammal Science*, **14**, 888-894.

Helle, E. (1980) Lowered reproductive capacity in female ringed seals (*Pusa hispida*) in the Bothnian Bay, northern Baltic Sea, with special reference to uterine occlusions. *Annales Zoologici Fennici*, **17**, 147-158.

Irvine, A.B. (1983) Manatee metabolism and its influence on distribution in Florida. *Biological Conservation*, **25**, 315-334.

Jepson, P.D., Brew, S., MacMillan, A.P., Baker, J.R., Barnett, J., Kirkwood, J.K., Kuiken, T., Robinson, I.R. and Simpson, V.R (1997) Antibodies to *Brucella* in marine mammals around the coast of England and Wales. *Veterinary Record*, **141**, 513-515.

Kanwisher, J. and Ridgway, S.H. (1983) The physiological ecology of whales and porpoises. *Scientific American*, **248**, 110-120.

Kooyman, G.L. (1989) *Diverse Divers: Physiology and Behavior.* Zoophysiology, Vol. 23. Springer-Verlag, Berlin and Heidelberg.

Kooyman, G.L., Wahrenbrock, E.A., Castellini, M.A., Davis, R.W. and Sinnett, E.E. (1980) Aerobic and anaerobic metabolism during voluntary diving in Weddell seals: evidence of preferred pathways from blood chemistry and behavior. *Journal of Comparative Physiology*, **138**, 335-346.

Lahvis, G.P., Wells, R.S., Kuehl, D.W., Stewart, J.L., Rhinehart, H.L. and Via, C.S. (1995) Decreased lymphocyte responses in free-ranging bottlenose dolphins (*Tursiops truncatus*) are associated with increased concentrations of PCBs and DDT in peripheral blood. *Environmental Health Perspectives*, **103** (Suppl. 4), 67-72.

Laist, D.W. (1987) Overview of the biological effects of lost and discarded plastic debris in the marine environment. *Marine Pollution Bulletin*, **18**(6B), 319-326.

Le Boeuf, B.J., Naito, Y., Huntley, A.C. and Asaga, T. (1989) Prolonged, continuous, deep diving by northern elephant seals. *Canadian Journal of Zoology*, **67**, 2514-2519.

Lenfant, C., Johansen, K. and Torrance, J.D. (1970) Gas transport and oxygen storage capacity in some pinnipeds and the sea otter. *Respiratory Physiology*, **9**, 277-286.

Loughlin, T. R. (Ed.) (1994) *Marine Mammals and the Exxon Valdez.* Academic Press, Inc., San Diego and London.

Marine Mammal Commission. (1998) *Annual Report to Congress 1997.* Marine Mammal Commission, 4340 East-West Highway, Bethesda, MD 20814.

Markussen, N.H. and Have, P. (1992) Phocine distemper virus infection in harp seals, *Phoca groenlandica. Marine Mammal Science*, **8**, 19-26.

Monnett, C. and Rotterman, L. (1988) Sex-related patterns in the post-natal development and survival of sea otters in Prince William Sound, Alaska. In: *Population Status of California Sea Otters.* (Ed. by D.B. Siniff and K. Ralls), pp. 162-190. U.S. Department of Interior,

Minerals Management Service, Pacific OCS Region, Los Angeles, CA. OCS Study MMS 88-0021.

Morrison, R., Rosenmann, M. and Estes, J.A. (1974) Metabolism and thermoregulation in the sea otter. *Physiological Zoology*, **47**, 218-229.

Neff, J.M. (1990) Composition and fate of petroleum and spill-treating agents in the marine environment. In: *Marine Mammals and Oil: Confronting the Risks* (Ed. by J.R. Geraci and D.J. St. Aubin), pp. 1-33. Academic Press, San Diego, CA.

Ortiz, C.L., Costa, D. and Le Boeuf, B.J. (1978) Water and energy flux in elephant seal pups fasting under natural conditions. *Physiological Zoology*, **51**, 166-178.

Osborn, K. and Williams, T.M. (1990) Postmortem examination of sea otters. In: *Sea Otter Rehabilitation Program: 1989 Exxon Valdez Oil Spill* (Ed. by T.M. Williams and R.W. Davis), pp. 134-146. International Wildlife Research.

O'Shea, T.J. (1999) Environmental contaminants and marine mammals. In: *Biology of Marine Mammals* (Ed. by J.E. Reynolds III and S.A. Rommel.), pp. 485-563. Smithsonian Institution Press, Washington, D.C.

O'Shea, T.J., Rathbun, G.B., Bonde, R.K., Buergelt, C.D. and Odell, D.K. (1991) An epizootic of Florida manatees associated with a dinoflagellate bloom. *Marine Mammal Science*, 7, 165-179.

Ralls, K., Brownell, R.L., Jr. and Ballou, J. (1980) Differential mortality by sex and age in mammals, with specific reference to the sperm whale. *Report of the International Whaling Commission*, (special issue **2**), 233-243.

Read, A.J. and Gaskin, D.E. (1988) Incidental catch of harbor porpoises by gill nets. *Journal of Wildlife Management*, **52**, 517-523.

Reijnders, P.J.H. (1986) Reproductive failure in common seals feeding on fish from polluted coastal waters. *Nature* (London), **324**, 456-457.

Ridgway, S.H. (1972) Homeostasis in the aquatic environment. In: *Mammals of the Sea: Biology and Medicine* (Ed. by S.H. Ridgway), pp. 590-747. Charles C Thomas Publisher, Springfield, IL.

Ross, H.M., Foster, G., Reid, R.J., Jahans, K.L. and MacMillan, A.P. (1994) *Brucella* species infection in sea-mammals. *Veterinary Record*, **134**, 359.

Ross, P.S., De Swart, R.L., Timmerman, H.H., Reijnders, P.J.H., Vos, J.G., Van Loveren, H. and Osterhaus, A.D.M.E. (1996) Suppression of natural killer cell activity in harbour seals (*Phoca vitulina*) fed Baltic Sea herring. *Aquatic Toxicology*, **34**, 71-84.

Safe, S. (1984) Polychlorinated biphenyls (PCBs) and polybrominated biphenyls (PBBs): biochemistry, toxicology and mechanism of action. *CRC Critical Reviews in Toxicology*, **13**, 319-395.

St. Aubin, D.J. and Geraci, J.R. (1986) Adrenocortical function in pinniped hyponatremia. *Marine Mammal Science*, **2**, 243-250.

St. Aubin, D.J. and Geraci, J.R. (1989) Adaptive changes in hematologic and plasma chemical constituents in captive beluga whales, *Delphinapterus leucas. Canadian Journal of Fisheries & Aquatic Sciences*, **46**, 796-803.

Smith, A.W., Brown, R.J., Skilling, D.E. and DeLong, R. (1974) *Leptospira pomona* and reproductive failure in California sea lions. *Journal of the American Veterinary Medical Association*, **165**, 996-998.

Stewart, B.S. and DeLong, R.L. (1991) Diving patterns of northern elephant seal bulls. *Marine Mammal Science*, 7, 369-384.

Thomson, C.A. and Geraci, J.R. (1986) Cortisol, aldosterone, and leukocytes in the stress response of bottlenose dolphins, *Tursiops truncatus. Canadian Journal of Fisheries and Aquatic Sciences,* **43**, 1010-1016.

Trillmich, F., Ono, K.A., Costa, D.P., DeLong, R.L., Feldkamp, S.D., Francis, J.M., Gentry, R.L., Heath, C.B., Le Boeuf, B.J., Majluf, P. and York, A.E. (1991) The effects of El Niño on pinniped populations in the Eastern Pacific. In: *Pinnipeds and El Niño: Responses to Environmental Stress.* (Ed. by F. Trillmich and K.A. Ono), pp. 247-270. Springer-Verlag, Berlin.

Van Bressem, M.-F., Van Waerebeek, K., Fleming, M. and Barrett, T. (1998) Serological evidence of morbillivirus infection in small cetaceans from the Southeast Pacific. *Veterinary Microbiology*, **59**, 89-98.

Watkins, W.A., Moore, K.E. and Tyack, P. (1985) Sperm whale acoustic behaviors in the southeast Caribbean. *Cetology*, **49**, 1-15.

Webster, R.G., Bean, W.J., Gorman, O.T., Chambers, T.M. and Kawaoka, Y. (1992) Evolution and ecology of influenza A viruses. *Microbiological Review*, **56**, 152-179.

Whittow, G.C. (1987) Thermoregulatory adaptations in marine mammals: interacting effects of exercise and body mass. A review. *Marine Mammal Science*, **3**, 220-241.

Woods, R., Cousins, D.V., Kirkwood, R. and Obendorf, D.L. (1995) Tuberculosis in a wild Australian fur seal (*Arctocephalus pusillus doriferus*) from Tasmania. *Journal of Wildlife Diseases*, **31**, 83-86.

Chapter 11

Living Together: The Parasites of Marine Mammals

F. JAVIER AZNAR, JUAN A. BALBUENA, MERCEDES FERNÁNDEZ
and J. ANTONIO RAGA[*]
*Department of Animal Biology, Cavanilles Institute of Biodiversity and Evolutionary Biology,
University of Valencia, Dr. Moliner 50, 46100 Burjassot, Valencia, Spain
E-mail (JAR): Toni.Raga@uv.es*

1. INTRODUCTION

The reader may wonder why, within a book of biology and conservation of marine mammals, a chapter should be devoted to their parasites. There are four fundamental reasons. First, parasites represent a substantial but neglected facet of biodiversity that still has to be evaluated in detail (Windsor, 1995; Hoberg, 1997; Brooks and Hoberg, 2000. Perception of parasites among the public are negative and, thus, it may be hard for politicians to justify expenditure in conservation programmes of such organisms. However, many of the reasons advanced for conserving biodiversity or saving individual species also apply to parasites (Marcogliese and Price, 1997; Gompper and Williams, 1998). One fundamental point from this conservation perspective is that the evolutionary fate of parasites is linked to that of their hosts (Stork and Lyal, 1993). For instance, the eventual extinction of the highly endangered Mediterranean monk seal *Monachus monachus* would also result in that of its host-specific sucking louse *Lepidophthirus piriformis* (Fig. 1b). Second, parasites cause disease, which may have considerable impact on marine mammal populations (Harwood and Hall, 1990). Scientists have come to realise this particularly after the recent die-offs caused by morbilliviruses

[*] Authorship order is alphabetical and does not reflect unequal contribution.

(see Domingo *et al.*, this volume). However, these epizootic outbreaks represent the most dramatic, but by no means the only, example of parasite-induced mortality in marine mammal populations (see Section 3 below). Third, parasites are elegant markers of contemporary and historical ecological relationships, providing information on host ecology, biogeography, and phylogeny (Gardner and Campbell, 1992; Brooks and McLennan, 1993; Hoberg, 1996, 1997). From the perspective of conservation and management, parasites have proved especially useful as biological tags for host social structure, movements, and other ecological aspects of marine mammal populations (Balbuena *et al.*, 1995). Finally, some parasites of marine mammals have public health and

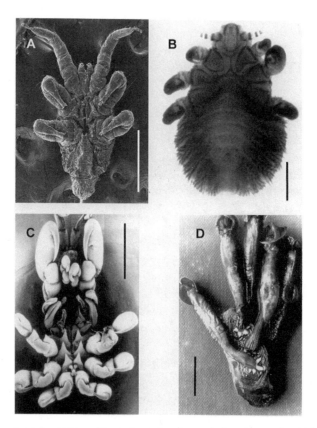

Figure 1.   Representative arthropod species from marine mammals: A) hexapod larva of *Orthohalarachne attenuata*, a mite from southern fur seal. B) *Lepidophthirus piriformis*, a sucking louse from Mediterranean monk seal. C) *Isocyamus delphini*, a whale louse from long-finned pilot whale. D) *Xenobalanus globicipitis*, a barnacle from striped dolphin. Scale bars: A and B, 0.5 mm; C, 3 mm; D, 6mm.

economic importance (Oshima and Kliks, 1987). For instance, the nematode *Anisakis simplex* uses fish and squid as intermediate hosts to infect cetaceans. Consumption of raw or lightly cooked seafood can result in human infections, causing a severe clinical condition known as anisakidosis. In addition, recent studies have shown that *A. simplex* in fish fillets can produce allergic reactions in humans (Fernández de Corres *et al.*, 1996; Armentia *et al.*, 1998). Besides, parasites of marine mammals cause important losses to the fish processing industry. Costs resulting from the removal of larvae of the sealworm *Pseudoterranova decipiens* from cod fillets alone were estimated to be in excess of CA$ 29 million in 1982 in Atlantic Canada (Burt, 1994).

In this chapter, we review different aspects of the biology of marine mammal parasites, organising the concepts under four main sections: (1) the parasite diversity; (2) the interaction between host and parasite populations; 3) the origin and evolution of the parasite faunas; and (4) the structure of parasite communities. There are obvious limitations of space and, therefore, we can only outline major ideas and provide selected examples that illustrate the conceptual framework. The chapter is not intended as a major review of the old literature and therefore readers keen to trace back the history of ideas to their very beginning, should look up the references provided herein.

2. PARASITE DIVERSITY

Biodiversity results from the complex interaction of phylogeny, ecology, geography, and history as determinants of the evolution and distribution of organisms (Hoberg, 1997; Brooks and Hoberg, 2000). In this section, we present the outcome of this interaction, *i.e.* a descriptive outline of patterns of association between marine mammals and their parasites. The processes leading to the observed patterns will be treated in some detail in Sections 4 and 5. It is important to distinguish between generalist organisms that can occur not only in marine mammals but in many other habitats (*e.g. Clostridium botulinum*, *Escherichia coli*) from those which have arisen specifically in these hosts. The latter are more meaningful from the biodiversity perspective.

We consider here the term 'parasite' in its widest sense (Price, 1980), and, for convenience, we divide this section in one part devoted to microparasites (*e.g.* viruses, bacteria, protozoans) and another, to macroparasites (commonly helminths and arthropods) (see Anderson, 1993, for a definition of these concepts). For completeness, other symbiotic associations, such as phoresis or commensalism are also included.

for a definition of these concepts). For completeness, other symbiotic associations, such as phoresis or commensalism are also included.

As a cautionary note, bear in mind that the differences in coverage of micro- and macroparasites of marine mammals reflect the degree to which both groups have been studied. Studies of metazoan parasites of sea mammals date back to Linnaean times, and gained momentum in the last 50 years, particularly thanks to the work of Soviet helminthologists (Delyamure, 1955; Delyamure and Skriabin, 1985). In contrast, research on microorganisms is a fairly recent addition to the field of marine mammal parasitology. For instance, not a single virus had been isolated from any marine mammal prior to 1968, and information on viral diseases was scanty until the 1980s (Lauckner, 1985a). In addition, the data presented here are influenced by the degree of sampling effort of the host groups, and, therefore, they do not necessarily reflect an even, complete representation of the richness and diversity of parasites in marine mammals.

2.1   Microparasites

Virological studies of marine mammals are currently in progress. Visser *et al.* (1991) reported 12 different viruses belonging to ten families, isolated from wild and captive pinnipeds; van Bressem *et al.* (1999) recorded nine virus families in some 25 cetacean species, and Duignan *et al.* (1995) recently found morbillivirus antibodies in sirenians (Table 1). We will focus on the genus *Morbillivirus* (family Paramyxoviridae) because this is one of the most studied, and several types show a specific relationship with marine mammals. The first morbillivirus of marine mammals was isolated in 1988 and was christened phocine distemper virus (PDV). The PDV caused mass mortality in the North Sea populations of harbour seals *Phoca vitulina* and, to a lesser extent, grey seals *Halichoerus grypus* (Domingo *et al.*, this volume). Other morbilliviruses have been discovered in other marine mammals, causing high mortality (Osterhaus *et al.*, 1998; Domingo *et al.*, this volume; van Bressem *et al.*, 1999). Two types are known in cetaceans: the porpoise morbillivirus (PMV) and the dolphin morbillivirus (DMV), which are considered as strains of the same species (Kennedy, 1998; van Bressem *et al.*, 1999; Domingo *et al.*, this volume).

Genomic and antigenic data suggest that the PDV is closely related to the canine distemper virus (CDV) of land carnivores, whereas the PMV-DMV seems more similar to morbilliviruses of ruminants (Barrett *et al.*, 1993; Visser, 1993; Kennedy, 1998). This should not be viewed as a result of parallel evolution (*i.e.* co-evolution, see Section 4) between the hosts and their viruses. Given the differences in the rate of evolutionary diversification between mammals and viruses, it seems more likely that common viral ancestors were able to infect phylogenetically related taxa because these are biologically similar (Visser, 1993).

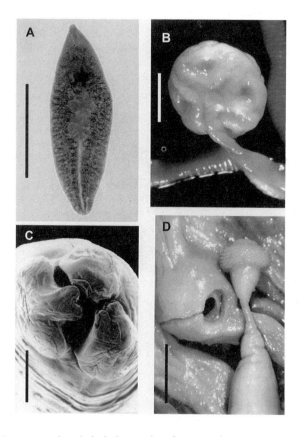

Figure 2. Representative helminth species from marine mammals: A) *Campula oblonga*, a trematode from harbour porpoise. B) Scolex of *Strobilocephalus triangularis*, a tapeworm from striped dolphin. C) Mouth of *Anisakis simplex*, a nematode from long-finned pilot whale. D) Anterior region of *Bolbosoma capitatum*, an acanthocephalan from false killer whale. Scale bars: A, B, D, 3 mm; C, 0.3 mm.

Table 1. Occurrence of virus, bacterium and protozoan families in marine mammals

	Host group				References
	Pinnipedia	Cetacea	Sirenia	Sea otter	
Viruses					
Adenoviridae	x	x			1, 2
Caliciviridae	x	x			1, 2
Coronaviridae	x				1
Hepadnaviridae		x			2, 3
Herpesviridae	x	x			1, 2
Orthomyxoviridae	x	x			1, 2
Papovaviridae		x			2
Paramyxoviridae	x	x	?		1, 2, 4, 5, 28
Picornaviridae	x	x			1
Poxviridae	x	x			1, 2
Retroviridae	x				1
Rhabdoviridae	x	x			1, 2
Bacteria					
Bacteroidaceae	x				6
Corynebacteriaceae	x				7, 8
Enterobacteriaceae	x	x			6, 7, 9, 10, 11
Leptospiraceae	x			?	7, 12
Micrococcaceae	x	x	x		9, 7, 13, 28
Mycobacteriaceae	x		x		13, 14, 15
Neisseriaceae	x	x			7, 9
Nocardiaceae	x				7
Pasteurellaceae	x	x		?	6, 7, 9, 11, 16, 28
Pseudomonadaceae	x	x	x		7, 9, 13
Streptococcaceae	x				7, 17, 18
Vibrionaceae	x	x			6, 7, 9, 19
*Abiotrophia**		x			20
*Alcaligenes**	x				7
*Arcanobacterium**	x				21
*Brucella**	x	x			22
*Cetobacterium**	x				23
*Clostridium**	x			?	6, 7, 10, 28
*Erysipelothrix**	x	x			7, 9
*Mycoplasma**	x				24, 25
Protozoa					
Eimeriidae	x		x		7, 13, 26
Sarcocystidae	x	x	x		7, 9, 26, 27

*Incertae sedis. References: 1 Visser *et al.* (1991), 2 van Bressem *et al.* (1999), 3 Bossart *et al.* (1990), 4 Duignan *et al.* (1995), 5 Kennedy (1998), 6 Baker *et al.* (1998), 7 Lauckner (1985b), 8 Pascual-Ramos *et al.* (1998), 9 Dailey (1985), 10 Baker and Ross (1992), 11 Foster *et al.* (1998), 12 Gulland *et al.* (1996), 13 Lauckner (1985c), 14 Romano *et al.* (1995), 15 Thorel *et al.* (1998), 16 Foster *et al.* (1996), 17 Goh *et al.* (1998), 18 Swenshon *et al.* (1998), 19 Fujioka *et al.* (1988), 20 Lawson *et al.* (1999), 21 Pascual-Ramos *et al.* (1997), 22 Jahans *et al.* (1997), 23 Foster *et al.* (1995), 24 Giebel *et al.* (1991), 25 Ruhnke and Madoff (1992), 26 Beck and Forrester (1988), 27 Domingo *et al.* (1992), 28 Lauckner (1985a).

Bacteria of about 30 genera have been isolated in marine mammals (Table 1 and references therein). Howard *et al.* (1983), Lauckner (1985a, 1985b, 1985c), Dailey (1985), and Dunn (1990) have reviewed the bacterial infections of both wild and captive marine mammals. The majority of these bacteria are widely distributed in many environments. However, new genera or species apparently specific to marine mammals have also been isolated recently (*e.g.* Foster *et al.*, 1995, 1998; Pascual-Ramos *et al.*, 1997, 1998). Bacterial infections certainly play a central role in the morbidity and mortality of marine mammals (Lauckner, 1985b), but it is often difficult to assess actual mortality as animals having bacterial diseases may be more susceptible to predation and thus may not be noticed (Thornton *et al.*, 1998). The pathological significance of bacterial infections is also difficult to interpret, particularly in dead animals, where natural or pathogenic endogenous bacteria may grow quickly (Howard *et al.*, 1983). In addition, it is difficult to determine whether a bacterium is a primary or a secondary pathogen in hosts physiologically compromised or suffering from other diseases (Geraci and Lounsbury, this volume).

Mycotic infections in marine mammals are relative rare. Many fungi infecting these hosts are secondary or opportunistic invaders (particularly in captive animals), as they normally grow as saprophytes in the soil or in organic debris (Migaki and Jones, 1983). *Loboa loboi*, however, has been reported as an agent causing disease (lobomycosis or Lobo's disease) in two species of dolphins, and also humans, of the same geographical area (Migaki and Jones, 1983).

Marine mammals seem to be remarkably free of protozoan parasites and thus protozoan diseases. Whether this impression is due to a lack of adequate sampling or to the actual absence of these organisms is not yet clear. Coccidia of the genera *Eimeria*, *Sarcocystis* and *Toxoplasma* are the most frequent protozoans reported from pinnipeds and cetaceans (Dailey, 1985; Lauckner, 1985b). *Eimeria phocae,* a typical apicomplexan of seals, is responsible for severe coccidiosis and associated death of harbour seals on the Atlantic coast of North America. This parasite causes acute and diffuse haemorrhagic and necrotising colitis, but hosts in good physical condition can survive (Lauckner, 1985b). Toxoplasmosis has been reported in the Californian sea lion *Zalophus californianus*, North Pacific harbour seal, Mediterranean striped dolphin *Stenella coeruleoalba*, and West Indian manatee *Trichechus manatus* (Buergelt and Bonde, 1983; Lauckner, 1985b; Domingo *et al.*, 1992).

Aside from some planktonic species of diatoms found accidentally on the skin of cetaceans, there are species from four genera obligately associated to these hosts: *Bennettella* (3 spp.), *Epipellis* (1 sp.), *Plumosigma* (2 spp.) and *Stauroneis* (2 spp.) (Nagasawa *et al.*, 1990).

2.2   Macroparasites

Some 313 species of helminths and arthropods, belonging to 104 genera and 40 families, have been reported in marine mammals (Table 2). Nevertheless, since many marine mammal species have not been thoroughly surveyed, and the above figures exclude sibling species, such as those described for anisakids (*e.g.* Mattiucci *et al.*, 1997), the actual number of parasite taxa might be far larger. Representative species of arthropods and helminths of marine mammals are shown in Figures 1 and 2.

Except for the occurrence of coronulid barnacles on both cetaceans and manatees, no parasite taxa up to the family level is shared between sirenians and other marine mammals (Table 2). In contrast, all families occurring in the sea otter *Enhydra lutris* appear also in pinnipeds, and seven are common to cetaceans and pinnipeds (Table 2). Apart from those of the Halarachnidae, all species from the shared families have indirect life cycles, being dispersed and transmitted through food webs. This strongly suggests that the carnivorous diet of cetaceans, pinnipeds, and sea otters have allowed exchange of parasites among these hosts (see Sections 4 and 5). This is particularly apparent in the sea otter, whose helminth fauna is made up chiefly of species acquired from pinnipeds (Margolis *et al.*, 1997).

The parasite fauna of sirenians, by contrast, seems more distinct, probably due, at least partly, to the host's herbivorous diet. The fauna is dominated by digeneans, particularly of the family Opisthotrematidae. Cestodes and arthropods are exceptional, and acanthocephalans are absent (Table 2). The families Opisthotrematidae and Rhabdiopoeidae (Digenea) are exclusive to sirenians (Table 2). These mammals harbour 18 genera, of which 15 (83%) are exclusive, and only 3 (17%) have representatives in either non-marine mammals or sea turtles. Therefore, the parasite fauna of sirenians seems fairly diverse (given the small number of extant host species), and very specific.

The bulk of parasite species infecting cetaceans belongs to the families Campulidae (Digenea), Diphyllobothriidae and Tetrabothriidae (Cestoda), Anisakidae, Pseudaliidae and Crassicaudidae (Nematoda), Polymorphidae (Acanthocephala) and Cyamidae (Amphipoda) (Table 2). Many of the numerically important families in pinnipeds are the same, *i.e.* Diphyllobothriidae and Tetrabothriidae (Cestoda), Anisakidae (Nematoda) and Polymorphidae (Acanthocephala), but the Heterophyidae (Digenea), Echinophthiriidae (Hexapoda), and Halarachnidae (Acarina) are not found in cetaceans and sirenians. There are obvious reasons for the absence of lice and acari in cetaceans, such as the lack of hair to glue the eggs, and a suitable surrounding climate. However, the absence of crustaceans on pinnipeds and their scarcity on sirenians (Table 2) raise interesting, unanswered questions about the processes leading to crustacean-cetacean associations (Fig. 3).

Figure 3. Whales, like this gray whale, are itinerant ecosystems for a great variety of crustacean ectoparasites and epizoits. (Photo: T. Walmsley)

Some families of parasites with species infecting cetaceans, pinnipeds, and the sea otter, are limited in distribution to, or occur mainly in (on), these hosts: Pseudaliidae, Crassicaudidae and Cyamidae in cetaceans, Echinophthiriidae in pinnipeds, and Campulidae and Diphyllobothriidae in cetaceans, pinnipeds, and the sea otter (Table 2). Additionally, cetaceans harbour two monospecific families: Pholeteridae and Brauninidae (Digenea). At the generic level, parasite specificity varies sharply between cetaceans and pinnipeds. Fifty-five genera of macroparasites or symbionts have been found in cetaceans, of which 43 (76%) are exclusive, 6 (11%) are shared with pinnipeds and/or other hosts, *e.g. Tetrabothrius* (Cestoda) with marine birds, *Pennella* (Copepoda) with fish, and *Balenophilus* (Copepoda) with sea turtles. In contrast, 18 out of 43 genera (42%) from pinnipeds and the sea otter are specific to these hosts, six (14%) are shared with cetaceans,

Table 2.   Occurrence of families of macroparasites and other symbionts in marine mammals

Parasite taxa	No. genera	No. spp.	Cetacea	Pinnipedia	Sea otter	Sirenia
Cestoda						
Diphyllobothriidae	8	48	a, d, j, k	l, m, n	+	-
Tetrabothriidae	5	23	a, b, c, d, g, j	l, n	-	-
Phyllobothriidae	2	2	a, b, d, g, h, j, k	-	-	-
Anoplocephalidae*	1	1	-	-	-	o
Digenea						
Campulidae	8	39	a, c, d, f, g, h, i, j, k	l, m, n	+	-
Notocotylidae	1	5	a	l, n	-	-
Heterophyidae	6	10	-	l, n	+	-
Opistorchiidae	4	5	-	m, n	+	-
Microphallidae	2	3	-	m, n	+	-
Pholeteridae	1	1	j, k	-	-	-
Brauninidae	1	1	j	-	-	-
Echinostomatidae	2	2	-	n	-	-
Paramphistomidae	2	3	-	-	-	o, p
Opisthotrematidae	6	11	-	-	-	o, p
Rhabdiopoeidae	4	4	-	-	-	p
Nudacotylidae	1	1	-	-	-	o
Labicolidae	1	1	-	-	-	p
Nematoda						
Anisakidae	4	18	a, b, d, e, f, g, h, i, j, k	l, m, n	+	-
Trichinellidae	1	1	i	m, n	-	-
Pseudaliidae	7	17	d, i, j, k	-	-	-
Crassicaudidae	2	11	a, b, d, h, j, k	-	-	-
Linhomoeidae	1	1	a, b	-	-	-
Filaroididae	1	4	-	l, n	-	-
Ancylostomatidae	1	2	-	l, n	-	-
Crenosomatidae	1	1	-	N	-	-
Dipetalonematidae	1	2	-	l, n	-	-
Dictyocaulidae	1	1	-	N	-	-
Dioctophymatidae	1	1	-	N	-	-
Capillariidae*	1	1	-	N	-	-
Ascarididae	2	3	-	-	-	o, p
Acanthocephala						
Polymorphidae	2	32	a, b, c, d, g, h, i, j, k	l, m, n	+	-
Arthropoda						
Echinophthiriidae	4	9	-	l, m, n	-	-
Halarachnidae	2	6	-	l, m, n	+	-
Sarcoptidae	1	1	-	N	-	-
Demodicidae	1	1	-	L	-	-
Ixodidae	1	1	-	M	-	-
Cyamidae	6	26	a, b, c, d, h, i, j, k	-	-	-
Coronulidae	6	7	a, b, c, d, g, j, k	-	-	o
Pennellidae	1	6	a, j	-	-	-
Balaenophilidae	1	1	a	-	-	-

a. Balaenopteridae, b. Balaenidae, c. Eschrichtiidae, d. Physeteridae, e. Platanistidae, f. Iniidae, g. Pontoporiidae, h. Ziphiidae, i. Monodontidae, j. Delphinidae, k. Phocoenidae, l. Otariidae, m. Odobenidae, n. Phocidae, o. Trichechidae, p. Dugongidae. *exceptional record.

and as many as 19 (44%) are shared with cetaceans and/or other hosts, particularly terrestrial carnivores and/or marine birds. The lower degree of parasite specificity in pinnipeds and the sea otter poses the question whether it is related to the more recent origin of these mammals as compared with sirenians and cetaceans. [Nevertheless, specificity is not necessarily related to the span of the host-parasite association (Hoberg and Adams, 1992; Hoberg *et al.*, 1997; Section 4)].

3. HOST-PARASITE INTERFACES – THE POPULATION PERSPECTIVE

The topic of the interactions between populations of parasites and marine mammals can be approached from either the parasite's or the host's perspective. In the first case, the emphasis is on the dynamics of parasite populations and how the host or other factors influence the parasite population (Smith, 1994). A few of these studies have focused on marine mammal parasites (*e.g.* Blair and Hudson, 1992; Aznar *et al.*, 1997; Marcogliese, 1997; Faulkner *et al.*, 1998). The second approach is host-based, the goal being either to ascertain the effect of parasites on the host population, or to gain information on the host population by analysing parasite data. We use here a mixed approach examining issues on host and parasite populations under the unifying framework of conservation and management of marine mammals. Three questions will be dealt with, namely, regulation by parasites of host populations, parasite control, and the use of parasites as population indicators.

3.1 Regulation

Regulation is a fundamental question for population ecologists because it deals with the determinants of population size, which, in turn, are crucial for conservation, management or control of wild animal populations.

A population is considered to be regulated if its size is maintained at some equilibrium value or set of values over time by density-dependent constraints on population growth (Smith, 1994). Evidence for parasite-mediated regulation can be obtained if parasite-induced effects on host mortality or reproduction increase with host population size and *vice versa* (Hassell *et al.*, 1982). However, such effects are difficult to demonstrate and estimate in wild populations, as this requires simultaneous comparison of host population size with and without the parasite. In addition, even if a significant parasite-induced, density-dependent effect on host abundance is shown, this does not necessarily imply regulation (density dependence is a necessary but not sufficient condition). If parasitism is concurrent with other factors, such as predation, hunting, shortage of resources or severe weather, its impact might be *compensatory*. Then, parasites would have no regulatory role because their removal would not rise, or otherwise affect, the trajectory of host population size. By contrast, regulation would occur if the effect of parasites is *additive*, that is, in addition to the other factors (Holmes, 1982).

Strong inference for regulation of vertebrate populations has been obtained by monitoring natural and human-induced introductions or removals of pathogens, or by manipulation of host-parasite systems [see examples in Dobson and Hudson (1995) and Hudson and Dobson (1995)]. Unfortunately, no marine mammal population has been sufficiently examined to establish whether parasite-mediated regulation occurs. Studies of these hosts are hampered by potential sampling biases, poor knowledge of many host and parasite species, and unsuitability for experimental work of many host-parasite systems (Raga *et al.*, 1997). For this reason, the evidence available is limited and should be interpreted with caution.

3.1.1   Microparasites – Viral Epizootics

Recent mass mortalities of marine mammals have received great media attention, raising concern among the public. Although the causes of such phenomena can be several (Domingo *et al.*, this volume), epizootics caused by morbilliviruses are perhaps the most important in terms of impact on the populations.

The PDV epizootic is by far the best documented die-off in marine mammals, both in terms of field data and epidemiology (Domingo *et al.*, this volume). There is absolutely no doubt that the epizootic had a significant impact on the population dynamics of the harbour seal. For instance, the high mortality in the Kattegat-Skagerrak area caused skewed sex and age distributions which will probably persist for decades (Heide-Jørgensen *et al.*, 1992). However, this does not answer the question over whether viral epizootics can regulate the harbour seal population.

Usually, regulation is achieved if the pathogen persists in the population. The classical epidemiology theory stipulates that the population of susceptible individuals has to exceed a given threshold to maintain the infection (May, 1991). Persistence of the PDV in harbour seals is unlikely because the threshold for transmission seems far larger than the susceptible population size (Grenfell *et al.*, 1992; Hall, 1995; Swinton *et al.*, 1998).

However, morbilliviruses have a remarkable ability for interspecific transmission between marine mammal populations (Harwood, 1998; Kennedy, 1998; van Bressem *et al.*, 1999; Domingo *et al.*, this volume; Geraci and Lounsbury, this volume). For the PDV die-off, in particular, it has been proposed that it was caused by a southward migration of harp seals from the Barents and Greenland Seas (de Swart *et al.*, 1995), where the virus is believed to be endemic (Stuen *et al.*, 1994). Therefore, regulation of the harbour seal population may be achieved by the occasional transmission from other seal species. Under this assumption, the size of the outbreaks would be proportional to the span after the last die-off, because the number of susceptible hosts increases with time (Grenfell *et al.*, 1992). [A similar regulatory mechanism by morbilliviruses has been proposed for cetacean populations (van Bressem *et al.*, 1999)]. Such density-dependent dynamics are substantiated by the fact that the 1988 outbreak occurred after several years of drastic increases in the seal population (Dietz *et al.*, 1989), and probably the number of naive hosts was high. Moreover, historical records provide circumstantial evidence for a long-term regulatory effect. Data from the last 200 years before the PDV outbreak show four major and other minor die-offs of harbour seals in the UK, the symptoms described being similar to those shown by the PDV-infected seals (Harwood and Hall, 1990).

An additional question is whether contaminants, particularly polychlorinated biphenyls, can increase the incidence, severity, or periodicity of viral epizootics (de Swart *et al.*, 1995; Ross *et al.*, 1996). This would facilitate regulation by decreasing host immunocompetence, thereby increasing susceptibility. However, high concentrations of these substances in the hosts are not necessarily responsible for the outbreaks (Kennedy, 1998).

It is unlikely that viral epizootics can cause the extinction of populations of sizes comparable to those of the North Sea seals (Grenfell *et al.*, 1992). However, such events have raised concern about the conservation of small, highly-endangered marine mammal populations. In 1997, more than half of the 300-individual monk seal population from Western Sahara died. The aetiology of this die-off has been much debated [see Domingo *et al.* (this volume) for details]. However, whichever the agent is, Osterhaus *et al.* (1997, 1998) isolated two different morbilliviruses from monk seals of Western Sahara and Greece respectively, confirming that these pathogens can occur in very small populations (probably due to interspecific transmission). Thus, the risk that viral outbreaks can drive endangered marine mammal populations to extinction should be considered in conservation programmes.

3.1.2   Macroparasites – Helminths

According to Harwood and Hall (1990), microparasite, particularly viral, infections seem the most important factor influencing population size of marine mammals in the absence of human exploitation. This may be so, but this appreciation may also reflect the fact that microparasite infections are simpler to monitor and model than those by macroparasites.

The large number of publications reporting parasite-induced damage shows that macroparasites can cause severe diseases in marine mammals (Geraci and St. Aubin, 1987; Geraci and Lounsbury, this volume). Few studies, however, have attempted to relate pathological findings to consequences on the host populations (*e.g.* Geraci *et al.*, 1978; Onderka, 1989; Raga and Balbuena, 1993; Bergeron *et al.*, 1997). Perrin and Powers (1980) surveyed cranial lesions attributable to the nematode *Crassicauda* sp. in spotted dolphins *Stenella attenuata* caught accidentally in fisheries in the eastern tropical Pacific. Assuming that bone damage is irreversible, the authors developed a model to describe variations in the prevalence of lesions with dolphin age. Then, mortality attributable to *Crassicauda* sp. was estimated as 11-14% of the natural mortality. However, sampling biases could affect this estimate if, for instance, infected dolphins were more likely to be caught in the fishery than uninfected ones (that is, the proportion of animals with lesions would be overestimated). Therefore, the study illustrates how inferences for mortality may be drawn from age-structure data, but also points to the difficulties that beset such inferences.

A similar investigation examined the prevalence of *Crassicauda boopis* in the kidney of North Atlantic fin whales *Balaenoptera physalus* (Lambertsen, 1986). The mortality rate accounted for by the parasite was estimated as 4.4-4.9%, which is within the range of 4.0-7.0% conventionally used for the total mortality rates of large whales. Parasite-induced mortality seems to account for a large proportion of the total death toll because other mortality causes, such as predation or starvation, are assumed to be small in fin whales. Thus, the impact of *C. boopis* seems additive and, therefore, might be significant in terms of regulation.

Additional evidence for regulation has been obtained from the hookworm *Uncinaria lucasi* infecting northern fur seals *Callorhynus ursinus* from the Pribilof Islands, Alaska. This nematode is transmitted from mother to pup with the milk, and represents an important source of pup mortality (Olsen and Lyons, 1965). The population size at St. Paul Island peaked around 1940 and has declined more or less steadily thereafter. Based on previous surveys, Fowler (1990) showed that mortality rates caused by the hookworm increased with pup numbers, which means that recent mortality is less than past. These data suggest a density-dependent relationship and, therefore, *U. lucasi* seems a good candidate to regulate fur seal populations. However, we do not know yet whether mortality by hookworms is additive, and further data are needed because density dependence was inferred from four surveys only.

3.2   Control

Control of parasite populations has traditionally been an issue for species infecting humans or domestic animals, whereas control of wildlife parasites has generally been attempted only when either public health or economy were at stake. However, as some wild animal populations become endangered and the interest in conservation increases, efforts to protect them have also led to attempts at controlling disease. Control programmes in wild animals can be carried out by chemotherapy, vaccination, or culling of hosts [see Dobson and Hudson (1995) and Gulland (1995) for examples]. Although generally applied to livestock, vaccination and chemotherapy are increasingly being tried as control measures in wild populations. However, their application to marine mammals seems difficult and costly. The use of antihelminthics, such as invermectin, has been proposed to control the abundance of the sealworm in grey seals during the breeding season, but treatments are hampered by long fasting periods, the need to administer the drug orally, and other practical problems (Stobo *et al.*, 1990; Burt, 1994). Vaccination can be a possibility to protect endangered species or populations, and has been considered already for the Mediterranean monk seal (Osterhaus *et al.*, 1992, 1998). This seems technically feasible since a vaccine has been developed to protect harbour seals from the PDV (Osterhaus *et al.*, 1989; Visser *et al.*, 1989).

Culling is usually proposed when host populations form a reservoir for a parasite of economic or public health relevance. This approach is often controversial because economic considerations are confronted with the ethics of environmental conservation. The sealworm is a paradigmatic case. The grey seal is the main definitive host, but the parasite spends most of its life encapsulated in the musculature of fish, many of which are of commercial importance (Fig. 4). Increases in sealworm abundance in fish fillets have been related to parallel increases in grey seals (Zwanenburg and Bowen, 1990; Burt, 1994). This suggests a density-dependent (regulatory) process, which supports the view that culling of seals would contribute to limiting infections in fishes. However, the relationship between seal and worm abundance is still poorly understood because the sealworm has a complex life cycle and the transmission may depend on many parameters (Fig. 4).

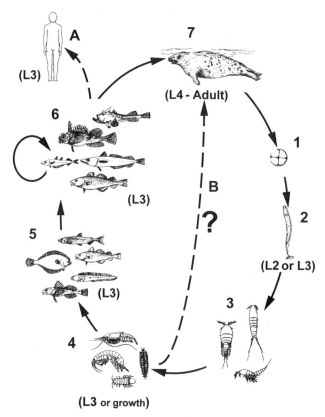

(L3 or growth)

Figure 4.   Life cycle of *Pseudoterranova decipiens*. (1) Eggs leave the host with the faeces and (2) hatch releasing a free-swimming larva, either second-stage (L2) (Measures and Hong, 1995) or third-stage (L3) (Køie *et al.*, 1995). Larvae sink and adhere to the substrate. (3) They are ingested by benthic copepods, (4) which are eaten by macroinvertebrates. Larvae either moult to L3 (Measures and Hong, 1995) or grow (Køie *et al.*, 1995). (5) Benthophagous fish feed on macroinvertebrates becoming infected. The L3's accumulate in the body cavity and muscle. (6) Demersal teleosts consume bentophagous fish, favouring parasite dispersion through the food web. (7) The seals become infected by preying on fish. The L3's moult to fourth-stage larvae (L4) and then to adult. Dashed arrows show unusual or potential routes: (A) humans can be infected (Pinel *et al.*, 1996) and (B) hypothetical infection of seals by eating macroinvertebrates (McClelland, 1990). (Reproduced and modified by permission from McClelland *et al.* 1990)

Mathematical models suggest that cod is a greater reservoir for the sealworm than seals because the parasite spends very little time of its life cycle in the latter (Dobson and May, 1987; des Clers, 1990; des Clers and Wootten, 1990). This suggests that control should concentrate primarily on the intermediate and paratenic hosts rather than on the definitive ones (Andersen *et al.*, 1995).

The 1988 PDV epizootic provided a unique natural experiment to test the validity of some conclusions drawn from mathematical models. Des Clers and Andersen (1995) examined the impact of the seal die-off on infection levels in fish in Hvaler, Norway. The mean abundance of the parasite in the cod decreased by one-half, but the reduction was less than that of the seal population (two-thirds of the original size). The decrease in worm abundance seemed short-lived. The cod caught in 1992 showed higher values than those caught in 1991, probably reflecting the recovery of seal populations in the area. The salient conclusion from the study is that the abundance of the sealworm in cod seems proportional to seal population size. However, it also suggests that culling might not be effective enough to bring down the number of worms in fish to an acceptable level unless a substantial part of the seal population is removed. This raises not only ethical questions but also concern on the impact of such reduction on the local environment.

An alternative to random culling would be to concentrate on the heavily infected seals (selective culling) (Stobo *et al.*, 1990), which requires additional work to identify the target individuals. In addition, caution should be exercised in designing such programmes because they can change the genetic makeup of the population in a few generations (Hartl *et al.*, 1995), with unpredictable long-term effects.

Clearly, additional data on sealworm biology should be incorporated into the models before they can be used as decision-making tools. In particular, current knowledge of transmission from invertebrates to fish (Marcogliese, 1996), the dynamics of worm populations (Marcogliese, 1997), and the role of non-commercial fish in transmission (Andersen *et al.*, 1995; Aspholm *et al.*, 1995; Jensen and Andersen, 1995) need to be considered.

3.3   Parasites as Host Population Tags

Both regulation and control issues focus upon the detrimental effect of parasites on either the host population or on human economy. However, the parasitic life-style has another, nonetheless important, dimension, which is that of two organisms living in close association. Because of this physical intimacy, the study of parasite populations can provide important information on host populations. This principle has been particularly applied in fisheries (MacKenzie and Abaunza, 1998), but it is also amenable to marine mammals (Balbuena *et al.*, 1995). In this case, parasites have proven useful to disclose information, relevant for conservation and management of the populations, particularly on stock identity (Delyamure *et al.*, 1984; Dailey and Vogelbein, 1991), social structure (*e.g.* Best, 1969;

Balbuena and Raga, 1991) and movements (or absence of them) of the hosts (Balbuena and Raga, 1994; Aznar *et al.*, 1995).

The technique consists of the comparison of infections with one or more parasite species between host groups, which are arranged according to ecological (migratory studies) or behavioural (social structure studies) criteria. Such comparison is usually based on differences in presence/absence, prevalence, abundance and intensity of the parasites. Two methodological approaches can be recognised (MacKenzie and Abaunza, 1998): single species studies, where a small number of parasite species are selected, the analyses being based on information of each species separately (*e.g.* Aznar *et al.*, 1995) or multispecies studies, where characteristics of a whole parasite assemblage are analysed with multivariate statistical methods (*e.g.* Balbuena and Raga, 1994).

A common problem in parasite studies of marine mammals is that of sampling biases. Animals obtained from strandings or as by-catch do not necessarily represent a random sample of the population. The effect of such biases is often difficult to avoid. However, it can be mitigated by dividing the sample into relatively homogeneous subsets of hosts or by using the multispecies approach, assuming that biases affecting one or few species are smoothed out by considering the entire set of parasites. In addition, sample sizes in marine mammal studies are usually low, which hampers statistical analyses. Larger samples can be obtained over time, but then it is difficult to separate spatial from temporal patterns (*e.g.* Aznar *et al.*, 1995). Comparison of presence/absence of parasite species somewhat alleviates this problem because this type of data is assumed not to change much over time (Balbuena *et al.*, 1995), but the analysis is rather crude. Collection and identification of the parasites should also be considered as a constraint, especially in studies of large whales. Generally, ectoparasites are better suited in such situations.

The analysis and interpretation of the results depends heavily on features of the parasite species selected as tags. Two of the most relevant are the parasite life span, which determines the time scale of study, and the source of infection, which is crucial to chart past host movements (Lester, 1990). The following example shows how both factors work together. *Anisakis typica* occurs in subtropical or tropical waters. Abril *et al.* (1986) found this species in a striped dolphin stranded on the French Atlantic coast, which suggests that the animal was a vagrant from southern latitudes. The life span of anisakids is in the order of a few weeks (McClelland, 1980), which provides information on the timing of the migration.

Another important parasite feature is the type of life cycle. Directly transmitted parasites are well suited for the study of host behaviour because transmission depends on contacts and other interactions between hosts. By contrast, recruitment of indirectly transmitted parasites is far more complex to follow because it depends on transmission rates between the intermediate and definitive hosts. However, since transmission is linked to the food web, such parasites can give information on past and present feeding grounds. For instance, the parasite fauna of the franciscana *Pontoporia blainvillei* exhibits substantial differences north and south of the River Plate Estuary in South America. Such differences suggest that franciscanas are sedentary, at least during the sampling period (spring), which corresponds with the breeding season (Aznar *et al.*, 1995). A boundary in feeding grounds can be established and, therefore, at least two franciscana stocks distinguished. Note, however, that such stocks are different from those unveiled by genetic methods. Parasite information indicates ecological stocks (spatial patterns of distribution and abundance in an ecological time scale) which can be related to the local food webs (see also Section 5), as opposed to genetic stocks (with a genetic basis). Both types may or may not have the same boundaries. For instance, individuals of a single genetic stock may be distributed in several ecological stocks. One possible, but by no means only, explanation is that the animals are sedentary, but the exchange of a small number of individuals is sufficient to over-ride genetic differences (Lester, 1990). Another possibility is that groups of young individuals from the same genetic stock migrate to different localities. Then they develop in each locality and do not move further, becoming ecologically isolated (*e.g.* Arthur and Albert, 1993).

4. ORIGIN AND EVOLUTION OF THE PARASITE FAUNAS

The term co-evolution includes systematic and ecological views in the study of a parasite-host association (Hoberg *et al.*, 1997). The systematic perspective is known as co-speciation and is evaluated by the congruence or incongruence between host and parasite phylogenies. The ecological perspective is known as co-adaptation and refers to the ways in which hosts and parasites interact. We will consider here only aspects related to the first approach.

How do a parasite species and its marine mammal host become associated? There are two possibilities, either (1) the association is inherited, *i.e.* it originated when the ancestors of both host and parasite interacted, or (2) the parasite, having emerged in another host, colonised the marine mammal species. Brooks and McLennan (1991, 1993) termed the first process *association by descent*, and the second, *association by colonisation*.

Tight association by descent implies co-evolution, *i.e.* a perfect congruence between the phylogenies of the parasite and the host at the taxonomic scale considered. Complete congruence can be considered as a sort of null hypothesis which is evaluated by comparing the phylogenies of the parasites and the hosts using matrix representation and co-speciation analysis (Brooks and McLennan, 1991, 1993; Hoberg *et al.*, 1997). This co-evolutionary hypothesis assumes a long-term association of parasites and hosts, and a high degree of co-speciation and co-adaptation between them [see Brooks and McLennan (1991) for more details]. When coincidence of host and parasite phylogenies is not exact (something that is fairly common), it might be due to (1) different speciation rates of hosts and parasites, (2) extinction of either host or parasite lineages, or (3) colonisation (host-switching) (Poulin, 1998).

Colonisation is continuously occurring: parasites are put in contact with unusual hosts through a variety of ecological ways. For instance, typical parasites of seals may end up in marine birds feeding on the same intermediate hosts (Hoberg, 1986a). Although, most of the time, parasites will not mature in the uncommon hosts, opportunities for successful colonisation arise. We cannot predict which of these encounters will eventually lead to new evolutionary associations, although two generalisations can be made. First, colonisation becomes more feasible as source and target hosts are phylogenetically closer because both provide similar conditions for the survival and reproduction of the parasite (Hoberg, 1987, 1996; Aznar *et al.*, 1998). Second, the chances also increase when there is a long-term contact between the parasite and the target host. Providing there is enough evolutionary time, this factor can over-ride the phylogenetic barrier (Hoberg, 1987; Hoberg, Gardner *et al.*, 1999; Hoberg, Jones *et al.* 1999). Accordingly, hosts with similar feeding habits have more chance of exchanging indirectly transmitted parasites than those with different diets. For instance, the cestode species of *Tetrabothrius* in cetaceans were probably acquired from marine birds (Hoberg and Adams, 1992). The results of colonisation typically are geographically delimited faunas, some degree of co-speciation and co-adaptation, depending on the age of the colonisation, and associations of variable temporal duration (Hoberg, 1992, 1997; Hoberg and Adams, 1992).

The span of an association (either by descent or colonisation) should be interpreted in relation to the host-parasite distribution, the patterns and processes involved in the geographical distribution of hosts (historical biogeography), and information about the local history and physical geology (Hoberg and Adams, 1992). The following examples illustrate how these processes are applied to understand the origin and evolution of the parasite fauna of marine mammals.

4.1   Sucking Lice of Pinnipeds: A Terrestrial Origin

The sucking lice (order Anoplura) are permanent ectoparasites of mammals, except the Monotremata, Cetacea, Sirenia, Pholidota, Edentata and Proboscida. Lice of the family Echinophthiriidae are found only on pinnipeds and the American river otter *Lutra canadensis* (Mustelidae) (Kim, 1985b) (Table 2). The origin of echinophthiriids is clearly terrestrial, and these lice had to develop very specific adaptations to cope with the marine conditions (Murray and Nicholls, 1965; Murray *et al.*, 1965). The genera of this family seem to be associated with specific host groups: *Proechinophthirus* with fur seals, *Lepidophthirus* with monk seals, *Echinophthirus* with phocids, and *Latagophthirus* with American river otters. The only exception is the genus *Antarctophthirus*, whose species occur in odobenids (walruses), phocids (seals), and otariids (sea lions and fur seals) (Fig. 5). Patterns of morphological and ecological adaptations, host specificity and biogeography of the echinophthiriids can be interpreted in different ways depending on the hypothesis on pinniped phylogeny considered.

The diphyletic hypothesis suggests that the Otarioidea (otariids and odobenids) have clear affinities with ursids and procyonids, having originated in the North Pacific, whereas the Phocoidea (Phocinae and Monachinae) are closely related to mustelids and originated in the North Atlantic (Timm and Clauson, 1985; Carroll, 1988). Accordingly, Kim (1985a, 1985b) suggested that the echinophthiriids were already present in the ancestors of carnivores, having co-evolved with the Otarioidea and Phocoidea, while becoming extinct in their terrestrial relatives. The (relictual) occurrence of *Latagophthirus* in otters (Mustelidae) would support this scenario (Kim, 1985b). However, it is not easy to reconcile this hypothesis with the cladogram provided by the author because of the most recent origin of *Latagophthirus* (Fig. 5).

Under the monophyletic hypothesis, the origin of the echinophthiriids should be more recent and intimately associated with pinnipeds. The presence of *Latagophthirus* in otters can be explained by colonisation from seals. The cladogram depicted in Figure 5, and the endemic character of the association between *Latagophthirus* and otters (Kim, 1985b), are *prima facie* evidence consistent with this scenario. Furthermore, the monophyletic origin of pinnipeds is supported by recent molecular and morphological findings (see Árnason *et al.*, 1995). However, much more research is necessary to test the congruence of host and parasite phylogenies.

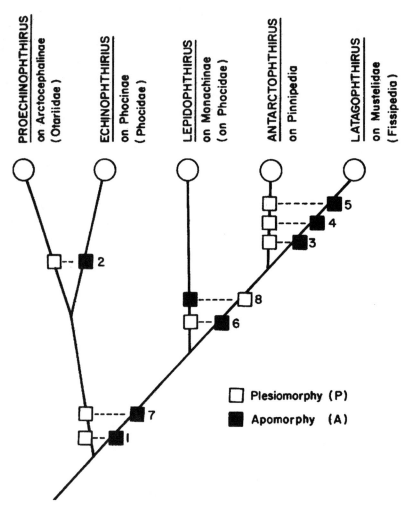

Figure 5.   Cladogram of the Echinophthiriidae genera, based in morphological characters indicated by numbers (see Kim (1985b) for details). The corresponding host taxa are indicated in parenthesis. Reprinted with permission from Coevolution of parasitic arthropods and mammals, K.C. Kim, ed. Copyright © 1985 John Wiley & Sons, Inc.

4.2   Tetrabothrids, Campulids and Marine Mammals: Examples of Marine Colonisation

Species of the order Tetrabothridea (Cestoda) parasitise cetaceans, pinnipeds, and some groups of marine birds. Hoberg (1987) postulated an origin via colonisation from elasmobranchs in the Tertiary, but recent phylogenetic evidence is compatible with diversification in the Cretaceous following a host switch from marine archosaurians to early seabirds (Hoberg *et al.*, 1997; Hoberg, Gardner *et al.* 1999; Hoberg, Jones *et al.*, 1999). Marine mammals, specifically cetaceans first and pinnipeds later, acquired tetrabothriideans after the initial radiation of orders of marine birds (Hoberg, 1996). Within the Tetrabothridea, a very interesting example of relatively recent colonisation, with subsequent radiation, is that of the genus *Anophryocephalus*. A phylogenetic reconstruction suggests that this genus appeared in pinnipeds by colonisation of tetrabothriids from odontocetes (Hoberg, 1989; Hoberg and Adams, 1992). Species of *Anophryocephalus* infect pinnipeds in the Arctic-Subarctic region, especially the Phocinae. There are also some representatives in otariids but none in odobenids (Hoberg and Adams, 1992; Hoberg, 1992, 1995). The phylogenies of hosts and parasites are highly incongruent, suggesting a history of frequent colonisation (Fig. 6). Assuming a monophyletic origin for the Pinnipedia, the Otariidae are the basal group, and the Odobenidae the sister-group of phocids. These relationships are not apparent when mapped onto the parasite tree (Fig. 6).

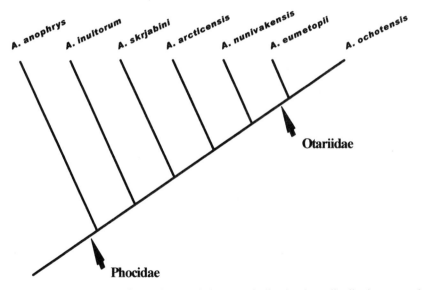

Figure 6. Cladogram of *Anophryocephalus* spp., indicating host distributions onto the parasite tree. [Based on data from Hoberg (1995)]

The origin and expansion of the Pinnipedia through the Holarctic area is well known and, along with the parasite and host phylogenies, allow reconstruction of the probable history of association between the species of

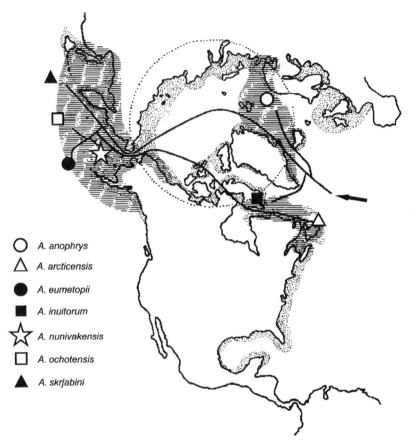

Figure 7.   Recapitulation of the distribution of *Anophryocephalus* spp. during the late
Pliocene and Pleistocene in the Arctic. Dotted area: emergent continental shelf at glacial
maxima. Striped area: current geographical limits of *Anophryocephalus* spp. The hypothetical
origin of *Anophryocephalus* sp. is proposed to be in *Phoca* sp. in the North Atlantic (arrow).
A. anophrys and *A. inuitorum* resulted from a vicariant or microallopatric process in *P.*
hispida of an ancestral parasite population. Later on, parasites dispersed with their hosts to
the North Pacific. Stadial isolation lead to *A. skrjabini*, *A. nunivakensis*, *A. ochotensis* and *A.*
eumetopii. *A. arcticensis* resulted from a secondary expansion of phocine hosts from the
North Pacific to the Atlantic. (Reproduced by permission from Hoberg, 1995)

Anophryocephalus and their hosts. The origin of *Anophryocephalus* is
assumed to have involved *Phoca* sp. in the North Atlantic. Subsequently,
Anophryocephalus expanded through the Arctic basin (Fig. 7), followed by
radiation of species in the Arctic and Subarctic area of the Pacific, where
colonisation of new hosts occurred [for more details, see Hoberg (1992,
1995) and Hoberg and Adams (1992)]. The putative mechanisms of these
patterns are intimately related to the physical geology of the Arctic during
that period. In the late Pliocene and Pleistocene, alternative episodes of
glaciation (stadials) and thaw (interstadials), with fluctuations in sea level,

occurred. During glacial maxima, the formation of faunal refugia was promoted due to the reduction in sea level, provoking isolation and speciation of hosts and parasites. During interstadials, expansion of parasites and hosts occurred, facilitating host switching (Hoberg and Adams, 1992; Hoberg, 1995). The latter might be accounted for by long-term trophic associations among hosts, all of which shared the same prey items.

Another group of marine parasites with a putative origin by colonisation is the Campulidae, a family of digeneans occurring world-wide in marine mammals, particularly cetaceans. The origin of the campulids has been traditionally discussed in the light of two exclusive hypotheses. First, the sister group of the Campulidae has been postulated to be the Fasciolidae (Brooks *et al.*, 1985). Since fasciolids are parasites of ruminants, this implies a terrestrial origin. Alternatively, the sister group might be the Acanthocolpidae, which are parasites of marine fish (see Brooks *et al.*, 1989). Here, a colonisation scenario seems plausible. Fernández, Littlewood *et al.* (1998) inferred a molecular phylogeny in which campulids appeared closer to acanthocolpids than fasciolids. The life cycle of acanthocolpids includes marine gastropods as the first intermediate host, and fishes as second intermediate and definitive hosts. Fernández, Littlewood *et al.* (1998) suggested that the campulids might have originated when their ancestor amplified the cycle by including cetaceans as new definitive hosts. Obviously, to promote this change, cetaceans should have fed predictably upon the ancient definitive hosts, *i.e.* fish.

5.   PARASITE COMMUNITIES: SCALE, HISTORY, AND 'LIMITED MEMBERSHIP'

The central question of community ecology may be formulated as follows: why does the pool of species that occur together constitute a limited subset of those that might occur together? (Roughgarden and Diamond, 1986; Roughgarden, 1989). We could consider three basic reasons for this 'limited membership': (i) some species cannot withstand the conditions of the physical environment (which includes the concept of host specificity in the case of parasites); (ii) some species never arrive in the community due to historical, biogeographical, or ecological reasons; and (iii) some species are excluded from the community through biotic interactions. The classical community theory considered the role of the latter factor, in particular contemporary competition, as the predominant force structuring communities and limiting diversity (Ricklefs and Schluter,

1993). Recent decades, however, have witnessed much effort in developing a more pluralistic approach to community structure (McIntosh, 1987).

Ecological processes operate on different temporal and spatial scales and, therefore, there is a range of scales to address the above question (Levin, 1992). To mention some examples, unique historical and biogeographical circumstances often determine fundamental aspects of the richness and composition of contemporary communities (Brooks and McLennan, 1991; Ricklefs and Schluter, 1993; Hoberg, 1996, 1997). Large-scale spatial processes can also account for major community patterns (Wiens, 1984; Holmes, 1987; Kennedy and Bush, 1994). For instance, species richness may depend solely on the regional pool of colonisers (Cornell and Lawton, 1992).

In this section, we analyse some of the causes of the 'limited membership' of species in parasite communities of marine mammals. Following the above emphasis upon scale, which is particularly obvious in the case of parasite communities (Poulin, 1998), we will adopt an hierachical perspective. A caveat is necessary concerning coverage. Investigation effort has been biased towards certain scales and groups (cetaceans and their helminths), and this bias is reflected herein. Also, the reader can find further arguments about the development of the helminth faunas of marine homeotherms (particularly seabirds) in Hoberg (1986a, 1986b, 1992, 1995, 1996, 1997) and Hoberg and Adams (1992).

5.1   The Consequences of Marine Colonisation

Several authors (Balbuena and Raga, 1993; Aznar *et al.*, 1994; see also Hoberg, 1987) have discussed the consequences of cetacean evolution from a terrestrial to a marine ecosystem upon the structure and evolution of their helminth communities. These authors speculated that the habitat shift by the host was paralleled by the extinction of most terrestrial parasites as they could not adjust their life cycles to the new environment. Therefore, the biodiversity clock was reset to nearly zero, and the helminth fauna observed today should ultimately have arisen from colonisation by marine parasites. This prediction seems substantiated at least for the three main groups of intestinal helminths, *i.e.* tetrabothriids (Cestoda), campulids (Digenea), and *Bolbosoma* species (Acanthocephala) (Table 2). [The origin of the latter is still unclear, but a similar scenario as that discussed in Section 4.2. for tetrabothriids and campulids has been suggested (Aznar, 1995)]. In contrast, lung-worms of cetaceans might have a terrestrial origin (Balbuena *et al.*, 1994), but data on parasite phylogeny is still needed to confirm this contention. What seems clear, however, is that extinction of terrestrial parasites should be less likely in marine mammals with higher dependency on land. For instance, a sizeable part of the nematode fauna of pinnipeds has obvious phylogenetic affinities with that of land carnivores. Not surprisingly, the cycle of these species is land-dependent and resembles that of their terrestrial counterparts (Anderson, 1992).

A further hypothetical outcome of marine colonisation is that marine mammals, during their subsequent evolution, had low probabilities of capturing parasites from other marine hosts because of the phylogenetic distance between them and the potential donors (Balbuena and Raga, 1993; see Section 4). Evidence for this hypothesis is lacking. However, Aznar *et al.* (1998) explored this question in marine turtles, which represent also a host lineage that have become ecologically isolated from their nearest broad phylogenetic relatives. Barriers for helminth exchange between marine turtles and other marine hosts seem clearly reflected at both ecological and evolutionary scales. We can therefore suggest that few successful host switchings occurred between marine mammals and other marine hosts, but, especially in cetaceans and sirenians, they should represent the main source of parasite diversity observed today. After these rare host captures, parasites would diversify by co-evolution with their mammalian hosts, allowing a certain rate of helminth exchange with other marine mammals (see Section 4). This fact is illustrated, for instance, by the specificity patterns shown in Section 2.2. In cetaceans, it has been proposed that the rarity of host captures coupled with the extinction of terrestrial parasites might have led to species-poor communities (Balbuena and Raga, 1993; Aznar *et al.*, 1994), but, as yet, this hypothesis has not been properly tested.

5.2   The Regional and Local Scales

The above factors limiting diversity allow global predictions about the composition of current communities. For instance, we can expect that the intestinal community of any cetacean individual in any locality will be composed of tetrabothriids, and/or campulids, and/or *Bolbosoma* species, rarely something else. This is a historical legacy. However, this scale is very coarse and does not allow us to predict community 'membership' at smaller geographical and host taxonomic scales. To do so, we should pose other questions: What are the patterns of specific host-parasite associations in relation to geographic regions? How important are the barriers for parasite dispersal? The answers to these questions may account for regional variation in parasite richness. Endemism could be explained simply by the fact that a host-parasite association is historically recent (*e.g. Anophryocephalus* and arctic seals, Section 4.2). Alternatively, environmental conditions and/or the use of hosts with limited geographical distribution may act as barriers for parasite dispersion (*e.g.* Hoberg, 1992,

1995). Note that the cycle of most helminths infecting marine mammals depends on definitive, intermediate, and paratenic hosts, and, if the cycle includes a free-living phase, on the physical environment. For instance, pelagic cetaceans are expected to have intestinal communities more diverse than coastal ones because speciation and dispersal of tetrabothriids and species of *Bolbosoma* seem restricted to the pelagic realm, perhaps due to the type of intermediate and paratenic hosts used (Aznar, 1995; Herreras *et al.*, 1997). Delyamure (1955) attempted to establish how the richness and composition of helminth communities of marine mammals vary among geographical regions depending on the regional history of host-parasite associations. However, information on parasite biogeography was limited at that time and powerful tools within the comparative method were still to be developed.

The host community is fundamental to understand parasite community structure at a local scale (*e.g.* Stock and Holmes, 1988; Edwards and Bush, 1989). In other words, ecological exchange of parasites will largely depend on how many, and how phylogenetically close, are the members of the local host community. For example, campulids of the genus *Hadwenius* are apparently specific to odontocetes (Fernández, Aznar *et al.*, 1998). Therefore, although baleen and toothed whales may co-occur in the same localities, it is very improbable that *Hadwenius* can successfully colonise baleen whales in ecological time (see also Aznar *et al.*, 1994).

Herreras *et al.* (1997) illustrated the interplay of regional and local factors based on a comparative geographical analysis of the gastro-intestinal helminth fauna of the harbour porpoise. Variation was mostly attributed to the geographical distribution of the parasite taxa and the community of sympatric cetaceans (and seals). Unfortunately, for most parasite groups, the relative importance of these factors remains to be analysed.

5.3   Parasite Infracommunities

So far, we have dealt with the factors limiting membership at the host population level (the so-called *component community* level). Factors acting at these scales determine the pool of helminth species that individual hosts of local marine mammal populations can draw from (Holmes, 1990). But what does determine membership at the level of the individual hosts (at the *infracommunity* level)? Two general mechanisms can be recognised: those determining the host-parasite encounter, and the interactions between the parasite species (Janovy *et al.*, 1992).

Comments on the former have to be focused largely upon features of the definitive host, because little is known about the intermediate and paratenic hosts of most marine mammal parasites. Aznar *et al.* (1994) compared infracommunity patterns of sedentary-coastal vs. migratory-pelagic cetaceans. Regardless of whether parasites are species-specific, the former host group appears to promote a more predictable infra-community structure, with frequently co-occurring helminths with larger population sizes. The reason seems to be simple: these hosts are exposed to the same parasites all the time (see Bush, 1990). Examples of this type of infracommunity might be those of the intestine of franciscanas (Aznar *et al.*, 1994), or the lungs of harbour porpoises (Balbuena *et al.*, 1994). Conversely, infracommunities in pelagic-vagrant cetaceans apparently represent random combinations drawn from the pool of locally available species. Examples are the intestinal infracommunities of long-finned pilot whales *Globicephala melas* (Balbuena and Raga, 1993), or those of the Pacific spotted dolphins (Dailey and Perrin, 1973). Obviously, these are extremes of a continuum and the predictive value of this heuristic rule is limited.

Due to inherent difficulties of the systems under study, detection of positive [*e.g.* 'hitch-hiking strategies' (Thomas *et al.*, 1997)] or negative interactions (competition) between marine mammal parasites have never been unequivocally demonstrated. Competition would become especially relevant if infracommunities are saturated with species. In this case, species richness would be limited at infracommunity level (Kennedy and Guégan, 1994). Saturation has been explored through the dependence of the local richness of species upon the regional richness [see concepts in Cornell and Lawton (1992) and, for an example of parasites, Barker *et al.* (1996)]. This relationship remains unexplored in parasite communities of marine mammals. However, our intuition is that interactions must be rare in these systems. We essentially view these communities as composed of those parasites that happened to reach marine mammals after many historical and ecological vagaries, there being much 'empty space' to be evolutionarily filled.

ACKNOWLEDGEMENTS

Thanks are due to David I. Gibson, Eric P. Hoberg, David J. Marcogliese, and Marie-Françoise van Bressem for critical comments and suggestions. We are also grateful to Geoff Foster for invaluable assistance with the literature and review of the section dealing with bacterial infections. This work was funded by grants from the DGES (project no. PB96-0801) and ICI of the Spanish Government. FJA holds an Associate Researcher Contract from the Ministry of Education and Culture of Spain.

REFERENCES

Abril, E., Almor, P., Raga, J.A. and Duguy, R. (1986) Parasitisme par *Anisakis typica* (Diesing, 1860) chez le dauphin bleu et blanc (*Stenella coeruleoalba*) dans le Nord-Est Atlantique. *Bulletin de la Societé Zoologique de France*, **111**, 131-133.

Andersen, K., des Clers, S. and Jensen, T. (1995) Aspects of the sealworm *Pseudoterranova decipiens* life-cycle and seal-fisheries interactions along the Norwegian coast. In: *Whales, seals, fish and man* (Ed. by A.S. Blix, L. Walløe and Ø. Ulltang), pp. 557-564. Elsevier Science, Amsterdam.

Anderson, R.C. (1992) *Nematode parasites of vertebrates – their development and transmission*, 1-578 pp. CAB International, Wallingford.

Anderson, R.M. (1993) Epidemiology. In: *Modern Parasitology* (Ed. by F.E.G. Cox), pp. 75-116, Blackwell, Oxford.

Armentia, A., Lombardero, M., Callejo, A., Martín Santos, J.M., Martín Gil, F.J., Vega, J., Arranz, M.L. and Martínez, C. (1998) Occupational asthma by *Anisakis simplex*. *Journal of Allergy and Clinical Immunology*, **102**, 831-834.

Árnason, Ú., Bodin, K., Gullberg, A., Ledje, C. and Mouchaty, S. (1995) A molecular view of pinniped relationships with particular emphasis on the true seals. *Journal of Molecular Evolution*, **40**, 78-85.

Arthur, J.R. and Albert, E. (1993) Use of parasites for separating stocks of Greenland halibut *(Reinhardtius hippoglossoides)* in the Canadian Northwest Atlantic. *Canadian Journal of Fisheries and Aquatic Sciences*, **50**, 2175-2181.

Aspholm, P.E., Ugland, K.I., Jødestøl, K.A. and Berland, B. (1995) Sealworm (*Pseudoterranova decipiens*) infection in common seals (*Phoca vitulina*) and potential intermediate fish hosts from the Outer Oslofjord. *International Journal for Parasitology*, **25**, 367-373.

Aznar, F.J. (1995) *Estudio biológico de la helmintofauna de la franciscana (Pontoporia blainvillei) (Cetacea) en aguas de Argentina*. Ph.D. Thesis, University of Valencia, Valencia, Spain. 181pp.

Aznar, F.J., Badillo, F.J. and Raga, J.A. (1998) Gastrointestinal helminths of loggerhead turtles (*Caretta caretta*) from the western Mediterranean: constraints on community structure. *Journal of Parasitology*, **84**, 474-479.

Aznar, F.J., Balbuena, J.A. and Raga, J.A. (1994) Helminth communities of *Pontoporia blainvillei* (Cetacea: Delphinidae) in Argentinean waters. *Canadian Journal of Zoology*, **72**, 443-448.

Aznar, F.J., Balbuena, J.A., Bush, A.O. and Raga, J.A. (1997) Ontogenetic habitat selection by *Hadwenius pontoporiae* (Digenea: Campulidade) in the intestine of franciscanas (Cetacea). *Journal of Parasitology*, **83**, 13-18.

Aznar, J., Raga, J.A. Corcuera, J. and Monzón, F. (1995) Helminths as biological tags for franciscana (*Pontoporia blainvillei*) in Argentinean and Uruguayan waters. *Mammalia*, **59**, 427-435.

Baker, J.R., Jepson, P.D., Simpson, V.R. and Kuiken, T. (1998) Causes of mortality and non-fatal conditions among grey seals (*Halichoerus grypus*) found dead on the coasts of England, Wales, and the Isle of Man. *Veterinary Record*, **142**, 595-601.

Baker, J.R. and Ross, H. (1992) The role of bacteria in phocine distemper. *Science of the Total Environment*, **115**, 9-14.

Balbuena, J.A. and Raga, J.A. (1991) Ecology and host relationships of the whale-louse *Isocyamus delphini* (Amphipoda: Cyamidae) parasitizing long-finned pilot whales

(*Globicephala melas*) off the Faroe Islands (Northeast Atlantic). *Canadian Journal of Zoology*, **69**, 141-145.

Balbuena, J.A. and Raga, J.A. (1993) Intestinal helminth communities of the long-finned pilot whale (*Globicephala melas*) off the Faeroe Islands. *Parasitology*, **106**, 327-333.

Balbuena, J.A. and Raga, J.A. (1994) Intestinal helminths as indicators of segregation and social structure of pods of pilot whales (*Globicephala melas*) off the Faroe Islands. *Canadian Journal of Zoology*, **72**, 443-448.

Balbuena, J.A., Aspholm, P.E., Andersen, K.I. and Bjørge, A. (1994) Lung-worms (Nematoda: Pseudaliidae) of harbour porpoises (*Phocoena phocoena*) in Norwegian waters: patterns of colonization. *Parasitology*, **108**, 343-349.

Balbuena, J.A., Aznar, F.J., Fernández, M. and Raga, J.A. (1995) The use of parasites as indicators of social structure and stock identity of marine mammals. In: *Whales, seals, fish and man* (Ed. by A.S. Blix, L. Walløe and Ø. Ulltang), pp. 133-139. Elsevier Science, Amsterdam.

Barker, D.E., Marcogliese, D.J. and Cone, D.K. (1996) On the distribution and abundance of eel parasites in Nova Scotia: local versus regional patterns. *Journal of Parasitology*, **82**, 697-701.

Barret, T., Visser, I.K.G., Mamaev, L., van Bressem, M.F. and Osterhaus, A.D.M.E. (1993) Dolphin and porpoise morbilliviruses are genetically distinct from phocine distemper virus. *Virology*, **193**, 1010-1012.

Beck, C. and Forrester, D.J. (1988) Helminths of the Florida manatee, *Trichechus manatus latirostris*, with a discussion and summary of the parasites of sirenians. *Journal of Parasitology*, **74**, 628-637.

Bergeron, E., Measures, L.N. and Huot, J. (1997) Lungworm (*Otostrongylus circumlitus*) infections in ringed seals (*Phoca hispida*) from eastern Arctic Canada. *Canadian Journal of Fisheries and Aquatic Sciences*, **54**, 2443-2448.

Best, P.B. (1969) The sperm whale (*Physeter catodon*) off the west coast of South Africa 3. Reproduction in the male. *Investigational Report of the Division of Sea Fisheries of South Africa*, **72**, 1-20.

Blair, D. and Hudson, B.E.T. (1992) Population structure of *Lankatrematoides gardneri* (Digenea: Opisthotrematidae) in the pancreas of the dugong (*Dugong dugon*) (Mammalia: Sirenia). *Journal of Parasitology*, **78**, 1077-1079.

Bossart, G.D., Brawner, T.A., Cabal, C., Kuhns, M., Eimstad, E.A., Caron, J., Trimm, M. and Bradley, P. (1990) Hepatitis B-like infection in a Pacific white-sided dolphin (*Lagenorhynchus obliquidens*). *Journal of the American Veterinary Medical Association*, **196**, 127-130.

Brooks, D. R., Bandoni, S. M., Macdonald, C. A. and O'Grady, R. T. (1989) Aspects of the phylogeny of the Trematoda Rudolphi, 1808 (Platyhelminthes: Cercomeria). *Canadian Journal of Zoology*, **67**, 2609-2624.

Brooks, D. R. and Hoberg, E.P. (2000) Triage for the biosphere: the need and rationale for taxonomic inventories and phylogenetic studies of parasites. *Comparative Parasitology*, **67**, 1-25.

Brooks, D. R., O'Grady, R. T. and Glen, D. R. (1985) Phylogenetic analysis of the Digenea (Platyhelminthes: Cercomeria) with comments on their adaptive radiation. *Canadian Journal of Zoology*, **63**, 411-443.

Brooks, D.R. and McLennan, D.A. (1991) *Phylogeny, ecology, and behavior. A research program in comparative biology*. University of Chicago Press, Chicago. 434 pp.

Brooks, D.R. and McLennan, D.A. (1993) *Parascript. Parasites and the language of evolution*. Smithsonian Institution Press, Washington D.C. 429 pp.

Buergelt, C.D. and Bonde, R.K. (1983) Toxoplasmic meningoencephalitis in a West Indian manatee. *Journal of the American Veterinary Medical Association*, **183**, 1294-1296.

Burt, M.D.B. (1994) The sealworm situation. In: *Parasitic infectious diseases; epidemiology and ecology* (Ed. by M.E. Scott and G. Smith), pp. 347-362, Academic Press, San Diego.

Bush, A.O. (1990) Helminth communities in avian hosts: determinants of pattern. In: *Parasite communities: patterns and processes* (Ed. by G.W. Esch, A.O. Bush and J.M. Aho), pp. 197-232. Chapman and Hall, London.

Carroll, R.L. (1988) *Vertebrate paleontology and evolution*. W.H. Freeman, New York, 698 pp.

Cornell, H.V. and Lawton, J.H. (1992) Species interactions, local and regional processes, and limits to the richness of ecological communities: a theoretical perspective. *Journal of Animal Ecology*, **61**, 1-12.

Dailey, M.D. (1985) Diseases of Mammalia: Cetacea. In: *Diseases of Marine Animals. Vol. IV, Part 2* (Ed. by O. Kinne), pp. 805-847. Biologische Anstalt Helgoland, Hamburg.

Dailey, M.D. and Perrin, W.F. (1973) Helminth parasites of porpoises of the genus *Stenella* in the eastern Tropical Pacific, with descriptions of two new species: *Mastigonema stenellae* gen. et sp. n. (Nematoda: Spiruroidea) and *Zalophotrema pacificum* sp. n. (Trematoda: Digenea). *Fishery Bulletin*, **71**, 455-471.

Dailey, M.D., and Vogelbein,W.K. (1991) Parasite fauna of three species of Antarctic whales with reference to their use as potential stock indicators. *Fishery Bulletin*, **89**, 355-365.

Delyamure, S.L. (1955) *Helminth fauna of marine mammals (ecology and phylogeny)*, Akademiya Nauk SSSR, Moscow. 517 pp. (Translated by Israel Program for Scientific Translation, Jerusalem, 1968, 522 pp.).

Delyamure, S.L. and Skriabin, A.S. (1985) Achievements of Soviet scientists in investigations of the helminthofauna of marine mammals of the world ocean. In: *Parasitology and pathology of marine organisms of the world ocean* (Ed. by W.J. Hargis Jr.), pp. 129-135. NOAA Technical Reports NMFS no. 25, Springfield.

Delyamure, S.L., Yurakhno, M.V., Popov, V.N., Shults, L.M. and Fay, F.H. (1984). Helminthological comparison of subpopulations of Bering Sea spotted seals, *Phoca largha* Pallas. In: *Soviet-American cooperative research on marine mammals: volume 1, Pinnipeds* (Ed. by F.H. Fay and G.A. Fedoseev), pp. 61-65. NOAA, Technical Reports NMFS no. 12, Springfield.

de Swart, R.L., Harder, T.C., Ross, P.S., Vos, H. W. and Osterhaus, A.D.M.E. (1995) Morbilliviruses and morbillivirus diseases of marine mammals. *Infectious Agents and Disease*, **4**, 125-130.

des Clers, S. (1990) Modelling the life cycle of the sealworm (*Pseudoterranova decipiens*) in Scottish waters. *Canadian Bulletin of Fisheries and Aquatic Sciences*, **222**, 273-288.

des Clers, S. and Andersen, K. (1995) Sealworm (*Pseudoterranova decipiens*) transmission to fish trawled from Hvaler, Oslofjord, Norway. *Journal of Fish Biology*, **46**, 8-17.

des Clers, S. and Wootten, R. (1990) Modelling the population dynamics of the sealworm (*Pseudoterranova decipiens*). *Netherlands Journal of Sea Research.*, **25**, 291-299.

Dietz, R., Heide-Jørgensen M.P. and Härkönen, T. (1989) Mass deaths of harbor seals (*Phoca vitulina*) in Europe. *Ambio*, **18**, 258-264.

Dobson, A.P. and Hudson, P.J. (1995) Microparasites: observed patterns in wild animal populations. In: *Ecology of infectious diseases in natural populations* (Ed. by B.T. Grenfell and A.P. Dobson), pp. 52-89. Cambridge University Press, Cambridge.

Dobson, A.P. and May, R.M. (1987) The effects of parasites on fish populations – theoretical aspects. *International Journal for Parasitology*, **17**, 363-370.

Domingo, M., Visa, J., Pumarola, M., Marco, A.J., Ferrer, L., Rabanal, R. and Kennedy, S. (1992) Pathologic and immunocytochemical studies of morbillivirus infection in striped dolphins (*Stenella coeruleoalba*). *Veterinary Pathology*, **29**, 1-10.

Duignan, P.J., House, C., Walsh, M.T., Campbell, T., Bossart, G.D., Duffy, N., Fernandes, P.J., Rima, B.K., Wright, S. and Geraci, J.R. (1995) Morbillivirus infection in manatees. *Marine Mammal Science*, **11**, 441-451.

Dunn, J.L. (1990) Bacterial and mycotic diseases of cetaceans and pinnipeds. In: *CRC handbook of marine mammal medicine: health, disease, and rehabilitation* (Ed. by L.A. Dierauf), pp. 89-96. CRC Press, Boca Raton.

Edwards, D.D. and Bush, A.O. (1989) Helminth communities in avocets: importance of the compound community. *Journal of Parasitology*, **75**, 225-238.

Faulkner, J., Measures, L.N., Whoriskey, F.G. (1998) *Stenurus minor* (Metastrongyloidea: Pseudaliidae) infection of the cranial sinuses of the harbour porpoise, *Phocoena phocoena*. *Canadian Journal of Zoology*, **76**, 1209-1216.

Fernández, M., Aznar, F.J., Latorre, A. and Raga, J.A. (1998) Molecular phylogeny of the families Campulidae and Nasitrematidae (Trematoda). *International Journal for Parasitology*, **28**, 767-775.

Fernández, M., Littlewood, D.T.J., Latorre, A., Raga, J.A. and Rollinson, D. (1998) Phylogenetic relationships of the family Campulidae (Trematoda) based on 18S rRNA sequences. *Parasitology*, **117**, 383-391.

Fernández de Corres, L., Audícana, M., Del Pozo, M.D., Muñoz, D., Fernández, E., Navarro, J.A., García, M. and Díez, J. (1996) *Anisakis simplex* induces not only anisakiasis: report on 28 cases of allergy caused by this nematode. *Journal of Investigational Allergology and Clinical Immunology*, **6**, 315-319.

Foster, G., Ross, H.M., Naylor, R.D., Collins, M.D., Pascual Ramos, C., Fernández Garayzabal, F. and Reid, R.J. (1995) *Cetobacterium ceti* gen. nov. sp., a new Gram-negative obligate anaerobe from sea mammals. *Letters in Applied Microbiology*, **21**, 202-206.

Foster, G., Jahans, K.L., Reid, R.J. and Ross, H.M. (1996) Isolation of *Brucella* species from cetaceans, seals and an otter. *Veterinary Record*, **138**, 583-586.

Foster, G, Ross, H.M., Patterson, I.A.P., Hutson, R.A. and Collins, M.D. (1998) *Actinobacillus scotiae* sp. nov., a new member of the family Pasteurellaceae Pohl (1979) 1981 isolated from porpoises (*Phocoena phocoena*). *International Journal of Systematic Bacteriology*, **48**, 929-933.

Fowler, C.W. (1990) Density dependence in northern fur seals (*Callorhinus ursinus*). *Marine Mammal Science*, **6**, 171-195.

Fujioka, R.S., Greco, S.B., Cates, M.B. and Schroeder, J.P. (1998) *Vibrio damsela* from wounds in bottlenose dolphins *Tursiops truncatus*. *Diseases of Aquatic Organisms*, **4**, 1-8.

Gardner, S.L. and Campbell, M.L. (1992) Parasites as probes for biodiversity. *Journal of Parasitology*, **78**, 596-600.

Geraci, J.R. and St. Aubin, D.J. (1987) Effects of parasites on marine mammals. *International Journal for Parasitology*, **17**, 407-414.

Geraci, J.R., Dailey, M.D. and St. Aubin, D. J. (1978) Parasitic mastitis in the Atlantic white-sided dolphin, *Lagenorhynchus acutus*, as a probable factor in herd productivity. *Journal of the Fisheries Research Board of Canada*, **35**, 1350-1355.

Giebel, J., Meier, J., Binder, A., Flossdorf, J., Poveda, J.B., Schmidt, R. and Kirchhoff, H. (1991) *Mycoplasma phocarhinis*, new species and *Mycoplasma phocacerebrale*, new species, two new species from harbor seals (*Phoca vitulina* L.). *International Journal of Systematic Bacteriology*, **41**, 39-44.

Goh, S.H., Driedger, D., Gillett, S., Low, D.E., Hemmingsen, S.M., Amos, M., Chan, D., Lovgren, M., Willey, B.M., Shaw, C. and Smith, J.A. (1998) *Streptococcus iniae*, a human and animal pathogen: Specific identification by the chaperonin 60 gene identification method. *Journal of Clinical Microbiology*, **36**, 2164-2166.

Gompper, M.E. and Williams, E.S. (1998) Parasite conservation and the black-footed ferret recovery program. *Conservation Biology*, **12**, 730-732.

Grenfell, B.T., Lonergan, M.E. and Harwood, J. (1992) Quantitative investigations of the epidemiology of phocine distemper virus (PDV) in European common seal populations. *Science of the Total Environment*, **115**, 15-29.

Gulland, F. (1995) The impact of infectious diseases on wild animal populations – a review. In: *Ecology of infectious diseases in natural populations* (Ed. by B.T. Grenfell and A.P. Dobson), pp. 20-51. Cambridge University Press, Cambridge.

Gulland, F.M.D., Koski, M., Lowenstine, L.J., Colagross, A., Morgan, L. and Spraker, T. (1996) *Leptospirosis* in California sea lions (*Zalophus californianus*) stranded along the central California coast, 1981-1994. *Journal of Wildlife Diseases*, **32**, 572-580.

Hall, A.J. (1995) Morbillivirus in marine mammals. *Trends in Microbiology*, **3**, 4-9.

Hartl, G.B., Klein, F., Willing, R., Apollonio, M. and Lang, G. (1995) Allozymes and the genetics of antler development in red deer (*Cervus elaphus*). *Journal of Zoology*, **237**, 83-100.

Harwood, J. (1998) Conservation biology –What killed the monk seals? *Nature*, **393**, 17-18.

Harwood, J. and Hall, A. (1990) Mass mortality in marine mammals: its implications for population dynamics and genetics. *Trends in Ecology and Evolution*, **5**, 254-257.

Hassell, M.P., Anderson, R.C., Cohen, J.E., Cvjetanović, B., Dobson, A.P., Gill, D.E., Holmes, J.C., May, R.M., McKeown, T., Perereira, M.S. and Tyrrel, D.A.J. (1982) Impact of infectious diseases on host populations group report. In: *Population biology of infectious diseases* (Ed. by R.M. Anderson and R.M. May), pp. 15-35. Springer-Verlag, Berlin, Heidelberg, New York.

Heide-Jørgensen, M.P., Härkönen, T. and Åberg, P. (1992) Long-term effect of epizootic in harbor seals in the Kattegat-Skagerrak and adjacent areas. *Ambio*, **21**, 511-516.

Herreras, M.V., Kaarstad, S., Balbuena, J.A., Kinze, C.C. and Raga, J.A. (1997) Helminth parasites of the digestive tract of the harbour porpoise *Phocoena phocoena* in Danish waters: a comparative geographical analysis. *Diseases of Aquatic Organisms*, **28**, 163-167.

Hoberg, E.P. (1986a) Aspects of ecology and biogeography of Acanthocephala in Antarctic seabirds. *Annales de Parasitologie Humaine et Comparée*, **61**, 199-214.

Hoberg, E.P. (1986b) Evolution and historical biogeography of a parasite-host assemblage: Alcatenia spp. (Cyclophyllidea: Dilepididae) in Alcidae (Charadriformes). *Canadian Journal of Zoology*, **64**, 2576-2589.

Hoberg, E.P. (1987) Recognition of larvae of the Tetrabothriidae (Eucestoda): implications for the origin of tapeworms in marine homeotherms. *Canadian Journal of Zoology*, **65**, 997-1000.

Hoberg, E.P. (1989) Phylogenetic relationships among genera of the Tetrabothriidae (Eucestoda). *Journal of Parasitology*, **75**, 617-626.

Hoberg, E.P. (1992) Congruent and synchronic patterns in biogeography and speciation among seabirds, pinnipeds, and cestodes. *Journal of Parasitology*, **78**, 601-615.

Hoberg, E.P. (1995) Historical biogeography and modes of speciation across high-latitude seas of the Holartic: concepts for host-parasite coevolution among the Phocini (Phocidae) and Tetrabothriidae (Eucestoda). *Canadian Journal of Zoology*, **73**, 45-57.

Hoberg, E.P. (1996) Faunal diversity among avian parasite assemblages: the interaction of history, ecology, and biogeography in marine systems. *Bulletin of the Scandinavian Society of Parasitology*, **6**, 65-89.

Hoberg, E.P. (1997) Phylogeny and historical reconstruction: host-parasite systems as keystones in biogeography and ecology. In: *Biodiversity II: understanding and protecting our resources* (Ed. by M. Reaka-Kudla, E.O. Wilson and D. Wilson), pp. 243-261, Joseph Henry Press, National Academy of Sciences, Washington, D.C.

Hoberg, E.P. and Adams, A.M. (1992) Phylogeny, historical biogeography, and ecology of *Anophryocephalus* spp. (Eucestoda: Tetrabothriidae) among pinnipeds of the Holartic during the late Tertiary and Pleistocene. *Canadian Journal of Zoology*, **70**, 703-719.

Hoberg, E.P., Brooks, D.R. and Siegel-Causey, D. (1997) Host-parasite cospeciation: history, principles and prospects. In: *Host-parasite evolution: general principles and avian models* (Ed. by D. Clayton and J. Moore), pp. 212-235. Oxford University Press, Oxford.

Hoberg, E.P., Gardner, S.L. and Campbell, R.A. (1999) Systematics of the Eucestoda: advances toward a new phylogenetic paradigm, and observations on the early diversification of tapeworms and vertebrates. *Systematic Parasitology*, **42**, 1-12.

Hoberg, E.P., Jones, A. and Bray, R.A. (1999) Phylogenetic analysis of the families of the Cyclophyllidea based on comparative morphology with new hypotheses for co-evolution in vertebrates. *Systematic Parasitology*, **42**, 51-73.

Holmes, J.C. (1982) Impact of infectious disease agents of the population growth and geographical distribution of animals. In: *Population biology of infectious diseases* (Ed. by R.M. Anderson and R.M. May), pp. 37-51. Springer-Verlag, Berlin, Heidelberg, New York.

Holmes, J.C. (1987) The structure of helminth communities. *International Journal for Parasitology*, **17**, 203-208.

Holmes, J.C. (1990) Helminth communities in marine fishes. In: *Parasite communities: patterns and processes* (Ed. by G.W. Esch, A.O. Bush and J.M. Aho), pp. 101-130. Chapman and Hall, London.

Howard, E.B., Britt, J.O., Matsumoto, G., Itahara, R. and Nagano, C.N. (1983) Bacterial diseases. In: *Pathobiology of Marine Mammal Diseases*, 1. (Ed. by E.B. Howard), pp. 69-118. CRC Press, Boca Raton.

Hudson, P.J. and Dobson, A.P. (1995) Macroparasites: observed patterns in naturally fluctuating animal populations. In: *Ecology of infectious diseases in natural populations* (Ed. by B.T. Grenfell and A.P. Dobson), pp. 144-176. Cambridge University Press, Cambridge.

Jahans, K.L., Foster, G. and Broughton, E.S. (1997) The characterisation of *Brucella* strains isolated from marine mammals. *Veterinary Microbiology*, **57**, 373-382.

Janovy, J. (Jr.), Clopton, R.E. and Percival, T.J. (1992) The roles of ecological and evolutionary influences in providing structure to parasite species assemblages. *Journal of Parasitology*, **78**, 630-640.

Jensen, T. and Andersen, K. (1995) The importance of sculpin (*Myoxocephalus scorpius*) as intermediate host and transmitter of the sealworm *Pseudoterranova decipiens*. *International Journal for Parasitology*, **22**, 665-668.

Kennedy, C.R. and Bush, A.O. (1994) The relationship between pattern and scale in parasite communities: a stranger in a strange land. *Parasitology*, **109**, 187-196.

Kennedy, C.R. and Guégan, J.F. (1994) Regional versus local helminth parasite richness in British freshwater fish: saturated or unsaturated parasite communities? *Parasitology*, **109**, 175-185.

Kennedy, S. (1998) Morbillivirus infections in aquatic mammals. *Journal of Comparative Pathology*, **119**, 201-225.

Kim, K.C. (1985a) Evolutionary relationships of parasitic arthropods and mammals. In: *Coevolution of parasitic arthropods and mammals* (Ed. by K.C. Kim), pp. 3-82. Wiley, New York.

Kim, K.C. (1985b) Evolution and host associations of Anoplura. In: *Coevolution of parasitic arthropods and mammals* (Ed. by K.C. Kim), pp. 197-232. Wiley, New York.

Køie, M., Berland, B. and Burt, D.B. (1995) Development of third-stage larvae occurs in the eggs of *Anisakis simplex* and *Pseudoterranova decipiens* (Nematoda, Ascaridoidea, Anisakidae). *Canadian Journal of Fisheries and Aquatic Sciences*, **52**, 134-139.

Lambertsen, R.H. (1986) Disease of the common fin whale (*Balaenoptera physalus*): Crassicaudosis of the urinary system. *Journal of Mammalogy*, **67**, 353-366.

Lauckner, G. (1985a) Diseases of Mammalia: Carnivora. In: *Diseases of marine animals, Vol. IV, Part 2* (Ed. by O. Kinne), pp. 645-682. Biologische Anstalt Helgoland, Hamburg.

Lauckner, G. (1985b) Diseases of Mammalia: Pinnipedia. In: *Diseases of marine animals, Vol. IV, Part 2* (Ed. by O. Kinne), pp. 683-793. Biologische Anstalt Helgoland, Hamburg.

Lauckner, G. (1985c) Diseases of Mammalia: Sirenia. In: *Diseases of marine animals, Vol. IV, Part 2* (Ed. by O. Kinne), pp. 795-803. Biologische Anstalt Helgoland, Hamburg.

Lawson, P.A., Foster, G., Falsen, E., Sjödén, B. and Collins, B.D. (1999) *Abiotrophia balaenopterae* sp. nov. isolated from the minke whale (*Balaenoptera acutorostrata*). *International Journal of Systematic Bacteriology*, **49**, 503-506.

Lester, R.J.G. (1990) Reappraisal of the use of parasites for fish stock identification. *Australian Journal of Marine and Freshwater Research*, **41**, 855-864.

Levin, S.A. (1992) The problem of pattern and scale in ecology. *Ecology*, **73**, 1943-1967.

Mackenzie, K. and Abaunza, P. (1998) Parasites as biological tags for stock discrimination of marine fish: a guide to procedures and methods. *Fisheries Research*, **38**, 45-56.

Marcogliese, D.J. (1996) Transmission of the sealworm, *Pseudoterranova decipiens* (Krabbe), from invertebrates to fish in an enclosed brackish pond. *Journal of Experimental Marine Biology and Ecology*, **205**, 205-219.

Marcogliese, D.J. (1997) Fecundity of sealworm (*Pseudoterranova decipiens*) infecting grey seals (*Halichoerus grypus*) in the Gulf of St. Lawrence, Canada: lack of density-dependent effects. *International Journal for Parasitology*, **27**, 1401-1409.

Marcogliese, D.J. and Price, J. (1997) The paradox of parasites. *Global Biodiversity*, **7**, 7-15.

Margolis, L., Groff, J.M., Johnson, S.C., McDonald, T.E., Kent, M.L. and Blaylock, R.B. (1997) Helminth parasites of sea otters (*Enhydra lutris*) from Prince William Sound, Alaska: comparisons with other populations of sea otters and comments on the origin of their parasites. *Journal of the Helminthological Society of Washington*, **64**, 161-168.

Mattiucci, S., Nascetti, G., Cianchi, R., Paggi, L., Arduino, P., Margolis, L., Brattey, J., Webb, S., D'Amelio, S., Orecchia, P. and Bullini, L. (1997) Genetic and ecological data on the *Anisakis simplex* complex, with evidence for a new species (Nematoda, Ascaridoidea, Anisakidae). *Journal of Parasitology*, **83**, 401-416.

May, R.M. (1991) The dynamics and genetics of host-parasite associations. In: *Parasite-host associations coexistence or conflict?* (Ed. by C.A. Toft, A. Aeschlimann and L. Bolis), pp. 102-128. Oxford University Press, Oxford.

McClelland, G. (1980) *Phocanema decipiens*: growth, reproduction, and survival in seals. *Experimental Parasitology*, **49**, 175-187.

McClelland, G. (1990) Larval sealworm (*Pseudoterranova decipiens*) infections in benthic macrofauna. *Canadian Bulletin of Fisheries and Aquatic Sciences*, **222**, 47-65.

McClelland, G., Misra, R.K. and Martell, D.J. (1990) Larval anisakine nematodes in various fish species from Sable Island Bank and vicinity. *Canadian Bulletin of Fisheries and Aquatic Sciences*, **222**, 83-118.

McIntosh, R.P. (1987) Pluralism in ecology. *Annual Review of Ecology and Systematics*, 18, 321-341.

Measures, L.N. and Hong, H. (1995) The number of moults in the egg of sealworm, *Pseudoterranova decipiens* (Nematoda: Ascaridoidea): an ultrastructural study. *Canadian Journal of Fisheries and Aquatic Sciences*, **52**, 156-160.

Migaki, G. and Jones, S.R. (1983) Mycotic diseases. In: *Pathobiology of marine mammal diseases. Volume II.* (Ed. by E. B. Howard), pp. 1-28. CRC Press, Boca Raton.

Murray, M.D. and Nicholls, D.G. (1965) Studies on the ectoparasites of seals and penguins. I. The ecology of the louse *Lepidophthirus macrorhini* Enderlein on the southern elephant seal, *Mirounga leonina* (L.). *Australian Journal of Zoology*, **13**, 437-454.

Murray, M.D., Smith, M.S.R. and Soucek, Z. (1965) Studies on the ectoparasites of seals and penguins. II. The ecology of the louse *Antarctophthirus ogmorhini* Enderlein on the Weddell seal, *Leptonichotes weddelli* Lesson. *Australian Journal of Zoology*, **13**, 761-771.

Nagasawa, S., Holmes, R.W., and Nemoto, T. (1990) The morphology of the cetacean diatom genus *Plumosigma* Nemoto. *Scientific Reports of Cetacean Research*, 1, 85-91.

Olsen, O.W. and Lyons, E.T. (1965) Life cycle of *Uncinaria lucasi* Stiles, 1901 (Nematoda: Ancylostomatidae) of fur seals, *Callorhinus ursinus* Linn., on the Pribilof Islands, Alaska. *Journal of Parasitology*, **51**, 689-700.

Onderka, D. (1989) Prevalence and pathology of nematode infections in the lungs of ringed seals (*Phoca hispida*) of the western Arctic of Canada. *Journal of Wildlife Diseases*, **25**, 218-224.

Oshima, T. and Kliks, M. (1987) Effects of marine mammal parasites on human health. *International Journal for Parasitology*, **17**, 415-421.

Osterhaus, A., Groen, J., Niesters, H., van de Bildt, M., Martina, B., Vedder, L., Vos, J., van Egmond, H., Abou Sidi, B. and Ely Ould Barham, M. (1997) Morbillivirus in monk seal mass mortality. *Nature*, **388**, 838-839.

Osterhaus, A.D.M.E., UytdeHaag, F.G.C.M., Visser, I.K.G., Veder, E.J., Reijnders, P.J.M., Kuiper, J. and Brugge, H.N. (1989) Seal vaccination success. *Nature*, **337**, 21.

Osterhaus, A.D.M.E., van de Bildt, M., Vedder, L., Martina, B., Niesters, H., Vos, J., van Egmond, H., Liem, D., Baumann, R., Androukaki, E., Kotomatas, S., Komnenou, A., Sidi, B.A., Jiddou, A.B. and Barham, M.E.O. (1998) Monk seal mortality: virus or toxin? *Vaccine*, **16**, 979-981.

Osterhaus, A.D.M.E., Visser, I.K.G., de Swart, R.L., van Bressem, M.F., van de Bildt, M.W.G., Örvell, C., Barrett, T. and Raga, J.A. (1992) Morbillivirus threat to Mediterranean monk seals? *Veterinary Record*, **130**, 141-142.

Pascual-Ramos, C., Foster, G. and Collins, M.D. (1997) Phylogenetic analysis of the genus *Actinomyces* based on 16s rRNA gene sequences: description of *Arcanobacterium phocae* sp. nov., *Arcanobacterium bernadiae* comb. nov., and *Arcanobacterium pyogenes* comb. nov. *International Journal of Systematic Bacteriology*, **47**, 46-53.

Pascual-Ramos, C., Foster, G., Alvarez, N. and Collins, M.D. (1998) *Corynebacterium phocae* sp. nov., isolated from the common seal (*Phoca vitulina*). *International Journal of Systematic Bacteriology*, 48, 601-604.

Perrin, W.F. and Powers, J.E. (1980) Role of a nematode in natural mortality of spotted dolphins. *Journal of Wildlife Management*, **44**, 960-963.

Pinel, C., Beaudevin, M., Chermett, R., Grillot, R. and Ambroise-Thomas, P. (1996) Gastric anisakidosis due to *Pseudoterranova decipiens* larva. *The Lancet*, **347**, 1829.

Poulin, R. (1998) *Evolutionary ecology of parasites*. Chapman and Hall, London. 212pp.

Price, P.W. (1980) *Evolutionary biology of parasites*. Princeton University Press, Princeton. 237pp.

Raga, J.A. and Balbuena, J.A. (1993) Parasites of the long-finned pilot whale, *Globicephala melas* (Traill, 1809), in European waters. *Report of the International Whaling Commission* (special issue **14**), 391-406.

Raga, J.A., Balbuena, J.A., Aznar, F.J. and Fernández, M. (1997) The impact of parasites on marine mammals: a review. *Parassitologia*, **39**, 293-296.

Ricklefs, R.E. and Schluter, D. (1993) *Species diversity in ecological communities: Historical and geographical perspectives*. University of Chicago Press, Chicago. 414pp.

Romano, M.I., Alito, A., Bigi, F., Fisanotti, J.C. and Cataldi, A. (1995) Genetic characterization of mycobacteria from South American wild seals. *Veterinary Microbiology*, **47**, 89-98.

Ross, P., de Swart, R., Addison, R., van Loveren, H., Vos, J. and Osterhaus, A. (1996) Contaminant-induced immunotoxicity in harbour seals: wildlife at risk? *Toxicology*, **112**, 157-169.

Roughgarden, J. (1989) The structure and assembly of communities. In: *Perspectives in ecological theory* (Ed. by J. Roughgarden, R.M. May and S.A. Levin), pp. 203-226. Princeton University Press, Princeton.

Roughgarden, J. and Diamond, J. (1986) Overview: The role of species interactions in community ecology. In: *Community ecology* (Ed. by J. Diamond and T.J. Case), pp. 333-343. Harper and Row, New York.

Ruhnke, H.L. and Madoff, S. (1992) *Mycoplasma phocidae*, new species isolated from harbor seals (*Phoca vitulina* L.). *International Journal of Systematic Bacteriology*, **42**, 211-214.

Smith, G. (1994) Parasite population density is regulated. In: *Parasitic and infectious diseases; epidemiology and ecology* (Ed. by M.E. Scott and G. Smith), pp. 47-63, Academic Press, San Diego.

Stobo, W.T., Beck, B. and Fanning, L.P. (1990) Seasonal sealworm (*Pseudoterranova decipiens*) abundance in grey seals (*Halichoerus grypus*). *Canadian Bulletin of Fisheries and Aquatic Sciences*, **222**, 147-162.

Stock, T.M. and Holmes, J.C. (1988) Functional relationships and microhabitat distributions of enteric helminths of grebes (Podicipedidae): the evidence for interactive communities. *Journal of Parasitology*, **74**, 214-227.

Stork, N.E. and Lyal, H.C. (1993) Extinction or co-extinction rates? *Nature*, **366**, 307.

Stuen, S., Have, P., Osterhaus, A.D.M.E., Arnemo, J.M. and Moustgaard, A. (1994) Serological investigations of virus infections in harp seals (*Phoca groenlandica*) and hooded seals (*Cystophora cristata*). *Veterinary Record*, **134**, 502-503.

Swenshon, M., Lammler, C. and Siebert, U. (1998) Identification and molecular characterization of beta-hemolytic streptococci isolated from harbor porpoises (*Phocoena phocoena*) of the North and Baltic seas. *Journal of Clinical Microbiology*, **36**, 1902-1906.

Swinton, J., Harwood, J., Grenfell, B.T. and Gillian, C.A. (1998) Persistence thresholds for phocine distemper virus infection in harbour seal *Phoca vitulina* metapopulations. *Journal of Animal Ecology*, **67**, 54-68.

Thomas, F., Mete, K., Helluy, S., Santalla, F., Verneau, O., de Meeus, T., Cezilly, F. and Renaud, F. (1997) Hitch-hiker parasites or how to benefit from the strategy of another parasite. *Evolution*, **51**, 1316-1318.

Thorel, M.F., Karoui, C., Varnerot, A., Fleury, C. and Vicent, V. (1998) Isolation of *Mycobacterium bovis* from baboons, leopards, and a sea-lion. *Veterinary Research*, **29**, 207-212.

Thornton, S. M., Nolan, S. and Gulland, F. (1998) Bacterial isolates from California sea lions (*Zalophus californianus*), harbor seals (*Phoca vitulina*), and northern elephant seals (*Mirounga angustirostris*) admitted to a rehabilitation center along the central California coast, 1994-1995. *Journal of Zoo and Wildlife Medicine*, **29**, 171-176.

Timm, R.M. and Clauson, B.L. (1985) Mammals as evolutionary partners. In: *Coevolution of parasitic arthropods and mammals* (Ed. by K.C. Kim), pp. 101-154. Wiley, New York.

Van Bressem, M.F., van Waerebeek, K. and Raga, J.A. (1999) A review of virus infections of cetaceans and the potential impact of morbilliviruses, poxviruses and papillomaviruses in host population dynamics. *Diseases of Aquatic Organisms*, **38**, 53-65.

Visser, I.K.G. (1993) *Morbillivirus infections in seals, dolphins and porpoises*. Ph.D. Thesis, University of Utrecht, Utrecht. 167pp.

Visser, I.K.G., Teppema, J.S. and Osterhaus, A.D.M.E. (1991) Virus infections of seals and other pinnipeds. *Reviews in Medical Microbiology*, **2**, 105-114.

Visser, I.K.G., van de Bildt, M.W.G., Brugge, H.N., Reijnders, P.J.H., Vedder, E.J., Kuiper, J., de Vries, P., Groen, J., Walwoort, H.C., UytdeHaag, F.G.C.M. and Osterhaus, A.D.M.E. (1989) Vaccination of harbour seals (*Phoca vitulina*) against phocid distemper virus with two different inactivated canine distemper virus (CDV) vaccines. *Vaccine*, **7**, 521-526.

Wiens, J.A. (1984) On understanding a non-equilibrium world: myth and reality in community patterns and processes. In: *Ecological communities: conceptual issues and the evidence* (Ed. by D.R. Strong, Jr., D. Simberloff, L.G. Abele and A.B. Thistle), pp. 439-457. Princeton University Press, Princeton.

Windsor, D.A. (1995) Equal right for parasites. *Conservation Biology*, **9**, 1-2.

Zwanenburg, K.C.T. and Bowen, W.D. (1990) Population trends of the grey seal (*Halichoerus grypus*) in eastern Canada. *Canadian Bulletin of Fisheries and Aquatic Sciences*, **222**, 185-197.

Chapter 12

Marine Mammal Mass Mortalities

[1]MARIANO DOMINGO, [2]SEAMUS KENNEDY, and
[3,4]MARIE-FRANÇOISE VAN BRESSEM

[1]*Department of Veterinary Pathology, Autonomous University of Barcelona, E-08193 Bellaterra, Barcelona, Spain;* [2]*Veterinary Sciences Division, Department of Agriculture and Rural Development, Stoney Road, Stormont, Belfast BT4 3SD, Northern Ireland, UK;* [3]*Peruvian Center for Cetacean Research (CEPEC), Jorge Chávez 302, Pucusana, Lima 20, Peru;* [4]*Department of Vaccinology-Immunology, Faculty of Veterinary Medicine, University of Liège, Sart Tilman, 4000 Liège, Belgium. E-mail (MD): domingo@cc.uab.es*

1. INTRODUCTION

Marine mammal mass strandings and large-scale mortalities have attracted considerable scientific and public attention in recent years. Investigations of the cause or causes of mass mortalities have also expanded knowledge of harmful agents capable of threatening marine mammal populations. To establish an integrated, explanatory framework for such phenomena, acceptable to all types of expertise in marine mammal science is an extraordinarily difficult task. Divergence in the interpretation of the importance of noxious effects and their interrelationships in several recent mortalities has been the rule rather than the exception. In this chapter we will (i) discuss the main causes of recent mass mortalities, and (ii) describe some recent events affecting pinnipeds and cetaceans, with special focus on those episodes in which a morbilliviral aetiology was demonstrated as a primary cause of the die-offs. The roles of other deleterious agents, such as chemical pollutants, parasites, fisheries, and noise pollution on marine mammal populations are considered in other chapters of this book.

Marine Mammals: Biology and Conservation, edited by
Evans and Raga, Kluwer Academic/Plenum Publishers, 2002

2.   CAUSES OF MARINE MAMMAL MORTALITIES

Large-scale marine mammal mortalities should be distinguished from mass strandings of live, mainly healthy animals. Mass strandings are usually recorded in cetacean species, and are defined as events involving three or more animals (Klinowska, 1985). Their patterns and causes may be numerous (Robson, 1984; Geraci and Lounsbury, 1993) and include (i) topographic and oceanographic conditions, (ii) prey behaviour or escape from predators, (iii) social cohesive behaviour (the group follows a disoriented or sick member, the 'key whale'), (iv) errors of navigation while following geomagnetic contours, and (v) disturbance of echolocation in shallow waters. In general, these events are geographically and temporally restricted. Regardless of the triggering factors, the stress associated with stranding, inability to thermoregulate, and compromise of circulatory and respiratory functions often lead ultimately to death several hours after stranding if first aid is not provided.

By contrast, mass mortalities are different in several ways: they usually occur over several months, many specimens are found ashore dead or sick, and the mortality usually occurs over a large area.

Recent mass mortalities that have been investigated in depth have been ultimately shown to have been caused by biotoxins, viruses, bacteria, or the 'El Niño' event (see Table 1), and will be further discussed here. We also provide details on some die-offs for which one or more causes could not be definitively established. The impact of fisheries, an obvious cause of mortality, is discussed by Hall and Donovan in Chapter 14 of this volume.

2.1   Biotoxins

Approximately twenty of the more than 1,000 known dinoflagellate species produce toxins which cause mortality in fish, birds and mammals (Steidinger and Baden, 1984).

Brevetoxins and saxitoxins, the respective causes of neurotoxic shellfish poisoning and paralytic shellfish poisoning in humans, have been implicated in die-offs of marine mammals (see Table 1). Biotoxins are produced in large quantities in dinoflagellate blooms, and accumulate in filtering shellfish and several fish species. Birds and mammals are poisoned by feeding on organisms which concentrate the toxins.

In the spring of 1978, increased mortality was observed in Hawaiian monk seals (*Monachus schauinslandi*), affecting young and very old animals (Gilmartin *et al.*, 1987). Several of these seals were emaciated and had gastric ulceration with nematodes embedded in the stomach wall. Two animals tested for ciguatoxin and maitotoxin, neurotoxins produced by

Table 1. Review of recent mass mortalities of marine mammals worldwide

Cause(s)	Year	Place	Species	Mortality	References
Biotoxins					
Biotoxins	1962	St.George Island, USA	*Callorhinus ursinus*	300	Cited from St.Aubin, 1991
Possibly ciguatoxin and maitotoxin	1978	Hawaii Islands, USA	*Monachus schauinslandi*	>50	Gilmartin et al., 1980
Brevetoxins	1982	Florida, USA	*Trichechus manatus*	>40	O'Shea et al. 1991
Saxitoxin	1987-1988	Cape Cod Bay, USA	*Megaptera novaeangliae*	14	Geraci et al.1989
Brevetoxin	1996	Florida, USA	*Trichechus manatus*	>149	Bossart et al.1998
Viruses					
Influenza virus	1979-1980	New England, USA	*Phoca vitulina*	>445	Geraci et al., 1982
Morbillivirus (PMV or DMV)	1982	Florida, USA	*Tursiops truncatus*	40-50	Hersh et al., 1990; Duignan et al., 1996
Morbillivirus (CDV)	1987	Baikal Lake	*Phoca sibirica*	2000	Grachev et al., 1989
Morbillivirus (PMV/DMV)	1987-1988	East coast USA	*Tursiops truncatus*	>742	Lipscomb et al., 1994; Taubenberger et al., 1996
Morbillivirus (PDV)	1988	North Sea	*Phoca vitulina*	18,000	Kennedy et al., 1988a; Osterhaus and Vedder, 1988
Morbillivirus (PDV)	1988	North Sea	*Halichoerus grypus*	300	Harwood et al., 1989
Morbillivirus (DMV)	1990-1992	Mediterranean Sea	*Stenella coeruleoalba*	>4000	Domingo et al., 1990; Van Bressem et al., 1991
Morbillivirus (PMV)	1993-1994	Gulf of Mexico	*Tursiops truncatus*		Krafft et al., 1995; Taubenberger et al., 1996
Morbillivirus	1994	Black Sea	*Delphinus delphis ponticus*	47	Birkun et al.,1999
Bacteria					
Leptospira sp.	1970	California and Oregon, USA	*Zalophus californianus*		Vedros et al. 1971
Leptospira sp.	1984	California	*Zalophus californianus*	>160	Dierauf et al. 1985
El Niño					
id.	1982-1983	Peru to California	Several species of pinnipeds		Trillmich and Ono 1991
id.	1992-1993	California	Several species of pinnipeds		Vidal and Gallo-Reynoso 1996
id.	1997-1998	Peru	Pinnipeds and cetaceans		Echegaray et al., 1998; Majluf et al., 1998
Undetermined					
id.	1918	Iceland	*Phoca vitulina*	>1000	Harwood and Hall, 1990
id. (morbillivirus-CDV?)	1955	Antarctica	*Lobodon carcinophagus*	>2000	Laws and Taylor 1957
id.	1978	Alaska	*Odobenus rosmarus*	Nearly 1000	Harwood and Hall, 1990
id.	1995	Upper Gulf of California, Mexico	Several species of marine mammals	>425	Vidal and Gallo-Reynoso 1996
id. (Biotoxins? /DMV?)	1997	Mauritania	*Monachus monachus*	About 200	Aguilar et al., 1998
id. (Morbillivirus-CDV?)	1997	Caspian Sea	*Phoca caspica*	Thousands	Forsyth et al., 1998
id.	1998	New Zealand	*Phocarctos hookeri*	>2300	Duignan, 1999

benthic dinoflagellates associated with tropical reefs, were positive (Gilmartin, 1987; Gilmartin *et al.*, 1987). Ciguatera was known to be present in at least one fish species eaten by monk seals. It was concluded that either the biotoxins or gastric parasites were responsible for the death of these animals (Gilmartin *et al.*, 1987).

Mass mortality, apparently associated with biotoxins, occurred in humpback whales (*Megaptera novaeangliae*) in 1987-88 (Geraci *et al.*, 1989). During a 5-week period, 14 humpback whales of both sexes stranded dead along the beaches of Massachusetts. Carcasses had abundant hypodermal fat and limited histopathologic examination carried out on three animals did not reveal any significant lesions. Geraci *et al.* (1989) concluded that the whales had died from saxitoxin poisoning after consumption of Atlantic mackerel (*Scomber scombrus*) containing the toxin.

Between February and April 1982, a total of 41 Florida manatees (*Trichechus manatus*) were found dead in southwestern Florida (O'Shea *et al.*, 1991). The epidemic was coincident with a *Gymnodinium breve* bloom as well as with death and morbidity of doubled-crested cormorants (*Phalacrocorax auritus*) and several species of fishes. Both sexes and a wide range of size classes were represented. The deaths of two of the 39 recovered manatees were attributed to collisions with boats and considered to be unrelated to the outbreak. The other carcasses were in good body condition and few gross lesions were detected at necropsy. Most manatees had fed recently. Ascidians (filter-feeding tunicates) were found in the intestinal contents in nearly 80% of the necropsied manatees, and tunicate ingestion was believed to have been the cause of the poisoning. Histologically there were no significant lesions (O'Shea *et al.*, 1991). Brevetoxins cause acute neurotoxicity, and no consistent macroscopic or microscopic lesions are to be expected from the action of these toxins, apart from those related to circulatory and respiratory failure (congestion and oedema in lungs and several other organs) (Steidinger and Baden, 1984).

From March to May 1996, another mass mortality affected the Florida manatee population, causing the death of at least 149 individuals (Bossart *et al.*, 1998). This mortality event also has been attributed to brevetoxin poisoning. There was no evidence of morbillivirus infection in tissues of affected manatees (Kennedy and Lipscomb, unpubl.).

2.2　Viruses

Knowledge of viruses of cetaceans, pinnipeds and sirenians is very recent (see Aznar *et al.*, this volume). Recent marine mammal mass mortalities have provided the impetus for the investigation of viruses in these species (Visser *et al.*, 1993a). Members of the genera *Influenza* and

Morbillivirus may cause large-scale mortalities in cetaceans and pinnipeds, as in terrestrial mammals (Appel *et al.*, 1981; Fenner *et al.*, 1993).

Figure 1.   Carcasses of Florida manatee recovered in one afternoon at the height of the 1982 mass mortality attributed to brevetoxins. A trench was dug and several tons of ice laid over the bodies. Some of the ice has been removed for the photograph. (Photo: S. Wright)

2.2.1   Influenza A Viruses

Viruses of the genus *Influenza* A and B (family *Orthomyxoviridae*) are pleomorphic enveloped virions, 80-120 nm in diameter, with a segmented single-stranded RNA of negative sense polarity (Fenner *et al.*, 1993). Influenza A strains may infect humans and other mammalian species, as well as a large variety of birds, and cause epidemics of respiratory disease (Murphy and Webster, 1996). Natural transmission is by aerosol, but may also be water-borne, and may occur between species of different orders and classes (Fenner *et al.*, 1993).

Among marine mammals, the first influenza A virus-associated epidemic was reported in harbour seals (*Phoca vitulina*) along the coast of New England (Geraci *et al.*, 1982). Between December 1979 and October 1980, more than 445 harbour seals, mostly immature animals, died along the New England coast (Geraci *et al.*, 1982). Affected seals were weak and had respiratory problems. Post-mortem examination revealed pneumonia characterised by necrotising bronchitis and bronchiolitis, and haemorrhagic alveolitis. An influenza A virus was isolated from the lungs and brains of the dead seals. A mycoplasma was also consistently recovered from the lungs of these animals. The virus was considered to be the cause of the outbreak of acute pneumonia while the mycoplasma was thought to have increased the severity of the disease (Geraci *et al.*, 1982). The seal influenza A virus was of avian origin and zoonotic, as indicated by the development of conjunctivitis in an accidentally infected person (Geraci *et al.*, 1982). Since this outbreak, another epidemic of pneumonia associated with influenza A virus was described in harbour seals off the New England coast in 1982-83. Influenza viruses were isolated from harbour seals which died during periods of increased strandings along the Cape Cod peninsula, Massachusetts, in January 1991, and from September 1991 to April 1992. These animals had pathological lesions of acute interstitial pneumonia and subcutaneous emphysema (Callan *et al.*, 1995). However, the heightened mortality of September 1991 to April 1992 was associated with a higher prevalence of phocine distemper virus (PDV) antibodies in those animals and coincided with morbillivirus disease in seals from the northeast Atlantic (Callan *et al.*, 1995; Duignan *et al.*, 1995a) and the actual cause of the mortality is therefore unclear.

2.2.2   Morbilliviruses

Viruses of the genus *Morbillivirus* (family *Paramyxoviridae*) are pleomorphic, enveloped virions about 150 nm in diameter with a single-stranded RNA of negative sense polarity (Fenner *et al.*, 1993). This genus comprises a group of viruses which are highly pathogenic in their respective hosts (Appel *et al.*, 1981). Up to 1988, only four members of this genus were known: measles virus (MV) of humans and other primates, canine distemper virus (CDV) of carnivores, and the two ruminant morbilliviruses, rinderpest virus (RPV) and peste-des-petits ruminants virus (PPRV). Since then, two other members of the genus have been identified - phocine distemper virus (PDV), which affects pinnipeds (Cosby *et al.*, 1988, Mahy *et al.*, 1988, Curran *et al.*, 1990, Visser *et al.*, 1991), and cetacean morbillivirus. The latter includes two strains: the porpoise (PMV) and dolphin (DMV) morbilliviruses which are closely related (Trudgett *et al.*, 1991; Welsh *et al.*, 1992; Barrett *et al.*, 1993; Visser *et al.*, 1993b).

The first morbillivirus of marine mammals was discovered in 1988 (see Table 1) when a die-off occurred among harbour seals and grey seals (*Halichoerus grypus*) on the coast of northern Europe (Kennedy *et al.*, 1988a; Osterhaus and Vedder, 1988). This mass mortality was caused by a previously unknown morbillivirus, christened phocine distemper virus (Cosby *et al.*, 1988). Nucleotide sequence analysis of PDV subsequently showed its distinctness from CDV (Cosby *et al.*, 1988; Mahy *et al.*, 1988; Curran *et al.*, 1990).

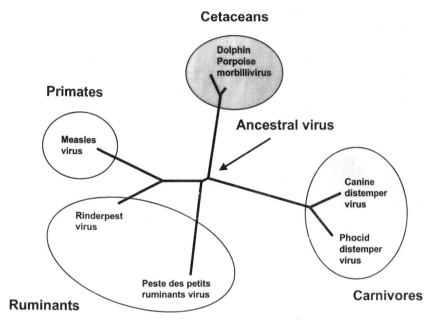

Figure 2.   Phylogenetic tree showing the relationships between the different morbilliviruses based on partial sequence of the P gene. A universal primer set was used to amplify a 429 base pair DNA fragment as described by Barrett *et al.*, 1993. The tree was derived using PHYLIP DNADIST and FITCH programmes (Felsenstein, J. 1989 Phylip 3.2. Manual. University of California, Herbarium, Berkeley, California). The branch lengths are proportional to the mutational differences between the viruses and the hypothetical common ancestor that existed at the nodes in the tree. Reproduced with permission from Barrett and Rima.

It subsequently became known that an epidemic had killed over 2,000 Baikal seals (*Phoca sibirica*) in Lake Baikal between autumn 1987 and October 1988 (Grachev *et al.*, 1989; Osterhaus *et al.*, 1989a; Heide-Jørgensen *et al.*, 1992). Seals had clinical signs resembling those observed in harbour seals during the PDV outbreak. However, the epidemic was caused by CDV (Osterhaus *et al.*, 1989b) and not by PDV as initially suspected.

These two incidents showed that (1) a newly recognised member of the genus *Morbillivirus* had been introduced to European pinnipeds causing an epidemic, and (2) epidemics in marine mammals could also be caused by morbilliviruses of terrestrial mammals (in this case CDV), if the virus jumps between the terrestrial and marine mammal species during interspecific contact.

Figure 3.   A stranded striped dolphin during the morbillivirus epidemic in the Mediterranean that started in autumn 1990. (Photo: J.A. Raga)

The next identified epidemic of morbilliviral disease in marine mammals affected the Mediterranean striped dolphin (*Stenella coeruleoalba*) (Fig. 3) from 1990 to 1992 and was caused by DMV. This epidemic brought attention to a morbillivirus previously identified in cetaceans, which caused a fatal infection in six harbour porpoises (*Phocoena phocoena*) from the Irish Sea in 1988 (Kennedy *et al.*, 1988b, Kennedy *et al.*, 1991). Characterisation of the porpoise and dolphin morbilliviruses revealed that they were previously unrecognised members of the genus *Morbillivirus*, antigenically and genetically very similar. They probably represent different strains of the same viral species for which the name cetacean morbillivirus was suggested (McCullough *et al.*, 1991; Trudgett *et al.*, 1991; Osterhaus *et al.*, 1992; Barrett *et al.*, 1993; Visser *et al.*, 1993b; Blixenkrone-Moller *et al.*, 1996). They are distinct from other morbilliviruses, and more closely related to the ruminant morbilliviruses (RPV, PPRV) than to the carnivore morbilliviruses (CDV and PDV) (Fig. 2; Barrett *et al.*, 1993; Visser *et al.*, 1993b).

The study of the role of morbilliviruses in previous cetacean mass mortalities then focused on the 1987-88 bottlenose dolphin (*Tursiops*

truncatus) die-off on the east coast of the USA. Results of this inves
showed that morbilliviral disease was prevalent in affected bottl
dolphins, and that both strains of cetacean morbillivirus, PMV and DM
had a primary role in the die-off (Lipscomb *et al.*, 1994; Krafft *et al.*, 1995,
Lipscomb *et al.*, 1996; Taubenberger *et al.*, 1996). Cetacean morbillivirus
also caused outbreaks of mortality in bottlenose dolphins from the
Indian/Banana river system (Florida) and the Gulf of Mexico in 1982 and
1993-94, respectively (Krafft *et al.*, 1995; Duignan *et al.*, 1996; Lipscomb *et
al.*, 1996; Taubenberger *et al.*, 1996). Finally, a morbillivirus is suspected to
have caused an increased mortality in common dolphins (*Delphinus delphis
ponticus*) from the Black Sea in 1994 (Birkun *et al*, 1998).

2.3   Bacteria

From late August to early November 1970, an epidemic of lethal disease
spread among subadult male California sea lions (*Zalophus californianus*)
along the coasts of northern California and Oregon. One hundred and
twenty-five dead sea lions were seen in California, while 230 sick and 5
dead sea lions were observed in one day on Shell Island, Oregon (Vedros *et
al.*, 1971). Diseased animals from California were depressed, extremely
thirsty, and had fever. All necropsied specimens had interstitial nephritis
and large numbers of spirochetes, identified as *Leptospira* sp., in the kidney
lesions. Leptospires were also seen by dark-field examination in kidney
suspensions from all specimens, and were isolated from kidneys, urine and
blood. Ten affected sea lions tested for serum antibodies against *Leptospira*
sp. had high titres against *L. pomona* and *L. autumnalis*. These results
strongly indicated that *Leptospira* sp. was the aetiological agent of this
epidemic (Vedros *et al.*, 1971). Between May and December 1984, a further
epidemic of leptospirosis struck California sea lions along the west coast of
the USA from Monterey County, California, to Seattle, Washington
(Dierauf *et al.*, 1985). More than 160 animals died in northern California.
This epidemic primarily affected juvenile and subadult males migrating
northwards from breeding rookeries of southern California's Channel
Islands (Dierauf *et al.*, 1985). *Leptospira pomona* has been associated with
abortions in California sea lions and with the 'multiple haemorrhagic
perinatal complex' in this species and in northern fur seals (*Callorhinus
ursinus*) from the Bering Sea (Smith *et al.*, 1974, 1977).

2.4   El Niño

El Niño is a meteorological and oceanographic phenomenon that occurs
at regular intervals in the eastern tropical Pacific. It is characterised by the

...nce, for 6 to 18 months, of anomalously warm water ...atorial ocean off Peru and Ecuador (Barber and ...ompanied by large reductions in zooplankton, fish ...nally rich waters of the eastern equatorial Pacific ...eduction of the primary production (Barber and ...,. During the El Niño of 1982-83, high mortalities of pinnipeds were observed from Peru to California (Trillmich and Ono, 1991). The seal populations in the area of Galapagos and Peru suffered most of the impact of this event (Trillmich and Dellinger, 1991). Nearby all of the Galapagos fur seal (*Arctocephalus galapagoensis*) pups born in 1982 at Fernandina Island, Ecuador, were dead after five months (Trillmich and Dellinger, 1991). A 100% mortality was also registered among dependent juveniles (1- and 2-years-old animals) between September 1982 and March 1983 (Trillmich and Dellinger, 1991). The pup and dependent juvenile death probably resulted from lack of foraging success of the females which stayed at sea for longer than in a normal year and were unable to provide enough milk to prevent pups from starving (Barber and Chavez, 1983). In addition, all adult territorial male fur seals starved to death after the breeding period, and an estimated 50% of the adult females also perished during this El Niño (Trillmich and Dellinger, 1991). Only 4% of the Galapagos sea lion (*Zalophus californianus wollebaeki*) pups born on Fernandina in 1982 survived to the age of one year, and a high yearling and adult mortality was registered in this species throughout the Galapagos Islands during the 1982-83 El Niño (Trillmich and Dellinger, 1991). A high mortality was also observed among pup, juvenile, and adult South American fur seals (*Arctocephalus australis*) and South American sea lions (*Otaria byronia*) at Punta San Juan, Peru, during this El Niño (Majluf, 1991). Besides, the increase in sea level and heightened storm activity which accompanied the El Niño at temperate latitudes, there was also an increase in the mortality rate in northern elephant seal (*Mirounga angustirostris*) pups at Año Nuevo and San Nicolas Islands (Trillmich *et al.*, 1991).

Trillmich and Ono (1991) provide an exhaustive account of the effects of El Niño on pinnipeds. Vidal and Gallo-Reynoso (1996) reported the mortality of several species of pinnipeds along the coast of California for the 1992-93 El Niño. During the El Niño of 1997-98, a high mortality was observed among South American sea lions and fur seals in Peru. At least 3,000 adult sea lions and 600 fur seals died between January and March 1998 around Punta San Juan (Majluf, 1998). This author estimated that total mortality for this event approached 40% for the sea lions and 10-20% for the fur seals. Premature births were observed among sea lions at the Ballestas Islands and Punta San Juan in November and December 1997, respectively. All these pups died within a few hours of birth (Majluf, 1998; K. Soto, *pers. comm.* to MFB). Besides, 85% to 90% of the pups normally born from December 1997 until February 1998 died of starvation because their mothers spent a longer time at sea foraging (K. Soto, *pers. comm.* to MFB). Cetaceans also suffered from this event. The death of 24.5% of 94 small cetaceans including dusky dolphins (*Lagenorhynchus obscurus*), bottlenose dolphins, common dolphins (*Delphinus* spp.) and Burmeister's porpoises (*Phocoena spinipinnis*) stranded on beaches around Pisco, Peru, between December 1997 and March 1998 was attributed to starvation due to El Niño (Echegaray *et al.*, 1998).

2.5   Undetermined Causes

Besides the events previously described, there are many other reports of mass mortalities in marine mammals, but the cause of most of them has not been investigated or could not be determined. Dietz *et al.* (1989) and Harwood and Hall (1990) briefly detailed some of these outbreaks. Here are described five well-documented die-offs of which the cause(s) could not definitively be established.

2.5.1   Mass Mortality in Crabeater Seals (*Lobodon carcinophagus*) in the Antarctic

From September until November 1995, a mass mortality was observed in crabeater seals wintering in the Crown Prince Gustave Channel (east coast of Graham Land) in the Antarctic. The disease spread rapidly among those seals but not to the intermingled Weddell seals (*Leptonychotes weddelli*). The average mortality in the 2,500 crabeater seal population was estimated at 85% (Laws and Taylor, 1957). Dead seals were in good body condition. Bleeding of the nose and mouth were often observed and some corpses had a swollen neck. Numerous abortions were observed concurrently with this mortality (Laws and Taylor, 1957). Histologically, there was acute congestion of the lungs, with areas of collapse, consolidation and emphysema, and evidence of nephritis. The die-off was considered to have been caused by a virus (Laws and Taylor, 1957). The lesions observed during the 1955 die-off and the finding of CDV-specific serum antibodies in 33% of 96 crabeater seal sera of all age classes, collected in 1989 near the Antarctic Peninsula (between 63°47'S, 56°27' and 67°31'S, 67°10'), led Bengston *et al.* (1991) to suggest that CDV was the cause of the mortality.

2.5.2   Die-off of Marine Mammals in the Gulf of California, Mexico

Between January and February 1995, a minimum of 366 dolphins, three fin whales (*Balaenoptera physalus*), two minke whales (*Balaenoptera acutorostrata*), one Bryde's whale (*Balaenoptera edeni*), two unidentified whales and 51 California sea lions died in the Upper Gulf of California (Vidal and Gallo-Reynoso, 1996). Most of the dead odontocetes were long-beaked common dolphins (*Delphinus capensis*) while the other identified specimens were bottlenose dolphins. Concurrently with this event, a mass-mortality was observed in seabirds from the same region (Vidal and Gallo-Reynoso, 1996). These authors reached the conclusion that biotoxins, viruses, or heavy metals were unlikely to have caused the die-off, and suggested that the animals died because an unknown toxic substance was discharged near the areas where they used to feed.

2.5.3   Mediterranean Monk Seal Die-off in Mauritania in 1997

From May to July of 1997, a mass mortality affected predominantly adult Mediterranean monk seals in the Mauritanian colony. It has been estimated that about 200 individuals from an initial population of 310 seals inhabiting the two main caves, died in this episode (Aguilar *et al.*, 1998; Fig. 4). Results of the investigations in several laboratories are contradictory, and, at the moment, there is no unified explanation of the cause of the die-off. Laboratory data indicate the presence of biotoxins including decarbamoyl saxitoxin, neosaxitoxin and gonyautoxin 1 in all tissues of the eight seals sampled and in the 18 fishes examined (Costas and Lopez-Rodas, 1998; Hernandez *et al.*, 1998). These authors suggested that the mortality was caused by poisoning by these biotoxins.

Other workers found evidence of a morbillivirus (related to the dolphin morbillivirus) infection in affected seals (Osterhaus *et al.*, 1997). These authors detected serum antibodies in an ELISA to canine distemper virus (CDV) in 7 out of 17 heart blood samples collected from cadavers. A still uncharacterised virus was isolated from seal tissues, and a positive result was obtained with a RT-PCR technique with primers specific for the genus *Morbillivirus*.

In a limited investigation of seal tissues, neither distemper-like lesions nor morbillivirus antigen were found (S. Kennedy and M. Domingo, unpubl. observs.). Although the number of seals investigated was low, these results do not support the participation of a morbillivirus in the die-off. An aetiological role for morbillivirus in this die-off has not therefore been established.

2.5.4   Mass Mortality in Caspian Seals (*Phoca caspica*)

In the spring of 1997, a mass mortality was observed among Caspian seals in Azerbaijan, on the western shore of the Caspian Sea (Forsyth *et al.*,

Figure 4.   Dead Mediterranean monk seal in West Africa. About two-thirds of the Mauritanian population were involved in a mass die-off during summer 1997. (Photo: A. Aguilar)

1998). Several thousand seals were believed to have died in the vicinity of the Aspheron peninsula. Histopathological examination of four seals revealed pulmonary congestion but no pneumonia nor evidence of brain lesions. However, lesions could have been masked by autolysis. Although no lesions were seen in any tissue examined, CDV nucleic acid was detected in the brain of one of four seals examined by the reverse transcriptase-polymerase chain reaction technique (RT-PCR) using 'universal' morbillivirus primers and a primer set specific for the CDV fusion (F) protein, but also by sequencing of the PCR products. These findings indicate that CDV was present in Caspian seals at the time of the outbreak, and may have played a role in this mortality. The virus may have been introduced by terrestrial carnivores such as dogs and wolves (*Canis lupus*) (Forsyth *et al.*, 1998).

2.5.5   New Zealand Sea Lion Mass Mortality in 1998

During January and February of 1998, a mass mortality affected the New Zealand sea lion population of the Auckland Islands. At least 1,600 to 2,300 pups died, and as many as 85 carcasses of adult individuals were collected at the beaches. Sick sea lions had locomotive disorders and dragged the hind limbs. The neck region was swollen, and there were raised cutaneous lesions. The results of investigations into the cause or causes of the mortality are not concluded. In ten pups examined, poor body condition and dehydration were prominent gross findings. Histological examination revealed interstitial pneumonia and bronchopneumonia, moderate lymphoid depletion, and suppurative encephalitis, but these lesions were not consistently found in all investigated animals. Adults had firm, red skin lesions, but internal abnormalities were not found. Histologically, these lesions consisted in suppurative dermatitis and vasculitis. Other microscopic lesions were bronchoneumonia, fibrinopurulent lymphadenitis and tonsillitis. Pleomorphic, gram negative bacteria were found associated with the pulmonary and dermal lesions. Virus isolation yielded negative results, and no evidence of a viral aetiology could be obtained by use of PCR or serological techniques. It was concluded that a fulminant septicaemia with vasculitis associated with bacterial infection was the most probable cause of the die-off (Duignan, 1999).

3.   EPIDEMICS CAUSED BY MORBILLIVIRUSES

Within a period of ten years, two recently recognised morbilliviruses have caused outbreaks of lethal disease in at least three cetacean species and two pinniped species. Scientists were confronted with the need to explain the sudden emergence of new highly pathogenic morbilliviruses distinct from previously known viruses. Although the lack of signs prior to the appearance of these epidemics suggested recent changes in one of the known morbilliviruses and subsequent spread to marine mammals, examination of archival sera and molecular virological data points to a longterm presence and circulation of these viruses in several species of marine mammal (Örvell and Sheshberadaran, 1991; Blixenkrone-Möller *et al.*, 1992; Visser *et al.*, 1993b; Blixenkrone-Möller *et al.*, 1996; for a review, see also Kennedy, 1998; Van Bressem *et al.*, 1999). Furthermore, molecular data indicate that divergence of these "new" viruses from other morbilliviruses occurred many years ago. The existence of immunologically naive marine mammal groups, and the introduction of morbilliviruses to them from other aquatic or terrestrial mammals are perhaps the critical factors involved in triggering an epidemic.

3.1   Morbillivirus Epidemic in Harbour Seals and Grey Seals in North-western Europe in 1988

The most serious morbillivirus mortality documented in pinnipeds occurred in the harbour seals, and to a minor extent, in the grey seals of north-western Europe in 1988. It has been estimated that approximately 18,000 harbour seals and over 300 grey seals died as a consequence of the PDV epidemic in northern European waters (Kennedy *et al.*, 1988a; Cosby *et al.*, 1988; Kennedy, 1990). The first affected seals were found on 12 April 1988 on the Danish island of Anholt. By early May, the epidemic had spread to the eastern Kattegat and into the Danish and Dutch Wadden Sea (Fig. 5). In June, deaths were also noted in the Skagerrak and the German Wadden Sea, and by July in the Oslo Fjord and the Southern Baltic (Heide-Jørgensen *et al.*, 1992). Thousands of dead seals were found in the Wadden

Figure 5.   Spread of the seal epidemic from April to October 1988 in the North Sea (modified from Harwood, 1989, 1990).

Sea. The epidemic spread south- and westwards, and reached England (east coast) and Ireland in late July and August, respectively. By September, sick seals were seen in most localities in the UK where these animals were known to occur regularly. Deaths were observed in seals in the UK until December 1988 (Heide-Jørgensen *et al.*, 1992; Hall *et al.*, 1992). Total mortality was estimated at approximately 40 to 60% in the Kattegat-Skagerrak and Wadden Sea, but there was wide variation between different areas (Heide-Jørgensen *et al.*, 1992).

Clinical signs observed in seals were those typical of canine distemper, and included respiratory, digestive, and nervous problems. Sick seals had mucopurulent oculonasal secretions, dyspnoea, coughing, diarrhoea, and nervous signs such as abnormal posture and muscle twitching (Kennedy, 1998). Abnormal buoyancy, due to subcutaneous emphysema, and behavioural alterations including increased tolerance to the presence of humans, were also noted. Abortions were observed at the beginning of the epidemic in the Kattegat.

3.2   Morbillivirus Epidemic in Baikal Seals

Between the autumn of 1987 and October 1988, several thousand Baikal seals died in Lake Baikal, Siberia, with typical signs of distemper. It was estimated that approximately 10% of this seal population had succumbed to the disease (Grachev *et al.*, 1989). Clinical signs were very similar to those of canine distemper in dogs, and to those reported later in seals which died during the distemper outbreak of 1988 in north-western Europe. The Lake Baikal epidemic was first attributed by Russian researchers to canine distemper virus (CDV), but when information on a new morbillivirus of European harbour seals was published, the opinion that PDV could have caused the epidemic gained acceptance. However, further characterisation of the virus isolated from Baikal seals showed that, in fact, it was a strain of CDV (Osterhaus *et al.*, 1989a, 1989b).

3.3   The 1990-92 Morbillivirus Epidemic in Mediterranean Striped Dolphins

A morbillivirus epidemic, caused by a newly recognised member of the genus *Morbillivirus*, the dolphin morbillivirus (DMV), ravaged the striped dolphins in the Mediterranean Sea in 1990-92 (Domingo *et al.*, 1990; Van Bressem *et al.*, 1991, 1993). The first dolphins affected by the disease were found moribund or dead in the vicinity of Valencia at the beginning of July 1990. In August, diseased dolphins were observed in Catalonia, the Balearic Islands, and south-eastern Spain (Aguilar and Raga, 1993). The epidemic subsequently expanded from the Spanish Mediterranean to the coasts of France and Italy in September and October 1990, respectively. Many striped dolphins stranded on the coasts of Morocco from September to November 1990. The number of diseased dolphins decreased significantly during the winter. However, between June 1991 and February 1992, the epidemic again hit striped dolphins, this time in waters around southern Italy and Greece. Cases were also reported in Turkey in 1992 (Bompar *et al.*, 1991; Bortolotto *et al.*, 1992; Webb, 1991; Aguilar and Raga, 1993; Van Bressem *et al.*, 1993; Cebrian, 1995; Osterhaus *et al.*, 1995). The age and sex distribution of the stranded specimens have been detailed elsewhere (Calzada *et al.*, 1994). Both sexes were equally affected but juvenile dolphins appeared to be only slightly affected by the epidemic.

The following data mainly refer to dolphins stranded on the Catalonian coast (Domingo *et al.*, 1992), although similar results have been reported for dolphins from other parts of Spain (Duignan *et al.*, 1992).

An overview of the strandings of striped dolphins along the Catalonian coast from August 1990 to August 1996 showed that strandings in this area decreased progressively from September to December 1990, to reach a figure of two to three striped dolphins per month, which seems to be the background level of mortality for this area after the DMV epidemic. From these data, it can be concluded that the epidemic subsided along the Catalonian coast in December 1990. The total number of dolphins found stranded along the Spanish Mediterranean coast during 1990 was 446. As many as 123 striped dolphins were found dead or moribund on the Catalonian coast in 1990. At least 26 of 147 (18%) dolphins beached alive on the Catalonian and Balearic Island coasts. However, most of them died within a few minutes or hours after stranding (Piza, 1991), in spite of rescue attempts. The longest survival time was 32 hours. Mortality after stranding was 100%. Affected dolphins appeared quiet and alert, or showed restlessness and disorientation. The mouth was closed and contracted, and vomiting was observed in one dolphin. Lymphopenia and leukocytosis were manifest in three blood samples analysed by the authors. Fifty-seven dolphin carcasses from the Catalonian Mediterranean waters, suitable for pathological examination, were necropsied from August to December 1990.

Morbillivirus antigen was detected by immunocytochemistry (using a monoclonal antibody directed against the haemagglutinin protein of PDV) in 37 of the 57 (65%) examined dolphins (Fig. 6).

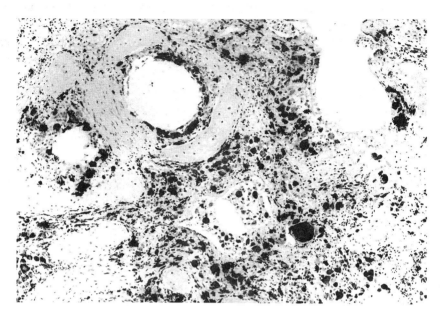

Figure 6.   Lung tissue of a striped dolphin, immunohistochemical staining with a monoclonal antibody against PDV (cross-reacting with DMV). Inflammatory infiltrates and syncytial cells contain large amounts of morbilliviral antigen. (Photo: M. Domingo). Reproduced from *Veterinary Pathology* 29:1-10, 1992, with permission from American College of Veterinary Medicine.

3.4   The 1987-88 and 1993-94 Morbillivirus Epidemic in Bottlenose Dolphins from the Eastern USA and Gulf of Mexico

From June 1987 through May 1988, at least 740 bottlenose dolphins washed ashore along the Atlantic coast of USA, from New Jersey to Florida (Lipscomb *et al.*, 1994). It was estimated that over 50% of the inshore population of bottlenose dolphins in the affected area had died (see Lipscomb *et al.*, 1994). The mortality was initially attributed to the brevetoxicosis. Brevetoxin was found in 8 out of 17 beached dolphins and in fish collected in the area. Opportunistic bacterial (*Vibrio* spp.), viral (poxvirus), and fungal (*Aspergillus* spp.) infections complicated the picture. Morbillivirus-specific antibodies were detected in the sera of six of 13 live-captured dolphins from the same region. However, neither clinical signs nor lesions attributable to a morbillivirus infection were detected in stranded dolphins, and attempts to isolate virus failed (Geraci, 1989). Therefore, it was concluded that although exposure of dolphins to a morbillivirus had taken place prior to the incident, it was unrelated to the mass mortality (Geraci, 1989).

After the Mediterranean striped dolphin epidemic of 1990-92, attention focused on the possible role of morbilliviruses in the 1987-88 bottlenose dolphin die-off. Lipscomb *et al.* (1994) performed histopathological examination of lung and lymph node samples from 79 dolphins. They also examined these samples by an immunoperoxidase technique using a monoclonal antibody to the haemagglutinin protein of PDV known to cross-react with DMV and PMV. They found lesions attributable to morbillivirus disease and detected morbilliviral antigen in 42 of 79 dolphins tested. The histopathological and immunoperoxidase findings were similar to those reported in harbour porpoises and Mediterranean striped dolphins (see below; Lipscomb *et al.*, 1994).

Figure 7.   Bottlenose dolphins that died along the U.S. central Atlantic coast in 1987-88 came ashore with a variety of conditions. Most dolphins were emaciated and had skin lesions associated with secondary bacterial and viral infections. (Photo: U.S. Department of Agriculture)

These studies strongly indicate that a morbilliviral infection was the cause of the mortality in bottlenose dolphins (Lipscomb *et al.*, 1994). Further investigations using RT-PCR and southern blot hybridisation techniques, as well as sequence analysis, have demonstrated that the two cetacean morbilliviruses, PMV and DMV, had a primary role in the die-off of 1987-88 (Krafft *et al.*, 1995; Lipscomb *et al.*, 1996; Taubenberger *et al.*, 1996). Some years later, from June 1993 until April 1994, an increased mortality of bottlenose dolphins was observed along the Gulf of Mexico coasts of Florida, Alabama, Mississippi and Texas (Fig. 7). Histological lesions characteristic of morbilliviral disease, morbillivirus antigen and a cetacean morbillivirus-specific amplification product were detected in tissues of these dolphins (Krafft, 1995; Lipscomb *et al.*, 1996). Sequencing of the morbillivirus phosphoprotein gene amplified from lung tissues of seven of these animals demonstrated that they had been infected by PMV (Taubenberger *et al.*, 1996). Collectively these results demonstrate that PMV was the aetiological agent of this epidemic.

3.5   Before and After a Morbillivirus Epidemic

What conditions should prevail for an epidemic to occur in a susceptible population? What kind of relationship between the virus and the population is established after the epidemic? Answers to these questions are relevant for understanding data collected in the field. It is generally accepted that the generation of epidemics by highly virulent, extremely contagious viruses such as morbilliviruses depends largely on the availability of sufficient numbers of non-protected individuals in the susceptible population. Predictions on the basis of this hypothesis are easy to perform. If a representative sample of a given population (for example, monk seals) has no antibodies against any morbillivirus, then the population is at risk. If antibodies do exist in most individuals, then the population has been exposed, although the effects of the exposure may have gone unnoticed.

There are several forms of interaction between a virus and the host species. How this relationship is established depends in turn on the ability of the virus to develop chronic infection in at least some individuals, with continuous or intermittent excretion of infective virus to the environment. Morbilliviruses do not possess this ability, so the virus depends, for its maintenance in the population, on the availability of susceptible, unexposed individuals. If infection occurs by close contact, or aerogenously over short distances, population density becomes a crucial factor.

Two possible scenarios could be imagined, both illustrated by examples from terrestrial mammals:

(A) The epidemic may be severe with high mortality. At any given time, there are no more susceptible hosts to be infected (all have been killed or, if survivors, they were infected previously and therefore are now immune) and the epidemic subsides in that area. If the virus is not re-introduced to the population, new susceptible individuals are born, but a substantial time is needed for a new epidemic to occur (the ratio between exposed: unexposed individuals is the key factor). Rinderpest has in the past caused major epidemics in cattle in Europe before disappearing from affected areas (for a historical review, see Schwabe, 1984).

(B) The epidemic may be more or less severe, but susceptible hosts (usually young individuals) are always available for infection. The epidemic can then become endemic. New cases of the disease are now usually restricted to young individuals at the time of decay of maternal immunity. Epidemics cease in that population because most, if not all, individuals are exposed at an early age. Those surviving remain immune for life. Such a situation now exists for canine distemper in urban dogs, in spite of vaccination programs, and also for measles in humans in many parts of the world.

For marine mammals affected by morbillivirus epidemics, both options may occur.

Option A could be valid for the Mediterranean striped dolphins. No active, systemic morbillivirus infection has been detected after the epidemic. A few cases of localised DMV infection in the central nervous system, considered to be cases of chronic morbillivirus infection, have been observed in striped dolphins (Domingo *et al.*, 1995), and in one bottlenose dolphin (Tsur *et al.*, 1997) after the epidemic. However, these chronic lesions are similar in many aspects to cases of subacute sclerosing panencephalitis in humans (Graves, 1984), and appear to be "closed" cases from the epidemiological point of view, *i.e.* affected dolphins did not excrete virus to the environment.

On the other hand, option B seems more appropriate for the situation in seals and pilot whales (*Globicephala melas* and *G. macrorhynchus*) from the North Atlantic (Ross *et al.*, 1992; Visser *et al.*, 1993c; Duignan *et al.*, 1995b; Van Bressem *et al.*, 1998). In these regions, sporadic cases of morbilliviral disease and high prevalences of morbillivirus specific antibodies have been reported. Single lethal cases of morbillivirus infection have been detected in a harp seal (*Phoca groenlandica*) (Daoust *et al.*, 1993), and in harbour seals (Duignan *et al.*, 1993) from the east coasts of Canada and the USA. Mortality due to morbilliviral disease also has been observed in a fin whale stranded on the Belgian coast in 1997 (Jauniaux, *et al.*, 1998). These findings suggest a continuous presence of the causative morbilliviruses in the North Atlantic, indicating that the disease has become endemic in many species.

3.6   Findings in Morbilliviral Disease

Macroscopic lesions due to morbilliviruses are not easy to recognise at necropsy. However, the diagnosis of morbilliviral disease by microscopic investigation is not a difficult task, provided that the case material is fresh and that adequate tools (*e.g.* morbillivirus-specific antibodies) for immunohistochemistry are available. The histo-pathological findings are remarkably similar in terrestrial and aquatic mammal species (Appel *et al.*, 1981; Kennedy *et al.*, 1989; Kennedy *et al.*, 1991; Kennedy *et al.*, 1992; Domingo *et al.*, 1992, Duignan *et al.*, 1992, Dungworth, 1993, Barker *et al.*, 1993, Schulman *et al.*, 1997).

For practical purposes, the following factors should be considered:

(i) Macroscopic lesions are not specific for morbillivirus infection.

(ii) Microscopic lesions due to morbillivirus infection are highly characteristic.

(iii) Immunocytochemical demonstration of morbillivirus antigens, in association with microscopic lesions, is one of the best confirmatory tests of morbilliviral disease.

Pulmonary congestion and alveolar oedema are usually observed in seals infected by PDV. Interlobular, subpleural, mediastinal and subcutaneous emphysema, the latter resulting in swelling of the neck and thorax, are also a common finding (Kennedy, 1998). Pulmonary abscesses, parasitic granulomas, lung abscesses, and heavy lungworm infection in the airways have also been frequently reported (Kennedy, 1998), but these lesions are not due to morbillivirus infection.

In striped dolphins infected by DMV, lung lesions were the only macroscopic lesions unequivocally attributable to the virus. These lesions consisted of multiple areas of collapsed, dark, lung tissue, alternating with inflated, paler zones, but were only evident in severe cases (Domingo *et al.*, 1992). Consistency of the tissue was not apparently modified due in part to the marked cartilaginous reinforcement of the airways present in cetacean lungs. Many striped dolphins had focal deep ulcerative stomatitis that frequently involved the gingival margin. A few dolphins showed vesicular-erosive stomatitis. Although most oral ulcerative lesions were not investigated histologically or by immunocytochemistry, DMV was demonstrated in the epithelium of the tongue of one dolphin, indicating a primary role for DMV in some of the ulcers. Other members of the genus *Morbillivirus*, rinderpest and peste-des-petits ruminants viruses, cause erosions in the mucous membranes of the upper alimentary tract (Scott, 1990a,b).

The microscopic lesions of morbillivirus infection in aquatic mammals are highly characteristic and very similar to those in terrestrial mammals. Dolphins, porpoises, and seals with morbillivirus infection usually had pneumonia, encephalitis, and lymphoid depletion. The presence of multinucleate syncytial cells was a frequent finding in cetaceans, and has also been reported in seals (Kennedy *et al.*, 1989; Kennedy *et al.*, 1991; Baker, 1992; Domingo *et al.*, 1992; Duignan *et al.*, 1992). These lesions could appear alone or in different combinations. The concurrence of such lesions in the lung, brain, and lymphoid organs constitutes a triad highly suspicious of morbillivirus infection (Appel *et al.*, 1981; Barker *et al.*, 1993; Dungworth, 1993).

Pneumonic lesions have been classified as bronchiolo-interstitial pneumonia. The interstitial pneumonia was apparently more acute in seals than in cetaceans, as denoted by the presence of hyaline membranes on the alveolar walls. Viral inclusions are frequently recognised in the nuclei and cytoplasm of syncytial and mononuclear cells (Kennedy *et al.*, 1989; Domingo *et al.*, 1992).

Mononuclear cell encephalitis is also usual in viral infections of the central nervous system. Perivascular infiltration by mononuclear cells, neuronophagia, and neuronal necrosis may be found in the cortical grey matter. Focal areas of encephalitis are also observed in the white matter, frequently around vessels, with malacia (necrosis) of the surrounding tissue. Demyelination is also a frequent finding. All morbilliviruses, except the ruminant morbilliviruses, can cause encephalitis in their respective hosts.

Besides pulmonary and encephalitic lesions, lymphoid depletion completes the triad of lesions characteristic of morbillivirus infection. Although macroscopically unremarkable, the cortical areas of lymph nodes are severely depleted of lymphocytes, and large syncytial cells may replace lymphoid tissue. The medullary cords are depopulated of cells, and the medullary sinuses appear almost empty.

The use of polyclonal and monoclonal antibodies and immuno-peroxidase techniques to demonstrate antigens of the various morbilliviruses causing disease in marine mammals, have been shown to be powerful diagnostic tools in recent morbillivirus epidemics (Kennedy *et al.*,1989; Kennedy *et al.*, 1991; Domingo *et al.*, 1992; Duignan *et al.*, 1992; Duignan *et al.*, 1993; Lipscomb *et al.*, 1994, Schulman *et al.*, 1997). The RT-PCR applied either to formalin-fixed, paraffin-embedded tissues or to frozen tissues used together with southern blot and sequencing, has also proven to be useful in detecting morbillivirus infection in marine mammals (Krafft *et al.*, 1995; Lipscomb *et al.*, 1996, Taubenberger *et al.*, 1996; Schulman *et al.*, 1997).

In all morbillivirus epidemics, viral antigens have been demonstrated primarily in the lung, lymph nodes, and brain tissue. In the lung, immunostaining was observed in epithelial cells lining the bronchiolar walls and alveoli. Exuded mononuclear cells and multinucleate syncytia in alveolar lumina also contained morbillivirus antigen. Immunostaining in the lymph nodes was usually widespread in the cortical areas, with many lymphoid cells and multinucleate syncytia, when present, showing a strong diffuse cytoplasmic reaction. In the brain of cetaceans, the cerebral cortical layer of neurones was frequently affected. Neurones, astrocytes, and microglial cells were intensely stained. The cerebellum showed immunostaining only in a few cases. In the cerebral white matter, cells of the inflammatory infiltrates around vessels were positively stained. In striped dolphins, the brain was the most frequently affected organ, with as much as 17/22 (77%) of dolphins showing DMV antigen. This was followed by the lung, with 16/23 (70%) positive (Domingo *et al.*, 1992). Other organs and tissues affected included the epithelium of the renal pelvis, trachea, intestine, biliary ducts, and mammary gland. Mononuclear inflammatory cells exuded in response to other inciting agents in any organ (for example, parasites) were often stained in infected dolphins.

Suspicion of morbillivirus infection in aquatic mammals should always be confirmed by histology and immunocytochemistry, or serology. Virus isolation and RT-PCR based assay together with Southern blot hybridisation and sequencing of the PCR products are highly recommended for a definitive diagnosis. The most reliable organs to be collected (into 10% buffered formalin) for histopathological and immunocytochemical investigations of morbillivirus infection in aquatic mammals are lungs, lymph nodes, spleen, brain, kidney, urinary bladder, and gastrointestinal tract. Samples for virus isolation should be relatively fresh and kept at -80°C. The preferred tissues are basically the same as for histology, although viruses have been more often isolated from the lungs, spleen and lymph nodes (Visser *et al.*, 1990; Van Bressem *et al.*, 1993). Frozen samples are the best material for molecular assays but formalin-fixed and paraffin-embedded samples have been successfully used for the diagnosis of cetacean morbillivirus infection in bottlenose dolphins from the eastern coast of the United States (Krafft *et al.*, 1995; Taubenberger *et al.*, 1996). Blood samples, even haemolysed, are useful for detecting the presence of morbillivirus-specific antibodies by indirect ELISA and virus neutralisation (Duignan *et al.*, 1995a,b; Van Bressem *et al.*, 1993, 1998) and should be stored at -20°C.

3.7   Why Do New Virus Infections Emerge?

Humans and other animal species are affected from time to time by apparently "new" viral diseases. Examples of recent viral pandemics include canine parvovirus infection, haemorrhagic viral disease of rabbits, porcine respiratory and reproductive syndrome, and human immunodeficiency virus infection. Where were these viruses prior to their appearance (often as an epidemic)? How did they originate? If they were silently present in reservoir species, which factors triggered or caused the emergence of devastating diseases? Explanations for the emergence of new viral diseases have been established on the basis of epidemiological and molecular biological investigations of diseases (Nathanson and Murphy, 1997). The emergence of a new viral disease can occur by the following mechanisms:

(i) Recognition of a previously unknown disease. The relationship between a virus and a disease is not recognised until interest in the disease leads to renewed investigative efforts and to the discovery of the 'new' viral agent.

(ii) An increase in the ratio of cases to infections is an important mechanism for the emergence of viral diseases, particularly in an epidemic form. Altered host resistance and increased virulence of a virus are often the basis of an epidemic. Influenza viruses, which possess elaborate mechanisms for reassortment of genetic material, offer clear examples of the emergence of viral strains with increased virulence for the normal host.

(iii) A sudden increase in the number of infected individuals by a particular virus may lead to the recognition of a new disease. This can occur in several ways. For instance, a virus can reinfect a population from which it has been absent for a long time. The lack of protective immunity in most of the individuals of this population favours rapid spread of the infection, sometimes resulting in disastrous epidemics. Measles and canine distemper viruses, the human and canine morbilliviruses respectively, often cause epidemics through extension to isolated populations. A similar situation may have occurred in the recent morbillivirus epidemics in seals and cetaceans.

Another way to increase the number of infections is to cross the species barrier. There are several examples that document this phenomenon. Feline parvovirus jumped to the domestic dog in 1978, causing a devastating pandemic with high mortality. Clearly, the entire dog population of the world was unprotected against this new pathogen. Small genetic changes in the genome of the feline parvovirus underlay its ability to cross the species barrier (Parrish, 1994). That marine mammals could be affected by viruses derived from terrestrial mammals was evidenced by the distemper outbreak in Baikal seals, caused by canine distemper virus, which presumably crossed from dogs to seals. However, this simple explanation is invalid for other epidemics in seals and in cetaceans, as the molecular analysis of the genomic sequence of the morbilliviruses isolated in these epidemics revealed that they are previously unrecognised members of the genus *Morbillivirus*, which diverged in the past from the terrestrial morbilliviruses.

REFERENCES

Aguilar, A., and Raga, J.A. (1993) The striped dolphin epizootic in the Mediterranean Sea. *Ambio*, **22**, 524-528.

Aguilar, A., Gonzalez, L.M., and Lopez-Jurado, L.F. (1998) The Mediterranean Monk Seal die-off in the western Sahara. *Workshop on the Biology and Conservation of the World's Endangered Monk Seals*, Monaco, 19-20 January 1998.

Appel, M.J.G., Gibbs, E.P.J., Ter Meulen, S.J.M., Rima, B.K., Stephenson, J.R., and Taylor, W.P. (1981) Morbillivirus disease of animals and man. In: *Comparative Diagnosis of Viral Diseases*, Vol. 4 (Ed. by E. Kustak and C. Kustak), pp. 235-297. Academic Press, New York.

Baker, J.R. (1992) The pathology of phocine distemper. *Science of the Total Environment*, **115**, 1-7.

Barber, R.T. and Chavez F.P. (1983) Biological consequences of El Niño. *Science*, **222**, 1203-1210.

Barker, I.K., Van Dreumel, A.A., and Palmer, N. (1993) The alimentary system. In: *Pathology of Domestic Animals*, 4rd ed, Vol 2 (Ed. by J.V.K. Jubb, P.C. Kennedy, and N. Palmer), pp. 1-317. Academic Press, San Diego, CA.

Barrett, T., Visser, I.K.G., Mamaev, L.V., Goatley, L., Van Bressem, M.-F., and Osterhaus, A.D.M.E. (1993) Dolphin and porpoise morbilliviruses are genetically distinct from phocine distemper virus. *Virology,* **193**, 1010-1012.

Bengtson, J.L., Boveng, P., Franzén, U., Have, P., Heide-Jørgensen, M.-P. and Härkönen, T. (1991) Antibodies to canine distemper virus in Antarctic seals. *Marine Mammal Science*, **7**, 85-87.

Birkun, A., Kuiken, T., Krivokhizhin, S., Haines, D.M., Osterhaus, A.D.M.E., Van de Bildt, M.W.G., Joiris, C.R., and Siebert, U. (1998) Epizootic of morbilliviral disease in common dolphins (*Delphinus delphis ponticus*) from the Black Sea. *Veterinary Record*, **144**, 85-92.

Blixenkrone-Möller, M., Sharma, B., Varsanyi, T.M., Hu, A., Norrby, E., and Kövamees, J. (1992) Sequence analysis of the genes encoding the nucleocapside protein (NP) and phosphoprotein (P) of Phocid Distemper Virus, and editing of the P gene transcript. *Journal of General Virology,* **73**, 885-893.

Blixenkrone-Möller, M., Bolt, G., Jensen, T.D., Harder, T., and Svansson, V. (1996) Comparative analysis of the attachment protein gene (H) of dolphin morbillivirus. *Virus Research*, **40**, 47-56.

Bompar, J.-M., Dhermain, F., Poitevin, F., and Cheylan, M. (1991) Les dauphins méditerranéens victimes d'un virus mortel. *La Recherche*, **22**, 506-508.

Bortolotto, A., Casini, L., and Stanzani, L.A. (1992) Dolphin mortality along the southern Italian coast (June-September 1991). *Aquatic Mammals*, **18**, 56-60.

Bossart, G.D., Baden, D.G., Ewing, R.Y., Roberts, B., and Wright, S.C. (1998) Brevetoxicosis in manatees (*Trichechus manatus latirostris*) from the 1996 epizootic: gross, histologic, and immunohistochemical features. *Toxicological Pathology*, **26**, 276-282.

Callan, R.J., Early, G., Kida, H. and Hinshaw, V.S. (1995) The appearance of H3 influenza viruses in seals. *Journal of General. Virology*, **76**, 199-203.

Calzada, N., Lockyer, C.H. and Aguilar, A. (1994) Age and sex composition of the striped dolphin die-off in the Western Mediterranean. *Marine Mammal Science*, **10**, 299-310.

Cebrian, D. (1995) The striped dolphin *Stenella coeruleoalba* epizootic in Greece, 1991-1992. *Biological Conservation*, **74**, 143-145.

Cosby, S.L., McQuaid, S., Duffy, N., Lyons, C., Rima, B.K., Allan, G.M., McCullough, S.J., and Kennedy, S. (1988) Characterisation of seal morbillivirus. *Nature* (London), **336**, 115-116.

Costas, E., and Lopez-Rodas, V. (1998) Paralytic phycotoxins in monk seal mass mortality. *Veterinary Record*, **142**, 643-644.

Curran, M.D., O'Loan, D., Rima, B.K., and Kennedy, S. (1990) Nucleotide sequence analysis of phocine distemper virus reveals its distinctness from canine distemper virus. *Veterinary Record*, **127**, 430-431.

Daoust, P.Y., Haines, D.M., Thorsen, J., Duignan, P.J., and Geraci, J.R. (1993) Phocine distemper in a harp seal (*Phoca groenlandica*) from the Gulf of St. Lawrence, Canada. *Journal of Wildlife Diseases*, **29**,114-117.

Dierauf, L.A., Vandenbroek, D., Roletto, J., Koski, M., Amaya, L. and Gage, L. (1985) An epizootic of leptospirosis in California sea lions. *Journal of the American Veterinary Medical Association*, **187**, 1145-1148.

Dietz, R., Heide-Jörgensen, M.-P. and Härkönen, T. (1989) Mass deaths of harbor seals (*Phoca vitulina*) in Europe. *Ambio*, **18**, 258-264.

Domingo, M., Ferrer, L., Pumarola, M., Marco, A., Plana, J., Kennedy, S., McAliskey, M., and Rima, B.K. (1990) Morbillivirus in dolphins. *Nature* (London), **348**, 21.

Domingo, M., Visa, J., Pumarola, M., Marco, A., Ferrer, L., Rabanal, R., and Kennedy, S. (1992) Pathologic and immunocytochemical studies of morbillivirus infection in striped dolphins (*Stenella coeruleoalba*). *Veterinary Pathology*, **29**, 1-10.

Domingo, M., Vilafranca, M., Visa, J., Prats, N., Trudgett, A., and Visser, I. (1995) Evidence for chronic morbillivirus infection in the Mediterranean striped dolphin (*Stenella coeruleoalba*). *Veterinary Microbiology*, **44**, 229-239.

Duignan, P., Geraci, J.R., Raga, J., and Calzada, N. (1992) Pathology of morbillivirus infection in striped dolphins (*Stenella coeruleoalba*) from Valencia and Murcia, Spain. *Canadian Journal of Veterinary Research*, **56**, 242-248.

Duignan, P., Sadove, J.S., Saliki, J.T., and Geraci, J.R. (1993) Phocine distemper in harbour seals (*Phoca vitulina*) from Long Island. New York. *Journal of Wildlife Diseases*, **29**, 465-469.

Duignan, P.J., Saliki, J.T., St Aubin, A.D., Early, G., Sadove, S., House, J.A., Kovacs, K. and Geraci, J.R. (1995a) Epizootiology of morbillivirus infection in North American harbor seals (*Phoca vitulina*) and grey seals (*Halichoerus grypus*). *Journal of Wildlife Diseases*, **31**, 491-501

Duignan, P.J., House, C., Geraci, J.R., Early, G., Copland, H.G., Walsh, M.T., Bossart, G.D., Cray, C., Sadove, S., St. Aubin, D.J., and Moore, M. (1995b) Morbillivirus infection in two species of pilot whales from the Western Atlantic. *Marine Mammal Science*, **11**, 150-162.

Duignan, P.J., House, C., Odell, D.K., Wells, R.S., Hansen, L.J., Walsh, M.T., St Aubin, D.J., Rima, B.K. and Geraci, J.R. (1996) Morbillivirus in bottlenose dolphins: evidence for recurrent epizootics in the Western Atlantic and Gulf of Mexico. *Marine Mammal Science*, **12**, 495-515.

Duignan, P.J. (1999) Gross pathology, histopathology, virology, serology and parasitology. In: *Unusual mortality of the New Zealand sea lion, Phocarctus hookeri, Auckland Islands, January February 1998*. Report of the Workshop, Wellington, 8-9 June 1998.

Dungworth, D.L. (1993) The respiratory system. In: *Pathology of Domestic Animals*, 4rd ed, Vol. 2, (Ed. by J.V.K. Jubb, P.C. Kennedy, and N. Palmer), pp. 539-699. Academic Press, San Diego, CA.

Echegaray, M.,Reyes, J.C., De Paz, N. and Vinces, M. (1998) Cuanto afecta El Niño a los cetáceos menores en Peru. *8a Reunião de Trabalho de Especialistas em Mamíferos Aquáticos da América do Sul, 2do Congresso da Sociedade Latinoamericana de Especialistas em Mamíferos Aquáticos-Solamac*, 25-29 October 1998, Olinda, Brazil p 73 (Abstract)

Fenner F.J., Gibbs, E.P.G., Murphy, F.A., Rott, R., Studdert, M.J. and White, D.O. (1993) *Veterinary Virology*. Academic Press, Inc., London, 666pp.

Forsyth, M.A., Kennedy, S., Wilson, S., Eybatov, T. and Barrett, T. (1998) Canine distemper virus in a Caspian seal. *Veterinary Record*, **143**, 662-664.

Geraci, J.R., St.Aubin, D.J., Barker, I.K., Webster, R.G., Hinshaw, V.S., Bean, W.J., Ruhnke, H.L., Prescott, J.H., Early, G., Baker, A.S., Madoff, S. and Schooley, R.T. (1982) Mass mortality of harbor seals: pneumonia associated with Influenza A virus. *Science*, **215**, 1129-1131.

Geraci, J.R. (1989) *Clinical Investigations of the 1987-88 Mass Mortality of Bottlenose Dolphins along the U.S. Central and South Atlantic Coast: Final report*. National Marine Fisheries Service, April 1989.

Geraci, J.R., Anderson, D.M., Timperi, R.J., St. Aubin, D.J., Early, G A., Prescott, J.H., and Mayo, C.A. (1989) Humpback whales (*Megaptera novaeangliae*) fatally poisoned by a dinoflagellate toxin. *Canadian Journal of Fisheries and Aquatic Science*, **46**, 1895-1898.

Geraci, J.R. and Lounsbury, V.J. (1993) *Marine Mammals Ashore: A Field Guide for Strandings*. Sea Grant Program, Texas A&M University Press, Galveston.

Gilmartin, W.G. (1987) Hawaiian monk seal die-off response plan, a workshop report, 1980 (Ed. by W.G. Gilmartin), pp. 1-7. San Diego, National Marine Fisheries Service.

Gilmartin, W.G., Delong, R.L., Smith, A.W., Griner, L.A., and Dailey, M.D. (1987) An investigation into unusual mortality in the Hawaiian monk seal, *Monachus schauinslandi*. In: Hawaiian monk seal die-off response plan, a workshop report, 1980 (Ed. by W.G. Gilmartin), pp. 32-41. San Diego, National Marine Fisheries Service.

Grachev, M.A., Kumarev, V.P., Mammev, V.P., Zorin, V.L., Baranova, L.V., Denikina, N.N., Belicov, S.I., Petrov, E.A., Kolsnik, V.S., Kolsnik R.S., Beim, A.M., Kudelin, V.N., Nagieva, F.G., and Sidorovo, V.N. (1989) Distemper virus in Baikal seals. *Nature* (London), **338**, 209.

Graves, M.C. (1984) Subacute Sclerosing Panencephalitis. *Neurological Clinics*, **2**, 267-280.

Hall, A.J., Pomeroy, P.P., and Harwood, J. (1992) The descriptive epizootiology of phocine distemper in the UK during 1988/89. *Science of the Total Environment*, **115**, 31-44.

Harwood, J. (1989) Lessons from the seal epidemic. *New Scientist*, 18 February, 38-42.

Harwood, J. (1990) A short history of the North Sea seal epidemic. *Ocean Challenge*, **1**, Spring, 4-8.

Harwood, J. (1998) What killed the monk seals? *Nature* (London), **393**, 17-18.

Harwood, J. and Hall, A. (1990) Mass mortality in marine mammals: its implications for population dynamics and genetics. *Trends in Ecology and Evolution*, 5, 254-257.

Heide-Jørgensen, M.-P., Harkonen, T., Dietz, R., and Thompson, P. (1992) Retrospective of the 1988 European seal epizootic. *Diseases of Aquatic Organisms*, **13**, 37-62.

Hernandez, M., Robinson, I., Aguilar, A., Gonzalez, L.M., Lopez-Jurado, L.F., Reyero, M. I., and Cacho, E. (1998) Did algal toxins cause monk seal mortality? *Nature*, London, **393**, 28.

Hersh, S.L., Odell, D.K., and Asper, E.D. (1990) Bottlenose dolphin mortality patterns in the Indian/Banana river system of Florida. *The Bottlenose Dolphin* (Ed. by S. Leatherwood and R.R. Reeves), pp. 155-164. Academic Press, San Diego.

Jauniaux, T., Charlier, G., Desmecht, M., and Coignoul, F. (1998) Lesions of morbillivirus infection in a fin whale (*Balaenoptera physalus*) stranded along the Belgian coast. *Veterinary Record*, **143**, 423-424.

Kennedy, S., Smyth, J.A., McMullough, S.J., Allan, G.M., and McNeilly, F. (1988a) Confirmation of cause of recent seal deaths. *Nature* (London), **335**, 464.

Kennedy, S., Smyth, J.A., Cush, P.F., McCullough, S.J., Allan, G.M., and McQuaid, S. (1988b) Viral distemper now found in porpoises. *Nature* (London), **336**, 21.

Kennedy, S., Kuiken, T., Ross, H.M., McAliskey, M., Moffett, D., McNiven, M., and Carole, M. (1992) Morbillivirus infection in two common porpoises (*Phocoena phocoena*) from the coasts of England and Scotland. *Veterinary Record*, **131**, 286-290.

Kennedy, S. (1998) Morbillivirus infections in aquatic mammals. *Journal of Comparative Pathology*, **119**, 201-225.

Klinowska, M. (1985) Cetacean live stranding sites relate to geomagnetic topography. *Aquatic Mammals*, **1**, 27-32.

Krafft, A., Lichy, J.H., Lipscomb, T.P., Klaunberg, B.A., Kennedy, S. And Taubenberger J.K. (1995) Postmortem diagnosis of morbillivirus infection in bottlenose dolphins (*Tursiops truncatus*) in the Atlantic and Gulf of Mexico epizootics by polymerase chain reaction-based assay. *Journal of Wildlife Diseases*, **31**, 410-415.

Laws, R.M. and Taylor, R.J.F. (1957) A mass dying of crabeater seals, *Lobodon carcinophagus* (Gray). *Proceedings of the Zoological Society of London*, **129**, 315-325.

Lipscomb, T.P., Schulman, F.Y., Moffett, D., and Kennedy, S. (1994) Morbilliviral disease in Atlantic bottlenose dolphins (*Tursiops truncatus*) from the 1987-1988 epizootic. *Journal of Wildlife Diseases*, **30**, 567-571.

Lipscomb, T.P., Kennedy, S., Moffett, D., Krafft, A., Klaunberg, B.A., Lichy, J.H., Regan, G.T., Worthy, G.A.J., and Taubenberger, J.K. (1996) Morbilliviral epizootic in bottlenose dolphins of the Gulf of Mexico. *Journal of Veterinary Diagnostic Investigation*, **8**, 283-290

Mahy, B.W., Barret, T., Evans, S., Anderson, E.C., and Bostock C.J. (1988) Characterisation of seal morbillivirus. *Nature* (London), **336**, 115.

Majluf, P. (1991) El Niño Effects on Pinnipeds in Peru. *Pinnipeds and El Niño, Responses to Environmental Stress* (Ed. by F. Trillmich and K.A. Ono), pp.55-65. Springer-Verlag, Berlin.

Majluf, P. (1998) Effects of the 1997/98 El Niño on pinnipeds in Peru. *8a Reunião de Trabalho de Especialistas em Mamíferos Aquáticos da América do Sul, 2do Congresso da Sociedade Latinoamericana de Especialistas em Mamíferos Aquáticos-Solamac*, 25-29 October 1998, Olinda, Brazil, p. 120.

McCullough, S.J., McNeilly, F., Allan, G.M., Kennedy, S., Smyth, J.A., Cosby, S.L., McQuaid, S., and Rima, B.K. (1991) Isolation and characterisation of a porpoise morbillivirus. *Archives of Virology*, **118**, 247-252.

Murphy, B.R. and Webster, R.G. (1996) Orthomyxoviruses. In: *Fields Virology* (Ed. B.N. Fields, D.M. Knipe, and P.M. Howley), pp. 1397-1445. Lippincott-Raven Publishers, Philadelphia..

Nathanson, N., and Murphy, F.A. (1997) Evolution of viral diseases. In: *Viral Pathogenesis* (Ed. Neal Nathanson *et al.*), pp. 353-369. Lippincott-Raven Publishers, Philadelphia.

Örvell, C., and Sheshberadaran, H. (1991) Phocine distemper virus is phylogenetically related to canine distemper virus. *Veterinary Record*, **129**, 267-269.

O'Shea, T.J., Rathbun, G.B., Bonde, R.K., Buergelt, C.D., and Odell, D.K. (1991) An epizootic of Florida manatees associated with a dinoflagellate bloom. *Marine Mammal Science*, **7**, 165-179.

Osterhaus, A.D.M.E., and Vedder, E.J. (1988) Identification of virus causing recent seal deaths. *Nature*, London, **335**, 20

Osterhaus, A.D.M.E., Groen, J., UydeHaag, F.G.C.M., Visser, I.K.G., Van de Bildt, M.W.G., Bergman, A., and Kligeborn, B. (1989a) Distemper virus in Baikal seals. *Nature*, **338**, 209-210.

Osterhaus, A.D.M.E., Broeders, H.W.J., Groen, J., UydeHaag, F.G.C.M., Visser, I.K.G., van de Bilt, M.W.G., Örvell, C., Kumarev, V.P., and Zorin, V.L. (1989b) Different morbilliviruses in European and Siberian seals. *Veterinary Record*, **125**, 647-648.

Osterhaus, A.D.M.E., Visser , I.K.G., De Swart, R.L., Van Bressem, M-F., Van de Bildt, M.W.G., Örvell, C., Barrett, T. and Raga, J.A. (1992) Morbillivirus threat to Mediterranean monk seals? *Veterinary Record*, **130**, 141-142.

Osterhaus, A.D.M.E., De Swart, R.L., Vos, H.W., Ross, P.S., Kenter, M.J.H. and Barrett, T. (1995) Morbillivirus infections of aquatic mammals: newly identified members of the genus. *Veterinary Microbiology*, **44**, 219-227.

Osterhaus, A., Groen, J., Niesters, H., Van de Bildt, M., Martina, B., Vedder, L., Vos, J., Egmond, H., Sidi, B.A., and Barhan, M.E.O. (1997) Morbillivirus in monk seal mass mortality. *Nature* (London), **388**, 838-839.

Parrish, C.R. (1994) The emergence and evolution of canine parvovirus- an example of recent host range mutation. *Seminars in Virology*, **5**, 121-132.

Piza, J. (1991) Striped dolphin mortality in the Mediterranean. *Proceedings of the Mediterranean striped dolphin mortality*. 4-5 Nov, Palma de Mallorca, Spain.

Robson, F.D. (1984) *Strandings: ways to save* whales. The Science Press, Johannesburg. 124pp.

Ross, P.S., Visser, I.K.G., Broeders, H.W.J., Van de Bildt, M.W.G., Bowen, W.D. and Osterhaus, A.D.M.E. (1992) Antibodies to phocine distemper virus in Canadian seals. *Veterinary Record*, **130**, 514-516.

Schulman, F.Y., Lipscomb, T.P., Moffett, D., Krafft, A.E., Lichy, J.H., Tsai, M.M., Taubenberger, J.K., and Kennedy, S. (1997) Histologic, immunohistochemical, and polymerase chain reaction studies of bottlenose dolphins from the 1987-1988 United States Atlantic coast epizootic. *Veterinary Pathology*, **34**, 288-295

Schwabe, C.W. (1984) *Veterinary Medicine and Human Health*, pp. 16-22. Williams and Wilkins, Baltimore London. 3rd edition.

Scott, G.R. (1990a) Peste-des-petits-ruminants (goat plague) virus. In: *Virus Infections of Vertebrates*, 3: Virus Infections of Ruminants (Ed. by Z. Dinter and B. Morein), pp. 355-361. Elsevier Science Publisher, B.V., Amsterdam.

Scott, G.R. (1990b) Rinderpest virus. In: *Virus Infections of Vertebrates*, 3: Virus Infections of Ruminants (Ed. by Z. Dinter and B. Morein), pp. 341-354, Elsevier Science Publisher, B.V., Amsterdam.

Smith, A.W., Brown, R.J., Skilling, D.E., and De Long, D.E. (1974) *Leptospira pomona* and reproductive failure in California sea lions. *Journal of the American Veterinary Medicine Association*, **165**, 996-998.

Smith, A.W., Brown, R.J., Skilling, D.E., Bray, H.L., and Keyes, M.C. (1977) Naturally-occuring leptospirosis in Northern fur seals (*Callorhinus ursinus*). *Journal of Wildlife Diseases*, **13**, 144-148.

St.Aubin, D. (1991) Review of recent mass mortalities of marine mammals in North America. *European Research on Cetaceans*, **5**.

Steidinger, K.A., and Baden, D.G. (1984) Toxic marine dinoflagellates. In: *Dinoflagellates*. (Ed. by D.L. Spector), pp. 201-261, Academic Press, New York.

Taubenberger, J.K., Tsai, M., Krafft, A.E., Lichy, J.H., Reid, A.H., Schulman, F.Y., and Lipscomb, T.P. (1996) Two morbilliviruses implicated in bottlenose dolphin epizootics. *Emerging Infectious Diseases*, **2**, 213-216.

Trillmich, F. and Dellinger, T. (1991) The effects of El Niño on Galapagos pinnipeds. In: *Pinnipeds and El Niño* (Ed. by F. Trillmich and K.A. Ono), pp. 66-74. Springer-Verlag, Berlin.

Trillmich, F. and Ono, K.A. (1991) *Pinnipeds and El Niño: Responses to Environmental Stress*. Springer-Verlag, Berlin. 293pp.

Trillmich, F., Ono, K.A., Costa, D.P., Delong, R.L., Feldkamp, S.D., Francis, J.M., Gentry, R.L., Heath, C.B., Leboeuf, B.J., Majluf, P. and York, A.E. (1991) The effects of El Niño on Pinniped Populations in the Eastern Pacific. *Pinnipeds and El Niño, Responses to Environmental Stress* (Ed. by F. Trillmich and K.A. Ono), pp. 247- 270. Springer-Verlag, Berlin.

Trudgett, A., Lyons, C., Welsh, M.J., Duffy, N., McCullough, S.J., and McNeilly, F. (1991) Analysis of a seal and a porpoise morbillivirus using monoclonal antibodies. *Veterinary Record*, **128**, 61.

Tsur, I., Goffman, O., Yakobsen, B., Moffett, D., and Kennedy, S. (1997) Morbillivirus infection in a bottlenose dolphin (*Tursiops truncatus*) from the Mediterranean Sea. *European Journal of Veterinary Pathology*, **2**, 83-85.

Van Bressem, M.F., Visser, I.K.G., Van de Bilt, M.W.G., Teppema, K.S., Raga, J.A., and Osterhaus, A.D.M.E. (1991) Morbillivirus infection in Mediterranean striped dolphins (*Stenella coeruleoalba*). *Veterinary Record*, **129**, 471-472.

Van Bressem, M.F., Visser, I.K.G., De Swart, R.L., Örvell C., Stanzani, L., Androukaki, E., Siakavara, K., and Osterhaus, A.D.M.E. (1993) Dolphin morbillivirus infection in different parts of the Mediterranean Sea. *Archives of Virology*, **129**, 235-242.

Van Bressem, M-F., Jepson, P., and Barrett, T. (1998) Further insight on the epidemiology of cetacean morbillivirus in the Northeastern Atlantic. *Marine Mammal Science*, **14**, 605-613.

Van Bressem, M.-F., Van Waerebeek, K. and Raga, J.A. (1999) A review of virus infections of cetaceans and the potential impact of morbilliviruses, poxviruses and papillomaviruses on host population dynamics. *Diseases of Aquatic Organisms*, **38**, 53-65.

Vedros, N.A., A.W. Smith, J. Schonewald, G. Migaki, and R.C. Hubbard. (1971) Leptospirosis epizootic among California sea lions. *Science*, **172**, 1250-1251.

Vidal, O., and Gallo-Reynoso, J-P. (1996) Die-offs of marine mammals and sea birds in the Gulf of California, México. *Marine Mammal Science*, **12**, 627-635.

Visser, I.K.G., V.P. Kumarev, Örvell, C., De Vries, P., Broeders, H.W.J., Van de Bildt, M.W.G., Groen, J., Teppema, J.S., Burger, M.C., UytdeHaag, F.G.C.M., and Osterhaus, A.D.M.E. (1990) Comparison of two morbilliviruses isolated from seals during outbreaks of distemper in North West Europe and Siberia. *Archives of Virology*, **111**, 149-164.

Visser, I.K.G., Teppema, J.S., and Osterhaus, A.D.M.E. (1991) Virus infections of seals and other pinnipeds. *Review Medical Microbiology,* **2**, 105-114

Visser, I.K.G., Van Bressem, M.F., Barrett, T., and Osterhaus, A.D.M.E. (1993a) Morbillivirus infections in aquatic mammals. *Veterinary Research*, **24**, 169-178.

Visser, I.K.G., Van Bressem, M.F., De Swart, R.L., Van de Bildt, M.W.G., Vos, H.W., Van der Heijden, R.W.J., Saliki, J., Örvell, C., Kitching, P., Barrett, T., and Osterhaus, A.D.M.E. (1993b) Characterisation of morbillivirus isolated from dolphins and harbour porpoises in Europe. *Journal of General Virology*, **74**, 631-641.

Visser, I.K.G., Vedder, E.J., Vos, H.W., Van de Bildt, M.W.G. and Osterhaus, A.D.M.E. (1993c) Continued presence of phocine distemper virus in the Dutch Wadden Sea seal population. *Veterinary Record*, **133**, 320-322.

Webb, J. (1991) Dolphin epidemic spreads to Greece. *New Scientist*, **131**, 18.

Welsh, M.J., Lyons, C., Trugett, A., Rima, B.K., McCullough, S.J., and Örvell, C. (1992) Characteristics of a cetacean morbillivirus isolated from a porpoise (*Phocoena phocoena*). *Archives of Virology*, **125**, 305-311.

E. Conservation and Management

The genesis and extinction of species is a natural process throughout the history of life on Earth. Indeed, our planet has witnessed several mass extinctions that have resulted in the collapse of a substantial portion of the ancient biotas. Human activities, however, have added a new facet to the extinction process. In particular, the irrational use of technology and resources during the twentieth century, especially associated with rapid industrial development, has generated an enormous impact upon the environment. As a result, many species, including marine mammals, have become endangered.

In the case of marine mammals, conservation problems first arose through direct exploitation. In their hunger for meat, blubber or skin, humans wiped out the Steller's sea cow (*Hydrodamalis gigans*), from the North Pacific and Bering Sea, as early as 1768, and, more recently, the West Indian monk seal (*Monachus tropicalis*), from the Caribbean around 1950. Subsequently, the lucrative benefits derived from large-scale whaling also led many cetacean species to the verge of extinction. Fortunately, several factors have worked to improve the situation. First, whale-derived products progressively lost their value in favour of industrial substitutes. Second, the International Whaling Commission (IWC) was created in 1946 following the International Convention for the Regulation of Whaling. Formerly devoted to regulate whale captures, this institution has increasingly become involved in the preservation and recovery of cetacean populations, an aspect that eventually resulted in a moratorium for commercial whaling in 1982 that has been largely in effect ever since (although not entirely observed, as evidenced graphically by the recent disclosures that the former Soviet Union continued hunting the protected blue and right whales during the 1950s-1960s, deliberately misreporting zero catches - Yablokov, 1994; Zemsky *et al.*, 1995; Tormosov *et al.*, 1998).

Throughout its history, the IWC has dealt with several proposals for stock management in commercial whaling, eventually adopting the revised management procedure (RMP) that its Scientific Committee proposed in 1991 (IWC, 1994; Cooke, 1994; Gambell, 1999). In many ways this was a model of how to incorporate the precautionary principle into the science of managing exploited populations. Catch limits were established usually at less than 2% of estimated abundance (occasionally up to 5%), and target levels of exploited stocks were set at 72% of initial abundance. A major goal of the RMP was to allow depleted whale stocks to recover and ensure that no additional stocks would fall below 54% of their initial abundance. Of course, one inevitable weakness in any management plan is our poor knowledge of initial abundance for any whale stock, and the continued difficulty of obtaining precise estimates of current abundance and catch rates. Even if these problems could be solved, there is always the economic drive towards over-exploitation of resources which lack ownership (and therefore lack direct management responsibility), particularly in situations where it may be more beneficial financially to exploit to extinction, and re-invest the profits elsewhere, than to harvest on a sustainable basis (Clark, 1976).

In Chapter 13, Enrique Crespo and Martín Hall discuss several aspects of such management problems with particular emphasis on ecological and economic interactions. They highlight the ecological problems faced by poorer countries around the world, emphasising how so many resources are exploited in those countries for export to the developed world, threatening net losses of regional biological production. The upwelling ecosystem of western South America, for example, transfers annually 12-15 million metric tonnes of biomass to other ocean basins. At the same time, they point out that the seas adjacent to Japan receive 3.25 million metric tonnes annually.

Fishing activities may also have important repercussions upon cetacean and pinniped populations due to direct competition for resources. Although the effects are often difficult to demonstrate, Steller's sea lion populations of the North Pacific, for example, have been dramatically affected by this problem (see Chapter 13).

Today, although the fishing industry targets mostly fish and marine invertebrates, there is still a substantial indirect mortality of cetaceans and other vertebrates because of incidental capture by different fishing gears - particularly driftnets, widely used in all oceans, and purse seines, typically employed by the tuna fishery in the eastern tropical Pacific Ocean. In particular, gillnet fisheries have led the vaquita (*Phocoena sinus*) to become the most threatened cetacean species (it is included as "critically endangered" in the World Conservation Union list). Martín Hall and Greg Donovan address issues of fisheries bycatch and their management in Chapter 14. They introduce the concept of potential biological removal (PBR), which was developed in the United States during the 1990s under the Marine Mammal Protection Act. Its aim was to estimate the number of individuals that could be 'safely' removed annually from a marine mammal population whilst maintaining an optimum sustainable population level (Wade, 1998). In practice, the goal has been to maintain populations above their Maximum Net Productivity Level, assumed to lie between 50-70% of carrying capacity. It has been applied particularly to the bycatch question, using information on abundance, bycatch and population growth rates.

Over recent decades, other human-related factors have been causing a great impact upon some cetacean species. Habitat degradation has led the

Asian river dolphins - the baiji (*Lipotes vexillifer*), the Ganges river dolphin (*Platanista g. gangetica*) and the Indus river dolphin (*P. g. minor*), to the verge of extinction (Reeves and Leatherwood, 1994). On the other hand, pollution, caused by a number of elements, such as heavy metals, organohalogenated compounds, oil spills, and acoustic pollution, has reached alarming levels in some geographic areas, and may be having detrimental effects upon marine mammal population growth. Various aspects of the problem of pollution are dealt with in detail in a book by Geraci and St. Aubin (1980), in the special issue of the IWC devoted to cetaceans (Reijnders *et al.*, 1999), and in a review chapter on environmental contaminants and marine mammals by Thomas O'Shea (1999). Here in Chapter 15 of this book, Ailsa Hall considers the special case of organohalogenated contaminants in marine mammals, and highlights the difficulties of obtaining direct experimental evidence of detrimental biological effects of pollutants.

A special form of pollution is that of sound disturbance. For around ten million years, marine mammals have evolved in a relatively quiet world accompanied only by the sounds of other animals, the occasional earthquake or some other natural perturbation. Only in the last hundred years have our oceans and rivers been disturbed in any substantial way with additional sounds created by mankind. In particular, man's use of the marine environment for commercial, industrial and recreational purposes has led to a considerable increase in the level of background sound (Urick, 1986), and created potential problems for many marine mammal species (Richardson *et al.*, 1995).

Following the more recent philosophy that human development must be sustainable, whale watching, an economically profitable business that contrasts with the old whaling tradition, has proliferated. This type of eco-tourism industry has experienced a recent boom in many areas of the world often without clear regulations. However, the current large-scale nature of this activity begs sound control and the creation of a universal ethical code for people approaching the animals (IFAW *et al.*, 1996; Evans, 1996).

In Chapter 16, Bernd Würsig and Peter Evans address the topic of sound disturbance and its effects upon cetaceans, the most vulnerable of marine mammal taxa, putting this in the context of other conservation threats which in many cases may be more serious. They point out that the effects of industrial, military, ocean science and directed human tourism can each be quite different, and so require separate examination. Sometimes those effects may cause physical damage to hearing, and of course the vessels generating some of the sounds can pose the additional threat of direct harm through ship strikes (a problem facing particularly the slower moving species like sperm whale and pilot whales, and which is a major threat to the survival of northern right whales and Florida manatees). Disturbance can have obvious short-term effects - avoidance of the sound source and other changes in behaviour may often be noted; but it is much more difficult to identify whether they have long-term implications upon life history parameters like survival and fecundity.

The final chapter by Bernd Würsig, Randall Reeves and J.G. Ortega-Ortiz examines a new conservation problem recognised recently - the possible effect of Global Climatic Change upon marine mammal populations (see also Tynan and Demaster, 1997; IWC, 1997; Würsig and Ortega-Ortiz, 1999). As a result of various human activities (emissions of ozone-depleting substances, increasing use of hydrocarbons for energy and fuel, and large-scale deforestation and desertification), the world is experiencing climate change to the extent that by the end of this century, temperatures are estimated to rise by $1.3\text{-}5.8^0\,C$, and overall sea level to rise by anywhere between 9-88 cm (Intergovernmental Panel on Climate Change, 2001). Obvious consequences will be the melting of polar ice, drowning of coastal plains and low lying islands, and changes to shallow seas. This will affect breeding sites for various pinniped species, and polar bears, coastal cetaceans and sirenians may find it difficult to adjust to loss of important feeding and breeding habitat. Shifts in areas of primary productivity may lead to distribution changes for many marine mammal species, whilst an increase in the frequency and velocity of storms, and more extreme seasonal fluctuations in local climate (*e.g.* El Niño Southern Oscillation events) may have further more indirect impacts.

In conclusion, a variety of human activities negatively influence marine mammal populations, to the extent that we can be certain that some species will not survive much longer. Although the introduction of international legislation protecting aspects of the marine environments (*e.g.* in the eastern North Atlantic and Europe: OSPAR Convention, HELCOM, ASCOBANS, and ACCOBAMS) is a welcome new trend, the lack of scientific information or the fundamentally different perspective that some governments take with regard to environmental policies still hamper the implementation of workable management and recovery plans for marine mammal populations. This applies particularly in the case of species that undergo long migrations or have extensive geographical ranges. To further compound the problem, there are jurisdictional and administrative obstacles at international and even national scales that, in practice, waste many conservation plans that otherwise appeared to be theoretically sound.

But there are a few examples that illustrate the success of a strategy of protection and recovery. One such is the case of the North Pacific elephant seal (*Mirounga angustirostris*). Only a tiny rookery of around 100 individuals of this species remained on an island off the Californian coast at the end of the nineteenth century; after a considerable conservation effort, its population size is now above 100,000 individuals, distributed along the Pacific coast of USA and Mexico. There is still room for hope.

REFERENCES

Clark, C.W. (1976) *Mathematical Bioeconomics: The Optimal Management of Renewable Resources*. Wiley Interscience, New York, NY.

Cooke, J.G. (1994) The management of whaling. *Aquatic Mammals*, **20**, 129-135.

Evans, P.G.H. (1996) Human disturbance of cetaceans. In: *The Exploitation of Mammals - principles and problems underlying their sustainable use*. (Ed. by N. Dunstone & V. Taylor), pp. 376-394. Cambridge University Press. 415pp.

Gambell, R. (1999). The International Whaling Commission and the contemporary whaling debate. In: *Conservation and Management of Marine Mammals* (Ed. by J.R. Twiss Jr and R.R. Reeves), pp. 179-198. Smithsonian Institution Press, Washington DC. 471pp.

Geraci, J.R. and St. Aubin, D.J. (Eds.) (1980) *Sea Mammals and Oil: Confronting the Risks*. Academic Press, San Diego, CA.

IFAW, Tethys Research Institute, and Europe Conservation. (1996) *Report of the Workshop on the Scientific Aspects of Managing Whale Watching*. Montecastello de Vibio, Italy, Mar 30-Apr 4, 1995. 40pp.

Intergovernmental Panel on Climate Change. (2001) Climate Change 2001: The Scientific Basis. *Third Assessment Report of the IPCC Working Group* I. Available from IPCC Secretariat, c/o World Meteorological Organization, 7bis Avenue de la Paix, C.P. 2300, CH-1211 Geneva. http://www.ipcc.ch/

International Whaling Commission. (1994) Report of the Scientific Committee. Annex H: The Revised Management Procedure (RMP) for Baleen Whales. *Report of the International Whaling Commission*, **44**, 145-152.

International Whaling Commission. (1997) Report of theIWC workshop on climate change and cetaceans. *Report of the International Whaling Commission*, **47**, 293-313.

Reijnders, P.J.H., Aguilar, A. and Donovan, G.P. (Eds.) (1999) Chemical Pollutants and Cetaceans. *The Journal of Cetacean Research and Management* (Special Issue **1**), 1-273.

Richardson, W.J., Greene, C.R. Jr., Malme, C.I. and Thomson, D.H. (1995) *Marine Mammals and Noise*. Academic Press, San Diego. 576pp.

O'Shea, T.J. (1999) Environmental contaminants and marine mammals. In: *Biology of Marine Mammals* (Ed. by J.E. Reynolds III and S.A. Rommel), pp. 485-564. Smithsonian Institution Press, Washington DC. 578pp.

Reeves, R.R. and Leatherwood, S. (1994) *Dolphins, porpoises and whales*. An Action plan for conservation of biological diversity:1994-1998. IUCN, Gland, Switzerland. 91 pp.

Tormosov, D.D., Mikhaliev, Y.A., Best, P.B., Zemsky, V.A., Sekiguchi, K., and Brownell, R.L. Jr (1998) Soviet catches of southern right whales *Eubalaena australis*, 1951-1971. Biological data and conservation implications. *Biological Conservation*, **86**, 185-197.

Tynan, C.T. and Demaster, D.P. (1997) Observations and predictions of Arctic climatic change: potential effects on marine mammals. *Arctic*, **50**, 308-322.

Urick, R.J. (1986) *Ambient Noise in the Sea*. Peninsula Publishing, Los Altos, CA.

Wade, P.R. (1998) Calculating limits to the allowable human-caused mortality of cetaceans and pinnipeds. *Marine Mammal Science,* **14**, 1-37.

Würsig, B. and Ortega-Ortiz, J. (1999) Global climate change and marine mammals. *European Research on Cetaceans*, **13**, 351-355.

Yablokov, A.V. (1994) Validity of whaling data. *Nature* (London), **367**, 108.

Zemsky, V.A., Berzin, A.A., Mikhaliev, Y.A., and Tormosov, D.D. (1995) Soviet Antarctic pelagic whaling after WWII: review of actual catch data. *Report of the International Whaling Commission*, **45**, 131-135.

Chapter 13

Interactions Between Aquatic Mammals and Humans in the Context of Ecosystem Management

[1]ENRIQUE A. CRESPO and [2]MARTÍN A. HALL
[1]Centro Nacional Patagónico and Universidad Nacional de la Patagonia, Blvd. Brown s/n., 9120 Puerto Madryn, Chubut, Argentina. E-mail: kike@cenpat.edu.ar; [2]Inter-American Tropical Tuna Commission, 8604 La Jolla Shores Dr., La Jolla, CA 92037, USA. E-mail: mhall@iattc.org

1. INTRODUCTION AND SCOPE

Aquatic mammals will be treated in the present chapter from an ecological point of view and in a broad sense. Even though aquatic mammals exhibit a wide variety of such ecological characteristics as size, behaviour, diet, and habitat, they also have some common basic characteristics. They also share a long history of ecological interactions with humans and, at present, a wide range of conflicts.

After a brief review of these ecological characteristics, we will briefly cover the relationship between aquatic mammals and man over the years. The first period goes from the origin of mankind through the first half of the 20th century. The second period will cover the important changes that took place during the decade of the 1960s, caused by the rapid expansion of the world's fisheries, which was accompanied, in some countries, by changes in attitudes toward aquatic mammals. Many habitat conflicts started or increased in this period, too. Finally, we will review the present situation.

Marine Mammals: Biology and Conservation, edited by
Evans and Raga, Kluwer Academic/Plenum Publishers, 2002

2. ECOLOGICAL CHARACTERISTICS OF AQUATIC MAMMALS

In most cases, aquatic mammals are close to the top of their food webs. They are considered to be K-strategists (Estes, 1979; McLaren and Smith, 1985). Their characteristics include large body size, long life spans, low reproductive rates, parental care of the offspring, delayed reproduction, juvenile survival that is less predictable than that of the adults, and predictable adult survival. In some cases they appear to show density-dependent responses in their population parameters (Fowler, 1981, 1984; Doidge *et al.*, 1984; Evans and Stirling, this volume). Some of these parameters for the cetaceans have been tabulated by Lockyer (1984) and by Evans (1987). Among the mammals, they can be placed at one of the extremes of the r-K continuum; at the other end, we find rodents and insectivores.

The adult size ranges from around 1 m and 15 kg in the sea otter (*Enhydra lutris*), to 30 m and 100 MT in the blue whale (*Balaenoptera musculus*). Aquatic mammals are long-lived species, ranging from around 20 years for small dolphins and sea lions to 40 or 60 years in many species of seals, large delphinids, and whales.

The most prolific species produce a calf per year in the best cases. Examples include seals, sea lions, and otters (Estes, 1979; Riedman, 1990). Lower reproductive rates are found in the larger whales, which rear one calf every several years (Evans, 1987). The breeding cycle in the southern right whale (*Eubalaena australis*) lasts between 3 and 4 years (Payne, 1986; Payne *et al.*, 1990*)*.

Parental care includes long periods of lactation of the calves, and the teaching of swimming, techniques for prey capture, *etc.* In many delphinids and in sperm whales (*Physeter macrocephalus*), calves remain in herds close to relatives for several years. Seals, on the other hand, have a shorter period of parental care, with the hooded (*Cystophora cristata*) and elephant seals (*Mirounga* spp.), at one extreme, having a period of care of less than one month, which is in part offset by the high rates of energy transfer between mother and calf during lactation (Riedman, 1990; Evans and Stirling, this volume; Lockyer, this volume).

Sexual maturity tends to come late in life. Most of the seals and sea lions, and several species of small cetaceans, attain sexual maturity four to six years after birth. Some larger delphinids and ziphiids begin to reproduce after nine years or more (Lockyer, 1984; Perrin and Reilly, 1984).

Survival is relatively uncertain between weaning and sexual maturity. This period, while the individual is learning different feeding and reproductive behaviours, is usually critical, and mortality is frequently the price for a mistake. But later, once the basic skills are mastered, life becomes more predictable (Caughley, 1966; Barlow and Boveng, 1991).

Population parameters, such as age at sexual maturity and reproductive rates, seem to be density dependent, but there is still some ongoing controversy on the subject (Eberhart, 1977; Fowler, 1981, 1984; DeMaster, 1984; McLaren and Smith, 1985). If the populations respond to changes in density by adjustments in age at sexual maturity, reproductive rates, juvenile mortality rates, and other parameters, then some of the management models proposed are valid. If this is not the case, many of our currently used models probably produce misleading results, and the population changes observed may be tracking environmental, or other, ecosystem changes.

Many species, such as large whales, otters, and some sea lions, have been driven to the edge of extinction. Several of them have recovered, or seem to be recovering. In some baleen whales, the response to the reduction in negative impacts has been quick (Best, 1993), but for others it has been argued that interspecific competition may slow down the process (see discussion in Clapham and Brownell, 1996). This leaves open questions concerning the existence of density dependence, and the role that other environmental effects may play in that recovery. The lesson, from the ecosystem management point of view, is that even resilient ecosystems may take decades to return to their previous conditions after major disturbances, and that there is no certainty that they will return to the same conditions. It is possible to eliminate or mitigate negative impacts on an ecosystem, but it is not possible to predict the trajectory of the response, or its endpoint.

3. THE USE OF AQUATIC MAMMALS AS A RESOURCE BY HUMANS

3.1 Primitive Societies

Humans have utilised aquatic mammals since prehistoric times, and archaeological records show the existence of tools and weapons made out of bone or stone for the apparent purpose of hunting mammals. In the cold regions of the planet, where the scarcity of animal protein of terrestrial origin, and of alternative food sources, led their inhabitants to the coastal zone or the sea, they were especially important. There were several characteristics that made them very valuable:

a)   The protein, and other nutrients contained in aquatic mammals came in large "packages," compared to fish, shellfish, *etc.*

b)   In primitive societies, the technology to catch fish schools in the ocean was not available in most cases, while the technology to capture aquatic mammals was.

c)   Because aquatic mammals are rich in fats, they provide a high-value diet in cold environments.

d)   They were predictable in their location in some cases (*i.e.* rookeries), or in their seasonal migrations in others.

e)   They were vulnerable because of their territorial and/or social behaviour, and in many cases they were a relatively easy (*i.e.* non-risky) prey.

f) They offered an opportunistic bounty when stranded; in cold weather, the preservation of food was easier, so the opportunity was not wasted.

g)   Because they were warm-blooded animals surviving in extremely cold conditions in many cases, they had developed adaptations to cold (*e.g.* fur, waterproof leather, and blubber layers) that could be used by humans exposed to similar conditions.

h)   They were also an energy source: the blubber could be burned into oil and used for light and warmth during the cold nights.

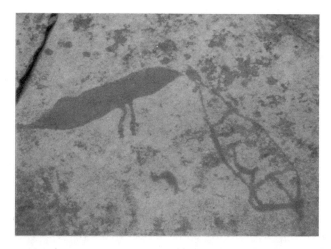

Figure 1.   Neolithic rock carving in Alta, northern Norway, shows that whales were killed by coastal arctic communities as long as 4,000 years ago. (Photo: P.G.H. Evans)

i) Because of their size and other characteristics, they had symbolic value for many cultures, and were integrated into rituals and ceremonies.

Some of these characteristics were of particular advantage with the pinnipeds or coastal small cetaceans, which are sufficiently large to provide an ample reward without serious problems or risks in hunting them (Bonner, 1982). Since Palaeolithic days, the hunting of seals and sea lions was a matter of survival and an activity of vital interest for many coast-dwelling communities in northern Europe, Asia, South America, and other parts of the world. In some cases the contribution of the aquatic mammals to the energy intake of some of these communities was very large. In aboriginal populations of southern South America more than 90% of the total energy intake was provided by fur seals and small cetaceans (Orquera and Piana, 1987; Piana, 1984; Schiavini, 1990).

A harpoon with a detachable head was developed and used in the Arctic and in southern South America (Piana, *pers. comm.*), by Inuit and Fuegian aboriginal people. The head, made of hard bone or ivory, was connected to a long stick. When the animal was harpooned, the head separated from the stick, but remained joined to a line and, in the case of the Inuit, to a float (avutang). The animal bled to death in the water, and was eventually retrieved by means of the line and float. The harpoons varied from single sticks, with stone knives at the tip, to complicated technologies in the Inuit culture, but they all followed the same basic principles (Bonner, 1982).

The economies of the Inuit people, the Fuegian groups, and many others were extremely dependent on seal and sea lion products. The blubber was eaten and used as fuel for lamps. According to Bonner (1982), the Inuit could have survived without the heat of the blubber lamps, but not without their light during the long Arctic nights.

3.2   Modern Societies: History of Whaling

The history of whaling as an industrial activity is much more recent (Evans, 1987), although it remained a subsistence harvest for several centuries in islands such as Japan, Tonga, *etc.* The earliest recorded European whaling was carried out by Norsemen of north-western Europe and by Basques of Spain who hunted whales in the Bay of Biscay around the 8th and 10th centuries (Allen, 1980). From the Bay of Biscay whaling spread to the rest of Europe, Greenland, and North America. At that time, during the 16th and 17th centuries, whaling became a major commercial operation quite different from subsistence activities. Northern right (*Eubalaena glacialis*), bowhead (*Balaena mysticetus*), and Atlantic gray whales (*Eschrichtius robustus*) were the main targets at that time, and the first stocks to be depleted. The Atlantic gray whale was driven to extinction.

During the 18th century, whaling became established as an industrial activity, and spread through the east coast of North America, the Indian and the Pacific Oceans, and the subantarctic islands of the Southern Ocean. During this period, whaling was carried out by means of hand harpoons and rowboats, and the whales were processed by the sides of the ships.

However, the industrial revolution resulted in a shift from a subsistence harvest to one based on monetary gain. The technological innovations brought about in those years, and the growing capital investment, resulted in more efficient ways of hunting that led to the decline of several stocks. Right whales, fur seals, and sea otters, among others, started to decline during that period. Large rorquals and other species followed them later.

Figure 2.   Scottish whaling ship with its catch at the peak of Southern Ocean whaling activities, South Georgia. During the twentieth century, around 360,000 blue whales were killed in the Antarctic alone. (Photo: J. Moncrieff).

Modern whaling started at the end of the 19th century when the harpoon gun and the explosive harpoon were introduced, and mounted on steam-driven vessels (Allen, 1980). Fast-swimming whales like the rorquals, which were too fast for the rowboats, could now be taken with the new technology. This kind of whaling, with relatively minor changes, is still being used in the limited exploitation that takes place today.

Increasing demand and conceptual errors in management, together with an intense competition for the resource, resulted in substantial declines in many stocks. The history of whale catches shows that each target species was replaced by another one, as the exploited stock was being depleted (Allen, 1980; Evans, 1987). The impacts of these harvests were probably very intense in some ecosystems, such as the Antarctic Ocean, but they also affected other ecosystems visited by the whales during their migrations.

The long migrations of some aquatic mammals and other highly migratory species raise the issue of the ecosystem boundaries. Should the management schemes for all ecosystems crossed by a major migration be harmonised with respect to the migrating species and its impacts in each region's prey, competitors, *etc*? Or should the definition of ecosystem, and therefore its management boundaries, be extended to include the whole migratory circuit? Practical reasons will probably decide the issue toward the first approach, but achieving harmony for all regions involved will not be easy, given the differences in priorities, economic standards, *etc*. The forum for this process may be, in the first stage, a combination of regional or national organisations when they exist, and international bodies such as the International Whaling Commission.

4. HUMANS AND AQUATIC MAMMALS AS COMPETITORS

The direct exploitation of aquatic mammal resources, both for the Stone Age communities and the modern industrial man, can be considered as a predator-prey relationship. In contrast to other relationships of this type, however, the reduction in prey abundance did not exert a negative effect on the predator population. Technology increased the efficiency of the whalers, making up for the decreased abundance of whales. What differentiates humans from other animal predators is the ability of humans to "evolve" through fast technological change, at a rate that the prey species cannot match. Because of this, management is needed to make a sustainable use of the resources. Until recently, there were no substantiated science-based procedures that could allow for a cautious management scheme.

In the last 30 or 40 years, the harvesting of whales and other aquatic mammals has been questioned (Barstow, 1990; Stroud, 1996), and this ended the direct exploitation of many of these populations. In the same period, however, a large increase in the amount of marine products extracted by the world's fisheries has resulted in other interactions between humans and aquatic mammals. The competition between aquatic mammals and man has intensified to the point that today it is probably more significant than the direct exploitation of aquatic mammals. This competition can be viewed at two levels (Trites *et al.*, 1997): (1) they share some prey items such as some fish and squid species that are taken by both fishermen and mammals (Pauly *et al.,* 1995*)*; and (2) since they are both part of the same food webs, ultimately they are competing for the primary productivity of the region where they occur.

Figure 3.   Franciscana on a quayside in southern Brazil. (Photo: E. Secchi).

In many cases, fisheries affect marine mammals by decreasing the abundance of their prey species. This problem has been well documented in the coastal waters of the southwestern Atlantic for the franciscana (*Pontoporia blainvillei*). The interactions between the fishery operating in this region and the small cetaceans, have been reviewed by Pinedo (1994) and Secchi *et al.*, (1997). The principal prey of this coastal and river dolphin is the sciaenids, which have been severely depleted by coastal fisheries. Two studies conducted for the same area (Rio Grande do Sul) in 1982 and 1997 showed preliminary evidence of a shift in the diet of the franciscana (Pinedo, 1982; Bassoi, 1997).

It is also believed that some Northwest Pacific fisheries have affected the Steller sea lion (*Eumetopias jubatus*), whose numbers have decreased in recent years (Alverson, 1992; Trites and Larkin, 1996). However, the interactions can be much more complex; fisheries also remove aquatic mammal predators and competitors by direct or incidental capture. Since the biomass of aquatic mammals may become very large, the interactions can be significant. In the tropical eastern Pacific, for example, the biomass of odontocetes almost equals the biomass of the three principal commercial tuna species (Table 1) (Michael D. Scott and Patrick K. Tomlinson, *pers. comm.*, IATTC).

Table 1. Estimated abundance (1986-1990), average weight and biomass of some eastern Pacific marine mammals, and biomass of the principal commercial tuna species.

Marine Mammals and tuna species		Number	Average Weight per indiv. (kg)	Biomass (metric tons)
Dolphins				
NE Spotted	*Stenella attenuata*	730,900	55	40,200
SW Spotted	*Stenella attenuata*	1,298,400	55	71,412
Coastal Spotted	*Stenella attenuata*	29,800	73	2,175
Eastern Spinner	*Stenella longirostris*	631,800	46	29,063
Whitebelly Spinner	*Stenella longirostris*	1,019,300	46	46,888
Striped	*Stenella coeruleoalba*	1,918,000	96	184,128
Northern Common	*Delphinus delphis*	476,300	83	39,533
Central Common	*Delphinus delphis*	406,100	83	33,706
Southern Common	*Delphinus delphis*	2,210,900	83	183,505
Fraser's	*Lagenodelphis hosei*	289,300	128	37,030
Bottlenose	*Tursiops truncatus*	243,500	173	42,126
Risso's	*Grampus griseus*	175,800	246	43,247
Rough-toothed	*Steno bredanensis*	145,900	173	25,241
Subtotal dolphins				**778,254**
Toothed Whales				
Pilot	*Globicephala* sp.	160,200	984	157,637
Melon-headed	*Peponocephala electra*	45,400	126	5,720
Pygmy killer	*Feresa attenuata*	38,900	139	5,407
False Killer	*Pseudorca crassidens*	39,800	828	32,954
Killer (Orca)	*Orcinus orca*	8,500	2,280	19,380
Cuvier's beaked	*Ziphius cavirostris*	20,000	1,810	36,200
Beaked whales	*Mesoplodon* spp.	25,300	600	15,180
Pygmy sperm	*Kogia simus*	11,200	167	1,870
Sperm	*Physeter macrocephalus*	22,700	27,170	616,759
Subtotal toothed whales				**891,107**
Total odontocetes				**~1,700,000**
Tunas				
Yellowfin tuna	*Thunnus albacares*			500,000
Skipjack tuna	*Katsuwonus pelamis*			1,000,000
Bigeye tuna	*Thunnus obesus*			500,000
Total principal commercial tuna species				**~2,000,000**

The ratio *annual consumption/biomass* for these species is approximately, 17 for tunas (range 13-21), 16 for dolphins, and 8 for toothed whales (Olson, Robert J., and Watters, George, *pers. comm.*, IATTC). The annual biomass consumed by these groups is:

Tunas	34 million MT
Dolphins	12 million MT
Toothed whales	7 million MT
Total odontocetes	19 million MT

The annual catch of these tuna species from the eastern Pacific is usually less than 500,000 MT (IATTC Annual Report for 1996), so the proportion of the biomass consumed by the tunas that reaches the human component of the food chain is less than 2%. As Trites *et al.,* (1997) have shown for the overall Pacific Ocean, the biomass consumed by cetacean populations in the eastern Pacific area, is greater than the catch of all the region's fisheries combined. However, as the same authors pointed out, the prey taken by the cetaceans and by the tunas is not necessarily the same as that consumed by humans. In any case, they are competing for the production of the lower levels. It is very clear that if we attempt to manage the ecosystem, we cannot make decisions for tunas while ignoring the dolphins, or *vice versa*. To increase fish production from this system may require that we develop new techniques, to capture species lower in the food chain, and new markets to utilise them. A decision of this type implies that humans will be replacing both tunas and dolphins (and sharks, billfishes, *etc.*) at the top of the food chain, rather than preying on them (incidental mortality may be considered a form of predation). A decision to grant total protection to aquatic mammals, on the other hand, implies severe constraints to the expansion of these fisheries lower in the food chain, and also actions to mitigate or eliminate incidental mortality when it occurs. Any intelligent decision must be based on establishing ecosystem objectives and constraints that allow for a sustainable use of the resources, while allowing the maintenance of the basic structure of the ecosystem. These decisions must be based on answers to very difficult questions concerning our societies' objectives with regard to aquatic mammals and target species (Hall and Donovan, this volume).

This very simplistic analysis omitted, among other things, the respective roles of the pinnipeds and baleen whales in the same ecosystem; the omission does not mean that their role is negligible, only that the main points can be made even with the more limited group used.

If we want to conserve the aquatic mammals as an "important" or "significant" part of the ecosystem, we must include them in every resource allocation scheme as other users. The definition of "important" or "significant" is not obvious, and it is loaded with cultural and economic ingredients. Are we ready to set aside total allowable catches (TACs) for aquatic mammals? The predation of aquatic mammals on any exploited species is, of course, included in the natural mortality of the species, and that is part of the traditional management models. But, if we set as an objective the recovery of the mammal population to some level higher than the current one, then that decision should be reflected in any allocation of catches of the target species in question. This is not easy, in a world where excessive fishing capacity is the norm rather than the exception, and where competition among nations and fleets is quite intense. Besides, our knowledge of ecosystem interactions is not good enough to predict which species will benefit from prey not harvested.

International co-operation is a must to achieve conservation goals for the vast majority of aquatic mammals that live in, and travel across, many national jurisdictions. But that co-operation requires a harmonisation of objectives that, up to now, has not often been achieved. Some sectors of the public, especially in more prosperous industrialised countries, assign a very high "moral" value to many aquatic mammals, and would like to see total protection for them. On the other hand, most inhabitants of developing countries have a different set of priorities, and for them the struggle against poverty, hunger, and unemployment is the primary concern. Culling of aquatic mammal populations could increase the resource base available to fishermen in a region, reducing the competition, but the expectation that there will be a direct and clear response ignores the complexities in the food web (Earle, 1996). For example, there are models of food webs developed for the Benguela Current ecosystem (Yodzis, 1994; Punt and Leslie, 1995; Punt and Butterworth, 1995) which show that, under a variety of assumptions, the effect of culling the Cape fur seal (*Arctocephalus pusillus pusillus*) population from the region could lead to reductions in the fisheries catch. It is not easy to see an intermediate point between total protection and culling.

It should also be noted that the strict environmental standards of some developed countries sometimes result in their fleets or industries being moved to developing nations, which in their eagerness to attract investments and to create jobs are more willing to overlook the negative environmental impacts of those activities. The impacts do not disappear; they are simply transferred to other areas. Although those investments bring opportunities to the developing nations, the end result is frequently overfishing and overcapitalisation, facilitated by the lack of regulations, and by the poor enforcement that takes place when the resources available for this purpose are very limited, so that its priority is not deemed high. For aquatic mammals, this results in higher levels of uncertainty about the impacts on populations for which there are less data available on population structure, abundance, *etc*, and less human resources trained to deal with these impacts.

5.   FISHERIES AND ECOSYSTEM PRODUCTIVITY

But there is another aspect of the interactions of fisheries with aquatic mammal populations that is increasing in importance with the globalisation of the economy. In primitive times, the influence of humans on the ecosystem was mostly at the local scale. Primitive groups consumed aquatic mammals and fishes from coastal populations, and the organic matter obtained in that way was "recycled" within a limited geographical space.

With the availability of modern transportation systems, and the economic inequalities in the world, a net flux has developed from some ocean basins, to other ones. Tunas caught in the eastern and western Pacific may be transported to Asia for canning, and finally to the United States or the European Union for consumption. Fishmeal produced on the west coast of South America, hake from the Argentinean shelf, salmon from Chilean farms, and many other marine products frequently find their way to distant markets. If the product is of high value, and many fish products are, it is more likely to end up in the markets of the industrialised nations of the Northern Hemisphere.

Fishmeal provides an example of a high-volume trade product. Chile and Peru produced 3.3 million MT of fishmeal in 1995, and exported close to 3 million MT of that (FAO, 1995). The biomass, extracted from the west coast of South America, needed to produce this amount of fishmeal is probably 4 to 5 times the total exported, or 12 to 15 million MT. It is sent to other ocean basins where it is consumed, and finally recycled. To make a quick comparison, that particular ecosystem "exports" every year the equivalent of 4-5 times the total biomass of odontocetes and commercial tunas present in the eastern Pacific at a given time. In the same year, Japan and China imported more than 1 million MT of fishmeal (FAO, 1995). If we consider several groups of fish products used by FAO, Japan had a net import in 1995 of over 3.25 million MT; Italy of over 570,000 MT; Spain of over 540,000 MT; on the other hand, Peru had a net export of over 2 million MT, and Chile of over 1.4 million MT (FAO, 1995). Argentina's exports went from 0.2 million MT in 1992 to 0.8 million MT in 1997 (PROMSA, 1998).

These very large biomasses were previously consumed by predators or scavengers in the respective regions, or sank to the bottom to feed the benthic components, *etc.* Now, they are transported to distant ocean basins, to feed humans, pets, or livestock. At these destinations they are consumed, and eventually recycled, but the recycling happens in a basin that is not the original one. The upwelling ecosystem of the west coast of South America is transferring annually 12 to 15 million MT of biomass to other basins. The seas adjacent to Japan receive 3.25 million MT annually. How much of this biomass is returned to their original ocean basin? Diffusion and advection could eventually balance the concentration, but over long periods of time. In return for the biomass exports, the countries of origin receive many high value-added products with no organic matter content (cars, computers, TVs, VCRs, *etc.*). Of course, there are many other products that are also being transferred among regions (timber, agricultural products, *etc.*).

The aquatic mammals at both ends of this process are exposed to opposite long-term trends. In the "extraction" ecosystems, they will face temporal or permanent reductions in biomass of many prey items, and probably changes in the size spectrum of the ecosystem. In the "consumption" ecosystems, the input of organic matter and nutrients may lead to higher productivity, eutrophication, *etc.*

Another way in which fisheries affect oceanic production is through the discards of dead individuals which are not marketable because of their size or other characteristics (Alverson *et al.*, 1994). The discards introduce what can be a major food subsidy to the species that can utilise it in either the water column or the benthos (Kennelly, 1995), while at the same time there is a removal of live individuals at different levels of the food chain, depending on the fishery. In the case of surface or midwater fisheries, there is a transfer from the surface or midwater layers of the water column to the bottom.

Some of these discards are consumed in the water column by scavengers (Britton and Morton, 1994), which in many cases have adapted to take advantage of fishing activities. This group includes seabirds (Hudson and Furness, 1988) and sharks, but also several species of aquatic mammals that are commonly seen following fishing boats or searching for food in fishing gear (Wickens and Sims, 1994; Kennelly, 1995). This subsidy may alter the competitive interactions among different components of the ecosystem, favouring those that utilise the fisheries over the others. When the discards reach the benthos, they can be utilised by different species (Hill and Wasenberg, 1990; Britton and Morton, 1994), depending on the depth. In shallow areas, the recycling may be fast, and even aquatic mammals may benefit from them. In deep areas, however, the discards will be out of reach for most components of the epipelagic ecosystem. Whether the discards are consumed by deep-sea scavengers, or decompose in situ, for surface fisheries, the completion of the cycle, bringing back its components to the epipelagic system, may be very lengthy (decades, or even centuries), depending on the circulation of the deep-water masses in the area. Deep-water currents may take these elements to other regions, before upwelling can bring them to the surface.

6.   HABITAT CONFLICTS

Some aquatic mammals live in oceanic areas, where their interactions with humans are limited to contacts with fishing, cargo, or other types of vessels, with active or lost fishing gear, or to indirect global effects (Johnston *et al.*, 1996) such as pollution, global warming, and prey depletion. But most aquatic mammals spend all or part of their lives in the coastal zones, and some in freshwater habitats, where the interactions with humans are much more frequent and intense.

The development of the marine coastal zone, or of the riverine habitats, leads to the construction of ports, piers, bridges, and breakwaters, to the closure of coastal lagoons for shrimp aquaculture, or to other changes, that in general, reduce the habitat available for aquatic mammals, or modify its characteristics (dredging, ambient noise, *etc.*).

Figure 4.   Dusky and Commerson's dolphin bycatches from midwater trawl activities in Argentina. (Photo: E.A. Crespo)

In some cases, the effect is direct and dramatic; *e.g.* an increasing human coastal population leads to more collisions with sport boats that are a major source of manatee mortality. In other cases, the effect is less obvious, but equally permanent, *e.g.* the building of a canal connecting two basins previously isolated, or of a dam isolating two areas previously connected, has consequences that may affect the survival or the genetic structure of a population. Yet canals and dams provide considerable benefits to the human population, and hydroelectric power has, until recently, been promoted as a clean alternative to fossil fuels or nuclear energy generation. Of the many and varied habitat conflicts we shall address only: (1) interactions with fishing gear and, very briefly, (2) the effects of dams; and (3) pollution.

6.1   Interactions of Aquatic Mammals with Fishing Gear

The primitive ways of harvesting aquatic mammals were harpoons, spears, clubs, and traps. Hook-and-line fishing seldom results in the capture of an aquatic mammal taking the hook. Other primitive ways of fishing, however, may have had some impacts on aquatic mammals. Plant extracts and other substances were used as poisons to catch fish under some circumstances (usually in freshwater environments). Traps of a variety of designs were used in most coastal cultures. It is possible that some of these systems, intentionally or not, may have caught aquatic mammals in some coastal environments with special characteristics (coastal lagoons, areas with narrow connections to the ocean or pronounced tidal changes, *etc.*)

The development of nets brought about an increase in the risk of entanglement and incidental mortality. Primitive nets, smaller and made of plant fibres, may not have been strong enough to capture aquatic mammals which become entangled in them.

In summary, the incidental impact of primitive fishermen will never be well estimated, but it is likely to have been significant only under some special conditions. Species of aquatic mammals living in, or entering, enclosed marine areas and freshwater systems may have been more vulnerable.

The industrial revolution allowed fishermen to operate further offshore, and to utilise much larger gear at a greater speed. Synthetic fibres made the nets less visible, and much harder to tear. At the same time, the increasing human population produced a demand for food that was, in part, filled by the rapid expansion of the world fisheries. In recent times, the impact of fisheries on marine mammals has escalated (Northridge, 1984, 1991). The entanglement in gillnets and other passive gear around the whole world in artisanal fisheries, in addition to that recorded in large pelagic fisheries like purse-seine and driftnet fisheries, became the major cause of mortality for stocks of dolphins, whales, and seals in the 1960-80 period, and it remains so today. The proliferation of synthetic gillnets emerged as a threat, not only to cetaceans and pinnipeds, but also to seabirds, sea turtles, non-target fish, *etc.*

FAO and other organisations promoted the use of gillnets in coastal areas because of their low cost, ease of use, and productivity. Soon they became the most common type of gear throughout the coastal regions of the world. The interactions between gillnets and cetaceans have been the object of a major world review (Perrin *et al.*, 1994). A variant of this type of gear was the driftnet, used in the open seas. Pelagic driftnets longer than 60 km were developed for fishing for squid, tuna, billfish, and salmon in areas where the densities of the target species were low. During the 1970s those fisheries increased in the North Pacific, the Mediterranean, and the North Atlantic. Due to concerns about the expansion and large-scale effects of pelagic driftnets, the United Nations General Assembly established a moratorium on the use of driftnets of more than 2.5 km in international waters in 1992 (Donovan, 1994).

Another type of gear that increased in use was the longline. The longlines can be deployed on the bottom or at different depths in the water column. There are very few published records (Pinedo, 1994; Bailey *et al.*, 1996) of aquatic mammals taken in longlines, although they are known to take some of the catches from the hooks (Nitta and Henderson, 1993; Yano and Dalheim, 1995). Killer whales, false killer whales, pilot whales, and bottlenose and common dolphins have been reported to do this, and occasionally they get caught (Heberer, ms.).

The impacts of incidental takes in trawl fisheries have been overlooked for many years, but recently several authors have addressed different aspects of the problem (Connelly *et al.*, 1997; Crespo *et al.*, 1994, Dans *et al.*, 1997; Lens, 1997; Morizur *et al.*, 1997).

Perhaps the best known example of an incidental catch of marine mammals in a fishery is that of the tuna-dolphin problem in the eastern Pacific Ocean, where several hundred thousand dolphins were taken each year during the 1960s in a purse-seine fishery targeting yellowfin tuna (*Thunnus albacares*) (Francis *et al.*, 1992). A long-term international program co-ordinated by the Inter-American Tropical Tuna Commission has succeeded in bringing the mortality levels down by more than 97% over the 1986-98 period, and the conservation problem has been solved without the sacrifice of the fishery (Hall, 1998; Hall and Donovan, this volume).

An issue that this fishery brought to the surface was that avoiding the mortality of aquatic mammals by shifting to other, alternative ways of fishing was not a sound ecological solution to the problem. The alternative ways of purse seining result in catches of undersize tunas and to large discards of tunas and of other species (Hall, 1998). Management has evolved from a single-issue approach–to reduce dolphin mortality–to more complex systems, where the consideration of the impact on dolphins is included in general considerations about the impacts on the target stocks and the other components of the ecosystem (Hall and Donovan, this volume). This evolution has been possible because of the joint efforts and co-operation of nations, fishermen, and environmentalists; the many lessons learned over more than 20 years of conflict can now be used by others dealing with similar problems (Hall, 1998).

Many of the incidental catches occur because some aquatic mammals have learned to take advantage of the fishing operations to obtain an easy meal, either by feeding from the catch or from the discards (Wickens and Sims, 1994; Kennelly, 1995). In these cases, the marine mammal population is receiving a "subsidy" from the fishery, at the same time that it is being negatively affected. While some of the negative effects (*i.e.* incidental mortality) can be assessed, the positive effects (*i.e.* how much higher is survival or reproduction because of the additional low-cost food input) are more difficult to estimate.

Figure 5.   Amazon river dolphin or boto. River dolphins are probably the most threatened group of marine mammals in the world, facing threats from habitat modification, fisheries, pollution and direct hunting. (Photo: T. Henningsen)

In other cases, the marine mammals manage to feed from a fishery or other human activity (*e.g.* aquaculture) while avoiding, in most cases, becoming entangled in the gear, pen, *etc.*, so only positive effects prevail. But the opposite case, where the aquatic mammals are killed by the fishing operations or other human activities, while receiving little benefit from them is also common (*e.g.* a right whale entangled in a lobster trap, a rough-toothed dolphin taken in a purse-seine set on a floating object, or a manatee hit by a boat's propeller). These different cases reflect different ecological interactions between the target of the fishery and the aquatic mammals, and management should take them into consideration (Hall, 1996; Hall and Donovan, this volume).

6.2   Dams

Dams and other barriers in large rivers have been constructed for hydroelectric power generation, irrigation, flood control, and trapping of floodwater for human use or fish culture. Examples all over the world show the significant impact of these projects on regional development. Nevertheless, the immediate and negative consequences in the river ecosystem for river dolphins, manatees, river otters, and other wildlife include, at least, the isolation of populations of aquatic mammals and their prey, above and below the dam; changes in biomass and diversity due to lowered nutrient availability; unnatural water flows, *etc.* (Perrin and Brownell, 1989; Smith, 1996).

Examples can be found in the major basins of the Indus, Brahmaputra-Ganges, Changjiang, Orinoco, and Amazon rivers. The population of the Indus River dolphin, *Platanista minor*, has been fragmented into six isolated sub-units (Khan and Niazi, 1989; Reeves and Leatherwood, 1994). The Ganges River dolphin, *Platanista gangetica,* in addition to threats caused by pollution, mining, and incidental and direct catches (Mohan, 1989), faces problems caused by dam construction.

In the Amazon and Orinoco basins, several major dams have been built, or are being planned (Best and da Silva, 1989), and the same applies to the Chang-jiang River in China, where the baiji, *Lipotes vexillifer,* one of the most endangered of aquatic mammals, lives (Chen and Hua, 1989; Zhou and Li, 1989). In many cases, the dams add to the other problems already affecting the river dolphins.

6.3   Pollution

Organochlorine compounds are carried by runoff from agricultural areas to coastal marine and estuarine areas. In the Ganges basin, for instance, at least 2,500 MT of pesticides and 1.2 million MT of fertilisers are used each year (Mohan, 1989).

Heavy metals and other by-products of industrial development accumulate in the environment, and in prey organisms, and are later concentrated by the top predators (Reijnders, 1996). Eutrophication, caused or enhanced by human activities, leads to low-oxygen areas that are abandoned by prey and predator alike. Excessive noise may impair the aquatic mammal ability to detect prey, communicate, *etc.*, which may also lead to loss of habitat (Gordon and Moscrop, 1996; Würsig and Evans, this volume).

Even though they have not been linked directly to the health of any population, some pollutants are believed to interfere with immune systems and cause reproductive abnormalities and diseases (Lockyer, this volume). Abnormal concentrations of pollutants were found in the striped dolphins (*Stenella coeruleoalba*) that died in the epizootic in the Mediterranean Sea between 1990 and 1992. Although the primary cause for the mortality was a morbillivirus, the dolphins were in poor nutritional condition, and had high levels of ectoparasitism and of epizoits (Aguilar and Raga, 1993).

7. ECONOMIC VALUE OF MARINE MAMMALS AT THE PRESENT DAY, AND THE NEED FOR INTEGRATED MANAGEMENT

With the moratorium on whale hunting and the changes in public attitude toward aquatic mammals, new economic activities have evolved in recent years, centring on the non-consumptive use of the species. Whale watching and ecotourism are now significant sources of income and employment in some areas of the world. Uncontrolled whale watching and tourism can also have adverse impacts on the populations, or on the habitat. A working group of the International Whaling Commission has started to consider this issue as another component of the management of the whale stocks (IWC, 1997). This non-consumptive utilisation of marine mammals may, in some cultures, replace the traditional harvest of the resource, but there are cultures still interested in the utilisation of the resource as a source of food and other products. The debate around the utilisation of marine mammals as a resource has two levels:

1) Those who accept the utilisation of the resource centre their concerns upon issues such as: Can we develop adequate and cautious management schemes that ensure the sustainable use of the resource? Can we monitor the compliance with the previous schemes? The Revised Management Procedure of the International Whaling Commission (IWC, 1994) is an example of a cautious and thoroughly scrutinised answer to the first question, which addresses a wide range of uncertainties. The answer to the second will be needed when the process is implemented.

2) Those who do not accept the utilisation of marine mammals as a resource, base their views on a moral judgement that is strongly culture-dependent (Barstow, 1990). The problem with moral issues is that they do not leave room for solutions that satisfy everyone, and that those involved in

the debate are frequently reluctant to accept alternative views. When some scientists adopt this moral judgment approach, it may influence their interpretation of scientific results. This creates confusion in the managers and the public, which expect a more objective approach, and cannot discriminate between legitimate scientific disagreements, and the pursuit of undeclared moral goals.

As can be seen, the interactions between aquatic mammals and man are related not only to ecology, but also to ethics and economics. The interactions flow in many directions, and the solution for conservation and management of aquatic mammal populations lies in an integrated management in which all the interest groups in conflict can define convenient courses of action that allow them to coexist. These parts should include the nations involved, fishermen, industry leaders, and environmentalists. Much of the success of conservation programmes lies in the ability of the leaders of the different sectors to understand and communicate with each other, and to find the common objectives that sometimes are lost in heated debates (Hall, 1996, 1998).

A quick example of these conflicting interests can be seen along the coast of Patagonia. Ecotourism in Patagonia is steadily increasing, based mainly on the observation of coastal wildlife. The southern right whale (*Eubalaena australis*) is the main attraction, followed by Magellanic penguins (*Spheniscus magellanicus*), South American sea lions (*Otaria flavescens*), and southern elephant seals (*Mirounga leonina*). It is clear, however, that the aquatic mammals of the area are the principal magnets for tourists.

The number of tourists visiting the area to watch whales has increased steadily in recent years, reaching 143,000 in the spring of 1997 (Organismo Provincial de Turismo del Chubut, unpubl. data). The revenue in northern Patagonia to observe wildlife is close to U$S 70 million per year in direct benefits, and it is increasing with the number of tourists. Indirect benefits can be estimated using a multiplier of 3.5, reaching a total benefit of close to U$S 250-300 million per year (Tagliorette and Losano, 1996; Barrera, 1996; A. Tagliorette, *pers. comm.*). It is impossible to predict today whether this industry will continue expanding for many years, or if it will soon reach some saturation point and then stabilise at that level or at a lower level. Will the visitors return time after time, or is it, for many of them, a one-time experience? The combination of economic and educational benefits for the coastal communities are clearly positive aspects (Constantine and Baker, 1997), and adequate regulations should make the activity sustainable in the long-term.

Figure 6.   Ecotourism attractions at Peninsula Valdés, Patagonia. Southern right whale (top) (Photo: B. Würsig); and South American sea lion rookery (bottom). (Photo: E.A. Crespo)

There is pressure from the private sector to open to tourism every attractive wildlife area along the coast and to allow the development of new activities, such as diving with dolphins, whales, and sea lions.

Other activities that produce significant economic benefits and employment in the region include fishing and oil exploitation. The rapid growth of fisheries over the last 25 years has resulted in overfishing of some of the target species, such as hake and squid, which are also an important prey for many aquatic mammals and seabirds (Crespo *et al.*, 1994, 1997; Koen Alonso *et al.*, 1998). The fishery also results in the incidental entanglement of South American sea lions (*Otaria flavescens*), dusky dolphins (*Lagenorhynchus obscurus*), Commerson's dolphins (*Cephalorhynchus commersonii*), and Magellanic penguins (*Spheniscus magellanicus*).

In addition, oil exploitation and transport of oil along the coast are major economic activities for the area, providing employment for many. However, these are also the most important sources of pollution for wildlife and the marine environment. Geraci and St. Aubin (1990) have reviewed some of the impacts of oil pollution on aquatic mammals. While the impacts of oil pollution are sometimes very visible (*i.e.* images of oil-covered penguins in the media), the impacts of fishing or tourism are not so obvious to the public, and are more difficult to trace to a direct cause-effect mechanisms.

8.   CONCLUSIONS

The conservation and management of aquatic mammals is a complex matter, but we have learned that it cannot be separated from the management of the other components of the ecosystem.

To approach the management of aquatic mammal resources from an ecosystem point of view requires that we consider, among other things:

a)   Habitat requirements: measures to achieve this include dealing with pollution and other forms of habitat degradation, fragmentation of habitats, and mitigation of impacts from fishing gear (while in use or after it has been lost).

b)   Environmental issues: including consideration of variability, short- and long-term changes (regime shifts), and trends in environmental parameters.

c)   Coexistence of different forms of utilisation.

d)   Socio-economic and cultural differences: including the development of an understanding and tolerance of those differences in all participants.

e)   As it is very difficult to conceive an ecosystem approach to management that completely protects some components of the ecosystem, the ecological approach suggests that the harvesting policies for all the components of the system be set following basic ecological principles, so as to retain the structure and function of the system (Hall, 1996). These policies may require that the harvests be spread in a balanced way over the whole food web. The term "harvesting" is used loosely, to include bycatches and other sources of mortality that can be controlled. Policies addressed to the protection of a single species or a group of species could be necessary when the threat of extinction is clear, but they should be avoided otherwise, and replaced with more holistic approaches. The unintended consequences of some of these "partial solutions" are clear in some cases (Hall, 1998).

Even though management, in most cases, must be international, each region presents different problems that need to be addressed. Both the global and its regional or local aspects of the status of a given population should be considered. A variety of alternative solutions, taking into account ecological, economic, "moral," cultural, and social variables should be considered.

ACKNOWLEDGEMENTS

The authors are indebted to many people who inspired the points of view exposed in this chapter. Thanks are given to Koen Van Waerebeek, Arne Bjørge, Bill Bayliff, and Adrián Schiavini, who kindly reviewed this chapter, and made important comments and suggestions for the improvement of the manuscript. Lori Pertini (Servicio Centralizado de Computación, CENPAT) kindly helped with the formatting of the chapter according to the rules provided by the editors.

REFERENCES

Aguilar, A. and Raga J.A. (1993) The striped dolphin epizootic in the Mediterranean Sea. *Ambio*, **22**, 524-528.

Allen, K.R. (1980) *Conservation and management of whales*. A Washington Sea Grant Publication. University of Washington Press and Butterworths, London. 107pp.

Alverson, D.L. (1992) Commercial fisheries and the Steller sea lion (*Eumetopias jubatus*): the conflict arena. *Reviews Aquatic Science*, **6**, 203-256.

Alverson, D.L., Freeberg, M.H., Murawski, S.A. and Pope, J.G. (1994) A global assessment of fisheries bycatch and discards. *FAO, Fisheries Technical Papers* 339, 235pp.

Anonymous (1998) *Inter-Amer. Tropical Tuna Commission, Annual Report for 1996*, 306pp.

Bailey, K. Williams, P.G., and Itano, D. (1996) Bycatch and discards in western Pacific tuna fisheries: a review of SPC data holdings and literature. *South Pacific Commission, Data Report*, 34. Noumea, New Caledonia.

Barlow, J. and Boveng, P. (1991) Modeling age-specific mortality for marine mammal populations. *Marine Mammal Science*, 7, 50-65.

Barrera, R.M. (1996) El desarrollo económico sustentable de Puerto Madryn. Informe de la Cámara de Industria y Comercio de Puerto Madryn (Argentina). 12 pp.

Barstow, R. (1990) Beyond whale species survival–peaceful coexistence and mutual enrichment as a basis for human-cetacean relations. *Mammal Review*, 20, 65-73.

Bassoi, M. (1997) Avaliação da dieta alimentar de toninhas, *Pontoporia blainvillei* (Gervais & D'orbigny, 1844), capturadas acidentalmente na costa sul do Rio Grande do Sul. Tese do Grau de Oceanología. FURG. 68pp.

Best, P.B. (1993) Increase rates in severely depleted stocks of baleen whales. *International Commision for the Exploration of the Sea, Journal of Marine Science*, 50, 169-186.

Best, R. and da Silva, V.F. (1989) Biology, status and conservation of *Inia geoffrensis* in the Amazon and Orinoco River basins. In: *Biology and Conservation of the River Dolphins* (Ed. by W.F. Perrin, R.L. Brownell Jr., K. Zhou and J. Liu), pp. 23-34. Occasional Papers of the IUCN Species Survival Commission, N° 3, Gland, Switzerland.

Bonner, W.N. (1982) *Seals and Man: a Study of Interactions*. University of Washington Press. 170pp.

Britton, J.C. and Morton, B. (1994) Marine carrion and scavengers. *Oceanography and Marine Biology: an Annual Review*, 32, 369-434.

Caughley, G. (1966) Mortality patterns in mammals. *Ecology*, 47, 906-918.

Chen P. and Hua Y. (1989) Distribution, population size and protection of *Lipotes vexillifer*. In: *Biology and Conservation of the River Dolphins* (Ed. by W.F. Perrin, R.L. Brownell Jr., K. Zhou and J. Liu), pp. 81-85. Occasional Papers of the IUCN Species Survival Commission, N° 3, Gland, Switzerland.

Clapham, P.J. and Brownell Jr., R.L. (1996) The potential for interspecific competition in baleen whales. *Report of the International Whaling Commission*, 46, 361-367.

Connelly, P.R., Goodson, A.D., Kaschner, K., Lepper, P.A., Sturtivant, C.R., and Woodward, B. (1997) Acoustic techniques to study cetacean behaviour around pelagic trawls. *International Council for the Exploration of the Sea, C.M. 1997/Q*, 15. 6 pp.

Constantine, R. and Baker, S. (1997) Monitoring the commercial swim-with-dolphins operations in the Bay of Islands. *Science for Conservation*, 56. 59 pp.

Crespo, E.A., Corcuera, J.F., and Cazorla, A.L. (1994) Interactions between marine mammals and fisheries in some coastal fishing areas of Argentina. *Report of the International Whaling Commission* (special issue 15), 269-281.

Crespo, E.A., Pedraza, S.N., Dans, S.L., Koen Alonso, M., Reyes, L., Garcia, N.A., Coscarella, M.A. and Schiavini, A.C.M. (1997) Direct and indirect effects of the high seas fisheries on the marine mammal populations in the northern and central Patagonian coast. *Journal of the Northwest Atlantic Fishery Science*, 22, 189-207.

Dans, S.L., Crespo, E.A., Garcia, N.A., Reyes, L.M., Pedraza, S.N. and Koen Alonso, M. (1997) Incidental mortality of Patagonian dusky dolphins in mid-water trawling: retrospective effects from the early 80's. *Report of the International Whaling Commission*, 47, 699-704.

DeMaster, D.P. (1984) A review of density dependence in marine mammals. In: Proceeding of the Workshop on biological interactions among marine mammals and commercial fisheries in the southeastern Bering Sea. *Alaska Sea Grant Report*, 84, 139-148, April 1984.

Doidge, D.W., Croxall, J.P. and Baker, J.R. (1984) Density–dependent pup mortality in the Antarctic fur seal *Arctocephalus gazella* at South Georgia. *Journal of Zoology* (London), 202, 449-460.

Donovan, G.P. (1994) Developments on issues relating to the incidental catches of cetaceans since 1992 and the UNCED Conference. *Report of the International Whaling Commission* (special issue **15**), 609-613.

Earle, M. (1996) Ecological interactions between cetaceans and fisheries. In: *The Conservation of Whales and Dolphins* (Ed. by M.P. Simmonds and J.D. Hutchinson), pp. 166-204. John Wiley and Sons, London.

Eberhardt, L.L. (1977) Optimal policies for conservation of large mammals with special reference to marine ecosystems. *Environmental Conservation*, **4**, 205-212.

Estes, J.A. (1979) Exploitation of marine mammals: r-selection of K-strategists? *Journal of the Fisheries Research Board of Canada*, **36**, 1009-1017.

Evans, P.G.H. (1987) *The Natural History of Whales and Dolphins*. Academic Press/Facts on File Publications, London & New York. 343pp.

F.A.O. (1995) *F.A.O. yearbook: Fishery statistics*. Commodities. Vol. 81. FAO, Rome. 183 pp.

Fowler, C.W. (1981) Density dependence as related to life history strategies. *Ecology*, **62**, 602-610.

Fowler, C.W. (1984) Density dependence in cetacean populations. *Report of the International Whaling Commission* (special issue **6**), 373-379.

Francis, R.C., Awbrey, F.T. , Goudey, C.A., Hall, M.A., King D.M., Medina, H., Norris, K.S., Orbach, M.K. Payne, R. and Pikitch, E. (1992) *Dolphins and the Tuna Industry*. National Academy Press, Washington, D.C. pp. i-xii, 1-176.

Geraci, J.R. and St. Aubin, D.J. (1990) *Sea mammals and oil: confronting the risks*. Academic Press. 282pp.

Gordon, J. and Moscrop, A. (1996) Underwater noise pollution and its significance for whales and dolphins. In: *The Conservation of Whales and Dolphins*. (Ed. by M.P. Simmonds and J.D. Hutchinson), pp. 281-319. John Wiley and Sons, London.

Hall, M.A. (1996) On bycatches. *Reviews in Fish Biology and Fisheries*, **6**, 319-352.

Hall, M.A. (1998) An ecological view of the tuna-dolphin problem: impacts and trade-offs. *Reviews in Fish Biology and Fisheries*, **8**, 1-34.

Heberer, C.F. m.s. A review of bycatches in tuna longline fisheries.

Hill, B.J. and Wasenberg, T.J. (1990) Fate of discards from prawn trawlers in Torres Strait. *Australian Journal of Marine and Freshwater Research*, **41**, 53-64.

Hudson, A.V. and Furness, R.W. (1988) Utilisation of discarded fish by scavenging seabirds behind whitefish trawlers in Shetland. *Journal of Zoology* (London), **215**, 151-166.

International Whaling Commission. (1994) Report of the Scientific Committee. Annex H: The Revised Management Procedure (RMP) for Baleen Whales. *Report of the International Whaling Commission*, **44**, 145-152.

International Whaling Commission. (1997) Report of the Scientific Committee. Annex Q: Report of the Whalewatching Working Group. *Report of the International Whaling Commission*, **47**, 250-256.

Johnston, P.A., Stringer, R.L., and Santillo, D. (1996) Cetaceans and environmental pollution: the global concerns. In: *The Conservation of Whales and Dolphins* (Ed. by M.P. Simmonds and J.D. Hutchinson), pp. 219-261. John Wiley and Sons, London.

Kennelly, S.J. (1995) The issue of bycatch in Australia's demersal trawl fisheries. *Reviews in Fish Biology and Fisheries,* **5**, 213-234.

Khan, K.M. and M.S. Niazi, (1989) Distribution and population status of the Indus river dolphin, *Platanista minor*. In: *Biology and Conservation of the River Dolphins* (Ed. by W.F. Perrin, R.L. Brownell Jr., K. Zhou and J. Liu), pp. 77-80. Occasional Papers of the IUCN Species Survival Commission, N° 3, Gland, Switzerland.

Koen Alonso, M., Crespo, E.A., Garcia, N.A., Pedraza, S.N., and Coscarella, M.A. (1998) Diet of dusky dolphins, *Lagenorhynchus obscurus*, in waters of Patagonia, Argentina. *Fishery Bulletin*, **96**, 366-374.

Lens, S. (1997) Interactions between marine mammals and deep-water trawlers in the NAFO Regulatory Area. *International Council for the Exploration of the Sea.*, *C.M.* 1997/Q:08. 6pp.

Lockyer, C. (1984) Review of baleen whale (Mysticeti) reproduction and implications for management. *Report of the International Whaling Commission* (special issue **6**), 27-50.

McLaren, I. and Smith T.G. (1985) Population ecology of seals: retrospective and prospective views. *Marine Mammal Science*, **1**, 54-83.

Mohan L. (1989) Conservation and management of the Ganges river dolphin, *Platanista gangetica*, in India. In: *Biology and Conservation of the River Dolphins* (Ed. by W.F. Perrin, R.L. Brownell Jr., K. Zhou and J. Liu). pp. 64-69. Occasional Papers of the IUCN Species Survival Commission, N° 3, Gland, Switzerland.

Morizur, Y., Tregenza, N., Heessen, H., Berrow, S., and Pouvreau, S. (1997) Incidental mammal catches in pelagic trawl fisheries of the Northeast Atlantic. *International Council for the Exploration of the Sea.*, *C.M.* 1997/Q:05. 9 pp.

Northridge, S.P. (1984) World review of interactions between marine mammals and fisheries. *F.A.O. Fisheries Technical Paper*, **251**, 1-190.

Northridge, S.P. (1991) An updated world review of interactions between marine mammals and fisheries. *F.A.O. Fisheries Technical Paper*, **251**, Suppl. 1, 1-58.

Orquera, L.A. and Piana, E.L. (1987) Human littoral adaptation in the Beagle Channel region: Maximum possible age. *Quarterly of South America and Antarctic Peninsula*, **5**, 133-162.

Pauly, D, Trites, A, Capuli, E, and Christensen, V. (1995) Diet composition and trophic levels of marine mammals. International Council for the Exploration of the Sea Council Meeting Papers., ICES, Copenhagen (Denmark), 22pp.

Payne, R. (1986) Long term behavioural studies of the southern right whale (*Eubalaena australis*). *Report of the International Whaling Commission* (special issue **10**), 161-167.

Payne, R., Rowntree, V., Perkins, J.S., Cooke , J.G., and Lankester, K. (1990) Population size, trends and reproductive parameters of right whales (*Eubalaena australis*) off Península Valdés, Argentina. *Reports of the International Whaling Commission* (special issue **12**), 271-278.

Perrin, W.F. and Brownell Jr., R.L. (1989) Report of the Workshop. Appendix 6: Report of the Subgroup on dams and dolphins. In: *Biology and Conservation of the River Dolphins* (Ed. by W.F. Perrin, R.L. Brownell Jr., K. Zhou and J. Liu). pp. 1-22. Occasional Papers of the IUCN Species Survival Commission, N° 3, Gland, Switzerland.

Perrin, W.F., Donovan, G.P., and Barlow, J. (eds.) (1994) Gillnets and cetaceans. *Report of the International Whaling Commission* (special issue **15**), 1-629.

Perrin, W.F. and Reilly, S.B. (1984) Reproductive parameters of dolphins and small whales of the family Delphinidae. *Report of the International Whaling Commission* (special issue **6**), 97-.134.

Piana, E.L. (1984) Arrinconamiento o adaptación en Tierra del Fuego. *En*: Antropología Argentina. (1984) Universidad de Belgrano, Buenos Aires. Pp. 12-110.

Pinedo, M.C. (1982) *Análise dos conteúdos estomacais de Pontoporia blainvillei (Gervais and D"Orbigny, 1844) e Tursiops gephyreus (Lahille, 1908) (Cetacea, Platanistidae e Delphinidae) na zona estuarial e costeira de Rio Grande, RS, Brasil.* M.A. Thesis Fundaçao Univ. do Rio Grande, Brazil, 95pp.

Pinedo, M.C. (1994) Review of small cetacean-fishery interactions in southern Brazil with special reference to the franciscana, *Pontoporia blainvillei*. *Report of the International Whaling Commission* (special issue **15**), 251-260.

PROMSA, (1998) Componente de Desarrollo Pesquero. Flota Pesquera Argentina, Capturas marítimas totales 1997. Secretaría de Agricultura, Ganadería, Pesca y Alimentación - BIRF-BID. March 1998, Buenos Aires, Argentina, 25pp.

Punt, A.E. and Butterworth, D.S. (1995) The effects of future consumption by the Cape fur seal on catches and catch rates of the Cape hakes. 4. Modelling the biological interactions between Cape fur seals *Arctocephalus pusillus pusillus* and the Cape hakes *Merluccius capensis* and *M. paradoxus*. *South African Journal of Marine Science*, **16**, 255-285.

Punt, A.E. and Leslie, R.W. (1995) The effects of future consumption by the Cape fur seal on catches and catch rates of the Cape hakes. 1. Feeding and diet of the Cape hakes *Merluccius capensis* and *M. paradoxus*. *South African Journal of Marine Science*, **16**, 37-55.

Reeves, R.R. and Leatherwood, S. (1994) *Dolphins, porpoises and whales*. An Action plan for conservation of biological diversity:1994-1998. IUCN, Gland, Switzerland. 91 pp.

Reijnders, P.J.H. (1996) Organohalogen and heavy metal contamination in cetaceans: observed effects, potential impact and future prospects. In: *The Conservation of Whales and Dolphins* (Ed. by M.P. Simmonds and J.D. Hutchinson), pp. 205-217. John Wiley and Sons, London.

Riedman, M. (1990) *The Pinnipeds: Seals, Sea Lions, and Walruses*. Univ. of California Press. 439pp.

Schiavini, A.C.M. (1990) *Estudio de la relación entre el hombre y los pinnípedos en el proceso adaptativo humano al Canal Beagle, Tierra del Fuego, Argentina*. Ph.D. Dissertation. University of Buenos Aires, 303pp.

Secchi, E.R., Zerbini, A.N., Basoi, M., Dalla Rosa, L., Möller, L.M. and Rocha-Campos, C.C. (1997) Mortality of franciscanas, *Pontoporia blainvillei*, in coastal gillnets in southern Brazil:1994-1995. *Report of the International Whaling Commission*, **47**, 653-658.

Smith, A. (1996) The river dolphins: the road to extinction. In: *The Conservation of Whales and Dolphins* (Ed. by M.P. Simmonds and J.D. Hutchinson), pp. 355-390. John Wiley and Sons, London.

Stroud, C. (1996) The ethics and politics of whaling. In: *The Conservation of Whales and Dolphins* (Ed. by M.P. Simmonds and J.D. Hutchinson), pp. 55-87. John Wiley and Sons, London.

Tagliorette, A. and Losano, P. (1996) Demanda turística en áreas costeras protegidas de la Patagonia. Plan de Manejo Integrado de la Zona Costera Patagónica. GEF/PNUD. WCS/FPN., Informe Técnico N°. 25: 1-29.

Trites, A.W. and Larkin, P.A. (1996) Changes in the abundance of Steller sea lions (*Eumetopias jubatus*) in Alaska from 1956 to 1992: how many were there? *Aquatic Mammals*, **22**, 153-166.

Trites, A.W., Christensen, V., and Pauly, D. (1997) Competition between fisheries and marine mammals for prey and primary production in the Pacific Ocean. *Journal of the Northwest Atlantic Fishery Science*, **22**, 173-187.

Wickens, P.A. and Sims, P.F. (1994) Trawling operations and South African (Cape) fur seals, *Arctocephalus pusillus pusillus*. *Marine Fisheries Review*, **56**, 1-12.

Yodzis, P. (1994) Local trophodynamics in the Benguela ecosystem: effect of a fur seal cull on the fisheries. Working Paper SAC94/WP 14. Third Meeting Scientific Advisory

Committee., Marine Mammal Action Plan, UNEP, Crowborough, UK, August 24-27, 1994. 50pp.

Zhou, K. and Li, Y. (1989) Status and aspects of the ecology and behaviour of the baiji, *Lipotes vexillifer*, in the lower Yangtze River. In: *Biology and Conservation of the River Dolphins* (Ed. by W.F. Perrin, R.L. Brownell Jr., K. Zhou and J. Liu), pp. 86-91. Occasional Papers of the IUCN Species Survival Commission, N° 3, Gland, Switzerland.

Chapter 14

Environmentalists, Fishermen, Cetaceans and Fish: Is There a Balance and Can Science Help to Find it?

[1]MARTÍN A. HALL and [2]G. P. DONOVAN
[1]*Inter-American Tropical Tuna Commission, 8604 La Jolla Shores Drive, La Jolla, CA. 92037, USA. E-mail: mhall@iattc.org*; [2]*International Whaling Commission, The Red House, 135 Station Road, Impington, Cambridge CB4 9NP, UK. E-mail: Greg@iwcoffice.org*

1. INTRODUCTION

1.1 Historical Background

It is only relatively recently that cetologists have considered anything other than direct exploitation as an important threat to cetaceans. Given the centuries old tradition of whaling and the history of attempts to regulate the industry (*e.g.* see review in Donovan, 1992), this is perhaps not surprising. The issue addressed in this chapter, that of non-deliberate or incidental captures of cetaceans in fishing gear, was not seriously considered a problem until the late 1960s when biologists on board fishing vessels observed high levels of incidental catches of dolphins in the tuna purse-seine fishery in the eastern tropical Pacific (Perrin, 1968). This fishery is now one of the best-studied examples (Joseph, 1994; Hall, 1998) and we will use it throughout the chapter as a case study to illustrate one approach to address a particular bycatch problem.

The tuna-dolphin problem led people to consider other fisheries, and the first attempts at a broad review were made in the mid-1970s, in particular under the auspices of the Scientific Committee of the International Whaling Commission (IWC) and its sub-committee on small cetaceans (Mitchell,

Marine Mammals: Biology and Conservation, edited by
Evans and Raga, Kluwer Academic/Plenum Publishers, 2002

1975a, b); it soon became apparent that many species of cetaceans were being killed incidentally around the world, particularly in gillnets.

Perhaps the most important scientific initiative regarding this problem was the holding of an IWC Symposium and Workshop on the Mortality of Cetaceans in Passive Fishing Gear and Traps held in California in 1990. Much of the content of this chapter concerns ideas arising out of discussions held during that workshop, and studies carried out in response to its recommendations (see Donovan, 1994; IWC, 1994b). Its particular strength was that it brought together not just cetologists but also experts in fisheries biology and management, gear technology, and fishermen themselves.

However, as will become apparent, the issue of cetacean (and indeed all marine mammal) bycatches in fisheries is a multi-layered, multi-disciplinary issue involving not just 'scientific' disciplines but also politico-ethical (Donovan, 1992) considerations, and the cetacean focus may not be the most appropriate in the wider context. The effects of incidental mortality of cetaceans in fisheries should be assessed taking into account the basic framework of ecological principles for conservation described in Mangel *et al.* (1996).

2.   PROBLEM - WHAT PROBLEM AND WHOSE PROBLEM?

Very often people refer to the 'problem' of incidental catches of cetaceans. Although it may appear to be self evident, in this section we would like to examine just what can be considered to be a problem and by whom. Let us define an incidental catch as the unintended mortality of a cetacean during the operation of the fishery. From the perspective of the cetacean population, the impact of this mortality, when considered with all other causes of mortality, may or may not be significant. The total mortality may upset the balance of births and deaths that has, up until the present, resulted in the persistence of the population in question. In ecological terms, the development of a fishery is equivalent to an invasion of an area by a new predator and/or competitor. Irrespective of whether the 'prey' is consumed, this new source of mortality was not part of the evolutionary context in which the population evolved, and succeeded in surviving. If mortality is very low in relation to population abundance, the reaction of the population may be only minor and perhaps very difficult to detect. If it is high, it will elicit more dramatic (*e.g.* density-dependent) responses. If these can compensate for the added losses, the population will stabilise at a new level, with a different age structure, and perhaps a different biomass. If they cannot compensate, then the population will decline over time towards extinction. The new selective pressure added to the system will also result in long-term genetic changes, favouring particular behavioural and ecological adaptations that improve the fitness of the individuals possessing them in the "new" ecosystem that now includes the fishery.

From the perspective of the fisherman, cetacean bycatches are almost always a problem, although the severity of the problem will vary depending on the circumstances, for example:

very severe problem - the gear is completely destroyed (*e.g.* in some instances with large whales, the gear may be towed away by the animal which subsequently dies, see Kraus, 1990);

severe problem - although not completely destroyed, operations must cease whilst repairs to the gear are made;

relatively minor problem - some time is lost as the entangled cetaceans must be cut free from the net;

no problem - if the commercial value of the cetacean is relatively high compared to the target species for the fishery, or if the cetacean can be used for bait (*e.g.* in some South American fisheries - see Read *et al.*, 1988; Bjørge *et al.*, 1994, pp. 99-100).

From the perspective of the cetacean population dynamicist, whether or not bycatches are a problem depends on the status of the cetacean population involved relative to the number of incidentally caught animals. Similarly, from the point of view of the resource manager, who has to keep one eye on the fishery and the other on the ecosystem, the marine mammal bycatch issue is essentially one of management and thus one in which science and biologists have a major role to play.

However, other issues superimposed on this complicate the issue. These are what might be termed the politico-ethical issues referred to earlier.

(1) Particularly in the USA and Western Europe, there are those who consider that cetaceans represent such a 'special' group of animals that they should not be killed, deliberately or incidentally, under any circumstances (*e.g.* Barstow, 1990). From this perspective, every incidental catch of a cetacean would be a severe problem. (Of course, by extension, it makes every single catch of a cetacean a severe problem for the fisherman too.)

(2) Notwithstanding the above, there is a general perception that bycatches (of any species) are a wasteful use of both life and resources and thus a problem. However, as we shall return to later, this relatively simple view is not necessarily always the case from an ecosystem perspective (*e.g.* Hall, 1996).

The weight given to points (1) and (2) above provide the political dimension, particularly when addressing the problem involves an international dimension, where the viewpoints of countries/cultures may be diametrically opposed.

The views of scientists on the politico-ethical issues are generally worth no more and no less than those of others. Given this, a potential danger arises if a scientist adopts a moral/ethical position that affects the interpretation or communication of scientific facts to managers or the public. The role of the scientist is to provide the best information possible to enable decision-makers to base their choices on the best available information and to enable the public to understand the issue as well as possible.

3.   MANAGEMENT

Simplistically, the cetacean bycatch problem can be viewed solely as a wildlife management issue. From the population biologist's perspective, whether or not the 'target' species is utilised is irrelevant (in this case, of course the cetacean, whilst being the management 'target', is not the fishery target). At this point, it is worth recalling that the history of cetacean management is not a particularly happy one (*e.g.* see review in Donovan, 1992). However, considerable advances have recently been made with respect to the management of large cetaceans, as summarised in Donovan (1995). This approach can be usefully considered when examining the cetacean bycatch problem, although as we shall see, there are a number of important conceptual differences as well as similarities.

3.1   Objectives

In attempting to develop a resource management scheme, the most important initial step is to define management objectives. It is relatively easy to define two 'extreme' objectives, for example:

a)   avoid extinction of the cetacean species incidentally caught;

b)   do not hinder the operation and profitability of the fishery in any way.

It is clear however that within these two general categories (interests of the bycaught species/interests of the fishermen), there are a number of potential options, and that there must be some form of trade-off between the objectives. The options chosen and the balance struck between them are generally taken as largely political rather than scientific decisions. However, there is a clear ecological change when the marine mammal population is stabilised at a level that is much lower than the initial one. Whether we see a fishery as a predator to some species, a competitor to others, or a mutualistic partner to others, the niche that the fishery occupies will not match exactly the one of those replaced/displaced by it. Furthermore, as recent history shows, fisheries lack the density-dependent mechanisms to regulate their evolution, and frequently their expansion is so fast that it overwhelms the capacity of the other species in the system to adapt to it.

In addition to the scientific objectives discussed below, however, it must be recognised that there are 'user' objectives that must be borne in mind. These include:

a)   the management strategy adopted must be as easily understood by the users (*i.e.* the fishermen) as possible and perceived as 'fair';

b)   the strategy should be practical given the circumstances of the fishery, *e.g.* a strategy that will work for an industrialised commercial fishery in a developed country may well not be applicable to an artisanal fishery in a developing country;

c)   it should be enforceable.

3.1.1   Possible Management Objectives with Respect to the Cetacean Bycatch

3.1.1.1   Zero Mortality

Choosing a 'zero mortality' option would mean that no cetaceans are allowed to be incidentally killed in a fishery under any circumstances. For those people who believe that it is immoral to kill cetaceans, this would always be the chosen management objective and would receive priority over all other considerations.

From a 'purely scientific' perspective, this may also be the chosen option under certain circumstances, for example, where the cetacean species or population being caught is in danger of extinction, *e.g.* the vaquita (see summaries in Bjørge *et al.*, 1994; Vidal, 1995).

There can be difficulties with the zero mortality option if it is considered the overriding management objective, whatever the status of the cetacean population involved. Fishermen confronted with such an objective perceive this as a direct threat to their livelihood, because for most fisheries, the only way to achieve this level is to shut them down. This may be counter-productive and result in little or no co-operation with biologists and/or managers. In most cases where cetacean bycatch levels have been reduced in fisheries, this has been achieved with the direct co-operation of fishermen, for example, in changing fishing gear or practices (IWC, 1994b).

There are likely to be similar problems in an international context if politico-ethical considerations based solely on the perceived values of cetaceans are used as the overriding management objective – not all nations, cultures, or socio-economic classes value species or groups of species in the same way (*e.g.* Donovan, 1992; Joseph, 1994).

3.1.1.2 Maintain a Trend in Population Abundance

This option appears to be intuitively appealing from a scientific perspective. For example, some level of cetacean bycatch may be acceptable if that population is (a) increasing in abundance or (b) not decreasing in abundance. Although superficially similar, these two options are quite different as they place the burden of proof on different sides of the problem. This was the focus of considerable debate in the IWC Scientific Committee in the early 1980s with respect to the management of large whale populations. The power of statistical tests for determining trends in cetacean populations is generally low (Gerrodette, 1987), and it may take decades to verify statistically that a whale or dolphin population is increasing. If that increase has to be demonstrated prior to any takes, the fishery in question will be closed for a long time. If on the other hand, a statistical decrease in abundance is required before taking action on a bycatch problem, the fisheries will operate for many years, and the cetacean populations affected may decrease considerably before the evidence becomes significant.

3.1.1.3 Maintain the Population of the Bycatch Species at Some Specified Level

This is a common general objective in managing wildlife populations. It of course implies that some level of bycatch level is sustainable. However, the major question is at what level should a population be kept. A number of different approaches have been proposed, normally in the context of directly harvested species.

3.1.1.3.1 Return the Population to its Pre-exploitation Level and Maintain it at that Level

This approach would allow the cetacean populations to recover to their 'pre-exploitation level'. There are at least two major problems with this approach: (i) the pre-exploitation levels of populations are rarely, if ever, known; and (ii) even if they are known, it is unlikely that the rest of the ecosystem, which has also been disturbed by human activities, will find a stable equilibrium point with a single population of an 'arbitrarily selected' key species.

In practice, even if it were possible to eliminate all human activities in an area, this would not guarantee that the ecosystem would return to its pristine condition; it is much more likely it would reach a new equilibrium and that this may take a very long time.

Figure 1. In several areas of the North Atlantic, especially the Bay of Fundy/Gulf of Maine, North Sea and Celtic Shelf, the bycatch of harbour porpoises is estimated to be above the values that would enable the populations to maintain their current level. (Photo: N. Tregenza)

3.1.1.3.2 Maintain the Population at the Current Level

Under this option, the bycatch is brought to a sustainable level, assuming that the current population level of the cetacean will reach an equilibrium with the other components of the system. It may, of course, be closely related to item 3.1.1.2, as one way to meet this objective would require the monitoring of the population size to ensure that it is not increasing or decreasing. This is not easy, as we have seen. Another way of trying to meet this objective is one that has been used in certain cases as a 'temporary' measure, and that is to make assumptions about the reproductive capacity of cetaceans (*e.g.* a maximum increase rate of 4%), and then take a conservative position such that the bycatch should not exceed half that rate (*i.e.* 2%), and that concern should be warranted if bycatches exceed 1% of the estimated population size (see also below). However, as is well recognised (*e.g.* IWC, 1996), this is a somewhat arbitrary approach. One apparent advantage of it is that it does not require estimating an 'optimum' level. It is also a relatively easy concept to explain. However, if the current population level is very low, there may, for example, be undesirable genetic consequences for the population, even before it is endangered or threatened, and/or the population will be more susceptible to environmental stochasticity, diseases, pollution, *etc.*

3.1.1.3.3   Maintain the Population at Some Maximum Productivity Level

Choosing an appropriate level is not a trivial exercise, and a number of such levels have been proposed. For example, the first attempts to develop serious scientific management procedures for fish and for large whales focused on the concept of the *maximum sustainable yield* or MSY[1]. Indeed, one of the problems with the IWC's so-called New Management Procedure developed in the mid-1970s was the difficulty in estimating MSY and related factors (Donovan, 1995).

Irrespective of the difficulties of establishing the particular level, such a strategy would allow incidental takes if the population is above or at the chosen level, but would reduce or eliminate them when the population is below it. To some extent, this is the approach adopted in the IWC's Revised Management Procedure, where the procedure is 'tuned' to a somewhat arbitrary target level (see Donovan, 1995). It is also relevant to the PBR (Potential Biological Removal) approach adopted by the USA to manage marine mammal populations where the goal is to maintain populations at or above their 'Maximum Net Productivity Level', which is assumed to lie between 50-70% of carrying capacity (Wade, 1998). This approach will be discussed in more detail later. However, given that the population experiencing the incidental mortality is not utilised as a resource, the rationale for maintaining it at a high productivity level is unclear. This was reflected in the discussion and adoption of objectives agreed by ASCOBANS (Agreement on the Conservation of Small Cetaceans of the Baltic and North Seas) at its 2nd Meeting of Parties in 1997. They chose as an interim objective to aim to restore populations to, or maintain them at, 80% or more of the carrying capacity, on the grounds that this 'is above the

---

[1]   In simple terms, at the initial equilibrium population size, the number of births balances the number of deaths by natural causes. As the population size is reduced, an excess of births over deaths (or animals reaching maturity) occurs as conditions (*e.g.* food supply) improve. The excess (or yield) reaches an absolute maximum number at a certain percentage of the initial population size. For whales, this was traditionally assumed to be at 60%.

level of maximum productivity and therefore more appropriate for a conservation agreement' (ASCOBANS, 1997).

3.1.1.3.4 Maintain Some Population Level that Results in a Known and Agreed-to Probability of Extinction

This approach emphasises the stochasticity of changes in abundance. Although obtaining the data to allow this to be explicitly determined would be extremely difficult, the principle underlying such an approach is similar to the approach adopted in developing the Revised Management Procedure of the IWC. That approach used simulation trials to determine how the management procedure would cope with a wide variety of scenarios, and in one sense looked at the probability of extinction under those scenarios. Table 1 gives some examples of the kinds of factors which the procedure had to deal with. How relevant these are to the bycatch issue will be referred to throughout the chapter.

Table 1. Some examples of the trials with which the IWC's Revised Management Procedure had to be able to cope

Several different population models and associated assumptions
Different starting population levels, ranging from 5% to 99% of 'initial' population size
Different MSY levels, ranging from 40% to 80%
Different MSY rates, ranging from 1% to 7% (including changes over time)
Uncertainty and bias in estimated population size
Changes in carrying capacity (including reduction by half)
Errors in historic catch levels
Catastrophes (irregular episodic events when the population is halved)
Various frequencies of surveys

3.1.1.3.5 Maintain Some 'Acceptable' Bycatch/Catch Ratio

If there is an ecological relationship (*e.g.* predator-prey) between the target species and the bycatch species, then it may be possible to establish a ratio between them that results in a minimisation of the impact on the system. If the exploitation removes the predator and prey in the 'correct' ratio, then the use may be deemed acceptable; if the ratio is very different from the optimal (in any direction), then changes to the fishing practices will be required. It is conceivable under this scenario that bycatches may have to be increased in some cases to maintain the ecological balance. Perhaps this option more than any other shows the necessity of considering the management of both catches and bycatches jointly, within an ecological framework. The major difficulty of such an approach is in successfully identifying ecological relationships in a predictive manner, a difficulty that affects all multi-species modelling and management.

3.2   Data Requirements

As has been well illustrated for other species, and particularly for the management of large whales, management objectives and schemes cannot be evaluated without consideration of data requirements and the ability of scientists and the industry to provide them (Donovan, 1995).

Several pieces of information are required if a quantitative assessment of the impact of bycatches on cetacean populations is to be made (Donovan, 1994; IWC, 1994b) irrespective of the management objectives (unless the zero option is chosen under all circumstances). These are:

(1)   reliable estimates of bycatch numbers;

(2)   knowledge of stock identity and migration;

(3)   reliable estimates of abundance.

These will provide an estimate of the bycatch level as a proportion of current population size. However, to interpret this in the context of management objectives (particularly sustainability), at least some knowledge is required of:

(4)   the dynamics of cetacean populations; and this should be viewed in the context of:

(5)   other factors affecting the population such as direct catches, habitat changes *etc.*, and:

(6)   the level of uncertainty of the basic information.

3.2.1   Reliable Estimates of Bycatch Numbers

Estimates of incidental mortality usually come from one of two sources: reports from fishermen, or observer records. Although both of these methods have errors and problems attached to them, the latter should provide results of higher quality, detail, and reliability (Northridge, 1996). In general, it should be possible to place observers on either all, or a representative sample, of boats in an 'industrial commercial' fishery. However, this has rarely been done (*e.g.* Hill *et al.*, 1990; Berrow *et al.*, 1994; Lennert *et al.*, 1994; Lowry and Teilmann, 1994; Vinther, 1999), with the notable exception of the tuna fishery in the eastern tropical Pacific, which now has complete observer coverage (Joseph, 1994; Lennert and Hall, 1996). This approach, however, is almost impossible where large numbers of small vessels are involved, as is the case for many artisanal fisheries in the developing world.

Other approaches, such as questionnaire surveys, are very difficult to interpret (Lien *et al.*, 1994), but may give some idea as to whether there is a potential problem.

Even observer schemes have sources of uncertainty that are not easy to estimate. Some may lead to underestimation of mortality, such as: (i) the observer may miss some of the mortality *e.g.* if the carcasses fall out of the net before it is hauled in; (ii) intentional under-reporting may occur if the observer asks fishermen to provide the information whilst he/she is not on watch, or if observers are bribed or intimidated; (iii) injuries caused by the fishing operations may lead to mortality some period later; (iv) stress caused by fishing operations may result in diminished survival or reproduction; (v) fishing operations may result in facilitated predation; and (vi) changes in the behaviour of the fishermen may be caused by the presence of the observer.

However, overestimation may also occur, for instance: (i) strained relationships between observers and crews may result in over-reporting; (ii) some of the injuries that are believed to result in death may not; and, probably the most important, (iii) some of the statistical techniques used for estimation, such as ratio estimates, tend to be biased at low sample sizes, and this may result in overestimates when observer or any other form of coverage is low (see Bravington and Bisack, 1996, and Vinther, 1999, for good discussions of the problems and approaches that should be considered when trying to estimate bycatch levels when complete coverage is not possible).

An extreme case of uncertainty takes place when an attempt is made to reconstruct a long time series for which there are no, or very few, valid observations. In the case of the eastern Pacific tuna fishery, a period of more than a decade (1959-71) when most of the mortality which occurred had no observer coverage; the only data available were from two letters of crew members, and a trip by a scientist. A U.S. National Academy of Sciences Committee (Francis *et al.*, 1992) reported:

'Thus, the data do not come from any sampling design. Trips were not selected at random or according to any pattern. The accuracy of the data is questionable because no standard procedure was used to collect the information, and interpretation of the data cannot be determined to be correct. ...... In summary, the mortality estimates for the period before 1973 .... have little or no statistical value, and the only conclusion that can be based on the data available is that mortality was very high.'

Similar problems can be found when trying to reconstruct long time series of catch statistics or CPUE (catch per unit of effort) data. It is problematic that these low-quality data are sometimes needed to determine the current status of the population with respect to some baseline condition.

In general, it must be recognised that for almost all fishery/cetacean interactions, we have, and will probably continue to have, only rough (often minimum) estimates of bycatch levels (see IWC, 1994b, Table 1). Management strategies developed must take this level of uncertainty into account, for example in a similar manner to the way that the RMP (Revised Management Procedure) takes into account uncertainty in the catch record (see Table 1). An important difference, however, is that the RMP assumes that once the management regime is underway, all future catches are known. This may well not be the case for incidental catches.

3.2.2   Knowledge of Stock Identity and Migration

Figure 2 illustrates just one of the problems that can arise when trying to manage a population with mistaken information on stock structure. Incidental catches are taken in the area bounded by the dotted line. The scientists believe that the area that they have surveyed (bounded by the thick black line) contains a single stock, when in fact it contains two stocks, an inshore and an offshore stock bounded by the thick grey line (an added complication is that such boundaries may vary temporally as well as geographically). They will thus underestimate the effect of the incidental catches on the inshore population.

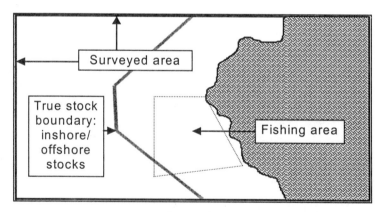

Figure 2.   Hypothetical illustration of one potential stock identity problem (see text).

Another example of the difficulties encountered when trying to associate a stock with a geographical region is the set of changes that accompany the El Niño-Southern Oscillation events. The "normal" habitat of a stock may shift to a new location, and it may shrink or expand, because of the oceanographic changes. There is an interesting example of a massive migration of individuals of common dolphins (*Delphinus delphis* and *D. capensis*) that perhaps could be linked to El Niño, (although it has lasted longer than the event), from the more tropical parts of the eastern Pacific to Californian waters (Anganuzzi *et al.*, 1993; Forney *et al.*, 1995; Barlow *et al.*, 1997).

The question of stock identity is a persistent problem in cetacean studies (see *e.g.* Donovan, 1991; Perrin and Brownell, 1994), and our knowledge of stock structure is poor for almost all small cetacean species in all areas. Despite the progress in biochemical techniques made in recent years (*e.g.* IWC, 1991; Dizon and Perrin, 1995), there are no simple unambiguous ways to address this problem. It is important that a suite of techniques is used (*e.g.* Donovan, 1991), and that information on movements is also obtained. The inevitable uncertainty must be taken into account in any management procedure developed (for example by following an approach such as the *Small Area* approach used in the RMP-International Whaling Commission, 1994a). There has been some criticism that the US PBR method (Wade, 1998) does not take problems relating to stock identity sufficiently into account (IWC, 1997).

3.2.3   Reliable Estimates of Abundance

The question of estimating the abundance of cetacean populations (notwithstanding the stock identity problems noted above) has been thoroughly addressed in recent years and is beyond the scope of this chapter. Therefore, we will not discuss it here other than to say that while the techniques exist, they are expensive. For example, the recent survey of the North Sea and adjacent waters cost over £1,000,000 (Hammond *et al.*, 1995). At present, we have relatively few estimates of abundance for cetacean populations affected by fisheries, particularly for developing countries (IWC, 1994b, Table 1). Again, despite the relatively advanced state of knowledge with respect to population estimation, it is important to recognise that some uncertainty is inevitable in obtaining any estimate, and this uncertainty must be taken into account in any management scheme, as is the case for the RMP (Table 1).

3.2.4   The Dynamics of Cetacean Populations

It is clear that our limited understanding of the population dynamics of most, if not all, cetacean species, and our inability to obtain sufficiently precise and unbiased estimates of biological parameters, make it impossible to obtain detailed predictions of the future (*e.g.* Reilly and Barlow, 1986). This was recognised in the development of the RMP where the sensitivity to various assumptions about population dynamics models and parameters were incorporated (see Table 1). The fact that in most cases we have only a wide range of possible values for a parameter, and that our estimates usually have large variances, makes the detection of changes very difficult.

3.2.4.1 Other Factors Affecting the Population Such as Direct Catches, Habitat Changes, *etc*

Clearly, incidental catches alone are not the only problem facing populations. Any management strategy must consider these other factors, such as directed catches, habitat degradation, stochastic variability, pollution, either explicitly or implicitly. Again, the simulation approach adopted during the development of the RMP provides a useful starting point for taking these factors into account (see Table 1).

3.3   Classifying the Problem

In the past, the information above has been used to try to scientifically evaluate whether the bycatches are potentially a problem *from the perspective of the cetacean population*. The following categories (IWC, 1994b; Hall, 1996) have been used to classify bycatches. However, whilst classification of a problem can be a stimulus to action, it is important that it is not seen as an end in itself; the classification simply suggests the degree of urgency, and the level of the actions to be taken. Apart from the first category below, where it is probable that the zero mortality option must be taken, then the other categories require action, preferably in the context of a management scheme and its objectives, as discussed above.

• *Critical bycatches*: bycatches of populations or species that are in danger of extinction (*e.g.* the 'vaquita' - Rugh *et al.*, 1993; Vidal, 1995). The problem here is that there is no universally accepted definition of 'in danger of extinction', and that may leave the door open to extreme approaches.

• *Non-sustainable bycatches*: in this case, the populations are not immediately at risk, but they will decline under the current levels of bycatch.

• *Sustainable bycatches*: bycatches that do not result in declines of the population.

• *Biologically-insignificant bycatches*: bycatches that are so low as to be considered negligible from the point of view of the dynamics of the population involved. These bycatches are also sustainable; the difference between the third and fourth categories is arbitrary, but it is an attempt to separate one that requires some control and monitoring, from one that is so low that it may not be worth the effort. The definition of biologically insignificant is arbitrary; a mortality level of 0.5% of a conservative estimate of population abundance has been implied for cetaceans (*e.g.* IWC, 1994b). Most, if not all, of the eastern Pacific dolphin populations fall under this category (Lennert and Hall, 1995).

• *Bycatches of unknown level*: when we lack the basic data on abundance or total mortality to determine if it is sustainable or critical. Given the lack of data for most populations, this is the category with the highest number of cases. IWC (1994b) created one additional class of unknown level, called 'Potential', which is a suggestion of possibly unsustainable level. This in effect reflects the 'gut feelings' of some biologists about the abundance of a population, and/or its total mortality, which in itself is problematic from both a scientific perspective and in terms of obtaining the confidence of fishermen and managers. However, it may be useful in deciding priority fisheries for study.

• *Charismatic bycatches*: this category is added to reflect the politico-ethical issues raised earlier and the fact that the way a society perceives the bycatch of a species may be independent of the level of the impact exerted on the species, or of its conservation status. The response of the public, which influences or determines management actions, reflects the value assigned to the species in question. Although not strictly an incidental catch, the case of the three ice-entrapped gray whales in Alaska might be seen as an example of where the ecological impact is minimal but where the public's perception and the political 'attractiveness' lead to disproportionate effort (*e.g.* Anonymous, 1989; Fraker, 1989; Scheffer, 1989).

The categories listed above are not mutually exclusive. More than one cetacean species may be affected by a fishery at different levels (*e.g.* eastern Pacific dolphins, Hall and Lennert, 1994).

3.4   An Ecological Perspective

We mentioned earlier that, from an ecological point of view, the problem of cetacean bycatches (and indeed any bycatches) may look very different from the perspective taken simply by the cetologist. It is a problem that has been highlighted by consideration of the tuna-dolphin problem in the eastern Tropical Pacific (*e.g.* Joseph, 1994; Hall, 1995, 1996, 1998) but which has implications for all fisheries. All fishing and agricultural activities have ecological costs associated with them, although these are often not recognised or acknowledged. An important role of scientists and managers is to try to identify ways of utilising resources at the lowest possible ecological cost.

However, this area of research is extremely difficult. It is not easy either to assess the impact of fishing operations, or to compare the impacts of different gears or modes of harvest. For example, one way of harvesting a resource may require vast amounts of energy; another one may cause some undesirable bycatches; a third may physically damage the habitat.

One option is to eliminate the harvest of the resource and all ecological impacts will disappear, but of course, this is no guarantee that an affected ecosystem will return to its 'pristine' state. However, most countries facing increasing population and other socio-economic problems will be forced to choose among a more limited set of options.

Given alternative ways of harvesting a resource, the most ecologically 'benign' options should be chosen, of course, providing they can be identified. The major difficulty is in coming up with an acceptable universal 'currency' to evaluate options even when the general ecological costs can be identified (which in almost all cases, our knowledge of whole ecosystems, or rather our lack of knowledge, precludes).

Some of the difficult types of questions that will arise include:

a)   How do we evaluate two different types of fishery, one which results in a small catch of a 'charismatic' species, and one which results in no 'charismatic' bycatches but large potentially damaging catches of less 'charismatic' species?

b)   How do we compare a way of fishing that causes physical disturbances on the bottom sediments and incidental mortality in benthic populations, with another one that causes bycatches among demersal or pelagic species but has no impact on the bottom?

The tuna-dolphin problem has identified just such a complex situation. There is a strong but poorly understood association of dolphins and adult tuna in the eastern tropical Pacific (and to a lesser extent, in other oceans). The fishing method for catching tuna has been well described elsewhere (*e.g.* Perrin, 1968, 1969). In simple terms, a large area of netting (up to 1 mile long and 600ft deep) is laid around a school of fish (Fig. 3). When the circle is completed, a cable that passes through the bottom of the net is pulled on board. This forms a 'purse'–hence the name 'purse-seining'. Each time this is done it is termed a 'set'.

There are three strategies used in purse-seining, largely determined by how the school of tuna is detected:

a)   by setting on dolphins, *i.e.* by exploiting the relationship between tuna and one of the three species of dolphins known to be associated with tuna (spotted dolphins, *Stenella attenuata*; spinner dolphins, *S. longirostris*; common dolphins, *Delphinus delphis*);

b)   by setting of free-swimming schools;

c)   by setting on floating objects ('logs').

Figure 3.   Purse-seine tuna net in operation in eastern tropical Pacific with diagrammatic representation of netting process (Photo: M. Hall).

The mortality of dolphins in sets on them was extremely high in the early days of the fishery (estimated at an annual average of some 350,000, although from very scanty data - Francis *et al.*, 1992), but as a result of a mixture of international co-operation, improved fishing techniques that included technological changes, training, and incentives for their performance (Hall, 1998, see below), and public pressure, has been reduced to around 3,000 in 1997 (Hall and Lennert, 1996) and mortality rates at less than 0.15% of the estimated dolphin population sizes.

Table 2 presents information on the total species composition of catches for the different types of sets, taken from data collected between 1993-1995.

It is not appropriate here to discuss the detail in the table or to debate the finer points of the extrapolation process but rather to highlight some of the general issues it illustrates. Firstly, it is clear that if dolphin catches themselves are excluded from consideration, then setting on dolphins reduces wasteful bycatches.

With respect to tuna, log fishing results in large catches of small tunas and a high level of discards (15%-20%). It also results in higher catches of skipjack, rather than the preferred yellowfin tuna. Setting on free-swimming schools ("school sets") is clearly better than setting on logs, with much lower discard levels (less than 5%), although the tuna caught are still smaller than for dolphin sets. However, some 40% of attempted sets produce virtually no yield (less than 0.5 tons), rendering the technique much less efficient. Dolphin sets result in almost exclusively market size tuna of the preferred species with discard levels of less than 1%.

On the assumption that the catches of dolphins are easily sustained by the populations, it would appear that setting on dolphins results in considerably less waste than setting on either free-swimming schools or logs, and may well be less ecologically damaging, depending on the status of the populations of the other bycaught species (Hall, 1998).

However, before leaving this issue, there is one more question that needs to be kept in mind, and that is the question of whether, from an ecosystem as opposed to a 'waste' (whether it be of yield or life) perspective, bycatches are necessarily a 'Bad Thing' and highly selective, targeted fisheries a 'Good Thing' (Hall, 1996)? To take a much simplified example, if you have a small unexploited lake and wish to begin a fishery, which is likely to alter the ecosystem more?–a fishery directed at a single species high in the food chain?–or a fishery that removes some proportion of all species present?

Table 2. Estimated average bycatch (in numbers of individuals unless stated) per 10,000 sets based on eastern tropical Pacific data from 1993-1998. (Sample sizes: n = 49,066 dolphin sets, n = 31,456 school sets, n = 23,870 log sets)

Bycatch	Dolphin sets	School sets	Log sets
Billfishes			
Sailfish	654	932	116
Swordfish	11	26	29
Black marlin	65	173	1,290
Striped marlin	86	189	313
Blue marlin	67	211	1,684
Shortbill spearfish	9	3	23
Unidentified billfish	108	20	119
Unid.entified marlin	35	48	234
Large bony fishes			
Mahi mahi	347	24,720	1,202,022
Wahoo	597	3,195	646,734
Rainbow runner	9	4,998	107,568
Yellowtail	2,297	60,838	92,563
Other large fish	45	49,908	57,766
Sharks and rays			
Silky shark	4,175	15,644	66,438
Whitetip shark	335	441	8,835
Hammerhead shark	160	1,141	1,883
Mantaray	614	4,050	260
Stingray	357	1,184	287
Other sharks	491	1,300	8,346
Unidentified sharks	928	1,606	12,200
Sea turtles			
Hawksbill turtle	0	1	1
Olive ridley	30	56	117
Loggerhead turtle	1	5	2
Green/black turtle	1	11	19
Unidentified turtle	12	19	38
Tuna discards (MT)			
Yellowfin	1,059	1,483	8,078
Bigeye	0	59	8,610
Skipjack	145	2,105	42,575
Other species	7	940	5,880

3.5   Action

3.5.1   Bycatch Reduction

On the assumption that cetacean bycatches are generally not a 'Good Thing' (either from the point of view of the cetacean population involved or the fisherman, for the reasons suggested earlier), in this section we will consider some of the practical measures that have been proposed to reduce cetacean bycatches. Despite our general lack of knowledge, one thing is clear. There is no single simple cause or solution to the incidental capture of cetaceans in fishing gear. Each case should be evaluated in the light of local conditions.

As suggested by Hall (1995), the bycatch can be considered as the product of the total effort and the bycatch per unit of effort. Reducing either or both of these simultaneously will reduce the bycatch.

Perhaps the most obvious way of reducing effort is to limit or ban effort by national or international regulations. And perhaps the best example of this was the ban on the high seas driftnet fishery from 1993, adopted by the UN General Assembly (Donovan, 1994). However, if such measures are to be successful, they must be perceived to be fair and to be enforceable. Complete bans should be seen as a last resort.

Another approach is to set a limit on the bycatch level, and then close the fishery when that level is reached. An advantage of this approach, provided again that it is enforceable, is that it encourages fishermen to develop ways to reduce the bycatch per unit effort (see below).

Both of these approaches require legislative action, but, whilst it is relatively easy to pass legislation and even to stress the need for enforcement, actually enforcing the law and monitoring the fishery can be logistically difficult, particularly in the case of artisanal fisheries with large numbers of small vessels (*e.g.* see Donovan, 1994; Van Waerebeek and Reyes, 1994, for a discussion of the situation in Peru).

There are a number of potential ways in which bycatch per unit effort can be reduced as discussed below.

3.5.1.1   Change in Fishing Practice/Modified Gear

This has proved particularly successful in the tuna-dolphin fishery, where careful fishing practice (the 'backdown' procedure) and modified gear (the Medina panel) have reduced bycatches dramatically to levels that should allow the populations of dolphins to increase (Joseph, 1994; Hall,

1998). There are encouraging signs that passive and acoustic modifications will be of value at least in some fisheries (*e.g.* Goodson *et al.*, 1994; Lien, 1994; Kraus *et al.*, 1995) and situations, although further work is needed to examine apparent differences between areas and seasons (IWC, 1997).

Figure 4. Harbour porpoise being released from fishing gear in the Bay of Fundy. Unfortunately only a small percentage of bycaught animals can be saved in this way. (Photo: J. Wang)

Modification of gear is one of many areas where co-operation between biologists, fishermen, and gear technologists is essential. For example, knowledge of the acoustic capabilities of the cetacean species involved is essential for determining suitable and effective gear modifications. Early efforts in this regard were hampered by general attempts to 'improve the detectability of nets' without reference to the behaviour and physiology of the species involved. An understanding of the 'biology of entrapment' is an important component in developing methods of bycatch reduction (IWC, 1994b). This information can be used to avoid setting nets at certain times of the day or year (*e.g.* sundown sets are prohibited in the tuna purse-seine fishery - Hall, 1995) or in certain localised geographical areas (*e.g.* see Gearin *et al.*, 1994). In certain fisheries for particular species (*e.g.* North Atlantic harbour porpoises), acoustic deterrents ('pingers') attached to nets have been shown to be effective in reducing bycatch levels, at least in the short-term (Fig. 5). This has been reviewed by the IWC Scientific Committee (IWC, 2000a).

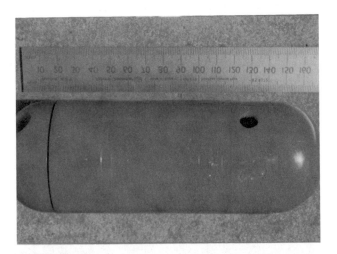

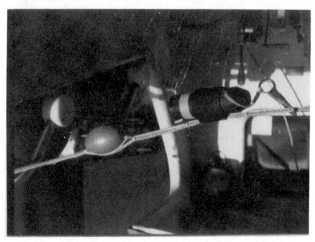

Figure 5.   Pingers have been employed very effectively in a number of North Atlantic gillnet fisheries to reduce porpoise bycatch. (Photos: P.G.H. Evans/N. Tregenza and F. Larsen)

3.5.1.2   Alternative Ways of Fishing

This involves completely changing the fishing method rather than modifying it (*e.g.* switching from gillnets to longlines - Corcuera, 1994). It has potential in some areas, but the effect of such changes must be evaluated and then monitored for several reasons: (a) as we have seen, reducing bycatches of cetaceans may result in increased bycatches of other, potentially more vulnerable, species; (b) the new methods may result in directed catches of cetaceans for bait (*e.g.* Félix and Samaniego, 1994; Van Waerebeek and Reyes, 1994); and/or (c) the new method may turn out to also result in incidental catches or, in the case of cetaceans stealing fish from longlines, result in direct kills by the fishermen.

An additional consideration that cannot be stressed too strongly is that for any bycatch reduction/elimination measures to be successful, fishing communities must be made aware of the reasons behind calls for a reduction in bycatches, and become involved in the process of finding solutions. The co-operation of fishing communities makes it much more likely that an equitable and observed system can be developed to mitigate problems. Orders from 'on high' (either at national or international levels) without involving the affected communities can often be counter-productive.

3.5.2   Management Procedures

Throughout this paper, we have suggested that in an ideal world, the cetacean bycatch issue should be approached in a similar manner to the development of the IWC's RMP. Indeed, for large baleen whales, the RMP may indeed be appropriate, even for populations not subject to commercial whaling.

The important features of such an approach (Donovan, 1995) are:

a)   scientists must accept their limitations and build the inevitable uncertainty explicitly into account;

b)   data and analysis requirements must be realistic and specified-resource users must recognise that unless requirements are met, they will not be able to utilise the resource;

c)   procedures should be rigorously tested using computer simulations;

d)   objectives must be explicitly stated and assigned priorities;

e)   a feedback mechanism to monitor the performance of the resource should be incorporated.

This approach, whilst achievable in the relatively straightforward situation of direct exploitation of baleen whales by commercial whaling operations from developed countries, still required a large commitment of effort and resources.   Whilst the principles remain good, it must be recognised that the cetacean bycatch problem is vastly more complex, even simply in terms of the number of species. When the fact that artisanal fisheries in developing countries involving large numbers of vessels represent a major part of the problem is taken into account, it is clear that it will not be possible to develop fully fledged management procedures applicable to the wide variety of situations that exists throughout the world. For these countries and fisheries, it is vital that governmental and non-governmental organisations from the developed world offer financial and logistical support for the necessary scientific work to be carried out, to at least identify priority fisheries that require attention (Donovan, 1994). However, for commercial fisheries in developed countries, it should be possible provided that the resources are made available.

At present, there has been only one procedure adopted that attempted to follow RMP principles and that is the PBR approach developed in the US referred to earlier (Wade, 1998). As noted earlier, the goal of this approach is to maintain populations at levels above their Maximum Net Productivity Level, assumed to lie between 50-70% of carrying capacity. It uses information on abundance, bycatch, and population growth rates to estimate a parameter known as the Potential Biological Removal (PBR) level. It accounts for precision of the abundance estimate in an explicit fashion, and bias and precision in other factors in an indirect manner. Bycatch levels that consistently exceed the PBR value are assumed to lead to a depletion of the stock.

The PBR is defined to be the product of three factors:

$$N_{min} \cdot 0.5 r_{max}.F_r \qquad\qquad\qquad \text{Eq. 1}$$

where N_{min} is a minimum population estimate for the stock, r_{max} is the maximum theoretical or estimated rate of increase of the stock at a small size, and F_r is a recovery factor, whose value lies between 0.1 and 1.0. In practice, r_{max} is either the maximum observed population growth rate for a stock or a default value (0.04 for cetaceans) if no specific estimate is available.

The procedure was tested using computer simulations similar to those used in the RMP (Wade, 1998). It is clearly an advance over previous *ad hoc* approaches, although there are some concerns that: (a) it may not adequately take into account all of the associated uncertainty, particularly with respect to case specific examples (IWC, 1997); (b) there is no clear scientific basis to maintain the populations at the Maximum Net Productivity Level, a concept developed for harvesting resources, rather than for ecosystem management; and (c) the level of caution is so high that in some fisheries, the take of one individual every several years would require management actions.

The level of caution used in the implementation of the PBR concept demonstrates clearly the high value placed on these species. It also illustrates the difficulties in separating the politico-ethical considerations from the science, because the consequences of applying extreme caution are very close to the zero mortality goal discussed earlier. Another factor to consider is that when the PBR value is high (*e.g.* 3,097 short beaked common dolphins - Barlow *et al.*, 1997), despite the fact that it is cautious, it may still "sound" too high to a large sector of the public. Management, which is the intersection between science and reality, reflects these perceptions, and deviates from its scientific basis to include them.

An important potential advance in developing a scientific management approach has taken place as a result of co-operation between ASCOBANS and the IWC Scientific Committee. It is to be hoped that this co-operation will result in an effective management scheme for harbour porpoises in the northeastern Atlantic (*e.g.* see Reijnders and Donovan, 1999; IWC, 1999; IWC, 2000b).

4. CONCLUSIONS

Bycatches result from a complex combination of environmental, ecological, biological, and gear factors as well as the motivation and ability of the fishermen themselves. Identifying these factors and their relative priority is of major importance in attempting to develop solutions to cetacean bycatch problems. Research programmes should be designed to allow management actions to be based on established scientific facts. Observer programmes should be designed to assist in the search for solutions as well as assessing bycatch levels and compliance with regulations. Wherever possible, regulatory actions should be part of an overall management scheme that encompasses both science and management. The *ad hoc* approach to management should become a thing of the past in the management of natural resources.

It must also be remembered that there is no universal solution to the incidental capture of cetaceans in fishing gear. Each case will need to be evaluated in the light of local conditions, whilst taking into account experience elsewhere.

The experience of the eastern Pacific tuna fishery shows that under certain circumstances bycatch problems can be tackled successfully, but that a number of conditions must be met (Hall, 1996). Many of these are of more general relevance and they are summarised below:

• Recognition by nations and industries/fishing communities that a problem exists and a commitment to its solution.

• Continued and constructive interaction among fishing communities/ industry, scientists, managers and environmentalists, based on the objective of finding a solution that achieves the desired conservation goals, while allowing the continuation of the fishery. This implies that the demands of the extreme fractions of all sectors involved will, most likely, not be met.

• The development of a scientific programme to understand why the bycatches happen, and the conditions that affect their level. A critical part of this programme is the flow of information to the fishermen concerning all factors affecting incidental mortality. It should also inform managers and the public of the full ecological consequences of any alternative proposal put forward to mitigate or eliminate the problem (*e.g.* new type of gear, area closure, *etc.*).

• The development of clear objectives with regards to the mitigation of impacts, with a schedule dictated by a realistic approach to the problem. Where appropriate, they should be defined within an international context and with the participation of all nations involved.

• All concerned should work towards the objectives in an iterative manner via realistic short-term goals that will encourage fishermen to achieve them.

• The development of a system of incentives, from the level of the nation down to that of the individual fisherman, with an emphasis on individual responsibility whenever possible. The system should serve as a selective force, encouraging the fishermen to develop gear, techniques, and decision-making skills that would allow them to continue using the resources while at the same time reducing the ecological impacts of their activity.

• The development of a fair system of regulations, based on scientific findings and statistical analyses. This should be done in close consultation with the fishermen. The system should allow for creativity and experimentation, and avoid micromanagement. For instance, if individual vessel limits are imposed on the participants, then the operational details that may affect mortality rates should be left to the discretion of the fishermen.

• The development of observer programmes designed to determine the factors that cause, or increase, incidental mortality as well as the estimation of bycatch numbers.

• Continued monitoring for unforeseen developments after an apparent solution has been found.

ACKNOWLEDGEMENTS

We would like to thank the University of Valencia, and in particular Antonio Raga, for the initiative that resulted in the courses, and in this volume. A number of scientists have contributed greatly to our discussions of this issue in recent years including Phil Hammond, Andy Read, Arne Bjørge, Dayton L. Alverson, Simon Northridge, Bill Perrin, Bob Brownell, Mark Bravington, and Finn Larsen. We would also like to thank the anonymous reviewers.

REFERENCES

Anganuzzi, A.A., Buckland, S.T. and Cattanach, K.L. (1993) Relative abundance of dolphins associated with tuna in the eastern Pacific ocean: analysis of 1991 data. *Report of the International Whaling Commission*, **43**, 459-465.

Anonymous. (1989) Whales freed in the Arctic. *Pollution Bulletin*, **20** (1), 8.

ASCOBANS. (1997) *Second Meeting of Parties to ASCOBANS: 17-19 November 1997, Bonn, Germany*, p. 43. ASCOBANS. 67pp.

Barlow, J., Forney, K.A., Hill, P.S., Brownell, R.L.J., Carretta, J.V., DeMaster, D.P., Julian, F., Lowry, S., Ragen, T. and Reeves, R.R. (1997) U.S. Pacific marine mammal stock assessments: 1996. *NOAA Technical Memorandum NMFS-SWFSC*, **248**, 1-223.

Barstow, R. (1990) Beyond whale species survival–peaceful coexistence and mutual enrichment as a basis for human-cetacean relations. *Mammal Review*, **20** (1), 65-73.

Berrow, S.D., Tregenza, N.J.C. and Hammond, P.S. (1994) Marine mammal bycatch on the Celtic Shelf., European Commission Document: DG XIV/C/1 Study Contract 92/3503.

Bjørge, A., Brownell, R.L.Jr., Donovan, G.P. and Perrin, W.F. (1994) Significant direct and incidental catches of small cetaceans. A report by the Scientific Committee of the International Whaling Commission to the United Nations Conference on Environment and Development (UNCED). *Report of the International Whaling Commission* (special issue **15**), 75-130.

Bravington, M.V. and Bisack, K.D. (1996) Estimates of harbor porpoise bycatch in the Gulf of Maine sink gillnet fishery, 1990-1993. *Report of the International Whaling Commission*, **46**, 567-574.

Corcuera, J. (1994) Incidental mortality of franciscanas in Argentine waters: the threat of small fishing camps. *Reports of the International Whaling Commission* (special issue **15**), 291-294.

Dizon, A.E. and Perrin, W.F. (1995) Report of a Workshop on analysis of genetic data to address problems of stock identity as related to management. Paper SC/47/Rep3 presented to the IWC Scientific Committee, May 1995 (unpublished). 66pp.

Donovan, G.P. (1991) A review of IWC stock boundaries. *Report of the International Whaling Commission* (special issue **13**), 39-68.

Donovan, G.P. (1992) The International Whaling Commission: Given its past, does it have a future? In: *Symposium «Whales: Biology-Threats-Conservation»* (Ed. by J.J. Symoens). Royal Academy of Overseas Sciences, Brussels, Belgium. 261pp.

Donovan, G.P. (1994) Developments on issues relating to the incidental catches of cetaceans since 1992 and the UNCED conference. *Report of the International Whaling Commission* (special issue **15**), 609-613.

Donovan, G.P. (1995) The International Whaling Commission and the Revised Management Procedure. In: *Additional Essays on Whales and Man* (Ed. by E. Hallenstvedt and G. Blichfeldt). High North Alliance, Lofoten, Norway.

Félix, F. and Samaniego, J. (1994) Incidental catches of small cetaceans in the artisanal fisheries of Ecuador. *Report of the International Whaling Commission* (special issue **15**), 475-480.

Forney, K.A., Barlow, J. and Carretta, J.V. (1995) The abundance of cetaceans in California waters. 2. Aerial surveys in winter and spring of 1991 and 1992. *Fishery Bulletin*, **93** (1), 15-26.

Fraker, M.A. (1989) A rescue that moved the world. *Oceanus*, **32** (1), 96-102.

Francis, R.C., Awbrey, F.T., Goudey, C.L., Hall, M.A., King, M.H., Norris, K.S., Orbach, M.K., Payne, R. and Pikitch, E. (1992) *Dolphins and the Tuna Industry*. National Academy Press, Washington,DC. pp. xii, 176.

Gearin, P.J., Melin, S.R., DeLong, R.L., Kajimura, H. and Johnson, M.A. (1994) Harbor porpoise interactions with a chinook salmon set-net fishery in Washington State. *Report of the International Whaling Commission* (special issue **15**), 427-438.

Gerrodette, T. (1987) A power analysis for detecting trends. *Ecology*, **68** (5), 1,364-1,372.

Goodson, A.D., Klinowska, M. and Bloom, P.R.S. (1994) Enhancing the acoustic detectability of gillnets. *Report of the International Whaling Commission* (special issue **15**), 585-595.

Hall, M.A. (1995) Strategies to reduce the incidental capture of marine mammals and other species in fisheries. In: *Whales, seals, fish and man: Proceedings of the International Symposium on the Biology of marine mammals in the North East Atlantic, Tromsø, Norway, 29 November-1 December 1994* (Ed. by A.S. Blix, L. Walløe and O. Ülltang) *Developments in Marine Biology*. 4. Elsevier, Amsterdam.

Hall, M.A. (1996) On bycatches. *Reviews in Fish Biology and Fisheries*, **6**, 319-352.

Hall, M.A. (1998) An ecological view of the tuna dolphin problem: impacts and trade-offs. *Reviews in Fish Biology and Fisheries*, **8**, 1-34.

Hall, M.A. and Lennert, C. (1994) Incidental mortality of dolphins in the eastern Pacific Ocean tuna fishery in 1992. *Report of the International Whaling Commission*, **44**, 349-351.

Hall, M.A. and Lennert, C. (1997) Incidental mortality of dolphins in the eastern Pacific Ocean tuna fishery in 1995, *Report of the International Whaling Commission*, **47**, 641-644.

Hammond, P.S., Benke, H., Berggren, P., Borchers, D.L., Buckland, S.T., Collet, A., Heide-Jorgensen, M.P., Heimlich-Boran, S., Hiby, A.R., Leopold, M.F. and Oien, N. (1995) Distribution and abundance of the harbour porpoise and other small cetaceans in the North Sea and adjacent waters., Final Report to the European Commission under contract LIFE 92-2/UK/O27. 242pp.

Hill, P.S., Jackson, A. and Gerrodette, T. (1990) Report of a marine mammal survey of the eastern tropical Pacific aboard the research vessel *McArthur* July 29-December 7, 1989., U.S. Dep. Commerc., NOAA-TM-NMFS-SWFC. 119pp.

International Whaling Commission. (1991) Report of the Workshop on the Genetic Analysis of Cetacean Populations, La Jolla, 27-29 September 1989. *Report of the International Whaling Commission* (special issue **13**), 3-21.

International Whaling Commission. (1994a) Report of the Scientific Committee, Annex H. The Revised Management Procedure (RMP) for Baleen Whales. *Report of the International Whaling Commission,* **44**, 145-152.

International Whaling Commission. (1994b) Report of the Workshop on mortality of cetaceans in passive fishing nets and traps, Annex G. List of cetacean species. *Report of the International Whaling Commission* (special issue **15**), 70-71.

International Whaling Commission. (1996) Report of the Scientific Committee, Annex H. Report of the sub-committee on small cetaceans. *Report of the International Whaling Commission,* **46**, 160-179.

International Whaling Commission. (1997) Report of the Scientific Committee, Annex H. Report of the sub-committee on small cetaceans. *Report of the International Whaling Commission,* **47**, 169-191.

International Whaling Commission. (1999) Report of the Scientific Committee. Annex I. Report of the Standing Sub-Committee on Small Cetaceans. *Journal of Cetacean Research and Management (Supplement)*, **1**, 211-225.

International Whaling Commission. (2000a) Report of the Scientific Committee. Annex I. Report of the sub-committee on small cetaceans. *Journal of Cetacean Research and Management (Supplement)*, **2**, 235-263.

International Whaling Commission. (2000b) Report of the Scientific Committee. Annex O. Report of the IWC-ASCOBANS Working Group on harbour porpoises. *Journal of Cetacean Research and Management (Supplement)*, **2**, 297-305.

Joseph, J. (1994) The tuna-dolphin controversy in the eastern Pacific Ocean: biological, economic, and political impacts. *Ocean Development and International Law*, **25**, 1-30.

Kraus, S.D. (1990) Rates and potential causes of mortality in North Atlantic right whales (*Eubalaena glacialis*). *Marine Mammal Science*, **6** (4), 278-291.

Kraus, S., Read, A., Anderson, E., Baldwin, K., Solow, A., Spradlin, T. and Williamson, J. (1995) A field test of the use of acoustic alarms to reduce mortality of harbor porpoises in gillnets. Draft Final Report available from New England Aquarium, Central Wharf, Boston, MA 02110.

Lennert, C. and Hall, M.A. (1995) Estimates of incidental mortality of dolphins in the eastern Pacific Ocean tuna fishery in 1993. *Report of the International Whaling Commission,* **45**, 387-390.

Lennert, C. and Hall, M.A. (1996) Estimates of incidental mortality of dolphins in the eastern Pacific Ocean tuna fishery in 1994. *Report of the International Whaling Commission,* **46**, 555-558.

Lennert, C., Kruse, S., Beeson, M. and Barlow, J. (1994) Estimates of incidental marine mammal bycatch in California gillnet fisheries for July through December, 1990. *Report of the International Whaling Commission* (special issue **15**, 449-463.

Lien, J. (1994) Entrapments of large cetaceans in passive inshore fishing gear in Newfoundland and Labrador (1979-1990). *Report of the International Whaling Commission* (special issue **15**), 149-157.

Lien, J., Stenson, G.B., Carver, S. and Chardine, J. (1994) How many did you catch? The effect of methodology on bycatch reports obtained from fishermen. *Report of the International Whaling Commission* (special issue **15**), 535-540.

Lowry, N. and Teilmann, J. (1994) Bycatch and bycatch reduction of the harbour porpoise (*Phocoena phocoena*) in Danish waters. *Report of the International Whaling Commission* (special issue **15**), 203-209.

Mangel, M., Talbot, L.M., Meffe, G.K., Agardy, M.T., Alverson, D.L., Barlow, J., Botkin, D.B., Budowski, G., Clark, T., Cooke, J., Crozier, R.H., Dayton, P.K., Elder, D.L.,

Fowler, C.W., Funtowicz, S., Giske, J., Hofman, R.J., Holt, S.J., Kellert, S.R., Kimball, L.A., Ludwig, D., Magnusson, K., Malayang III, B.S., Mann, C., Norse, E.A., Northridge, S.P., Perrin, W.F., Perrings, C., Peterman, R.M., Rabb, G.B., Regier, H.A., Reynolds III, J.E., Sherman, K., Sissenwine, M.P., Smith, T.D., Starfield, A., Taylor, R.J., Tillman, M.F., Toft, C., Twiss Jr, J.R., Wilen, J. and Young, T.P. (1996) Principles for the conservation of wild living resources. *Ecological Applications,* **6** (2), 338-362.

Mitchell, E.D. (1975a) *IUCN Monograph No. 3. Porpoise, Dolphin, and Small Whale Fisheries of the World: Status and Problems.* International Union for Conservation of Nature and Natural Resources, Morges, Switzerland. 129pp.

Mitchell, E. (1975b) Report of the Meeting on Smaller Cetaceans, Montreal April 1-11, 1974. *Journal of the Fisheries Research Board of Canada,* **32** (7), 889-983.

Northridge, S.P. (1996) *JNCC Report.* A review of marine mammal bycatch observer schemes with recommendations for best practice. *Joint Nature Conservation Committee, Aberdeen, UK.* 42pp.

Perrin, W.F. (1968) The porpoise and the tuna. *Sea Frontiers,* **14** (3), 166-174.

Perrin, W.F. (1969) Using porpoise to catch tuna. *World Fishing,* **18** (6), 4 unnumbered.

Perrin, W.F. and Brownell, R.L. (1994) A brief review of stock identity in small marine cetaceans in relation to assessment of driftnet mortality in the North Pacific. *Report of the International Whaling Commission* (special issue **15**), 393-401.

Read, A.J., Van Waerebeek, K., Reyes, J.C., McKinnon, J.S. and Lehman, L.C. (1988) The exploitation of small cetaceans in coastal Peru. *Biological Conservation,* **46**, 53-70.

Reijnders, P.J.H. and Donovan, G. (1999) Report of the Scientific Committee. Annex L. Furthering Scientific Cooperation Between ASCOBANS and the IWC. *Journal of Cetacean Research and Management (Supplement),* **1**, 245.

Reilly, S.B. and I arlow, J. (1986) Rates of increase in dolphin population size. *Fishery Bulletin,* **84** (3), 527-533.

Rugh, D.J., Breiwick, J.M., Dahlheim, M.E. and Boucher, G.C. (1993) A comparison of independent, concurrent sighting records from a shore-based count of gray whales. *Wildlife Society Bulletin,* **21** (4), 427-437.

Scheffer, V.B. (1989) How much is a whale's life worth, anyway? *Oceanus,* **32** (1), 109-111.

Van Waerebeek, K. and Reyes, J.C. (1994) Post-ban small cetacean takes off Peru: A review. *Report of the International Whaling Commission* (special issue **15**), 503-520.

Vidal, O. (1995) Population biology and incidental mortality of the vaquita, *Phocoena sinus. Report of the International Whaling Commission* (special issue **16**), 247-272.

Vinther, M. (1999) Bycatches of harbour porpoises (*Phocoena phocoenaL*) in Danish set-net fisheries. *Journal of Cetacean Research and Management,* **1** (2), 123-136.

Wade, P.R. (1998) Calculating limits to the allowable human-caused mortality of cetaceans and pinnipeds. *Marine Mammal Science,* **14**, 1-37.

NOTE ADDED IN PROOF

Since the completion of this chapter, a number of relevant papers have been published to which we would like to draw the attention of the reader.

Aguilar, A. (2000) Population biology, conservation threats and status of Mediterranean striped dolphins (*Stenella coeruleoalba*). *Journal of Cetacean Research and Management,* **2**, 17-26.

Crespo, E.A., Alonso, M.K., Dans, S.L., García, N.A., Pedraza, S.N., Coscarella, M. and González, R. (2000) Incidental catches of dolphins in mid-water trawls for Argentine anchovy (*Engraulis anchoita*) off the Argentine shelf. *Journal of Cetacean Research and Management*, **2**, 11-16.

Gearin, P.J., Gosho, M.E., Laake, J., Cooke, L., Delong, R.L. and Hughes, K.M. (2000) Experimental testing of acoustic alarms (pingers) to reduce bycatch of harbour porpoise, *Phocoena phocoena*, in the state of Washington. *Journal of Cetacean Research and Management*, **2**, 1-10.

Murray, K.T., Read, A.J. and Solow, A.R. (2000) The use of time/area closures to reduce bycatches of harbour porpoises: lessons from the Gulf of Maine sink gillnet fishery. *Journal of Cetacean Research and Management*, **2**, 135-141.

Rojas-Bracho, l. and Taylor,B.L. (1999) Risk factors affecting the vaquita (*Phocoena sinus*). *Marine Mammal Science*, **15**, 974-989.

Tolley, K.A., Rosel, P.E., Walton, M., Bjørge, A. and Øien, N. (1999) Genetic population structure of harbour porpoises (*Phocoena phocoena*) in the North Sea and Norwegian waters. *Journal of Cetacean Research and Management*, **1**, 265-274.

Chapter 15

Organohalogenated Contaminants in Marine Mammals

AILSA J. HALL

Sea Mammal Research Unit, University of St. Andrews, Gatty Marine Laboratory, St. Andrews, Fife, KY16 8LB. E-mail: a.hall@smru.st-andrews.ac.uk

1.   INTRODUCTION

The widespread use of synthetic chemicals by man has led to the accumulation of many persistent compounds in the marine environment. This contamination of the ocean, and the subsequent uptake of foreign compounds (xenobiotics) by marine biota, has led to a vast amount of research and related literature on their concentration in, and effect on, the marine environment at all levels of the food chain (*e.g.* Harding *et al.,* 1997; Tatsukawa, 1992).

Many of the domestic and industrial chemicals that remain in the world's seas and oceans do so because of their high molecular weight, their chemical stability and concomitant resistance to metabolic degradation (Menzer, 1991).   The adsorption of xenobiotics onto particulates and organic matter leads to uptake by plankton, crustaceans and fish and thus they ultimately find their way into top predators such as marine mammals (Kawano *et al.,* 1988; Mossner and Ballschmiter, 1997).

The fact that many compounds are bioconcentrated and are highly lipophilic (fat-soluble) means that levels, particularly in blubber, can be extremely high (Wells *et al.,* 1994; Martineau *et al.,* 1994; Jarman *et al.,* 1996).   This has heightened concern for the potential effects they may have at both the individual and population levels.   Now that it has been well established that a variety of chemicals are being taken up by seals and cetaceans worldwide (*e.g.* Borrell, 1993; Tatsukawa, 1992; Kemper *et al.,*

Marine Mammals: Biology and Conservation, edited by
Evans and Raga, Kluwer Academic/Plenum Publishers, 2002

1994; Muir *et al.,* 1992b; O'Shea, 1999), recent research is being directed towards a better understanding of their dynamics and toxicology. Although this area is still in its infancy, we are beginning to gain some insights into the toxicodynamics and metabolism of the more common compounds in marine mammals. Studies into the subsequent effects on the health of individual animals are understandably still few and far between, but in this chapter I will highlight the most significant findings to date and outline how marine mammals may deal with exposure to some of the most persistent of the ocean's contaminants.

The extent to which compounds are bioavailable to marine animals (*i.e.* are available for uptake into the blood and tissues) will depend not only on the physico-chemical properties of the compound, the route of exposure and the trophic level at which exposure occurs, but also on the physiological and biochemical make-up of the target species (James and Kleinow, 1994). Persistent ocean contaminants are known to cause a variety of toxic effects, including reproductive and immunological dysfunction (Safe, 1994; Jansen *et al.,* 1993) carcinogenesis, either directly or indirectly (Anderson *et al.,* 1994; Hemming *et al.,* 1993), and neurological effects (Shain *et al.,* 1991).

The endocrine disrupting properties of persistent ocean contaminants on humans and wildlife (Colborn and Clement, 1992), including marine organisms (Reijnders and Brasseur, 1992), has recently caused increased concern about their effects. However, many of the studies reporting these effects have been conducted on laboratory animals and whilst it is important to study such responses in as wide a context as possible, extrapolation from laboratory animals to free-ranging species, even from marine species to other marine species, may be invalid because of the physiological, biochemical, and genetic differences between them. Persistent ocean contaminants seldom cause specific lesions or health outcomes that are unambiguously diagnostic of exposure. Thus only correlative evidence between exposure to environmental contaminants and morbidity in marine mammals has been reported (see section on the effects of contaminants in marine mammals). Although the results from studies on animals in captivity have provided more robust evidence for the potential effects on free-living populations, even in these studies it is difficult to attribute effects observed to specific compounds (Addison, 1989).

It is beyond the scope of this chapter to summarise the vast body of published data on the concentrations of synthetic chemicals in seals and cetaceans. Indeed a recent annotated bibliography on marine mammals and pollutants by Aguilar and Borrell (1997) lists 541 references. There are a number of reviews which assess the importance of continued monitoring (Wagemann and Muir, 1984; Addison *et al.,* 1984; Bignert *et al.,* 1993;

Hutchinson and Simmonds, 1994; Addison and Smith, 1998), for example, temporal and spatial changes in tissue concentrations have pointed to potential links between contaminants and marine mammals epidemics (Hall *et al.,* 1992a; Aguilar and Borrell, 1994a; Kannan *et al.,* 1993). The impact and significance of these findings are now being investigated in a more integrated and unified context.

Before moving away from the monitoring aspects, there are a number of issues surrounding the measurement and interpretation of tissue residue data which should be borne in mind when comparing the results from different studies. Age, sex, condition and reproductive state are all important factors that influence the concentration of compounds found in tissue samples (Aguilar and Borrell, 1994b; Borrell *et al.,* 1995; Marsili *et al.,* 1997; Addison and Smith, 1974; Pomeroy *et al.,* 1996; O'Shea and Brownell, 1994), and these should be accounted for when reporting levels of contaminants in marine mammals. Similarly, when tissue samples have been collected from dead animals, the decomposition state can significantly affect the concentrations measured (Borrell and Aguilar, 1990).

It is also important to consider the main target organs for the contaminant of interest. For example, organohalogenated compounds are associated with fatty tissues such as the blubber, milk of lactating females and melon in cetaceans, whereas heavy metals are found at higher concentrations in kidney, liver and bone. The blubber is a popular tissue for lipophilic contaminant analysis because of its high fat content but the amount of lipid in the blubber can vary with location on the body. Similarly, fats are often stratified within the blubber layer (particularly in cetaceans), with the inner layers containing more high saturation fatty acids than the outer layers (Aguilar and Borrell, 1991; Gauthier *et al.,* 1997). These differences may account for observed variations in contaminant burdens. Although the fat-soluble contaminants are generally distributed in the target tissues on the basis of lipid content, the type of lipid is also important as these chemicals may have differential lipid affinities. For example, their affinity for phospholipids may be lower than for triglycerides or non-esterified fatty acids (Reijnders, 1980).

In this chapter I will concentrate on the one group of compounds which have caused the most concern for their potential effects on marine mammals, the organohalogenated compounds. However, readers should be aware that there is also a great deal of information on the occurrence of other contaminants in marine mammals, particularly heavy metals (*e.g.* Law *et al.,* 1992; Zeisler *et al.,* 1993; Meador *et al.,* 1993), radionuclides (Samuels *et al.,* 1970; Anderson *et al.,* 1990; Calmet *et al.,* 1992), and the butyl tins (Kannan *et al.,* 1996; Law *et al.,* 1998).

2.   ORGANOCHLORINES

Marine top predators may have particularly high body burdens of organochlorine (OC) contaminants (Luckas *et al.,* 1990; Jarman *et al.,* 1996*)*, largely because OCs are among the most fat soluble of the persistent ocean contaminants, and marine mammals have large stores of fat (over 90% of the lipid in the body is found in blubber, Tanabe *et al.,* 1981). OCs are sequestered into the blubber where they may remain in storage until the lipid is mobilised for energy or milk during lactation. For many species there is a regular dynamic turnover of blubber during annual and seasonal cycles such as during the breeding season or migrations off feeding grounds. It is during these periods, or when animals are stressed through disease or starvation, that the contaminants may exert extra toxic effects.

There is also a negative correlation between blubber thickness and contaminant concentration (Addison and Smith, 1974), and in many carnivores there is a predictable relationship between the number of adipocytes (fat cells) and body mass (Pond and Mattacks, 1985). The size of these cells increases as fat stores increase, although in some species (Colby *et al.,* 1993) new cells may be produced when increased fat storage is required. If the number of adipocytes in marine mammal blubber were predictable, the relationship between the proportion of lipid in the tissue and the blubber thickness would tend towards an asymptote. This has been clearly seen in samples from harbour porpoises (*Phocoena phocoena*), and less clearly for harbour seals found dead around the UK coast (SMRU unpublished, Hall *et al.,* 1999). Blubber is therefore a very important tissue in the toxicokinetics of OCs in marine mammals. However, the blood may play a more central role as it is the medium through which OCs are distributed to other, more sensitive target tissues and organs (see section on biotransformation and metabolism).

There are two major groups of OCs that have been the main focus of attention for the past thirty years or so: the chlorinated pesticides and the polychlorinated biphenyls.

2.1   Chlorinated Pesticides

DDT (dichlorodiphenyltrichloroethane) is one of the most widely reported OC's found in marine mammal tissue. It was first described in these species in the 1960s (Koeman and Van Genderen, 1966), and although the progressive reduction in the agricultural uses of DDT in the developed world led to its almost total withdrawal by 1983-86 (Newton *et al.,* 1989), it is still in use today, particularly in many parts of the developing world.

DDT is primarily a neurotoxin but can adversely affect many organs and systems such as the liver and reproductive tract. As an insecticide it was widely used with great success, but it had many unforeseen effects, particularly on ecosystems. Among these was its widespread occurrence in marine mammals at concentrations much higher than in their terrestrial counterparts (Kawano *et al.*, 1988). DDT is metabolised (transformed into a different form by a metabolic process) into two major compounds (metabolites) namely DDE (dichlorodiphenyldichloroethylene) and DDD (dichlorodiphenyldichloroethane). The sometimes high concentrations of total DDT (the sum of DDT and its metabolites) in marine mammals has led to many different interpretations of their significance to the health of the animal. It is clear, however, that the exposure of marine mammals to DDT is global, and ranges from coastal to open ocean species (Kleivane and Skaare, 1998; Salata *et al.*, 1995; Tanabe *et al.*, 1983). Concentrations range from (in extreme cases) 1,000 ppm or more in blubber to less than 10 ppm for most individuals (particularly in the baleen whales - O'Shea and Brownell, 1994). This skewed pattern of concentrations, with many animals having relatively low concentrations and only a few individuals having very high levels, is typical of that found in the many populations investigated and is similar for most of the OCs.

Other neurologically and reproductively toxic chlorinated insecticides found in marine mammals are the cyclodienes (*e.g.* chlordane, heptachlor, aldrin and dieldrin) and toxaphene. Although concentrations of these in marine mammals are often lower than the DDTs, there is still concern over the potential effects of dieldrin and toxaphene in particular (Boon *et al.*, 1998). As with DDT, they are distributed in marine mammals throughout the world, including in polar species (Kucklick *et al.*, 1993; Wade *et al.*, 1997; Muir *et al.*, 1992a).

2.2   Polychlorinated Biphenyls (PCBs)

Polychlorinated biphenyls (PCBs) comprise 209 possible congeners (different chemical configurations known as chlorobiphenyls or CBs), of which approximately 100 were used in commercial products and may therefore be found in the environment (Lemesh, 1992). They were produced in high tonnages by most industrialised nations up to about 1980. They have a high fire resistance and were used as capacitor impregnants and transformer coolants (commercial names include Aroclor and Clophen).

PCBs are chlorinated hydrocarbons with varying substitution of chlorine atoms for hydrogen atoms at different positions on each ring (Fig. 1). For easier identification of congeners, Ballschmiter and Zell (1980) classified the PCBs according to their structure, using a numbering system from 1 to

209. The potential effects of these contaminants were most notoriously seen when some rice oil in Japan and later Taiwan was accidentally contaminated with a PCB mixture. The resultant so called 'Yusho' disease was characterised by skin lesions and acne known as chloracne (Urabe *et al.*, 1979; Kimbrough, 1987). The chronic effects of PCB ingestion in laboratory animals include effects on reproduction, tumour promotion, oestrogenic activity and immuno-suppression (Bitman *et al.*, 1972; Vos, 1972).

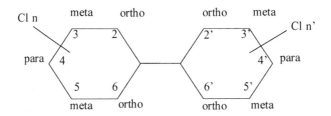

Figure 1.   The generalised structure of polychlorinated biphenyls.

In the older literature, PCBs were analysed simply as a single group, but with the development of more sophisticated analytical techniques using gas chromatography and mass spectrometry, the concentration of individual congeners are now more often reported. Although this has led to some difficulties in comparing concentrations between studies or investigating time trends, more detailed and accurate assessment of exposure, uptake, distribution, elimination, and potential toxic effects can now be made.

2.3   Other Organohalogenated Compounds

Other important contaminants in this group include HCH (hexachlorocyclohexane), HCB (hexachlorocyclobenzene), the poly-brominated biphenyls (PBBs) and the polybrominated diphenyl ethers (PBDEs). Concentrations of HCH (particularly the gamma isomer, known as Lindane) and HCB in marine mammals are often reported with the other chlorinated pesticides. Tanabe *et al.* (1996) carried out detailed investigations of the HCH isomers in 10 odontocete species. The PBBs (manufactured as flame retardants and shown to cause similar toxic effects to PCBs in laboratory animals, Kimbrough, 1987) have recently been reported in marine mammals (Kuehl and Haebler, 1995), including open ocean species such as the sperm whale (*Physeter macrocephalus*, DeBoer *et al.*, 1998). Attention is now focusing on this group because their concentration in the environment is increasing where levels of those chemicals whose production and use have been controlled, are decreasing (Norén and Meironyté, 1998).

3. THE BIOTRANSFORMATION AND METABOLISM OF ORGANOHALOGENATED COMPOUNDS

There are four main phases in the disposition of a foreign compound in a mammalian system:

ABSORPTION → DISTRIBUTION → METABOLISM → EXCRETION

The routes of absorption, distribution and excretion of xenobiotics from the body are shown schematically in Figure 2.

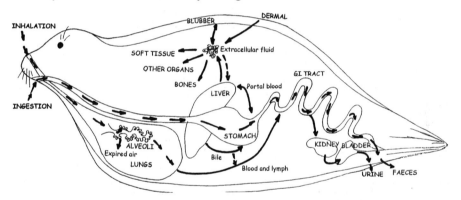

Figure 2. Schematic diagram showing the routes of absorption, distribution and excretion of xenobiotics from the body.

The absorption and distribution of a compound will depend on its lipophilicity or hydrophilicity, the presence or absence of acidic or basic groups, and the size of the molecules (Klaassen and Rozman, 1991). The central uptake process is passive diffusion (although carrier-mediated diffusion has been identified in some mammals), in which the solubility of the compound in water and lipid is important. Contaminants ingested in prey items by top predators may be presented to the gastro-intestinal tract in different forms and food matrices (James and Kleinow, 1994). These differences will affect the availability of the compound, as will other commonly overlooked factors such as gastric transit times (dependent in turn on the size and type of the meal and frequency of feeding) and lipid content and form (Bloedow and Hayton, 1976). Generally, lipophilic contaminants with high triglyceride solubility, molecular weights of <600 (Niimi and Oliver, 1988), low degrees of chlorine substitution, and molecular volumes <0.25nm are more readily absorbed (De Bruin, 1980).

Once absorbed, a substance will enter the circulation leading to its distribution throughout the animal. However, compounds are often segregated into non-sensitive tissues which become storage sites (*e.g.* blubber, connective tissues and bones). The materials in such depots are exchanged with the circulatory system. Serum proteins, such as albumin, function as major carriers in the blood. The binding between proteins and compounds (sequestration) must be broken if substances are to enter tissues by moving through biological membranes (Timbrell, 1991).

Having gained access to actively metabolising tissues, foreign substances are subjected to enzymatic attack and undergo biotransformation. Some compounds will be 'metabolically inert' and will be eliminated from the animal unchanged. However, most substances undergo some form of biotransformation, depending on the characteristics of the compound and the type of biochemical reaction involved. The metabolic products formed during this process are chemically distinct from their parent compounds and may be more or less toxic to the animal, *i.e.* the bioactivation or detoxication of xenobiotics occurs depending on whether active or stable metabolites are produced. The pathways of the metabolism of a foreign compound are shown in Figure 3. Largely the latter occurs; lipophilic compounds are converted to more polar, hydrophilic substances that are then excreted more easily because their affinity for proteins and fatty tissues has been reduced. Bioactivation, although rarer than detoxication during the handling of chemicals by mammals, is toxicologically very important.

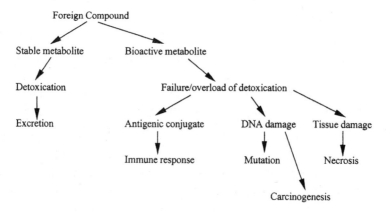

Figure 3.   Pathways in the metabolism of foreign compounds .

The metabolism of xenobiotics occurs in two phases (Williams, 1959). Phase I is the oxidation, reduction or hydroxylation of the compound; phase II occurs when metabolites are coupled to form conjugates. Imbalances between these two reactions are often the cause of tissue damage. Metabolism of foreign compounds largely occurs in the microsomal fraction of the endoplasmic reticulum in the liver (known as microsomal transformation), but other tissues and sites may also metabolise toxicants to a lesser extent (for example mitochondria and cytosol of other tissues, the dermis layer of the skin and blood plasma (Sipes and Gandolfi, 1991)).

3.1 Biotransformation Enzymes

The cytochrome P450 system (also known as the monooxygenase or mixed function oxidase (MFO) system) and the mixed function amine oxidase (a flavin-monooxygenase) are two oxidative enzyme systems which are the prominent phase I pathways. Basically, they add a hydroxyl group to a foreign substance. The cytochrome P450s are the most important enzymes and different forms (isoenzymes) catalyse different reactions. Although they occur mostly in the liver, extrahepatic biotransformation does occur, although the rate of transformation and capacity in these tissues is usually lower. Chronic exposure to low levels of toxicants, as often occurs in the marine environment, may have a marked effect on their ultimate deposition and extrahepatic biotransformation may be implicated in causing tissue damage at the site of metabolism (Stegeman and Hahn, 1994). Reaction rates involving these enzymes can be altered by inducing or inhibiting agents and chemicals, a phenomenon central to the biotransformation and toxicity of a wide range of aquatic contaminants. Over 70 distinct cytochrome P450 genes have been identified in various species, now classified into eight major families (CYP1 - CYP8; Nebert *et al.*, 1987; Nelson *et al.*, 1993). A capital letter identifies subfamilies. CYP1A1, CYP1A2 (also referred to as P448) and CYP2B are enzymes inducible by foreign compounds.

Since this system was originally identified as the drug metabolising system, groups of enzymes are also classified as phenobarbital (PB) inducible, 3-methyl colanthrene (3-MC), or ethanol inducible or mixed inducible (induced by both PB and 3-MC). The CYP1A subfamily has the same induction pattern as 3-MC inducible forms and CYP2B forms the same pattern as PB forms. However, there are some anomalies within this classification. Members of the P450 subfamilies CYP1A, 2B and 2E (Nelson *et al.*, 1993) are important in the metabolism of certain pollutants. CYP1A forms are inducible by polyaromatic hydrocarbons and by planar polyhalogenated aromatic compounds such as chlorobiphenyls, which may then be metabolised by these enzymes. Proteins of the CYP2B form are inducible in terrestrial mammals by barbiturates and by *ortho*-substituted (non-coplanar) polyhalogenated biphenyls, which in turn may be metabolised by these enzymes (De Bruin, 1980).

Alkoxy-derivates of the dye resorufin (Burke and Mayer, 1983) have been used extensively as diagnostic substrates for measuring the induction of microsomal enzymes. Resorufin is a multi-ring structure whose shape is similar to those of several P450 inducers. The induction of ethyoxyresorufin *O*-deethylase (EROD) activity in vertebrates is generally accepted as an indicator of CYP1A induction while pentoxyresorufin *O*-deethylase (PROD) activity generally marks CYP2B forms. There are, however, uncertainties in extrapolating from results in laboratory animal models to new species. For example, PROD activity can be slightly enhanced by CYP1A inducers (White *et al.*, 1994).

Figure 4 indicates the relationships between the distribution, disposition and toxicity of a contaminant in a mammalian system.

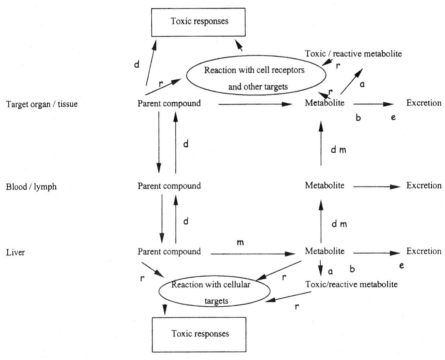

Figure 4.   The relationships between the distribution and toxicity of a foreign compound. The parent compound may move out of the blood (d) to other tissues and there react with receptors, causing toxic responses (r). Alternatively metabolism may occur (m) and the metabolites (a) will react with the targets. Metabolites may also distribute back into the blood (b) and be excreted (e). (After Timbrell, J.A. (1991) 'Principles of Biochemical Toxicology', Taylor and Francis Publishers, London)

Phase II reactions need energy to drive them. These reactions include glucuronidation, sulphation, glutathione conjugation, acytelation and amino acid conjugation. Glucuronidation is one of the most important, resulting in the elimination of glucuronides in the body by the urinary or biliary-faecal routes.

3.2   Induction of Enzymes

Biotransformation enzyme activities can be increased or enhanced (induction) by the exposure of organisms to chemicals. This results generally in an increased rate of their synthesis and was thought to be limited to the cytochrome P450 dependent monooxygenases, but certain conjugating enzymes may also be enhanced in this manner (Sipes and Gandolfi, 1991). As their rate of synthesis increases, a concomitant increase in the biotransformation rate of foreign compounds may also occur. Thus, many compounds induce their own metabolism, such that after repeated exposure the deposition or removal of the compound is more rapid or the bioactivation to a toxic metabolite more extensive.

The onset, magnitude and duration of induction vary with the agent, its dose, the species and sex of the affected animal, and the tissue in which the enzyme activity is measured. As mentioned earlier, the two most widely studied inducing agents are phenobarbital and 3-methyl colanthrene, which produce different morphological and biochemical effects in the liver. Substrates with a high affinity for the induced P450 isoenzymes show the greatest increases in rates of biotransformation. One of the 3-methyl colanthrene inducible forms of cytochrome P450 is also called P1450 or aryl hydrocarbon hydroxylase (AHH). A single P450 isoenzyme can metabolise a number of different substrates and catalyse a number of different reactions.

Major classes of marine environmental contaminants that are inducing agents include the OCs, steroids, and chlorinated dioxins. Selective induction may occur at low doses, and atrophy of the liver at much higher levels (Conney, 1982). In addition, cytochrome P450s play a critical role in the metabolism of endogenous compounds such as steroids, fatty acids, leukotrienes, and prostoglandins (Gonzalez, 1989). Furthermore, the sex hormones in particular appear to have a powerful influence on the metabolism of foreign compounds, the metabolic control of which also involves the hypothalamus and pituitary glands. The male hypothalamus produces a factor which inhibits the release of a hormone and leaves the liver in a 'male state', whereas in the female, the hypothalamus is inactive and hence the pituitary releases a feminising factor (possibly growth hormone) which changes the liver to the 'female state' (Timbrell, 1991).

These different states mean that the toxicity of certain compounds differs between the sexes. For example, the insecticides aldrin and heptachlor are metabolised more rapidly to toxic epoxides in males and are less toxic to females. Males are more sensitive to the renal toxicity of chloroform than females, and this difference can be removed by castration of the males, and restored by administration of androgens. These interactions between hormonal and contaminant metabolism are important in the understanding of how contaminants can disrupt mammalian endocrine systems. The thyroid, adrenal and pituitary hormones, and insulin mainly act directly on the liver to affect metabolism, whereas the sex hormones act via the pituitary and hypothalamus. The potential interactions are, however, complex and often unknown.

3.3   Bioactivation

Biotransformation enzymes can also produce reactive intermediates or metabolites which interact with amino acid or hydroxyl groups present on DNA, RNA or proteins (Timbrell, 1991). For many chemicals these intermediates are detoxified, provided there is a balance between rates of formation and detoxication. If this is the case, reactive intermediates or bioactive metabolites may not necessarily cause tissue injury. When this balance is disturbed (*e.g.* during enzyme induction which increases the overall rate of biotransformation), excess production of reactive intermediates may occur. Where doses are high, non-toxic pathways may become saturated and minor pathways which form these toxic intermediates become involved. Epoxides, free radicals, acid chlorides, and N-hydroxy derivatives are often encountered as reactive intermediates. Metabolic 'switching' may also occur. Because compounds are metabolised in different ways, one metabolic pathway may be inhibited by one compound, leading to the increased activity of another, minor pathway that may affect the toxicity of the chemical.

4.   METHODS FOR DETERMINING UPTAKE AND METABOLISM IN MARINE MAMMALS

The toxicokinetics and metabolism particularly of organochlorine contaminants in marine mammals have been investigated using a variety of methods. These techniques will be reviewed in the following section.

4.1   Polychlorinated Biphenyl Patterns

Some of the PCB congeners have the ability to attain a coplanar configuration, which mimic the structure of highly toxic contaminants such as 2,3,7,8-tetrachlorodibenzo-*p*-dioxin (TCDD) and 2,3,4,7,8-pentachloro-dibenzofuran (PCDF). Effects produced by exposure to these pollutants include hepatic damage, reproductive and immune function disorders, thymic atrophy and dermal disorders. They are also high inducers of 3-MC (CYP1A) type hepatic microsomal enzyme activity (Poland and Knutson, 1982; Tanabe *et al.*, 1987).

Figure 5.   Polychlorinated biphenyls (PCBs) are synonymous with industry, being used in heat transfer systems, capacitors, and plastics, as dielectric fluids in electrical transformers, and as hydraulics, inks, additives and lubricants. Though banned in most industrialised countries during the 1970s and 1980s, they are likely to continue entering the marine environment from transformers and other machinery through leakage, degradation and disposal. (Photo: P.G.H. Evans)

More accurate insight into the mechanisms of uptake, distribution and elimination of PCBs has been gained from studying individual CB congeners in marine mammal tissues rather than assessing an animal's total PCB burden (Norstrom and Muir, 1994). Comparisons between, and changes in, the patterns of individual CB congeners in tissues such as blubber and muscle, in species at various levels of the food chain, have been used to infer the metabolic capacity of animals at higher trophic levels.

Contaminants present in the prey, but not found in the predators, are assumed to be metabolised.

By comparing the ratio of each congener to the most persistent congener (such as CB153, which is found at consistently high levels in marine biota), patterns and levels can be standardised, making them directly comparable and independent of concentration. (Boon *et al.*, 1992). Highly significant differences occurred in patterns of fish, seal blood and seal faeces in an experiment comparing two groups of captive harbour seals (*Phoca vitulina*) fed contaminated fish from the Wadden Sea and cleaner fish from the Atlantic (Boon *et al.*, 1987). Congener ratios to CB153 in fish and seal blood were compared, the relative behaviour of each congener assessed, and contaminants grouped into similarly structured compounds that are either metabolised by seals or are inert. The group of compounds found at lower levels in seal blood than in fish all possessed vicinal hydrogen atoms (*i.e.* hydrogen atoms bound to adjacent carbon atoms of an aromatic ring). Those with vicinal H atoms at the *ortho* and *meta* positions all possessed only one *ortho*-chlorine atom (see Fig. 1). Compounds occurring at the same level in seal blood as fish, and therefore not metabolised, possessed either two or three *ortho*-chlorines.

Boon and Eijenraam (1988) reported that the contribution of the lower chlorinated congeners up to the pentachlorobiphenyls, to the total CBs present in a sample could be compared between fish, seal faeces and seal blood such that fish > faeces > blood. They concluded that enzyme mediated metabolism, via an arene oxide intermediate, was the probable cause of the relatively lower concentration of certain congeners in seal blood. CBs with a di-*ortho* chlorination were apparently not metabolised. Groups of such persistent CBs showed a constant pattern throughout all seal species studied (Boon *et al.*, 1992). These structural similarities have been used to classifiy CB congeners according to the way they are metabolised in different species (structure-activity relationships), and how they may therefore interact with cytochrome P450 enzymes (Parke, 1985). Highly consistent CB patterns within representative biota at the same trophic level was also observed by Duinker and Hillebrand (1983). Duinker *et al.*, (1989) found the same pattern of CBs in different areas of the ocean, assuming the harbour porpoise were eating fish with the same general CB pattern as the fish eaten by the harbour seals.

Tanabe *et al.* (1987) concluded from comparing CB patterns in prey and predators, that the metabolic capacity of marine mammals appeared to be lower than terrestrial mammals. Thus, the accumulation of CB congeners, especially coplanar forms with stereo-isomerism resembling 2,3,7,8,TCDD, may indicate that these animals are at higher risk of toxic effects following exposure than their terrestrial counterparts. In order to investigate the metabolic capacities of marine compared to terrestrial species further, Tanabe *et al.* (1988) proposed a 'metabolic index' for mammals, such that the accumulation of CBs from the diet relative to the recalcitrant congener CB180, were compared between different cetacean and terrestrial species. CB congeners in fish, as the main prey of marine mammals, are significantly affected by ambient water and less so by feeding habits (McKim and Nichols, 1994). Thus fish and cephalopods from the common feeding grounds of small cetaceans should show the same CB pattern. However, CB congener patterns do not reflect metabolic capacities alone, for example lower chlorinated CBs are found at higher levels in cetaceans from warmer waters feeding on prey with similar CB patterns to those in colder waters. To account for this, Tanabe's metabolic index was calculated as: $MIj = Log\ CR180 - Log\ CRj$

CRj = Concentration Ratio (between cetaceans and prey) isomer j

$CR180$ = Concentration Ratio of isomer CB180

Cetaceans appeared to have a very low capacity to metabolise CBs with vicinal *meta* and *para* hydrogen atoms (Tanabe *et al.*, 1988). Seals had intermediate capacities between birds and terrestrial mammals

These structure-activity relationships were refined and used by others (Gagnon *et al.,* 1990) to conclude that cetaceans indeed have a low capacity to metabolise *meta, para* unsubstituted CBs. This may indicate that they lack PB-type (CYP2B) metabolising enzymes found in many other mammalian species.

Muir *et al.* (1988) showed that the pattern of PCBs in Arctic cod was almost the same as a mixture of commercial PCBs, whereas the pattern became increasingly simplified in ringed seals (*Phoca hispida*) and polar bears (*Ursus maritimus*), the seals' main predator. Similarly Wolkers *et al.* (1998) observed that a decrease in the lower chlorinated CBs was associated with higher concentrations of CB153. The metabolic index confirmed the model of persistency, except with CB128 and CB138 (congeners with vicinal H atoms exclusively in the *ortho, meta* positions in combination with ≥ 2 ortho-chlorines). It appeared these could, to some extent, be metabolised in ringed seals (although co-elution of other congeners with CBs 128 and 138 may have resulted in an underestimation of the metabolic index for these congeners). The congener distributions in narwhal (*Monodon monoceros*) and Arctic beluga whales (*Delphinapterus leucas*) were almost identical, including a large number of tetrachloro- and pentachloro-congeners, a pattern very similar to that in Arctic cod (Norstrom and Muir, 1994). However, beluga whales in the St Lawrence had higher proportions of hexachloro- and heptachloro- congeners in their blubber. Walrus (*Obdobenus rosmarus*) appeared to be more efficient at metabolising CBs than ringed seals, with lower tetrachloro- and pentachloro-biphenyls in their blubber.

Figure 6.   Walruses (top); Ringed seal (bottom). The efficiency with which these pinnipeds metabolise chlorinated biphenyls may relate to their different diets, walruses taking bivalve prey that is likely to particularly concentrate contaminants. (Photos: I. Stirling)

However, little is known about the contaminant levels in the food of the walrus being studied (mainly bivalves). Thus, body burden pattern differences between these species may be reflecting dietary differences. Using this comparative pattern method, ringed seals also metabolise CBs with vicinal *meta, para* hydrogen atoms to some extent, but other contributory factors must also be involved because, for example, CB101 (2,5,2',4',5', pentachlorobiphenyl) and CB87 (2,5,2',3',4', pentachloro-biphenyl) both have one free *meta, para* position but only CB87 decreases in ringed seals compared to beluga whales and narwhal. Norstrom *et al.* (1992) placed narwhal and beluga whales in a special category (with a lower capacity to metabolise CBs with unsubstituted *meta, para* positions).

Selective metabolism in different species based on congener patterns was discussed by Reijnders (1988) and Boon *et al.* (1987). CB patterns in three harbour seals, one ringed seal and one harbour porpoise were virtually identical in blood, liver and blubber from the same animal. However, considerable differences occurred between animals of different species, and between animals of the same species, depending on physiological state. According to Tanabe *et al.*'s (1988) model, CB101 should be persistent in harbour porpoise. However, Reijnders (1988) found CB101:CB153 similar in fish and porpoise. The ratio of CB149:CB153 was higher than in fish prey and higher than in harbour seals. He recommended that the proposal that cetaceans lack PB metabolising enzymes (CYP2B) should be refined. Either harbour porpoise have a different metabolic capacity compared with other cetaceans, or the distinction is rather the substitution of two *ortho*-chlorines (still metabolisable) or three or more *ortho*-chlorines (persistent).

Ringed seals and harbour porpoises therefore have metabolic capacities between those of other cetaceans, and Duinker *et al.* (1989), using a similar comparative method, placed porpoises intermediate in capacity to other cetaceans. Indeed Bruhn *et al.* (1995), comparing CB congener to CB153 ratios in herring with those in harbour porpoise, also found these cetaceans appeared able to metabolise congeners with *meta, para* vicinal H atoms, even in the presence of two or more *ortho*-chlorines.

This model was further refined by Boon *et al.* (1997), using data on CB congener concentrations in marine mammals from different European monitoring programmes. The species included in the study were otter (*Lutra lutra*), harbour porpoise, common dolphin (*Delphinus delphis*), grey seal (*Halichoerus grypus*) and harbour seal. Three types of metabolism were defined and inferred from detailed comparisons between the ten individual congeners studied relative to CB153. These patterns were compared with ratios calculated for cod liver oil as a reference source for patterns in the diet. Type 1 metabolism was defined as the persistent congeners (not metabolised). Type 2 was inferred where biotransformation appeared to occur at a constant rate. Non-inducible, constitutive enzymes may mediate this type of metabolism. Type 3 metabolism was defined where the biotransformation of the congeners became stronger with increasing concentration of CB153, *i.e.* biotransformation being mediated by inducible enzymes.

In conclusion the authors reported that the ability to metabolise congeners with vicinal H atoms only in the *ortho* and *meta*-positions, and with one *ortho*-chlorine substituent, generally increased in the order: otter < cetacean < phocid seal, but the metabolism of congeners with vicinal H atoms in the *meta* and *para* positions with two *ortho*-chlorines increased in the order: cetacean < seal < otter. Both these groups of congeners are probably metabolised by different cytochrome P450 enzyme subfamilies, and levels differ between seal and cetacean species. Within-species patterns varied in a concentration-dependent manner, which suggested the induction of cytochrome P450 enzymes. However, confounding factors such as starvation could also have a similar effect on CB congener patterns.

4.2 Cytochrome P450 in Marine Mammals

The dynamics of contaminants in marine mammals can be more accurately defined by directly studying the activity of the enzymes which catalyse the metabolic reactions involved, such as those of the cytochrome P450 group.

The cytochrome P450 enzymes catalyse a variety of different, primarily oxidative, reactions in mammalian systems. They are only just beginning to be fully characterised in marine mammals and studies have been restricted to a few species of pinniped (Addison *et al.*, 1986; Addison and Brodie, 1984; Goksøyr *et al.*, 1992; Troisi and Mason, 1997; Wolkers *et al.*, 1998) and cetacean (White *et al.*, 1994; Watanabe *et al.*, 1989; Goksøyr *et al.*, 1986). Because tissues (mainly liver) have to be collected from freshly dead or recently killed animals, to ensure enzyme activity is not degraded, studies to date have been limited. Often, substantial quantities of cytochrome P420 (degraded P450) are found in samples, and this may bias the interpretation of results (White *et al.*, 1994; Troisi and Mason, 1997).

In a study of three species of Northern Hemisphere pinnipeds, higher P450 activities were found in harbour compared with harp (*Phoca groenlandica*) and hooded seals (*Cystophora cristata*, Goksøyr *et al.*, 1992). Female harp seals seemed to have lower cytochrome P450 activities than other species, and the low but detectable levels of PROD activity (surrogates for CYP2B, see section on biotransformation and metabolism) in harp seals were not seen in hooded seals. Monoxygenase activities were also monitored in harp and hooded seal pups. Levels were largely lower in pups than adults, although both harp seal EROD and MCOD (7-methoxycoumarin *O*-deethylase) and hooded seal MCOD activities were higher in pups than adults. During the same study, harp and hooded seal pups were treated with phenobarbital. EROD activities increased in both species but MCOD, ECOD and PROD were only induced in harp seals. The conclusion was that seals do respond to phenobarbital, but typical CYP2B activity (PROD) is lower than might be expected from studies in other mammal species.

Reagent antibodies rather than alkylresorufin substrates are now being used to determine CYP1A regulation in marine mammals. Goskøyr *et al.* (1988) used a polyclonal anti-cod P450c and a monoclonal anti-scup CYP1A1 to identify immunochemically related forms in the minke whale (*Balaenoptera acutorostrata*). White *et al.* (1994) used the same anti-scup CYP1A1 monoclonal in the beluga whale (*Delphinapterus leucas*). Watanabe *et al.* (1989) investigated the effects of rat cytochrome P450 antibodies to CYP1A1, 1A2 and 2B1 on monooxygenase activity in the liver microsomes of short-finned pilot whales (*Globicephala macrorhynchus*). CYP1A1 appeared to have an inhibitory effect on EROD and AHH activity. There was also a weaker effect of CYP1A2 and no effect of CYP2B1. Goksøyr *et al.* (1989) also reported some evidence for CYP3 isoenzymes in minke whale.

Results of initial studies using anti-CYP2B antibodies have supported Tanabe *et al.*'s (1988) hypothesis that CYP2B forms are reduced in some cetaceans compared with terrestrial mammals. Goskøyr *et al.* (1988, 1989) found that a polycloncal antibody to rat CYP2B1 and 2B2 did not detect these forms in minke whale liver samples. Stegeman and Hahn (1994) also reported the same findings, as did White *et al.* (1994) in the beluga whale. However, a rabbit polyclonal antibody against CYP2B4 did recognise a band in the beluga whale (White *et al.*, 1994), suggesting a type of CYP2B is present in these animals in a form more closely resembling the rabbit than the rat. CYP1A levels in the beluga whale were related to EROD, PROD and AHH, activities suggesting that this P450 form is a primary catalyst for these reactions in this species.

The fact that samples failed to cross-react with polyclonal antibodies raised against rat CYP2B1 and scup P450B may indicate the lack of a gene encoding a protein homologous to CYP2B forms or very low/absent expression of such a gene. Interestingly, CB congeners known to be inducers of CYP2B in terrestrial mammals were present in these beluga whales. PROD activity (although low) strongly correlated with CYP1A, suggesting in *this* species PROD activity is catalysed by CYP1A. However, the presence of protein bands that cross react with anti-rabbit CYP2B4 and anti-dog CYP2B11 is intriguing (White *et al.*, 1994).

Figure 7.   Killer whale (top); Beluga whale (bottom). Males of these two whale species may be particularly susceptible to cytochrome P450 enzyme induction by certain environmental contaminants. (Photos: T. Similå and T. Henningsen)

There were also sex differences in CYP1A content in beluga whales (White *et al.*, 1994). Levels were five times higher in males than females, and were also highly correlated with the blubber concentrations of non-*ortho* and mono-*ortho* CB congeners, compounds that induce CYP1A in other mammals. Sex differences in monooxygenase activities in the rat have been linked to hormonal regulation of different P450 isoenzymes (Waxman *et al.*, 1985, Dannan *et al.*, 1986). However, other studies have shown no sexual differentiation in CYP1A1 activity in rats, either treated or untreated with inducing agents (Waxman *et al.*, 1985). Thus, some rates of reaction (AHH and EROD transformation) are similar in marine mammals to those induced in rats, whereas others (PROD, aldrin epoxidase) appear to be lower. In certain cetacean species (killer whales (*Orcinus orca*) - Watanabe *et al.*, 1989 and beluga whales - White *et al.*, 1994), the presence of appreciable EROD activity in some individuals suggested induction by persistent environmental contaminants.

In seals, anti-rat CYP2B1 did not recognise any bands in hepatic microsomes of harbour seals (Hahn *et al.*, 1991) or fur seals (*Callorhinus ursinus*, Stegeman and Hahn, 1994). A single band was found in fur seals using the anti-rabbit CYP2B4 polyclonal. Goskøyr *et al.* (1992) investigated the cross reactivity of anti-cod CYP1A1 in harp and hooded seal liver samples. They found reactions with two protein bands. They also used an anti-dog CYP2B11, which recognised multiple bands in harp and hooded seals, but which was strongest in male pups. This form of P450 is reported to be an effective catalyst of CB153 metabolism (Graves *et al.*, 1990) which, to date, has only been apparent in dogs. These cross reactivities may indicate evolutionary relationships between species and their abilities to metabolise certain CBs.

Thus both pinnipeds and cetaceans possess proteins recognised by antibodies to mammalian and fish CYP1A1 forms. There is some evidence for the presence and expression of 2B enzymes in pinnipeds and cetaceans, but the metabolic capabilities of these forms and their relationship to other mammalian species still needs to be clarified (Stegeman and Hahn, 1994).

The regulation of CYP1A enzymes involves a ligand-activated transcription factor known as the Ah (aromatic hydrocarbon) receptor (Landers and Bunce, 1991). This receptor has been investigated in beluga whales (Hahn *et al.*, 1992) but knowledge of it in other species is sparse. The presence and properties of the Ah receptor could determine the susceptibility of animals to CYP1A induction and other effects resulting from exposure to polyaromatic hydrocarbons and halogenated hydrocarbons.

Stegeman and Hahn (1994) pose a list of questions that need to be answered regarding cytochrome P450 in marine mammals. These include:

a) what mechanisms regulate the genes for specific enzyme activities?
b) how are their functions modified by internal and external variables?
c) what are the temporal or physiological links between the function or
regulation of biotransforming enzymes and the function or regulation of
other molecular processes in particular cell systems? The regulation of the
P450 system may be as important as the function of the enzymes
themselves, particularly in mediating the effects of toxic chemicals. The
cellular targets may determine the nature of any systemic effects, but neither
the regulatory mechanism nor the cellular targets are known.

The catalytic properties of individual P450 isoenzymes are fundamental
to determining their role in xenobiotic metabolism. The ultimate objective
would be to determine the structure/function relationship or the substrate
structure/activity relationships for different P450s. For example, Benzo-*a*-
pyrene metabolism or AHH activity, catalysed primarily by CYP1A forms,
was one of the first activities strongly linked to an induction mechanism.
AHH activity induction in fish by hydrocarbons first indicated not only that
a similar catalyst occurs in fish and mammals, but also that induction of this
nature could be used as a marker of exposure to aromatic hydrocarbons.
The fact that phenobarbital treatment of fish does not increase the metabolic
rate of drug substrates as seen in mammals, indicated that differences
between these species exist, either in the enzymes themselves, or in their
regulation.

Different biochemical characteristics of the cytochrome P450 system
appear to exist between seals and cetaceans, and between seal species.
Reijnders (1994) suggests that caution is required when extrapolating results
for some compounds acting as P450 substrates in some species to other
compound and/or other species, since organisms can develop species- and
sometimes even organ-specific isoenzyme profiles.

4.3   Organochlorine Metabolites

Chlorobiphenyls and other organochlorine compounds can exert toxic
effects in various forms, either as parent compounds or as metabolised
products. Some CBs are not retained or biomagnified, but are metabolised
to hydroxylated or sulphonated CBs (Bergman and Haraguchi, 1994).
Jensen and Jansson (1976) reported the presence of a new class of CB and
DDT metabolites, the methyl sulphones, which are persistent in mammalian
systems and may be as toxic as their parent compounds. Mio *et al.* (1976)
also identified methylthio-metabolites and methylsulphonyl metabolites of
2,5,2',5'-tetrachlorobiphenyl. The measurement of these metabolites in
tissues and excreta of marine mammals will provide additional information
about the uptake and dynamics of toxicants.

CB and DDE methylsulphones (MeSO$_2$) are formed via nucleophilic attack by glutathione on an arene oxide intermediate formed during phase I metabolism (Bergman *et al.*, 1994a). These arene oxides are intermediates with a limited half-life. They are metabolised to aromatic compounds, but also to sulphur containing metabolites via the mercapturic acid pathway. The toxicological significance of MeSO$_2$ CBs have not been determined, although they have been linked to respiratory distress (Urabe *et al.*, 1979). MeSO$_2$-4,4',-DDE has been shown to cause severe necrosis in the adrenal cortex of mice at low doses (Lund *et al.*, 1988).

Methylsulphonyl metabolites of CBs and DDE were first detected in the blubber of grey seals from the Baltic (Jensen and Jansson, 1976), and reports now indicate methylsulphonyl (MeSO$_2$) CBs may be major pollutants in many marine mammals. MeSO$_2$-4,4'-DDE has also been shown to be present at high levels in some mammals (Brandt and Bergman, 1987) and are retained in certain tissues due to protein sequestration.

Samples have been analysed from beluga whale, grey seal and false killer whale (*Pseudorca crassidens*). Methylsulphonyl CBs were present in concentrations between 0.1-10 g/g, together with concentrations of CBs between 10 and 200 times higher than this (Bergman *et al.*, 1994a). The liver generally contained higher concentrations on a lipid weight basis than adipose tissues, indicating retention in the liver. In grey seal, mink, and otter only a few MeSO$_2$ CBs were found but concentrations were higher (the most abundant being 3-MeSO$_2$-2,5,6,2',4',5'-hexachlorobiphenyl). The structure of the parent compound is deduced from replacing the methylsulphonyl group with hydrogen. Seven CB congeners have been identified as precursors: CB49, CB70, CB101, CB87, CB149, CB132 and CB141. The small number of congeners yielding methylsulphonyl metabolites, compared with the number known to be easily metabolised, indicates selective formation rather than selective tissue retention. Those identified in wild mammals are CBs with a 2,5 dichloro- or a 2,3,6 trichloro-substitution. These patterns have unsubstituted *meta*, *para* positions adjacent to two chlorine atoms, which may stabilise the *meta*, *para* epoxide primary product from cytochrome P450 oxidation and allow easier reaction with glutathione. Levels of MeSO$_2$-CB congeners were higher than their CB precursors in the same sample, but were similar to the more recalcitrant congeners, indicating they have the potential to be highly persistent.

Haraguchi *et al.* (1992) also found MeSO$_2$-CBs in grey, harbour and ringed seal samples. As in other studies, the lower chlorinated CBs were not retained. Interestingly, adult ringed and adult female grey seals all lacked CB118 and ringed seals also retained CBs 49, 52 and 101 in larger amounts than other seal species (this may reflect differences in metabolism or feeding habits). There was no significant difference in MeSO$_2$-CB or

DDE in seals collected before the 1988 seal epidemic and those collected during it. The higher concentrations of CBs in adults and in seals from the Baltic compared to the Skagerrak/Kattegat were not reflected in the levels of metabolites. The ratio of CB:MeSO$_2$-CB varied, being largely between 10 and 30. Ratios of 4,4'-DDE:MeSO$_2$-DDE for all the species studied indicated that MeSO$_2$-CBs were more strongly retained than MeSO$_2$-DDE. This may be due to differences in the rate of metabolism of the parent compounds, or that MeSO$_2$-DDE is further metabolised at a higher rate.

Methylsulphone (MSF) metabolites of PCBs were recorded in a number of cetacean species by Troisi *et al.* (1998). They found PCB:MSF ratios varied considerably between species. Harbour porpoise had ratios of 1:10, pilot whale (*Globicephalus melas*) and Atlantic white-sided dolphin (*Lagenorhynchus acutus*) had ratios of 1:50 and common dolphin, Risso's dolphin (*Grampus griseus*), and striped dolphin had ratios of 1:100. The authors suggest these variations are due to interspecies differences in capacity to form methylsulphone metabolites.

Lund (1994) investigated MeSO$_2$-DDE toxicity in seals. MeSO$_2$-DDE is biotransformed to reactive intermediates that are cytotoxic. DDD is a well-known adrenocorticolytic agent in dogs and humans and can be bioactivated in mink and otter (Nelson and Woodard, 1949). Lund looked at the bioactivation of the DDT metabolites DDD and MeSO$_2$-DDE by seal adrenal *in vitro*. He incubated adrenal protein with radio-labelled hydrocarbon, and determined if the radioactivity became irreversibly bound to the protein. The involvement of cytochrome P450s was investigated by omitting NADPH (necessary for P450 activity) and by including selective P450 inhibitors. The adrenal was bioactivated by *o,p'*-DDE and MeSO$_2$-DDE. The irreversible binding of *o,p'*-DDD was 17 times higher than that of MeSO$_2$-DDE. The addition of cytochrome P450 inhibitors indicated that cytochrome P450s were responsible for the bioactivation. In addition, there is a good correlation between irreversible binding *in vitro* and the *in vivo* toxicity of compounds (Jönsson *et al.*, 1993). Lund concludes there may be two mechanisms for the adrenotoxic effects of DDD, DDT and PCBs; a) they are cytotoxic after the generation of reactive intermediates, or b) they inhibit glucocorticoid-synthesising enzymes.

The hydroxylated metabolites of CBs are excreted or conjugated to glucuronic acid or sulphate. Hydroxylated CBs have been measured in faeces from environmentally exposed seals (Jansson *et al.*, 1975). In rodents, metabolism of CB congeners can result in decreased toxicity of the parent compound and increased specific toxicity of the metabolites. For example, the hydroxylated metabolites of CB77 caused reduced plasma vitamin A (retinol) levels and reduced levels of thyroxine in rodents. This finding was reproduced in the seal feeding experiment when two groups of harbour seals were fed contaminated and clean fish (Brouwer *et al.*, 1988).

CB31, CB77 and CB105 fed to laboratory animals have shown that several isomers of hydroxylated metabolites, including dihydroxylated and dechlorinated metabolites, are retained in the blood (Klasson-Wehler *et al.,* 1989). Nine of 13 hydroxylated CB metabolites were found in grey seal blood and compared with standards (Bergman *et al.,* 1994b). It is interesting to note that the concentration of the hydroxylated metabolites of CB105 were higher than one of the most persistent congeners, CB153. A dominating hydroxylated CB in rat plasma and seal blood was formed following a 1,2 shift of a chlorine in the *para* position in the 2,3,4 trichlorobiphenyl ring. A similar arrangement was also seen in the 3,4-dichlorobiphenyl substituted rings of CB77, CB105, CB118 and CB156. Thus, all the major mono-*ortho* CBs can theoretically be metabolised to produce 4-hydroxylated CBs (4-OH-CBs) retained in the blood.

Highly selective retention of these OH-CBs may be related to their structural resemblance to thyroxine. 4-OH-3,5,2',3',4'-pentachloro-biphenyl competes for the thyroxine (T4) binding sites twice as well as 4-OH-3,4,3'-4'-tetrachlorobiphenyl. PCBs have been reported as inhibiting the transport of thyroid hormones in rat plasma by competing with T4 for the binding site on transthyretin (Brouwer and Van den Berg, 1988). Perinatal exposure to CB118 and CB153 markedly decreased serum T4 levels in young rats, whereas CB28 did not and CB118 and CB153 are more likely than CB28 to form hydroxylated metabolites in the blood (Ness *et al.,* 1993).

5. BIOMARKERS AND EFFECTS OF ORGANOHALOGENATED CONTAMINANTS ON MARINE MAMMALS

5.1 Effects

The most ecologically important aspect of the toxicology of contaminants in marine mammals is the adverse effect they may have on individuals and therefore populations. This issue really came to the forefront of marine mammal science following two major events. The first was the report that the population of harbour seals in the Wadden Sea was in serious decline, and this might be due to the reproductive effects of contaminant exposure (Reijnders, 1980, 1984). The second was a viral epidemic among harbour seals in Europe, the effects of which may have been greater in populations with high OC contaminant burdens because of their immunosuppressive effects (Hall *et al.,* 1992a; Kendall *et al.,* 1992).

Figure 8.   Harbour seals in European seas in 1988 were subject to an outbreak of phocine distemper virus which killed up to 50% of some populations. Subsequent studies suggested that exposure to PCBs may have compromised their immune systems, causing higher mortality than would otherwise have occurred. (Photo: A. Hall/SMRU)

Reijnders (1986) addressed the first of these in an elegant experiment using captive harbour seals. Two groups of female harbour seals were fed fish from different areas one contaminated with OCs the other much cleaner. Blood samples were monitored regularly over two years for progesterone and oestradiol 17ß, and breeding males were introduced into both groups. Reproductive success was significantly lower in the group fed contaminated fish, and failure was thought to occur at the implantation stage of pregnancy. That OCs are potential hormone disrupters is now well recognised and much emphasis is being given to investigating the mechanisms by which this may occur (see section on metabolites). Indeed, studies on Dall's porpoise indicated a negative relationship between testosterone levels in the blood and PCB and DDE concentrations in the blubber, although this was not seen for the other hormone (aldosterone) measured (Subramanian *et al.*, 1987).

Correlative evidence for the effect of OC exposure on reproduction in pinnipeds have been reported in grey and ringed seals from the Baltic (Helle *et al.*, 1976), and California sea lions (*Zalophus californianus*, DeLong *et al.*, 1973). In the former studies, outcomes such as uterine occlusions or stenosis have been reported, and in the latter, stillbirths and premature pupping was related to the uptake of DDT. However, the interpretation of these findings is fraught with difficulties because of the many other factors which may affect reproduction in pinnipeds (for further reviews and discussion, see Addison, 1989; Reijnders, 1984; and O'Shea and Brownell, 1998). It has been assumed that OC contaminants are inert when sequestered in blubber. However, this is questioned by, for example, Reijnders (1980), and needs further investigation particularly because recent studies in humans suggest that lipid is oestrogenic (Pedersen *et al.*, 1996). If this is also the case in marine mammals, it may not be assumed that these contaminants are only bioavailable during periods when blubber is mobilised. PCBs in particular are important in this respect because of their oestrogen-like structure. If they compete with oestrogens at receptor sites within the lipid they may be causing hormone disruption as a consequence of storage in the blubber. Further studies on the nature of marine mammal blubber and its endocrinological importance are clearly required.

In cetaceans, the evidence for reproductive effects is even more limited. Beluga whales in the St Lawrence River in Canada are the best studied species in this regard (Fig. 9). Martineau *et al.*, (1987) suggested that the observed population declines were due to high PCB levels in these animals, and Bland *et al.*, (1993) observed some reproductive pathologies in stranded animals. However, some of the potential confounding factors were not taken into consideration and when habitat degradation or body condition are considered, these animals may have similar OC burdens to belugas in other populations (Addison, 1989). Until further, well controlled, experimental studies are carried out in conjunction with observations on free-living animals, the interpretation of these findings remains open.

Bergman and Olsson (1985) also reported the occurrence of adrenocortical hyperplasia, hyperkeratosis, and other lesions in grey and ringed seals from the Baltic. The pathologies seen were indicative of a disease complex involving OCs and hormone disruption, a finding also demonstrated in laboratory animals (Fuller and Hobson, 1986). Also of note was a study by Freeman and Sanglang (1977) which demonstrated alterations in the synthesis of steroids after exposing grey seal adrenal cells *in vitro* to a commercial mixture of PCBs. Other lesions correlated with high exposure to PCBs include skull and bone lesions in grey seals (Bergman *et al.*, 1992; Zakharov and Yablokov, 1990) and harbour seals from the Baltic (Mortensen *et al.*, 1992). The relationship between adrenocorticohyperplasia and OC exposure was investigated in harbour porpoises by Kuiken *et al.* (1993), and although these changes were found in the adrenal tissue, there was no relationship between these and OC burdens.

Figure 9.   Beluga whales in the Gulf of St Lawrence have some of the highest contaminant burdens of any marine mammal yet it has not been possible to provide clear evidence for them having a primary effect upon life history parameters. (Photo: T. Henningsen)

During the outbreak of phocine distemper among harbour and some grey seals in European waters, differential mortality rates were reported among harbour seal populations around the UK coast (Hall *et al.,* 1992b). It was suggested that other mechanisms might have been involved in determining the outcome of the disease. This observation led to a study of the OC contaminant burdens among animals that were victims and survivors of the epidemic. The results suggested that animals that died of the disease had higher blubber levels of OCs than survivors, although it was not possible to control for all potential confounders (Hall *et al.,* 1992a).

Interestingly, this finding was also repeated in a study of contaminant burdens in striped dolphins following a similar outbreak of dolphin morbillivirus in the Mediterranean Sea in 1990 (Kannan *et al.,* 1993; Aguilar and Borrell, 1994a). The results of studies on laboratory animals had suggested that PCBs in particular are immunosuppressive. Seals with higher body burdens at the time the virus struck may have been unable to mount a sufficient immune response and this exacerbated the effects of the outbreak at the population level. Again, in order to test this hypothesis on seals themselves, captive feeding experiments were carried out. A study by Ross *et al.* (1995) and DeSwart *et al.* (1994) found evidence for immunosuppression in a group of captive harbour seals fed contaminated fish compared with animals fed clean fish. Natural killer cell activity in particular was depressed and lymphocyte function measured *in vitro* was lower in the exposed group.

Only a few studies have investigated the immunosuppressive effects of contaminants on wild populations. In a study of free-living grey seal pups, various haematological and biochemical parameters were no different among those born to mothers with high OC burdens (and thus exposed to higher contaminant levels through the milk) compared with those born to females with lower burdens (Hall *et al.*, 1997). However, the exposure of pups through the milk may not have been as high as calculated from the expected dose assumed from the lipid intake and growth rate of the pups (Pomeroy *et al.*, 1994), and differences between high and low exposed pups were not sufficiently large for effects to be detected. In a study of free-living bottlenose dolphins, proliferation responses of lymphocytes *in vitro* were correlated with higher OC concentrations (Lahvis *et al.*, 1995).

Figure 10.   Striped dolphins in the Mediterranean suffered mass mortality in 1990 caused by a morbillivirus epidemic. Animals that died during the outbreak had higher organochlorine levels in their blubber than those sampled prior to and after the outbreak. (Photo: J.M. Bompar)

5.2   Biomarkers

The National Academy of Sciences in the US defines biomarkers as 'xenobiotically-induced variation in cellular or biochemical components or processes, structure or functions that are measurable in a biological system or sample' (NRC, 1987). Some of these have already been considered in the preceding sections, notably cytochrome P540 enzymes. The relationship between alterations in biomarkers and specific health effects must be established for these to be useful in assessing the risk to the populations.

Stegeman and Hahn (1994) suggest that cytochrome P450 induction as a biomarker of exposure to inducers, indicating the degree and risk of environmental contamination, may provide a useful molecular tool. It has been repeatedly tested as a marker in many species (Payne *et al.*, 1987), and could be used in marine mammals given further confirmation of recent findings. Indeed Fossi *et al.* (1992), Marsili *et al.* (1998), and Fossi *et al.* (1997) have investigated MFO induction in skin samples from whales, dolphins and fur seals with promising results. If extra-hepatic induction is sufficiently high or very low protein levels can be measured, then more readily obtained tissues and samples from live animals will ensure this technique is not confined to potentially biased samples from dead animals. Troisi and Mason (1997) investigated using cytochrome P420 levels in harbour seal liver (the degraded form of P450) as a biomarker. Total PCB burdens were positively correlated with cytochrome P450, P420 and MFO activity levels, and the authors suggest cytochrome P420 biomarkers may enable poorer quality samples from by-caught animals to be used in monitoring studies.

However, how far induction can be interpreted as indicative of potential toxicity as it does exposure, is not clear. Induction is closely linked to the degree of contamination, but the significance of this finding still needs to be considered with some care. There are various signs that inducers or substrates of P450 are indeed toxic, and that generalised effects may occur as a result, but highly induced P450 is not necessarily associated with a greater risk of toxicity. The effect of long-term low level exposure to CYP1A inducers needs to be investigated further.

Reijnders (1994) defines biomarkers as physiological parameters (such as physical and chemical blood parameters) which provide more direct insight into the response of an organism to be used in a multiple response concept. Experimental studies investigating the effects of exposure on blood chemistry and biochemistry, and steroid hormone levels, can provide useful biomarkers, but these must be used in an integrated way, since data can diverge from normal values for several reasons. For example, in a study by Hall *et al.* (1998) of thyroid hormones in grey seals, significant age-related changes during the lactation period were reported in pups and females respectively, factors which must be accounted for when using these parameters as biomarkers. Nevertheless, many of these approaches outlined offer an important opportunity to determine the risk of contaminant exposure to marine mammals. Biomarkers should ideally be quantitative (dose-dependent), and responses should be distinguishable from natural variations, the more direct the relationship between biomarker, exposure to contaminant and adverse health effect, the more meaningful the marker will be.

The field of research into the potential effects of contaminants on marine mammal populations presents some significant challenges. We still need to develop new and novel ways of monitoring exposure and response. Effort must also be directed towards integrating studies using marine mammal cell lines and laboratory animals with those on captive and free-living marine mammals. We ultimately need to understand how individual responses to contaminants may affect the population dynamics of different species.

REFERENCES

Addison, R.F. and Smith, T.G. (1974) Organochlorine residue levels in Arctic ringed seals: variation with age and sex. *Oikos*, **25**, 335-337.

Addison, R.F. and Brodie, P. (1984) Characterisation of ethoxyresorufin *O*-deethylase in the grey seal, *Halichoerus grypus*. *Journal of Comparative Biochemistry and Physiology*, **79C**, 261-265.

Addison, R., Brodie, P., Edwards, A. and Sadler, M. (1986) Mixed function oxidase activity in the harbour seal *Phoca vitulina* from Sable Is., N.S. *Journal of Comparative Biochemistry and Physiology*, **85C**, 121-124.

Addison, R.F. (1989) Organochlorines and marine mammal reproduction. *Canadian Journal of Fisheries and Aquatic Sciences*, **46**, 360-368.

Addison, R.F. and Smith, T.G. (1998) Trends in organochlorine residue concentrations in ringed seals (*Phoca hispida*) from Holman, Northwest Territories, 1972-91. *Arctic*, **51**, 253-261.

Addison, R.F., Brodie, P.F., Zinck, M.E. and Sergeant, D.E. (1984) DDT has declined more than PCBs in Eastern Canadian seals during the 1970s. *Environmental Science and Technology*, **18**, 935-937.

Aguilar, A. and Borrell, A. (1994a) Abnormally high polychlorinated biphenyl levels in striped dolphins (*Stenella coeruleoalba*) affected by the 1990-1992 Mediterranean epizootic. *Science of the Total Environment*, **154**, 237-247.

Aguilar, A. and Borrell, A. (1994b) Reproductive transfer and variation of body load of organochlorine pollutants with age in fin whales (*Balaenoptera physalus*). *Archives of Environmental Contamination and Toxicology*, **27**, 546-554.

Aguilar, A. and Borrell, A. (1997) *Marine Mammals and Pollutants, An Annotated Bibliography*. Departmento de Biología Animal, Facultad de Biología, Universidad de Baercelona, España. 251pp.

Aguilar, A. and Borrell, A. (1991) Heterogenous distribution of organochlorine contaminants in the blubber of baleen whales – implications for sampling procedures. *Marine Environmental Research*, **31**, 275-286.

Anderson, L.M., Logsdon, D., Ruskie, S., Fox, S.D., Issaq, H.J., Kovatch, R.M. and Riggs, C.M. (1994) Promotion by polychlorinated biphenyls of lung and liver tumors in mice. *Carcinogenesis*, **15**, 2245-2248.

Anderson, S.S., Livens, F.R. and Singleton, D.L. (1990) Radionuclides in grey seals. *Marine Pollution Bulletin*, **21**, 343-345.

Béland, P., DeGuise, S., Girard, C., Lagacé, A., Martineau, S., Michaud, R., Muir, D.C.G., Norstrom, R.J., Pelletier, E., Ray, S. and Shugart, L.R. (1993) Toxic compounds and health and reproductive effects in St. Lawrence beluga whales. *Journal of Great Lakes Research*, **19**, 766-775.

Ballschmiter, K. and Zell, M. (1980) Analysis of polychlorinated biphenyls (PCB) by glass capillary gas chromatography. *Zeitschrift für Analytische Chemie*, **302**, 20-31.

Bergman, A. and Haraguchi, K. (1994) Analysis of PCB and DDE methyl sulfones - persistent metabolites of PCB and DDT in biota. *The Standard*, **1**, 1-3.

Bergman, A., Norstrom, R.J., Haraguchi, K., Kuroki, H. and Beland, P. (1994a) PCB and DDE methyl sulfones in mammals from Canada and Sweden. *Environmental Toxicology and Chemistry*, **13**, 121-128.

Bergman, A., Klasson-Wehler, E. and Kuroki, H. (1994b) Selective retention of hydroxylated PCB metabolites in blood. *Environmental Health Perspectives*, **102**, 464-469.

Bergman, A., Olsson, M. and Reiland, S. (1992) Skull-bone lesions in the Baltic grey seal (*Halichoerus grypus*). *Ambio*, **21**, 517-519.

Bergman, A and Olsson M. (1985) Pathology of Baltic grey seal and ringed seal females with special reference to adrenocortical hyperplasia: Is environmental pollution the cause of a widely distributed disease syndrome? *Finnish Game Research*, **44**, 47-62.

Bignert, A., Gothberg, A., Jensen, S., Litzen, K., Odsjo, T., Olsson, M. and Reutergardh, L. (1993) The need for adequate biological sampling in ecotoxicological investigations: a retrospective study of twenty years pollution monitoring. *Science of the Total Environment*, **128**, 121-139.

Bitman, J., Cecil, H.C. and Harris, S.J. (1972) Biological effects of polychlorinated biphenyls in rats and quail. *Environmental Health Perspectives*, **1**, 145-149.

Bloedow, D.C. and Hayton, W.L. (1976) Effects of lipids on bioavailability of sulfisoxazole acetyl, dicumarol and griseofulvin in rats. *Journal of Pharmacological Science*, **65**, 328-334.

Boon, J.P., Reijnders, P.J.H., Dols, J., Wensvoort, P. and Hillebrand, M.T.J. (1987) The kinetics of individual polychlorinated biphenyl congeners in female harbour seals (*Phoca vitulina*) with evidence for structure-related metabolism. *Aquatic Toxicology*, **10**, 307-324.

Boon, J.P. and Eijgenraam, F. (1988) The possible role of metabolism in determining patterns of PCB congeners in species from the Dutch Wadden sea. *Marine Environmental Research*, **24**, 3-8.

Boon, J.P., Sleiderink, H.M., Helle, M.S., Dekker, M., VanSchanke, A., Roex, E., Hillebrand, M.T.J., Klamer, H.J.C., Govers. B., Pastor, D., Morse, D., Wester, P.G. and DeBoer, J. (1998) The use of a microsomal in vitro assay to study phase I biotransformation of chlorobornanes (toxaphene R) in marine mammals and birds – Possible consequences of biotransformation for bioaccumulation and genotoxicity. *Journal of Comparative Biochemistry and Physiology*, **121C**, 385-403.

Boon, J.P., van Arnhem, E., Jansen, S., Kannan, N., Petrick, G., Duinker, J.C., Reijnders, P.J.H. and Goksyr, A. (1992) The Toxicokinetics of PCBs in Marine Mammals with Special Reference to Possible Interactions of Indiviual Congeners with the Cytochrome P450-dependent Monooxygenase System: an Overview. In: *Persistent Pollutants in the Marine Ecosystem* (Ed. by C.H. Walker, and D.R. Livingstone), pp. 119-159. Pergamon Press, Oxford.

Boon, J.P., Van der Meer, J., Allchin, C.R., Law, R.J., Lungsyr, J., Leonards, P.E.G., Spliid, H., Storr-Hansen, E., McKenzie, C. and Wells, D.E. (1997) Concentration dependent changes of PCB patterns in fish-eating mammals: Structural evidence for induction of cytochrome P450. *Archives of Environmental Contamination and Toxicology*, **33**, 298-311.

Borrell, A. (1993) PCBs and DDTs in blubber of cetaceans from the Northeastern North Atlantic. *Marine Pollution Bulletin*, **26**, 146-151.

Borrell, A. and Aguilar, A. (1990) Loss of organochlorine compounds in the tissues of a decomposing stranded dolphin. *Bulletin of Environmental Contamination and Toxicology*, **45**, 46-53.

Borrell, A., Bloch, D. and Desportes, G. (1995) Age trends and reproductive transfer of organochlorine compounds in long-finned pilot whales from the Faroe Islands. *Environmental Pollution*, **88**, 283-292.

Brandt, I. And Bergman, A. (1987) PCB methyl sulphones and related compounds: Identification of target cells and tissues in different species. *Chemosphere*, **16**, 1671-1676.

Brouwer, A., Blaner, W.S., Kukler, A. and Van den Berg, K.J. (1988) Study on the mechanisms of interference of 3,4,3',4',-tetrachlorobiphenyl with the plasma retinol-binding proteins in rodents. *Chemical and Biological Interactions*, **68**, 203-217.

Brouwer, A. and Van den Berg, K.J. (1986) Binding of a metabolite of 3,4,3'4'tetrachlorobiphenyl to transthyretin reduces serum vitamin A transport by inhibiting the formation of the protein complex carrying both retinol and thyroxin. *Toxicological Applications to Pharmacology*, **85**, 301-312.

Bruhn, R., Kannan, N., Petrick, G., Schulzbull, D.E. and Duinker, J.C. (1995) CB patterns in the harbor popoise – Bioaccumulation, metabolism and evidence for cytochrome P450 PB activity. *Chemosphere*, **31**, 3721-3732.

Burke, M.D.and Mayer, R.T. (1983) Differential effects of phenobarbitone and 3-methyl-colanthrene induction on the hepatic microsomal metabolism and cytochrome P450-binding of phenoxazone and a homologous series of its n-alkyl ethers (alkoxyresorufins). *Chemical and Biological Interactions*,**45**, 243-248.

Calmet, D., Woodhead, D. and André, J.M. (1992) 210Pb, 137Cs and 40K in three species of porpoises caught in the eastern Tropical Pacific Ocean. *Journal of Environmental Radioactivity*, **15**, 153-169.

Colborn, T. and Clement, C. (eds.) (1992) *Chemically-induced alterations in sexual and functional development: the wildlife/human connection.* Advances in Modern Environmental Toxicology, Vol XXI, Princeton Scientific. Publishing Company, Princeton, USA, 403pp.

Colby, R.H., Mattacks, C.A. and Pond, C.M. (1993) The gross anatomy, cellular structure and fatty acid composition of adipose tissue in captive polar bears (*Ursus maritimus*) *Zoological Biology*, **12**, 267-275.

Conney, A.H. (1982) Induction of microsomal enzymes by foreign chemicals and carcinogenesis by polycyclic aromatic hydrocarbons. *Cancer Research*, **42**, 4875-4917.

Dannan, G.A., Porubek, D.J., Nelson, S.D., Waxman, D.J. and Guengerich, F.P. (1986) 17--estradiol 2- and 4-hydroxylation catalysed by human cytochrome P4501A1: a comparison of the activities induced by 2,3,7,8-tetrachlorodibenzo-p-dioxin in MCF-7 cells with those from heterologous expression of the cDNA. *Archives of Biochemistry and Biophysics*, **293**, 342-345.

DeBoer, J., Wester, P.G., Klammer, H.J.C., Lewis, W.E. and Boon, J.P. (1998) Do flame retardants threaten ocean life? *Nature* (London), **394**, 28-29.

De Bruin, A. (1980). *Biochemical Toxicology of Environmental Agents*. Elsevier, Amsterdam.

DeLong, R.L., Gilmartin, W.G. and Simpson, J.G. (1973) Premature births in Californian sea lions: association with organochlorine pollutant residue levels. *Science*, **181**, 1168-1170.

DeSwart, R.L., Ross, P.S., Vedder, L.J., Timmerman, H.H., Heisterkamp, S., Van Loveren, H., Vos, J.G., Reijnders, P.J.H. and Osterhaus, A.D.M.E. (1994) Impairment of immune function in harbor seals (*Phoca vitulina*) feeding on fish from polluted waters. *Ambio*, **23**, 155-159.

Duinker, J.C. and Hillebrand, M.T.J. (1983) Composition of PCB mixtures in biotic and abiotic marine compartments (Dutch Wadden Sea). *Bulletin of Environmental Contamination and Toxicology*, **31**, 25-32.

Duinker, J.C., Hillebrand, M.T.J., Zeinstra, T. and Boon, J.P. (1989) Individual chlorinated biphenyls and pesticides in tissues of some cetacean species from the North Sea and the Atlantic Ocean; tissue distribution and biotransformation. *Aquatic Mammals*, **15**, 95-124.

Fossi, M.C., Marsil, L., Leonzio, C., Notarbartolo Di Sciara, G., Zanardelli, M. and Focardi, S. (1992) The use of non-destructive biomarker in Mediterranean cetaceans: preliminary data on MFO activity in skin biopsy. *Marine Pollution Bulletin*, **24**, 459-461.

Fossi, M.C., Marsili, L., Junin, M., Castello, H., Lorenzani, J.A., Casini, S., Sarelli, C. and Leonzio, C. (1997) Use of nondestructive biomarkers and residue analysis to assess the health status of endangered species of pinnipeds in the southwest Atlantic *Marine Pollution Bulletin*, **34**, 157-162.

Freeman, H.C. and Sanglang, G.B. (1977) A study of the effects of methyl mercury, cadmium, arsenic, selenium and a PCB (Aroclor 1254) on adrenal and testicular steroidgeneses in vitro by the gray seal (*Halichoerus grypus*). *Archives of Environmental Contamination and Toxicology*, **5**, 369-383.

Fuller, G.B. and Hobson, W.C. (1986) Effect of PCBs on reproduction in mammals. In: *PCBs and the Environment Vol II* J.(Ed. by S. Waid). Pp. 101-125. CRC Press, Boca Raton, Florida.

╭ ╌on, M.G., Dodson, J.J., Comba, M.E. and Kaiser, K.L.E. (1990) Congener-specific analysis of the accumulation of polychlorinated biphenyls (PCBs) by aquatic organisms in the maximum turbidity zone of the St. Lawrence Estuary, Quebec, Canada. *Science of the Total Environment*, **97/98**, 739-759.

Gauthier, J.M., Metcalfe, C.D. and Sears, R. (1997) Validation of the blubber biopsy technique for monitoring of organochlorine contaminants in balaenopterid whales. *Marine Environmetal Research*, **43**, 157-179.

Goksøyr, A., Andersson, T., Forlin, L., Snowberger, E.A., Woodin, B.R. and Stegeman, J.J. (1989) Cytochrome P-450 monooxygenase activity and immunochemical properties of adult and foetal piked (minke) whales, Balaenoptera acutorostrata. In: *Cytochrome P450: Biochemistry and Biophysics* (Ed. by I. Schuster), pp. 698-701. Proceedings of the 6th International Conference on Biochemistry and Biophysics of Cytochrome P450, July, 1988, London.

Goksøyr, A., Andersson, T., Forlin, L., Stenersen, J., Snowberger, E.A., Woodin, B.R. and Stegemn, J.J. (1988) Xenobiotic and steriod metabolism in adult and foetal piked (minke) whales, *Balaenoptera acutorostrata*. *Marine Environmental Research*, **24**, 9-12.

Goksøyr, A., Beyer, J., Larsen, H.E., Andersson, T. and Forlin, L. (1992) Cytochrome P450 in seals: monooxygenase activities, immunochemical cross-reactions and response to phenobarbital treatment. *Marine Environmental Research*, **34**, 113-116.

Goksøyr, A., Solbakken, J., Tarlebo, J. and Klungsoyr, J. (1986) Initial characterisation of the hepatic microsomal cytochrome P-450 system of the piked (minke) whale, *Balaenoptera acutorostrata*. *Marine Environmental Research*, **19**, 185-187.

Gonzalez, F.J. (1989) The Molecular Biology of the Cytochromes P450s. *Pharmacological Review*, **40**, 243-288.

Graves, P.E., Elhag, G.A., Ciaccio, P.J., Bourque, D.P. and Halpert, J.R. (1990) cDNA and deduced amino acid sequences of a dog hepatic cytochrome P450IIIB responsible for the metabolism of 2,2'4,4'5,5'-hexachlorobiphenyl. *Archives of Biochemistry and Biophysics*, **281**, 106-115.

Hahn, M.E., Poland, A., Glover, E. and Stegeman, J.J. (1992) The Ah receptor in marine animals: phylogenetic distribution and relationship to P450IA inducibility. *Marine Environmental Research*, **34**, 87-90.

Hahn, M.E., Steiger, G.H., Calambokidas, J., Shaw, S.D. and Stegeman, J.J. (1991) Immunochemical characterization of the cytochrome P450 in harbor seals (*Phoca vitulina*). Ninth Biennial Conference on the Biology of Marine Mammals, (Abstract), Society for Marine Mammalogy. 30pp.

Hall, A.J., Green, N.J.L., Jones, K.C., Pomeroy, P.P. and Harwood, J. (1998) Thyroid hormones as biomarkers in grey seals. *Marine Pollution Bulletin*, **36**, 424-428.

Hall, A., Pomeroy, P., Green N., Jones, K. and Harwood, J. (1997) Infection, haematology and biochemistry in grey seal pups exposed to chlorinated biphenyls. *Marine Environmental Research*, **43**, 81-98.

Hall, A.J., Duck, C.D., Law, R.J. and Allchin, C. (1999) Organochlorine contaminants in Caspian and harbour seal blubber. *Environmental Pollution*, **106** (2), 203-212.

Hall, A.J., Law, R.J., Wells, D.E., Harwood, J., Ross, H.M., Kennedy, S., Allchin, C.R., Campbell, L.A. and Pomeroy, P.P. (1992a) Organochlorine levels in common seals (*Phoca vitulina*) that were victims and survivors of the 1988 phocine distemper epizootic. *Science of the Total Environment*, **115**, 145-162.

Hall, A.J., Pomeroy, P.P. and Harwood, J. (1992b) The descriptive epizootiology of phocine distemper in the UK during 1988/89. *Science of the Total Environment*, **115**, 31-44.

Haraguchi, K., Athanasiadou, M., Bergman, A., Hovander, L. and Jensen, S. (1992) PCB and PCB methyl sulfones in selected groups of seals from Swedish waters. *Ambio*, **21**, 546-549.

Harding, G.C., LeBlanc, R.J., Vass, W.P., Addison, R.F., Hargrave, B.T., Pearre, S., Dupuis, A. and Brodie, P.F. (1997) Bioaccumulation of polychlorinated biphenyls (PCBs) in the marine pelagic food web, based on a seasonal study in the southern Gulf of St. Lawrence, 1976-1977. *Marine Chemistry*, **56**, 145-179.

Helle, E., Olsson, M. and Jensen, S. (1976) PCB levels correlated with pathological changes in seal uteri. *Ambio*, **5**, 261-263.

Hemming, H., Flodström, S., Warngard, L., Bergman, A., Kronevi, T., Nordgren, I. and Ahlborg, U.G. (1993) Relative tumour promoting activity of three polychlorinated biphenyls in rat liver. *European Journal of Pharmacology*, **248**, 163-174.

Hutchinson, J.D. and Simmonds, M.P. (1994) Organochlorine contamination in pinnipeds. *Review of Environmental Contamination and Toxicology*, **136**, 123-167.

James, M.O. and Kleinow, K.M. (1994) Trophic Transfer of Chemicals in the Aquatic Environment. In: *Aquatic Toxicology: Molecular, Biochemical and Cellular Aspects* (Ed. by D.C. Malins and G.K. Ostrander), pp. 1-35. Lewis Publishers, Boca Raton.

Jansen, H.T., Cook, P.S., Porcelli, J., Lui, T.C. and Hansen, L.G. (1993) Estrogenic and antiestrogenic actions of PCBs in female rat: in vitro and in vivo studies. *Reproductive Toxicology*, **7**, 237-248.

Jansson, B., Jensen, S., Olsson, M., Renberg, G., Sundstrom, G. and Vaz, R. (1975) Identification by GC-MS of phenolic metabolites of PCB and p,p'-DDE isolated from Baltic guillemot and seal. *Ambio*, **4**, 93-97.

Jarman, W.M., Norstrom, R.J., Muir, D.C.G., Rosenberg, B., Simon, M. and Baird, R.W. (1996) Levels of organochlorine compounds including PCDDs and PCDFs in the blubber of cetaceans from the west coast of North America. *Marine Pollution Bulletin*, **32**, 426-436.

Jensen, S. and Jansson, B. (1976) Methyl sulfone metabolites of PCB and DDE. *Ambio.*, **5**, 257-260.

Jönsson, C.-J., Lund, B.-O. and Brandt, I. (1993) Adrenocorticolytic DDT metabolites: Studies in mink, *Mustela vison* and otter, *Lutra lutra. Ecotoxicology*, **2**, 41-53.

Kannan, K., Corsolini, S., Focardi, S., Tanabe, S., Tatsukawa, R. (1996) Accumulation patterns of butyltin compounds in dolphin, tuna and shark collected from Italian coastal waters. *Archives of Environmental Contamination and Toxicology*, **31**, 19-23.

Kannan, K., Tanabe, S., Borrell, A., Aguilar, A., Focardi, S., Tatsukawa, R. (1993) Isomer specific analysis and toxic evaluation of polychlorinated biphenyls in striped dolphins affected by an epizootic in the western Mediterranean Sea. *Archives of Environmental Contamination and Toxicology*, **25**, 227-233.

Kawano, M., Inoue, T., Wada, T., Hidaka, H. and Tatsukawa, R. (1988) Bioconcentration and residue patterns of chlordane compounds in marine animals: invertebrates, fish, mammals and seabirds. *Environmental Science and Technology*, **22**, 792-797.

Kemper, C., Gibbs, P., Obendorf, D., Marvanek, S. and Lenghaus, C. (1994) A review of heavy metal and organochlorine levels in marine mammals in Australia. *Science of the Total Environment*, **154**, 129-139.

Kendall, M.D., Safieh, B., Harwood, J., Pomeroy, P.P. (1992) Plasma thymulin concentrations, the thymus and organochlorine contaminant levels in seals infected with phocine distemper virus. *Science of the Total Environment*, **115**, 133-144.

Kimbrough, R.D. (1987) Human health effects of polychlorinated biphenyls (PCBs) and polybrominated biphenyls (PBBs). *Annual Review of Pharmacological Toxicology*, **27**, 87-111.

Klaassen, C.D. and Rozman, K. (1991) Absorption, Distribution and Excretion of Toxicants. In: *Casarett and Doull's Toxicology. The Basic Science of Poisons* (Ed. by M.O. Amdur, J. Doull and C. Klaassen), pp. 50-87. Pergamon Press, New York.

Klasson-Wehler, E., Bergman, A., Darnerud, P.O. and Wachtmeister, C.A. (1989) 3,3',4,4'-Tetrachlorobiphenyl: excretion and tissue retention of hydroxylated metabolites in the mouse. *Drug Metabolism Disposal*, **17**, 441-448.

Kleivane, L. and Skaare, J.U. (1998) Organochlorine contaminants in northeast Atlantic minke whales (*Balaenoptera acutorostrata*). *Environmental Pollution*, **101**, 231-239.

Koeman, J.H. and Van Genderen, H. (1966) Some preliminary notes on residues of chlorinated hydrocarbon insecticides in birds and mammals in the Netherlands. *Journal of Applied Ecology*, **3**(Suppl), 99-106.

Kucklick, J.R., McConnell, L.L., Bidleman, T.F., Ivanov, G.P. and Walla, M.D. (1993) Toxaphene contamination in Lake Baikal's water and food web. *Chemosphere*, **27**, 2017-2026.

Kuehl, D. and Haebler, R. (1995) Organochlorine, organobromine, metal and selenium residues in bottlenose dolphins (*Tursiops truncatus*) collected during an unusual mortality event in the Gulf of Mexico. 1990. *Archives of Environmental Contamination and Toxicology*, **28**, 494-499.

Kuiken, T., Höfle, U., Bennett, P.M., Allchin C.R., Kirkwood, J.K., Baker, J.R., Appleby, E.C., Lockyer, C.H., Walton, M.J. and Sheldrick M.C. (1993) Adrenocortical hyperplasia disease and chlorinated hydrocarbons in harbour porpoises (*Phocoena phocoena*). *Marine Pollution Bulletin*, **26**, 440-446.

Lahvis, G.P., Wells, R.S., Kuehl, D.W., Steward, J.L., Rhinehart, H.L. and Via, C.S. (1995) Decreased lymphocyte responses in free-ranging bottlenose dolphins (*Tursiops truncatus*) are associated with increased concentrations of PCBs and DDT in peripheral blood. *Environmental Health Perspectives*, **103**(Suppl4), 62-67.

Landers, J.P. and Bunce, N.J. (1991) The Ah receptor and the mechanism of dioxin toxicity. *Biochemisry Journal*, **276**, 349-353.

Law, R.J., Jones, B.R., Baker, J.R., Kennedy, S., Milne, R. and Morris, R.J. (1992) Trace metals in the livers of marine mammals from the Welsh coast and the Irish Sea. *Marine Pollution Bulletin*, **24**, 296-304.

Law,R.J., Blake, S.J., Jones, B.R. and Rogan, E. (1998) Organotin compounds in liver tissue of harbour porpoises (*Phocoena phocoena*) and grey seals (*Halichoerus grypus*) from the coastal waters of England and Wales. *Marine Pollution Bulletin*, **36**, 241-247.

Lemesh, R.A. (1992) Polychlorinated biphenyls - An overview of metabolic toxicologic and health consequences. *Veterinary and Human Toxicology*, **34**, 256-260.

Luckas, B., Vetter, W., Fischer, P., Heidemann, G. and Plotz, J. (1990) Characteristic chlorinated hydrocarbon patterns in the blubber of seals from different marine regions. *Chemosphere*, **21**, 13-19.

Lund, B-O. (1994) In vitro adrenal bioactivation and effects on steroid metabolism of DDT, PCBs and their metabolites in the grey seal (*Halichoerus grypus*). *Environmental Toxicology and Chemistry*, **13**, 911-917.

Lund, B-O., Bergman, A. and Brandt, I. (1988) Metabolic activation and toxicity of a DDT-metabolite - 3-methylsulphonyl-DDE - in the adernal zona fasciculata in mice. *Chemical and Biological Interactions*, **65**, 25-40.

Marsili, L., Casini, C., Marini, L., Regoli, A., Focardi, S. (1997) Age, growth and organochlorines (HCB, DDTs and PCBs) in Medieterranean striped dolphins (*Stenella coeruleoalba*) stranded in 1988-1994 on the coasts of Italy. *Marine Ecology Progress Series*, **151**, 273-282.

Marsili, L., Fossi, M.C., diScaria, G.N., Zanardelli, M., Nani, B., Panigada, S. and Focardi, S. (1998) Relationship between organochlorine contaminants and mixed function oxidase activity in skin biopsy specimens of Mediterranean fin whales (*Balaenoptera physalus*). *Chemosphere*, **37**, 1501-1510.

Martineau, D., Bland, P., Desjardins, C. and Lagace, A. (1987) Levels of organochlorine chemicals in tissues of beluga whales (*Delphinapterus leucas*) from the St. Lawrence estuary, Quebec, Canada. *Archives of Environmental Contamination and Toxicology*, **16**, 137-147.

Martineau, D., DeGuise, S., Fournier, M., Shugart, L., Girard, C., Lagace A., and Bland, P. (1994) Pathology and toxicology of beluga whales from the St. Lawrence estuary, Quebec, Canada – Past, present and future. *Science of the Total Environment*, **154**, 201-215.

McKim, J.M. and Nichols J.W. (1994) Use of physiologically based toxicokinetic models in a mechanistic approach to aquatic toxicology. In: *Aquatic Toxicology: Molecular, Biochemical and Cellular Aspects* (Ed. by D.C. Malins and G.K. Ostrander), pp. 469-519. Lewis Publishers, Boca Raton.

Meador, J. P., Varanasi, U., Robisch, P.A. and Chan, S.L. (1993) Toxic metals in pilot whales (*Globicephala melena*) from strandings in 1986 and 1990 on Cape Cod, Massachusetts. *Canadian Journal of Fisheries and Aquatic Sciences*, **50**, 2698-2706.

Menzer, R.E. (1991) Water and Soil Pollutants. In: *Casarett and Doull's Toxicology. The Basic Science of Poisons* (Ed. by M.O. Amdur, J. Doull and C. Klaassen). Pp. 872-904. Pergamon Press, New York.

Mio, T., Sumino, K. and Mizutani, T. (1976) Sulfur-containing metabolites of 2,2'5,5'-tetra-chlorobiphenyl, a major component of commercial PCBs. *Chemical Pharmacology Bulletin*, **24**, 958-1960.

Mortensen, P. Bergman, A., Bignert, A., Hansen H-J., Hrknen, T and Olsson, M. (1992) Prevalence of skull lesions in harbour seals (*Phoca vitulina*) in Swedish and Danish museum collections, 1835-1988. *Ambio*, **21**, 520-524.

Mossner, S. and Ballshmiter, K. (1997) Marine mammals as global pollution indicators for organochlorines. *Chemosphere*, **34**, 1285-1296.

Muir, D.C., Norstrom, R.J. and Simon, M. (1988) Organochlorine contaminants in Arctic marine food chains: accumulation of specific polychlorinated biphenyls and chlordane-related compounds. *Environmental Science and Technology*, **22**, 1071-1079.

Muir, D.C.G., Wagemann, R., Hargrave, B.T., Thomas, D.J., Peakall, D.B. and Norstrom, R.J. (1992a) Arctic marine ecosystem contamination. *Science of the Total Environment*, **122**, 75-134.

Muir, D.C.G., Ford, C.A., Grift, N.P., Stewart, R.E.A. and Bidleman, T.F. (1992b) Organochlorine contaminants in narwhal (*Monodon monoceros*) from the Canadian Arctic. *Environmental Pollution*, **75**, 307-316.

National Research Council. (1987) Committee on Biological Markers. *Environmental Health Perspectives*, **74**, 3-9.

Nebert, D.W., Adesnik, M., Coon, M.J., Estabrook, R.W., Gonzalez, F.J., Guengerion, F.P., Gunsalus, I.C., Johnson, E.F., Kemper, B., Levin, W., Phillips, I.R., Sato, R. and Waterman, M.R. (1987) The P450 gene superfamily: recommended nomenclature. *DNA* , **6**, 1-8.

Nelson, A.A. and Woodard, G. (1949) Severe adrenal cortical atrophy (cytotoxic) and hepatic damage produced in dogs by feeding 2,2-bis(parachloro-phenyl)-1,1-dichloroethane (DDD or TDE). *Archives of Pathology*, **48**, 387-394.

Nelson, D.R., Kamataki, T., Waxman, D.J., Guengerich, F.P., Estabrook, R.W., Feyereisen, R., Gonzalez, F.J., Coon, M.J., Gunsalus, I.C., Gotoho, O., Okuda, K. and Nebert, D.W. (1993) The P450 superfamily - update on new sequences, gene mapping, accession numbers, early trivial names of enzymes and nomenclature. *DNA Cell* Biology, **12**, 1-12.

Ness, D., Schantz, S., Moshtaghian, J. and Hansen, L. (1993) Effects of perinatal exposure to specific PCB congeners on thyroid hormone concentrations and thyroid histology in the rat. *Toxicological Letters*, **68**, 311-323.

Newton, I., Bogan, J.A. and Haas, M.B. (1989) Organochlorines and mercury in the eggs of British Peregrines, *Falco peregrinus*. *Ibis*, **131**, 355-376.

Niimi, A.J. and Oliver, B.G. (1988) Influence of molecular weight and molecular volume on dietary absorption efficiency of chemicals by fishes. *Canadian Journal of Fisheries and Aquatic Sciences*, **45**, 222-227.

Norn, K. and Meironyt, D. (1998) Contaminants in Swedish human milk. Decreasing levels of organochlorine and increasing levels of organobromine compounds. *Organohalogen compounds*, **38**, 1-4.

Norstrom, R.J. and Muir, D.C.G. (1994) Chlorinated hydrocarbon contaminants in arctic marine mammals. *Science of the Total Environment*, **154**, 107-128.

Norstrom, R.J., Muir, D.C.G., Ford, C.A., Simon, M., MacDonald, C.R. and Bland, P. (1992) Indications of P450 monooxygenase activities in Beluga *Delphinapterus leucas* and narwhal *Monodon monoceros* from patterns of PCB, PCDD and PCDF accumulation. *Marine Environmental Research*, **34**, 267-272.

O'Shea, T.J. (1999) Environmental Contaminats and Marine Mammals. In: *Biology of Marine Mammals* (Ed. by J.E. Reynolds III and S.A. Rommel), pp. 485-563. Smithsonian Institution Press, Washington DC. 578pp.

O'Shea, T.J. and Brownell, R.L.Jr. (1998) California sea lion (*Zalophus californianus*) populations and DDT contamination. *Marine Pollution Bulletin*, **36**, 159-164.

O'Shea, T.J. and Brownell, R.L.Jr. (1994) Organochlorine and metal contaminants in baleen whales – A review and evaluation of conservation implications. *Science of the Total Environment*, **154**, 179-200.

Parke, A. (1985) The role of cytochrome P450 in the metabolism of pollutants. *Marine Environmental Research*, **17**, 97-100.

Payne, J.F., Fancey, L.L., Rahimtula, A.D. and Porter, E.L. (1987) Review and perspective on the use of mixed-function oxygenase enzymes in biological monitoring. *Journal of Comparative Pharmacology and Physiology*, **86C**, 233-245.

Pedersen, S.B., Hansen, P.S., Lund, S., Andersen, P.H., Odgaad, A. and Richelsen, B. (1996) Identification of estrogen receptors and estrogen receptor messenger RNA in human adipose tissue. *European Journal of Clinical Investigations*, **26**, 262-269.

Poland, A. and Knutson, J.C. (1982) 2,3,7,8-tetrachlorodibenzo-p-dioxin and related halogenated aromatic hydrocarbons: examinations of the mechanism of toxicity. *Annual Review of Pharmacology and Toxicology*, **22**, 517-520.

Pomeroy, P.P., Green, N., Hall, A.J., Walton, M., Jones, K. and Harwood, J. (1996) Congener specific exposure of grey seal (*Halichoerus grypus*) pups to chlorinated biphenyls during lactation. *Canadian Journal of Fisheries and Aquatic Sciences*, **53**, 1526-1534.

Pond, C.M. and Mattacks, C.A. (1985) Body mass and natural diet as determinants of the number and volume of adipocytes in eutherian mammals. *Journal of Morphology*, **185**, 183-193.

Reijnders, P.J.H. (1980) Organochlorine and heavy metal residues in harbour seals from the Wadden Sea and their possible effects on reproduction. *Netherlands Journal of Sea Research*, **14**, 30-45.

Reijnders, P.J.H. (1984) Man-induced environmental factors in relation to fertility changes in pinnipeds. *Environmental Conservation*, **11**, 61-65.

Reijnders, P.J.H. (1986) Reproductive failure in common seals feeding on fish from polluted coastal waters. *Nature* (London), **324**, 456-457.

Reijnders, P.J.H. (1988) Ecotoxicology perspectives in marine mammology: research principles and goals for a conservation policy. *Marine Mammal Science*, **4**, 91-102.

Reijnders, P.J.H. (1994) Toxicokinetics of chlorobiophenyls and associated physiological responses in marine mammals, with particular reference to their potential for ecotoxicological risk assessment. *Science of the Total Environment*, **154**, 229-236.

Reijnders, P.J.H. and Brasseur, S.M.J.M. (1992) Xenobiotic induced hormonal and associated developmental disorders in marine organisms and related effects in humans: An overview. In: *Chemically-induced alterations in sexual and functional development: the wildlife/human connection* (Ed. by T. Colborn and C. Clement). Advances in Modern Environmental Toxicology, Vol XXI, pp. 159-174. Princeton Scientific. Publishing Company, Princeton, USA.

Ross, P.S., DeSwart, R.L., Reijners, P.J.H., Loveren, H.V., Vos, J.G. and Osterhaus, A.D.M.E. (1995) Contaminant-related suppression of delayed-type hypersensitivity and antibody responses in harbour seals fed herring from the Baltic Sea. *Environmental Health Perspectives*, **103**, 162-167.

Safe, S.H. (1994) Polychlorinated biphenyls (PCBs): environmental impact, biochemical and toxic responses and implications for risk assessment. *Critical Reviews in Toxicology*, **24**, 87-149.

Salata, G.G., Wade, T.L., Sericano, J.L., Davis, J.W. and Brooks. J.M. (1995) Analysis of Gulf of Mexico bottlenose dolphins for organochlorine pesticides and PCBs. *Environmetal Pollution*, **88**, 167-175.

Samuels, E.R., Cawthorn, M., Lauer, B.H. and Baker, B.E. (1970) Strontium-90 and cesium-137 in tissues of fin whales (*Balaenoptera physalus*) and harp seals (*Phagophilus groenlandicus*). *Canadian Journal of Zoology*, **48**, 267-269.

Shain, W., Bush, B. and Seegal, R. (1991) Neurotoxicity of polychlorinated biphenyls: Structure activity relationship of individual congeners. *Toxicological Applications to Pharmacology*, **111**, 33-42.

Sipes, I.G. and Gandolfi, A.J. (1991) Biotransformation of Toxicants. In: *Casarett and Doull's Toxicology. The Basic Science of Poisons* (Ed. by M.O. Amdur, J. Doull and C. Klaassen), pp. 88-126. Pergamon Press, New York.

Stegeman, J.J. and Hahn, M.E. (1994) Biochemistry and Molecular Biology of Monooxygenases: Current Prespectives on Forms, Functions, and Regulation of Cytochrome P450 in Aquatic Species. In: *Aquatic Toxicology: Molecular, Biochemical and Cellular Aspects* (Ed. by D.C. Malins and G.K. Ostrander), pp. 87-206. Lewis Publishers, Boca Raton.

Subramanian, A., Tanabe, S., Tatsukawa, R., Saito, S. and Miyazaki, N. (1987) Reduction in the testosterone levels by PCBs and DDE in Dall's porpoises of northwestern North Pacific. *Marine Polltion Bulletin*, **18**, 643-646.

Tanabe, S., Kannan, N., Subramanian, A., Watanabe, S. and Tatsukawa, R. (1987) Highly toxic coplanar PCBs: occurrence, source, persistency and toxic implications to wildlife and humans. *Environmental Pollution*, **47**, 147-163.

Tanabe, S., Kumaran, P., Iwata, H., Tatsukawa, R. and Miyuazaki, N. (1996) Enantiomeric ratios of x-hexachlorocyclohexane in blubber of small cetaceans. *Marine Pollution Bulletin*, **32**, 27-31.

Tanabe, S., Mori, T., Tatsukawa, R. and Miyazaki, N. (1983) Global pollution of marine mammals by PCBs, DDTs and HCHs (BHCs). *Chemosphere*, **12**, 1269-1275.

Tanabe, S., Tatsukawa, R. Tanaka, H., Maruyama, K., Miyuazaki, N. and Fujiyama, T. (1981) Distribution and total burdens of chlorinated hydrocarbons in bodies of striped dolphins (*Stenella coeruleoalba*). *Agricultural Biology and Chemistry*, **45**, 2569-2578.

Tanabe, S., Watanabe, S., Kan, H. and Tatsukawa, R. (1988) Capacity and mode of PCB metabolism in small cetaceans. *Marine Mammal Science*, **4**, 103-124.

Tatsukawa, R. (1992) Contamination of chlorinated organic substances in the ocean ecosystem. *Water Science and Technology*, **25**, 1-8.

Timbrell, J.A. (1991) *Principles of Biochemical Toxicology*. Taylor and Francis, London.

Troisi, G.M., Haraguchi, K., Simmonds, M.P. and Mason, C.F. (1998) Methyl sulphone metabolites of polychlorinated biphenyls (PCBs) in cetaceans from the Irish and the Aegean Seas. *Archives of Environmental Contamination and Toxicology*, **35**, 121-128.

Troisi, G.M. and Mason, C.F. (1997) Cytochromes P450, P420 and mixed function oxidases as biomarkers of polychlorinated biphenyl (PCB) exposure in harbour seals (*Phoca vitulina*). *Chemosphere*, **35**, 1933-1946.

Urabe, H., Kogda, H. and Asahi, M. (1979) Present status of Yusho patients. *Annals of the New York Academy of Science*, **320**, 273-276.

Vos, J.G. (1972) Toxicity of PCBs for mammals and birds. *Environmental Health Perspectives*, **1**, 105-117.

Wade, T.L., Chambers, L., Gardinali, P.R., Sericano, J.L., Jackson, T.J., Tarpley, R.J. and Suydam, R. (1997) Toxaphene, PCB, DDT, and chlordane analysis of beluga whale blubber. *Chemosphere*, **34**, 1351-1357.

Wagemann, R. and Muir, D.C.G. (1984) Concentrations of heavy metals and organochlorines in marine mammals of Northern waters: overview and evaluation. *Canadian Technical Reports of Fisheries and Aquatic Sciences*, **1279**, 1-97.

Watanabe, S., Shimada, T., Nakamura, S., Nishiyama, N., Yamashita, N., Tanabe, S. and Tatsukawa, R. (1989) Specific profile of liver microsomal cytochrome P-450 in dolphin and whales. *Marine Environmental Research*, **27**, 51-65.

Waxman, D., Dannan, G.A. and Guengerich, F.P. (1985) Regulation of rat hepatic cytochrome P-450: Age-dependent expression, hormonal imprinting and xenobiotic inducibility of sex-specific isoenzymes. *Biochemistry*, **24**, 4409-4417.

Wells, D.E., Campbell, L.A., Ross, H.M., Thompson, P.M. and Lockyer, C.H. (1994) Organochlorine residues in harbour porpoise and bottlenose dolphins stranded on the coast of Scotland 1988-1991. *Science of the Total Envionment*, **151**, 77-99.

White, R.D., Hahn, M.E., Lockhart, W.L. and Stegeman, J.J. (1994) Catalytic and immunochemical characterization of hepatic microsomal cytochromes P450 in Beluga whale (*Delphinapterus leucas*). *Toxicological Applications to Pharmacology*, **126**, 45-57.

Williams, R.T. (1959) *Detoxication Mechanisms: The metabolism and detoxification of drugs, toxic substances and other organic compounds*. John Wiley, New York.

Wolkers, J., Burkow, I.C., Lydersen, C., Dahle, S., Monshouwer, M., Witkamp, R.F. (1998) Congener specific PCB and polychlorinated camphene (toxaphene) levels in Svalbard ringed seals (*Phoca hispida*) in relation to sex, age, condition and cytochrome P450 enzyme activity. *Science of the Total Environment*, **216**, 1-11.

Zakharov, V.M. and Yablokov, A.V. (1990) Skull asymmetry in the Baltic grey seal: effects of environmental pollution. *Ambio*, **19**, 266-269.

Zeisler, R., Demiralp, R., Koster, B.J., Becker, P.P., Burrow, M., Ostapczuk, P. and Wise, S.A. (1993) Determination of inorganic constituents in marine mammal tissues. *Science of the Total Environment*, **140**, 365-386.

Chapter 16

Cetaceans and Humans: Influences of Noise

[1]BERND WÜRSIG and [2]PETER G.H. EVANS

[1]*Marine Mammal Research Program, Texas A&M University, 4700 Ave. U, Galveston, TX, 77551, USA. E-mail: wursigb@tamug.tamu.edu;* [2]*Department of Zoology, University of Oxford, South Parks Road, Oxford, OX1 3PS, UK. E-mail: peter.evans@zoo.ox.ac.uk*

1. INTRODUCTION

Whales, dolphins and porpoises face a multitude of problems at the hands of humans. These include incidental kills in fishing nets and lines, direct kills for food or bait, debilitation (and death) due to pollution or other forms of habitat degradation (see Chapters 12-15 in this volume), and the mainly unknown potential effects of larger-scale ecological changes such as human-induced global climate change (Brownell *et al.*, 1989; Tynan and DeMaster, 1997). But besides these problems, there is another often difficult to measure cause of decline in health of habitat for marine mammals. This is human-induced noise, and its potential effects were not investigated until relatively recently, in the past thirty years or so. A rather thorough discussion of noise impacts on marine mammals is provided by Richardson *et al.* (1995); we here summarise the situation for cetaceans, and provide several examples and possible mitigation efforts.

2. CETACEAN USE OF SOUND AND THE MODERN UNDERWATER NOISE ENVIRONMENT

Cetaceans are all acoustic animals par excellence, and we therefore expect that underwater anthropogenic noises might be particularly

Marine Mammals: Biology and Conservation, edited by
Evans and Raga, Kluwer Academic/Plenum Publishers, 2002

disruptive to their lives (see Chapters 4 and 8). However, cetaceans are also large-brained, generally adaptive creatures, and while considering influences of habitat change we would be amiss not to take this adaptability into account. Despite all that has been written on sound influences, we still do not know very much about how sounds can affect long-term behaviour, well-being, reproduction, and - overall - the fitness of individuals and the health of populations.

As a general rule, the larger the mammal, the more probable it is to have sensitive low frequency hearing, usually at the expense of acute high frequency hearing. We humans hear reasonably well from about 20 Hz (cycles per second) up to as high as 15,000 - 20,000 Hz (or 15-20 kHz). The lowest frequency that we can hear is called the upper limit of infrasound; and the highest frequency is the lower limit of ultrasound. In general, human females have higher frequency hearing capability than males. Everybody loses some high frequency hearing with age, and men tend to do so more rapidly and drastically than women. The first author of this paper is fifty years old, and can hear to only about 10 kHz. Elephants and rhinoceroses have low frequency hearing, stretching into infrasound. Large whales, especially blue and fin whales, produce such sounds and are probably sensitive to them as well; we have no evidence for infrasound production or hearing in any toothed whale species (review by Richardson *et al.*, 1995). On the other hand, toothed whales have sensitive hearing at mid and very high frequencies but not at low frequencies. They make and receive clicks that reach far into ultrasound and that are used largely for echolocation. We have no good evidence that any baleen whales can make ultrasounds or can echolocate (for a more detailed description, see Chapter 4).

The higher the frequency, the more sound attenuates with distance. In other words, a sound of a particular intensity at 100 Hz might reach to a distance of one kilometre, while a sound of the same intensity but at 10 kHz might only reach for 150 metres (for a more thorough review, see Malme, 1995). So, we expect to find low frequency sounds being used as long-distance communication or contact calls, and higher frequency sounds being used for short-range communication and echolocation. This is exactly what happens, with blue and fin whales emitting moans that reach into infrasound and that can be heard for up to several hundred kilometres in some situations in deep water (Payne and McVay, 1971); and with dolphins whistling to each other and echolocating at relatively close distances within generally one kilometre (Au, 1993, Norris *et al.*, 1994). Large toothed whales, such as pilot whales, appear to have lower frequencies of sound projection and optimum hearing than do the smaller dolphins and porpoises, but there is much variability in this among species.

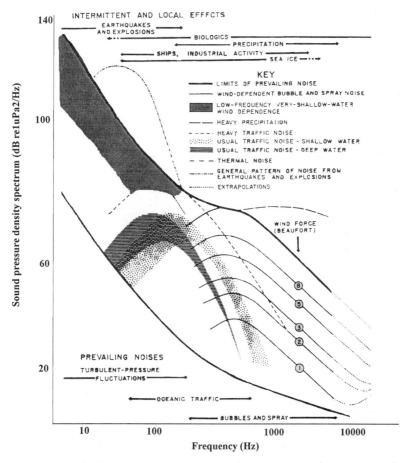

Figure 1. Generalised ambient noise spectra from a variety of sources: geophysical, weather-related, biological, and human-caused. This was originally compiled by Wenz (1962) and has been re-plotted to presently used units of sound intensity on the ordinate.

The sea has always been a noisy place, but is even more so today (Wenz, 1962; Fig. 1). In the pre-industrial ocean, there were wave noises; the din of rain on the surface; sounds of earthquakes and sea ice; and biological sounds of croaker fishes, pistol shrimps, and marine mammals. The general ambient, or background, noise due to these multiple sources usually would have been about 50-70 dB at 100Hz on a standardised measurement scale, but louder near the surface during rainstorms. However, now in many regions, there is the constant din of major shipping lanes, explosions and pile driving near shore, oil and gas exploration activities, and fishing and pleasure craft (Fig. 2). There is also an ever-greater reliance on acoustics for naval operations (sonar) and the gathering of some oceanographic data, such as synoptic basin-wide temperature measurements (Munk *et al.*, 1994).

Figure 2.   Examples of vessels which can disturb cetaceans through the sounds they produce: a) Seismic vessel; b) Supertanker; c) Warship on NATO exercise; and d) Whale watching vessel. (Photos: a, b, c: P.G.H. Evans; d: W. Rossiter)

All in all, the ocean is tremendously noisy especially in the northern hemisphere, where most shipping lanes occur. Indeed, Ross (1976) estimated that between 1950 and 1975, ambient noise had risen by 10 dB in areas where shipping noise dominates, and he predicted it would rise a further 5 dB by the end of the twentieth century as shipping traffic increased. Of course, there is more noise near large harbours and generally more noise in surface waters than in ocean depths. However, in deep water, there tends to be a mid-water depth where water temperature and pressure interact to create a sound speed minimum, about 1 km below the surface in the tropics but only 100 m or so below the surface in cold waters. This area of sound speed minimum channels sounds (by bending waves of sound into it), and therefore this is an especially noisy place for low frequency sounds that travel for great distances (Malme, 1995). It has been speculated that blue and fin whales might increase their communication distances by calling and listening while in this channel, utilising "cylindrical sound spreading" instead of the general "spherical sound spreading" of the deep ocean (Payne and McVay, 1971), but this assertion remains untested. Conversely, such species may also move out of these sound channels to avoid loud noises.

3.   SOUND DISTURBANCE

Anthropogenic noises can disrupt the lives of animals in several ways, and cetaceans are no exception: sounds can 1) frighten them or make them curious, but in either way change their behaviour; 2) compete with communication signals or echolocation, by sound masking, and thus decrease the efficiency of finding food, mating, caring for young, or avoiding predators; cause 3a) physical effects such as stress leading to changes in hormone levels and perhaps lowered immunity from diseases, 3b) a temporary loss of hearing (or temporary threshold shift) or permanent damage to hearing, or 3c) - in the worst of cases with explosions or other loud noises that also send shock waves - possible death (Richardson and Würsig, 1997). We have tentative information about No. 1 above, at least for changes in behaviour in the short term. We know even less about long-term behavioural changes, such as threshold of sound intensity that might cause abandonment of an area. Almost nothing is known about noise-induced stress in marine mammals. Information about hearing losses, and more debilitating chronic or catastrophic effects is limited but some relevant studies are underway. Let us begin with several examples of industrial sounds changing behaviour.

One of the most dramatic sound-induced behavioural changes of cetaceans occurs in springtime near ice, when beluga whales and narwhals hear icebreakers in the distance. One such study, which relied on thousands of animals of both species being censused from the air during a three-year research project, found that the animals reacted to ships passing by at distances as great as about 90 km (LGL and Greeneridge, 1986; Finley *et al.*, 1990; Cosens and Dueck, 1993). They moved away from the ships, and were almost totally absent within about 20 km of the ships' tracks. Belugas and narwhals are generally less sensitive in summer, but it is unclear whether this is due to less confining ice in the area or perhaps because by summer they have habituated to vessel noises. One of us (BW) has repeatedly seen bowhead whales in summer being "more skittish" of industrial noises when near a confining shoreline than when in open water, and perhaps the ice at least in part enhances such a nervousness factor for the belugas and narwhals.

Bowhead whales have been studied intensively relative to oil exploration and development activities, with many of these data reviewed in a book on bowheads by Burns *et al.* (1993), and with a summary in Richardson *et al.* (1995). Behavioural reactions can vary dramatically depending on whether the animals are migrating, feeding, socialising, resting; or with no obvious relationship to general patterns of behaviour. However, one rather consistent finding stands out: bowhead whales tend to have shorter surfacings, shorter dives, fewer respirations per surfacing, and, overall, somewhat higher respiration rates when there is strong industrial noise in close proximity. This is particularly dramatic for loud pulsive "seismic" sounds made to explore the ocean bottom for oil and gas. These short duration intense sounds (usually about 242-252 dB re 1μPa @ 1 m in the case of multiple airgun arrays, and 226 dB re 1μPa @ 1 m for single airguns), projecting from behind a medium-sized moving vessel, strongly affect respiration and dive-related behaviour at distances of at least 10 km, and become particularly strong when the seismic vessel is closer than 5 km (Ljungblad *et al.*, 1988). Recent (1996-98) studies show near-total avoidance out to longer ranges of 20 km for migrating bowheads passing a localised area of seismic work (W. John Richardson, LGL Ltd., Toronto, Canada, *pers. comm.*, March 1999). When a fast-moving vessel rapidly converges on a group, bowheads tend to scatter in "all directions", and the resultant social disruption can last for at least several hours.

Gray whale responses to disturbance have also been measured during migration and while they are feeding, and these are broadly similar to the findings in bowheads. A particularly instructive experiment was conducted off the California coast, with a vessel anchored near the path of northward-migrating gray whales while observers on shore tracked movements of whales and whale pods with the help of surveyor's transits, or theodolites (Malme and Miles, 1985). On the vessel, researchers projected scaled-down sounds of seismic airguns to approaching whales. Observers on shore did not know when the experiment was "on" or "off", and therefore could not have been biased in their descriptions of whale movements. Analyses found a dramatic shift in movement patterns beginning at a distance of about 1km from the single airgun. In terms of sound level, this corresponds to a distance of about 5km with a full array of airguns used during normal seismic exploration. Gray whales tended to slow when they heard the noises, and some reversed or shifted directions for a time before passing by the vessel several hundred metres to the side of their original track. The intervals between respirations tended to decrease, and there was some evidence of whales apparently purposefully entering near-shore sound shadows created by underwater topography. The majority of whales avoided the area where the received sound level was greater than 170 dB (re. 1µPa @ 1 m); less consistent but still measurable effects were obtained at about 140-160 dB, which would correspond to distances of about 5 to 10 km from a full seismic source (Malme and Miles, 1985). Recent studies of U.S. Navy low frequency active sonar hums (LFA) at 100-500 Hz indicate that migrating gray whales react in similar fashion as to the seismic noises, at about the same received sound levels (Tyack and Clark, 1998). However, there was little indication of disturbance when feeding blue and fin whales were exposed to the same sounds, although blue whales may have made slightly fewer of their own low-frequency calls when the LFA was on (Clark *et al.*, 1998).

Sperm whales, the largest odontocetes, appear to react to both seismic (Bowles *et al.*, 1994) and pulsed sounds of sonar (Watkins *et al.*, 1993), by moving away and by limiting or altogether stopping their own pulsed calls (or "clicks"). In one case in the southern Indian Ocean, sperm whales ceased calling in apparent response to a seismic airgun array greater than 300 km from the whales, with a received sound level only about 10-15 dB above ambient sound intensity (Bowles *et al.*, 1994). As well, sperm whales and pilot whales of the same study ceased calling within at least 10 km range while experimental low frequency (57 Hz) hums were projected from a research vessel, but hourglass dolphins (as well as Antarctic fur seals) may have been attracted to the hums, as they approached the ship during playbacks, and the dolphins rode the bow. Other reactions to noises by changing vocalisations are known. Lesage *et al.* (1999) reported that for beluga whales in the presence of ferries in the St. Lawrence River, calling rate declined overall from that of boat absence (but, with redundancy of some calls, see Mitigation section, below) and frequencies of calls shifted upwards. Narwhals have been known to become totally silent and leave an area ensonified by ice-breaking or other vessels (Finley *et al.*, 1990). An excellent summary of other examples, including shifts in vocalisations in response to noises or other stimuli in captivity, is found in Lesage *et al.* (1999).

One of us (BW) has observed that Hawaiian spinner dolphins that rest during daytime as a group of about 100 individuals just underneath the flight path of low-flying commercial jetliners at Kaneohe Airport, Kona, Hawaii, dive abruptly in response to the jets.   On the other hand, a helicopter flying low over a pod of 12 sperm whales in the Fair Isle Channel, North Scotland, in July 1998 elicited no response at all (N. Thompson *pers. comm.* to PE).   Data are badly needed for especially toothed whales, as few detailed studies have been done, and the many anecdotal accounts give often conflicting and certainly inconclusive results. For large whales, attention must be paid to the plethora of low frequency sounds, including those of large ships and ocean acoustic studies; for toothed whales, it seems especially important to investigate how higher frequency clicks and pulses, also used in sonar and other active interrogation of the undersea environment, affect behaviour and health of animals.

Table 1. Estimated threshold values for effects of sound impulses upon fish and cetaceans (assuming spherical spreading of sound; dB values are re. 1μPa @ 1 m)

(a) FISH	PHYSICAL DAMAGE	BEHAVIOURAL AVOIDANCE
Sound Intensity	180 - 220 dB	160 - 180 dB
Dist. from Multiple Seismic Array (248 dB re 1mPa @1m)	0.25 - 2.5 km	2.5 - 25.0 km
Dist. from Single Airgun (226 dB re 1mPa @1m)	0.02 - 0.2 km	0.2 - 2.0 km
(b) BALEEN WHALES		
Sound Intensity	?220 dB	130 - 170 dB
Dist. from Multiple Seismic Array (248 dB re 1mPa @1m)	0.25 km	7.9 - 25.0 km
Dist. from Single Airgun (226 dB re 1mPa @1m)	0.02 km	0.6 - 2.0 km

Source: Evans & Nice (1996), derived from values obtained from various studies detailed therein.

From the various experimental studies conducted on marine mammals and fish, it may be possible to draw some broad generalisations concerning the sound intensities likely to cause either physical damage or at least short term avoidance. Assuming sound transmission loss occurs by spherical spreading (a reasonable approximation for most situations except very shallow waters), one can derive some estimates of the distances at which seismic sound sources within the hearing range of such organisms are likely to have an effect (see Table 1, derived from Evans & Nice, 1996). The interesting implication is that some quite disparate types of marine organism may be affected similarly. However, when the sound intensities are lower at frequencies of high animal sensitivity, then the effects will also be less.

Short term behavioural reactions do not necessarily translate to measurable or important harm to individuals or populations. Such harm may be inferred if communication, food finding, predator avoidance, and - perhaps especially important - care of young are compromised by acoustic masking, by social scattering, or by causing the animals to abandon an area. None of these possibilities, to our knowledge, has been demonstrated with certainty. It is highly likely, however, that gray whales, for example, abandoned the use of San Diego Bay (Reeves, 1977) because of noise; and temporarily abandoned Laguna San Ignacio in Baja California because of traffic, dredging, and noise generated by a salt-production plant. When the plant and associated activities ceased operation, the whales came back (Withrow, 1983, Bryant *et al*., 1984). A similar scenario of abandonment of area may apply to Indo-Pacific humpbacked dolphins, which are no longer found in the waters immediately adjacent to Victoria Harbour, Kowloon, and north Hong Kong Island (Leatherwood and Jefferson, 1997). However, in that case, both excessive noise and pollution may have been responsible. On the other hand, bottlenose dolphins seem able to handle a prodigious amount of noise, as there are many cases where even if they show short term avoidance, they continue to feed, socialise, and rear calves directly in busy harbours or shipping lanes - a prime example is the Galveston Ship Channel, where the senior author has his base (Fertl, 1994), and another is the Sado Estuary, Portugal (dos Santos *et al.,* 1996).

4.   WHALE AND DOLPHIN WATCHING

Whale and dolphin watching has rapidly become a major industry-by-the-sea, with statistics indicating that in the mid-1990's, 5.4 million persons were involved within 65 countries of the world, yielding an annual direct income of $504 million (Hoyt 1996). The financial value may be at least twice as high in 1999 (E. Hoyt, *pers. comm*. to BW). Many former fishing villages, as well as some areas where whale or dolphin hunting used to be the norm, have utilised this new economic resource to great satisfaction. Overall, it is certainly more attractive to have such a sustainable use of our natural resources, and to "love" the animals instead of eating them. However, whale and dolphin watching does need to be controlled, as excessive noise and the mere physical presence of many vessels (and for swim-with-dolphin operations, people in the water) certainly have the potential to change behaviour and perhaps elevate levels of stress (IFAW *et al.*, 1996). There is one major difference between most industrial sounds and the sounds generated by whale watching operations: whereas the former is stationary or tends to "pass by" the animals in some standard or at least partially systematic fashion; in the latter, boats directly orient towards, and stay with, their subjects - the marine mammals themselves. As a result, the animals may not have the chance to habituate as well as with some industrial activities, and may become irritated (or "sensitised") to constant or near-constant day-time approaches. On the other hand, they may become habituated to "constant" human presence. We have few good data on sensitisation or habituation, and find that - overall - whales and dolphins who wish to avoid boats can generally do so in remarkably efficient fashion. Watkins (1986) found that baleen whales in Cape Cod Bay, an area with much industrial as well as commercial tourism activity, have become generally quite habituated to boats around them. They respond by diving or by increased surface activity (such as flipper slapping or leaping) when boats approach rapidly and "head-on", as well as when there are rapid shifts in engine speed and direction.

Much has recently been written about whale and dolphin watching and potential disturbance, especially for gray, humpback, right, and sperm whales; as well as for dusky, common, Hector's and bottlenose dolphins (for example, see recent reviews by Evans, 1996, Constantine & Baker, 1997, and Constantine, 1999). The general conclusion for the present appears to be that common sense should prevail when approaching marine mammals: do not have more than three boats within 100m of the animals, do not approach rapidly, avoid sudden changes in direction and engine speed, do not cut into a group so as to separate group members. Weinrich (1989) provided a useful figure of distances and boat operation to keep in mind (Fig. 3).

Whales and dolphins may also be in different "moods" relative to their general behavioural mode; and depending on such things as whether they have recently fed, are composed of a nursery unit of mothers and calves, or are sensitised by other human actions or by predators such as killer whales nearby (Würsig 1996). One of us (PE), observing the reactions of harbour porpoise to different vessels, has found marked differences in reaction depending upon the size of boat, its behaviour to the porpoises, and various ecological features of the porpoises (group size, presence of young, season, *etc.*) (Evans *et al.*, 1994). In other words, normally approachable animals can be quite "skittish" at times, and it is up to the experience of the whale watching skipper to recognise these traits (Fig. 4).

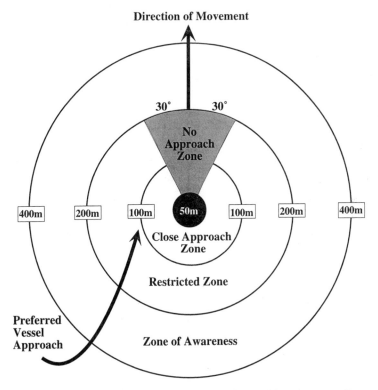

Figure 3. Vessel approach diagram. The 50 m "bull's eye" is a no approach zone. Note that circle diameters are not to scale. (Modified from Beach and Weinrich, 1989)

Constantine (1999) discussed long-term effects of whale and dolphin watching in New Zealand, where five species of dolphins and six of whales are targeted commercially. New Zealand tourism operations are generally quite well regulated and appear to be "sustainable" without chasing animals away from their near-shore haunts. However, Constantine suggested that perhaps too many permits were being issued too rapidly, resulting in the potential to harm individuals and populations in future.

There are certainly opposite extremes of potential disturbance: gray whale mothers and their newborns in Baja California calving lagoons are probably much more easily disturbed than when the same animals (with slightly older calves) are feeding along Vancouver Island, British Columbia. Dusky dolphins that rest and socialise in a bay south of Kaikoura, New Zealand, appear to be much less easily disturbed than are spinner dolphins of Hawaii that spend their daytime in deep rest in small bays (Norris *et al.*, 1994, Würsig, 1996). While tourism has not driven spinner dolphins out of these bays (perhaps because of potentially great importance of the bays to the dolphins), the dolphins are constantly forced by human presence to change their activity levels, from rest to a heightened level of alertness (Forest, 1998). We presently have no information on how this frequent change in behaviour may affect level of stress, and in turn how this might affect survivability.

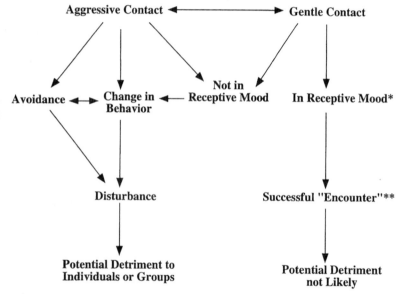

Figure 4.   A schematic of how to approach cetaceans for best "whale-watching results. The schematic includes dolphin and porpoise watching, and swimming-with-dolphins. Aggressive contact is defined as rapid approach by the vessel, or rapid changes in speed and direction. Gentle contact means careful appraisal of the animals, their behavioural state, and how best to approach them, usually slowly and not head-on. Note that even gentle contact can result in unacceptable potential detriment if the animals are not in a receptive mood. *Receptive means that the animals are in a behavioural state to likely cause least disturbance (see Fig. 4). **Successful Encounter refers to both the animals not being disturbed, and the humans being happy with the situation.

5.   GENERAL CONSIDERATIONS

We know something about short term reactions for whales and dolphins subjected to incidental ("industrial") and directed ("tourism") human activities, and we believe that most of these reactions are due to increased and variable noises. We surmise that auditory masking may be important at times, by decreasing information transfer and efficiency of group functions.

Behavioural and social disruption in the short-term might lead to long-term problems. There are some correlates of behavioural disruption and long-term stress in wild and domesticated terrestrial mammals (Richardson and Würsig, 1995 provide a summary of such studies), and by extrapolation to cetaceans we believe that stress, reproductive dysfunction, and other problems may result. However, requisite physiological studies (in the simplest case, involving hormone analyses from blood) of "stressed" and "unstressed" populations or sub-populations have not been carried out.

One consideration is certainly the importance to dolphins and whales of the area affected by humans, or the area affected versus the habitat available to the animals. Thus, if a particular bay, for example, is the only habitable bay where animals can rest away from deep water dangers, potential abandonment of that bay might be disastrous (in terms of inclusive fitness) to the group of animals that habitually use it. If there are many such bays and they can simply shift their habitat use patterns a bit to avoid humans, then no (or little) harm may take place. On the other hand, places "important" to marine animals are notoriously difficult for humans to define. Our behavioural observations may show little long-term effect in an area simply because dolphins or whales do not leave because they have nowhere else to go. When this is the case, subtle chronic effects may take place that might impact health, reproduction, and ultimately survivability: the animals may become stressed as they tolerate instead of habituate to the human presence.

It was mentioned previously that whales and dolphins may react differently not only by different types and cadences of sounds, but also by other factors of general behaviour, group disposition, *etc.* We expand on this concept in Figure 4, with the understanding that these are generalities gleaned from personal experience. Each statement in Figure 4 is subject to contrary examples; each point is open to argument. In general, however, we find that more gregarious large groups of whales and dolphins tend to be less easily disturbed than small grouped ones or "loners". River dolphins are particularly shy and skittish, as are animals close to shore or surrounded by ice or islands. Pelagic "open ocean" dolphins in general are least disturbed, but here caveats must be made. For example, we have found that striped dolphins are often very shy of vessels from even a large distance, but the congeneric spinner, spotted and clymene dolphins come up to ride the bow of vessels during a majority of encounters (Würsig *et al.*, 1998). Likewise, the white-beaked dolphin commonly bow-rides vessels on the European continental shelf but its close relative, the Atlantic white-sided dolphin, rarely does (Evans, 1990). It is also likely, although few data exist on this point, that whales and dolphins are most "skittish" or easily disturbed, when they are not in their usual, or most familiar, surroundings.

Thus, we might find that nearshore Hector's dolphins, for example, may actually become more nervous when they are in deeper oceanic waters than in their usual range. This possibility has, to our knowledge, not been investigated. Dolphins and whales react differently whether they are socialising, looking for food, resting, or taking care of young (Würsig and Würsig, 1980; Fig. 5).

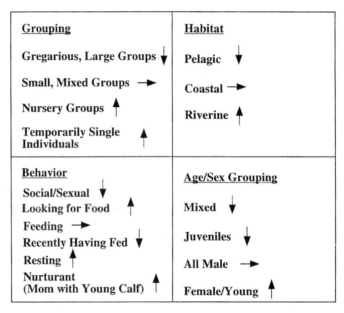

Figure 5. Generalised cetacean disturbance variables, with a down-pointing area indicating likely non-disturbance, a right-pointing arrow indicating the possibility of disturbance, and an up-pointing arrow indicating "near-certain" disturbance.

Striking examples of cetaceans that seem oblivious to disturbance while socialising are mating right and bowhead whales, which can be approached easily for photography, biopsy darting, or radio tagging (*e.g.*, Wartzok *et al.*, 1989). Different age and sex groupings also show different amounts of "skittishness". Mixed age and sex groups as well as juvenile groups tend to be least easily disturbed, whereas females and young are often quite shy, and either edge away from approaching activity, or - at times - actively race from it. Interestingly, it is our impression that all-male adult groups are often quite skittish. Males are also known to be especially prone to capture shock when captured for scientific study, tagging, or incidentally in nets (Norris *et al.*, 1994). Perhaps their stress hormone levels are already high, due to social interactions, and an extra dose of human-induced stress is not healthy.

6. MITIGATION

Mitigation of anthropogenic noises can occur both by the animals affected and by the humans producing the noise. Dolphins can at least partially mitigate against masking of their signals by adjusting their echolocation frequencies to overlap minimally with the noise (Au *et al.*, 1974). It has been suggested (Dahlheim, 1987) that gray whales exposed to industrial noises change their frequencies of calling to minimise overlap, but this assertion remains unproved. Lesage *et al.* (1999) found that beluga whales, subjected to vessel noises in the St Lawrence River, repeat certain calls more often, and shift their usual calling range from major energy at about 3.6 kHz without vessel noise to over 5.3 kHz with it. These higher frequencies are above the major energy of ferry noise common in the area. Lesage *et al.* postulate that repetition of signals may create a redundancy of information in order to communicate in the noisy environment, and the frequency shift is intended to overcome masking. Rendell and Gordon (1999) report that long-finned pilot whales in the Mediterranean increased their output of certain whistles during and immediately after military sonar signals centring around 4-5 kHz. It is not known whether this might be due to "nervousness" when the sonar signals were heard or due to an attempt to communicate in spite of the anthropogenic noise. In all of these cases cited above, the human sounds were long-term parts of the animals' environment, and one would expect that they would have habituated to them. Since they did not, we may assume either chronic disturbance, of potential but unknown detriment, an attempt to overcome the noise barrier to communication, or both. It is also possible that whales and dolphins increase their volume, or "talk louder", when other sounds threaten to obscure the message, but no information exists on that point.

At least toothed whales have highly directional hearing. There is some limited evidence of directional hearing in baleen whales as well (Richardson, 1995). It is likely that directional discrimination of noise versus useful biological signals may help in detecting communication and echolocation sounds. This is not unlike the well-known "cocktail effect" that allows humans at a party to converse despite a din of background noise that is, in its entirety, of greater volume than the point-to-point conversation. It is not thoroughly understood how we humans do this, but in our case, eye contact and lip movements as well as directional hearing are likely to be involved.

Habituation to noises with no negative consequences is known to take place throughout the animal kingdom, and cetaceans are no exception (*e.g.*, Watkins, 1986). Thus, a constant hum of activity not correlated with harm will soon be ignored (although it may still increase irritability and therefore at least minor stress); while animals pursued by whaling boat or tuna purse seiner become hypersensitive, or sensitised, to a similar sound. One of us (PE) was particularly struck by how much less approachable pods of spinner dolphins became in a bay in Dominica, West Indies, following the slaughter of a group of animals. This persisted for at least five years.

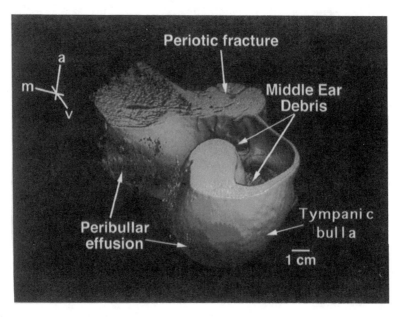

Figure 6.   A three-dimensional reconstruction from CT scans of an ear from a humpback whale (*Megaptera novaengliae*) with blast injuries shows multiple fractures throughout the periotoc, which are consistent with intracochlear blood. Blood, serum, and cellular debris of both intra-and extra-cochlear origin filled the middle ear and surrounding peribullar region (Reproduced with permission from D. Ketten; see also Ketten *et al.*, 1993).

Although behavioural habituation seems to be an obviously advantageous trait, it is also possible that it can result indirectly in chronic damage to inner ear hair cells or affect physiology in subtle but long term ways. Lien *et al.* (1995) reported ear damage during humpback whale post-mortem examinations from two individuals found dead in the vicinity of Trinity Bay, NE Newfoundland, where industrial noises of underwater drilling, blasting and dredging occurred at high sound levels, mainly between 20 and 400 Hz. The two humpbacks that had died in fishing gear near blasting both had damaged ears: ruptures of round windows, ossicular chain disruption, and haemorrhages, whilst two autopsied individuals similarly killed in gear from areas where there was no industrial activity, showed no signs of ear damage (Ketten *et al.*, 1993, Fig. 6). Besides the dangers of chronic hearing loss through long term exposure to loud sounds, a habituated whale or dolphin is unlikely to respond appropriately to real danger from a similar sound.

Finally, cetaceans can mitigate against noises by moving, or allowing themselves to be displaced from their traditional habitat. Although this would seem highly negative for the animals, it may not be if other suitable habitat is readily available. In such a case, the displacement may allow the animals to better communicate and be less stressed than staying where they had been. The point here is that habituation may not be totally positive and displacement may not be totally negative, and each case must be evaluated on its own merits.

But humans as well can mitigate against impacts of potentially adverse noise. We have already mentioned the obvious tour-operator rules of approaching slowly, changing engine speeds with care, staying at a respectable distance, *etc.* These mitigation techniques may be externally enforced, but are also self-trained since tourism on animals can only be conducted when animals are not scared, skittish, and evasive. However, industry and acoustic ocean science may not always have the same inherent reasons to act responsibly, and education and enforcement are, in our opinion, necessary. Richardson and Würsig (1995) spelled out basic noise mitigation techniques: 1) Design of equipment to be as silent as possible. Propeller shrouding that has been used to silence ships of war is an example; as are also acoustic uncoupling of generators from hulls, engine trains from drive shafts and propellers, and other engineering techniques. 2) Seasonal and hourly timing of activities can help to mitigate effects of industry. For example, if seismic oil exploration needs to be done, it would be advisable not to do it when gray whales migrate through the area in spring and fall; or during the daytime in a bay used by day-resting spinner dolphins. 3) Changes of locations can help to mitigate sounds, so that industrial supply vessels, for example, do not move directly through bowhead whale nearshore feeding grounds, but actually route around the main concentration of animals with only minimal increase in expense of fuel and time. 4) Adjustment of operational procedures can help to mitigate against adverse effects. One way to help is to monitor the area for marine mammals before blasting or projecting other loud sounds. If mammals are present, the activity has to be delayed. Such monitoring has been widely practiced, and is especially well organised for oil rig removal in the Gulf of Mexico (Klima *et al.*, 1988). Monitoring presently tends to rely on visual, not acoustic, methods, and may certainly miss animals. It is important to conduct both simultaneously since each has its own advantages and limitations. 5) Other operational changes include keeping vessel speed down, slowly ramping up sounds, staggering sound production so that it does not occur throughout the day, and providing lower-charge warning blasts before projecting intense sounds needed for the job. Except for vessel speed, these latter operational procedures appear to us to be questionable, and may even do more harm than good, if, for example, whales or dolphins are attracted to ramping up sounds, to low-level blasts, or to changes in duty cycles. Even reduced vessel speed may act conversely if animals are less disturbed by a vessel moving rapidly through an area in 20 minutes than if it lingers and takes twice as long.

One technique that has not been thoroughly investigated but which shows promise for the future is a way of shrouding sound once it has been projected into the water. The best method of reduction may be to create an impedance mismatch by a curtain of air bubbles. Air is about 800 times less dense than water, and air bubbles therefore effectively "swallow up" much sound energy moving from water to the bubbles. The technique has recently been investigated in some detail for shrouding around a stationary, very loud percussive hammering ("pile driving") activity for creating a wharf in Hong Kong (Würsig *et al.*, 2000). A curtain of bubbles was created by running air into a perforated rubber hose surrounding the pile driver. Sounds that were bubble-screened were reduced at 250 to 1,000 m distances in the broadband (from 100 Hz to 25.6 kHz) by about 3-5 dB, with greatest reduction at 400-6400 Hz. Indo-Pacific humpbacked dolphins that occurred in the area were therefore subjected to less noise than without the bubble curtain operating. Nevertheless, more experimental studies need to be carried out to ascertain if and how bubble screening can become a commonplace reality both for stationary and moving sources of noise.

There are occasions, however, when reduction of sounds can have detrimental effects. Cetaceans may avoid sound not just because it interferes with their communication or feeding, but because it signifies a potential threat through physical damage; there are a number of recorded instances where a vessel has collided and killed a whale. Propeller damage is another potential threat although it is generally difficult to be sure that wounds on the back or dorsal fin of dolphins are actually caused by this rather than by attacks from sharks or killer whales. For this reason, it is not advisable to suddenly accelerate, slow down or cut one's engine. Most sounds produced by small vessels are in the high frequency range above 1 kHz, particularly when there is cavitation of the propeller (as occurs at high speed or when the propeller is damaged in some way). These sounds do not carry over large distances. Using field trials in Cardigan Bay, West Wales, the second author calculated the distances at which a bottlenose dolphin would hear various small craft, and obtained values of 450 m (jet ski), 800 m (speedboat low speed), 1 km (inflatable high speed), 1.2 km (lobster boat low speed), 1.8 km (speedboat high speed), and 3.2 km (lobster boat high speed) (Evans *et al.*, 1992). Thus, although the area to which a dolphin might be exposed to sound from a jet ski would be small, the often erratic track that a jet ski takes imposes an obvious risk of physical damage particularly to young, "naive" animals that have little time to take avoiding action.

7.   CONCLUSIONS

We know that noises can be a problem, and we believe that industrial, military, ocean science, and directed human tourism effects can be quite different. We also know that reactions by animals are likely to be different depending on species, aggregations of groups, age and sex and "mood" variables, and unknown factors. After all, these are large-brained social mammals and - like us - are allowed to be unpredictable just when we believe that we have it all figured out. Noise and high amounts of human activity can hurt cetaceans. Therefore, we need to study the boundaries of such problems and advise operators and legislators of potential mitigation action. However, we do not want "the noise problem" to obscure what we perceive to be generally more evident problems to whales and dolphins. These are directed killing of some declining populations, accidental killing due to indiscriminate fishery practices, and the problems of mortality and lowered fitness due to catastrophic (oil or chemical spills) as well chronic pollution (DDT, PCB, organochlorine, heavy metal leakages) of cetacean prey and therefore the cetaceans themselves. But, we are at the least pleased that humans care enough about our own actions to address "the noise problem", and to attempt to better the well-being - and perhaps simply the quality - of the lives of our mammals of the sea.

ACKNOWLEDGEMENTS

We thank Suzanne Yin for careful preparation and editing of references, Jim Boran for his logistic help, and Paul Wu and Stefan Bräger for preparing figures. A special thanks to W. John Richardson and John Calambokidis for their insightful comments on a draft of this chapter.

REFERENCES

Au, W.W.L. (1993) *The Sonar of Dolphins*. Springer-Verlag, New York. 277pp.
Au, W.W.L., Floyd, R.W., Penner, R.H. and Murchison, A.E. (1974) Measurement of echolocation signals of the Atlantic bottlenose dolphin, *Tursiops truncatus* Montagu, in open waters. *Journal of the Acoustical Society of America*, **56**(4), 1280-1290.

Beach, D.W. and Weinrich, M.T. (1989) Watching the whales: Is an educational adventure for humans turning out to be another threat for endangered species? *Oceanus*, **32**(1), 84-88.

Bowles, A. E., Smultea, M., Würsig, B., DeMaster, D.P. and Palka, D. (1994) Relative abundance and behavior of marine mammals exposed to transmissions from the Heard Island Feasibility Test. *Journal of the Acoustical Society of America*, **96**(4): 2469-2484.

Brownell, R. L., Jr., Ralls, K. and Perrin, W.F. (1989) The plight of the 'forgotten' whales; It's mainly smaller cetaceans that are now in peril. *Oceanus*, **32**(1): 5-13.

Bryant, P. J., Lafferty, C.M. and Lafferty, S.K. (1984) Reoccupation of Laguna Guerrero Negro, Baja California, Mexico, by gray whales. In: *The Gray Whale Eschrichtius robustus* (Ed. by M. L. Jones, S. L. Swartz, and S. Leatherwood), pp. 375-387. Academic Press, Orlando, FL.

Burns, J. J., Montague, J.J. and Cowles, C.J. (eds.) (1993) *The Bowhead Whale*. Spec. Publ. No. 2, The Society for Marine Mammalogy, Lawrence, KS. 787pp.

Clark, C. W., Tyack, P. and Ellison, W.T. (1998) *Quick Look, phase 1: Responses of blue and fin whales to SURTASS LFA, Southern California Bight*. Unpublished report, Bioacoustics Research Program, Lab. of Ornithology, Cornell Univ., Ithaca, N.Y. 14850.

Constantine, R. (1999) *Effects of tourism on marine mammals in New Zealand*. Science for Conservation 106, Published by Department of Conservation, P.O.B. 10-420, Wellington, New Zealand, 60pp.

Constantine, R. and Baker, C.S. (1997) *Monitoring the commercial swim-with-dolphin operations in the Bay of Islands*. Science & Research Series, 104. Department of Conservation, Wellington, NZ.

Cosens, S.E. and Dueck, L.P. (1993) Icebreaker noise in Lancaster Sound, N.W.T., Canada: Implications for marine mammal behavior. *Marine Mammal Science*, **9**, 285-300.

Dahlheim, M. E. (1987) *Bio-acoustics of the gray whale (Eschrichtius robustus)*. Ph.D. Thesis, Univ. British Columbia, Vancouver, B. C. 315pp.

Evans, P.G.H. (1990) Whales, Dolphins and Porpoises. The Order Cetacea. In: *Handbook of British Mammals* (Ed. by G.B. Corbet and S. Harris), pp. 299-350. Blackwell, Oxford. 588pp.

Evans, P.G.H. (1996) Human disturbance of cetaceans. In: *The Exploitation of Mammals - principles and problems underlying their sustainable use*. (Ed. by N. Dunstone & V. Taylor), pp. 376-394. Cambridge University Press. 415pp.

Evans, P.G.H. and Nice, H.E. (1996) *Review of the Effects of Underwater Sound generated by Seismic Surveys on Cetaceans*. Sea Watch Foundation, Sussex. 50pp.

Evans, P.G.H., Carson, Q., Fisher, P., Jordan, W., Limer, R. and Rees, I. (1994) A study of the reactions of harbour porpoises to various boats in the coastal waters of SE Shetland. *European Research on Cetaceans*, **8**, 60-64.

Evans, P.G.H., Canwell, P.J. and Lewis, E.J. (1992) An experimental study of the effects of pleasure craft noise upon bottle-nosed dolphins in Cardigan Bay, West Wales. *European Research on Cetaceans*, **6**, 43-46.

Fertl, D. C. (1994) *Occurrence, movements, and behavior of bottlenose dolphins (Tursiops truncatus) in association with the shrimp fishery in Galveston Bay, Texas*. M. S. Thesis, Texas A&M University, College Station. 116pp.

Finley, K.J., Miller, G.W., Davis, R.A. and Greene, C.R. (1990) Reactions of belugas, *Delphinapterus leucas*, and narwhals, *Monodon monoceros*, to ice-breaking ships in the Canadian high arctic. *Canadian Bulletin of Aquatic Sciences*, **224**, 97-117.

Forest, A. (1998) *Spinner dolphin (Stenella longirostris) behavior: Effects of tourism*. M. S. Thesis draft. Texas A&M University, College Station.

Goodson, A.D., Klinowska, M. and Bloom, P.R.S. (1994) Enhancing the acoustic detectability of gillnets. *Report of the International Whaling Commission*, (special issue **15**), 585-596.

Goodson, A.D., Mayo, R.H., Klinowska, M. and Bloom, P.R.S. (1994) Field testing passive acoustic devices designed to reduce the entanglement of small cetaceans in fishing gear. *Report of the International Whaling Commission* (special issue **15**), 597-605.

Hoyt, E. (1996) *The Worldwide Value and Extent of Whale Watching 1995.* Report from Whale & Dolphin Conservation Society, Bath. 34pp.

IFAW, Tethys Research Institute, and Europe Conservation. (1996) *Report of the Workshop on the Scientific Aspects of Managing Whale Watching.* Montecastello de Vibio, Italy, Mar 30-Apr 4, 1995. 40pp.

Ketten, D.R., Lien, J. and Todd, S. (1993) Blast injury in humpback whale ears: Evidence and implications. *Journal of the Acoustical Society of America*, **94**(3), 1849-1850.

Kraus, S.D., Read, A., Solow, A., Baldwin, K., Spradlin, T., Anderson, E., and Williamson, J. (1997) Acoustic alarms reduce porpoise mortality. *Nature* (London), **388**, 525.

Klima, E.F., Gitschlag, G.R. and Renaud, M.L. (1988) Impacts of the explosive removal of offshore petroleum platforms on sea turtles and dolphins. *Marine Fisheries Review*, **50**(3), 33-42.

Leatherwood, S. and Jefferson, T.A. (1997) Dolphins and development in Hong Kong: a case study in conflict. *IBI Reports*, **7**, 57-69.

Lesage, V., Barrette, C., Kingsley, M.C.S., and Sjare, B. (1999) The effect of vessel noise on the vocal behavior of belugas in the St. Lawrence River estuary, Canada. *Marine Mammal Science*, **15**, 65-84.

LGL and Greeneridge. (1986) *Reactions of beluga whales and narwhals to ship traffic and ice-breaking along ice edges in the eastern Canadian High Arctic: 1982-1984.* Environmental Studies, 37. Indian & Northern Affairs Canada, Ottawa, Ont. 301pp.

Lien, J., Taylor, D.G. and Borggaard, D. (1995) Management of Underwater Explosions in Areas of High Whale Abundance. Marienv 1995. Proceedings of the International Conference on Technologies for Marine Environment Preservation. Vol. 2, 627-632.

Ljungblad, D.K., Würsig, B., Swartz, S.L. and Keene, J.M. (1988) Observations on the behavioral responses of bowhead whales (*Balaena mysticetus*) to active geophysical vessels in the Alaskan Beaufort Sea. *Arctic*, **41**(3), 183-194.

Malme, C.I. (1995) Sound propagation. In: *Marine Mammals and Noise* (Ed. by W. J. Richardson, C.R. Greene, Jr., C.I. Malme and D.H. Thomson), pp. 59-86. Academic Press, San Diego, CA.

Malme, C.I. and Miles, P.R. (1985) Behavioral responses of marine mammals (gray whales) to seismic discharges. In: *Proceedings of the Workshop on Effects of Explosives Use in the Marine Environment, Jan. 1985, Halifax.* Pp. 253-280. Nova Scotia Technical Report No. 5. Canadian Oil and Gas Lands Administration Environmental Protection Branch, Ottawa, Ontario, 398pp.

Miles, P.R. and Malme, C.I. (1983) *The acoustic environment and noise exposure of humpback whales in Glacier Bay, Alaska.* BBN Tech. Memo. 734. Rep. from Bolt Beranek & Newman Inc., Cambridge, MA for U. S. Natl. Mar. Fish. Serv., Seattle, WA. 81pp.

Munk, W.H., Spindel, R.C., Baggeroer, A. and Birdsall, T.G. (1994) The Heard Island feasibility test. *Journal of the Acoustical Society of America*, **96**(4), 2330-2342.

Norris, K.S., Würsig, B., Wells, R.S. and Würsig, M. with Brownlee, S. M., Johnson, C.M. and Solow, J. (1994) *The Hawaiian Spinner Dolphin.* Univ. Calif. Press, Berkeley, CA. 408pp.

Payne, R.S. and McVay, S. (1971) Songs of humpback whales. *Science*, **173**(3997), 585-597.

Perrin, W.F., Donovan, G.P. and Barlow, J. (editors) (1994) Gillnets and Cetaceans. *Report of the International Whaling Commission.* (special issue **15**), 1- 629.

Reeves, R.R. (1977) *The problem of gray whale (Eschrichtius robustus) harassment: At the breeding lagoons and during migration.* MMC-76/06. U. S. Mar. Mamm. Comm. 60pp. NTIS PB-272506.

Reeves, R. and Leatherwood, S. (1994) *Dolphins, Porpoises and Whales: 1994-1998.* Action Plan for the Conservation of Cetaceans. IUCN Species Survival Commission. Gland, Switzerland.

Rendell, L.E. and Gordon, J.C.D. (1999) Vocal response of long-finned pilot whales (*Globicephala melas*) to military sonar in the Ligurian Sea. *Marine Mammal Science*, **15**, 198-204.

Richardson, W.J. (1995) Marine mammal hearing. In: *Marine Mammals and Noise* (Ed. by W. J. Richardson, C.R. Greene, Jr., C.I. Malme and D.H. Thomson), pp. 205-240. Academic Press, San Diego, CA.

Richardson, W.J. and Würsig, B. (1995) Significance of responses and noise impacts. In: *Marine Mammals and Noise* (Ed. by W. J. Richardson, C.R. Greene, Jr., C.I. Malme and D.H. Thomson), pp. 387-424. Academic Press, San Diego, CA.

Richardson, W.J. and Würsig, B. (1997) Influences of man-made noise and other human actions on cetacean behaviour. *Marine and Freshwater Behaviour and Physiology*, **29**, 183-209.

Richardson, W.J., Greene, C.R. Jr., Malme, C.I. and Thomson, D.H. (1995) *Marine Mammals and Noise.* Academic Press, San Diego, CA. 576pp.

Ross, D. (1976) *The Mechanics of Underwater Noise.* Pergamon, New York.

dos Santos, M.E., Ferreira, A.J., Ramos, J., Ferreira, J.F. and Bento-Coelho, J.L. (1996) The acoustic world of the bottlenose dolphins in the Sado Estuary. *European Research on Cetaceans*, **9**, 62-64.

Tyack, P.L. and Clark, C.W. (1998) *Quick Look, phase II, Playback of low frequency sound to gray whales migrating past the central California coast - January, 1998.* Unpublished report, Department of Biology, Woods Hole Oceanographic Inst., Woods Hole, MA 02543.

Tynan, C.T. and DeMaster, D.P. (1997) Observations and predictions of arctic climatic change: Potential effects on marine mammals. *Arctic*, **50**, 308-322.

Wartzok, D., Watkins, W.A., Würsig, B. and Malme, C.I. (1989) Movements and behaviors of bowhead whales in response to repeated exposure to noises associated with industrial activities in the Beaufort Sea. Rep. from Purdue Univ., Fort Wayne, IN, for Amoco Production Co., Anchorage, AK. 228pp.

Watkins, W.A. (1986) Whale reactions to human activities in Cape Cod waters. *Marine Mammal Science*, **2**(4), 251-262.

Watkins, W. A., Draher, M.A., Fristrup, K.M., Howald, T.A. and Notarbartolo di Sciara, G. (1993) Sperm whales tagged with transponders and tracked underwater by sonar. *Marine Mammal Science*, **9**(1), 55-67.

Wenz, G.M. (1962) Acoustic ambient noise in the ocean: Spectra and sources. *Journal of the Acoustical Society of America*, **34**(12), 1936-1956.

Withrow, D.E. (1983) Gray whale research in Scammon's Lagoon (Laguna Ojo de Liebre). *Cetus*, **5**(1), 8-13.

Würsig, B. (1996) Small cetaceans and humans: a perspective of world problems. *Delfines y otros mamiferos acuaticos de Venezuela*, Caracas, pp. 15-22.

Würsig, B. and Würsig, M. (1980) Behavior and ecology of the dusky dolphin, *Lagenorhynchus obscurus*, in the south Atlantic. *Fishery Bulletin*, **77**(4), 871-890.

Würsig, B., Cipriano, F., Slooten, E., Constantine, R., Barr, K. and Yin, S. (1997) Dusky dolphins (*Lagenorhynchus obscurus*) off New Zealand: Status of present knowledge. *Report of the International Whaling Commission*, **47**, 715-722.

Würsig, B., Greene, C.R. and Jefferson, T.A. (2000) Development of an air bubble curtain to reduce underwater noise of percussive piling. *Marine Environmental Research*, **49**, 79-93.

Würsig, B., Lynn, S.K., Jefferson, T.A. and Mullin, K.D. (1998) Behavior of cetaceans in the northern Gulf of Mexico relative to survey ships and aircraft. *Aquatic Mammals*, **24**(1), 41-50.

Chapter 17

Global Climate Change and Marine Mammals

[1]BERND WÜRSIG, [2]RANDALL R. REEVES, and [1]J. G. ORTEGA-ORTIZ
[1]Marine Mammal Research Program, Texas A&M University, 4700 Ave. U, Galveston, TX, 77551, USA. E-mail: wursigb@tamug.tamu.edu; [2]Okapi Wildlife Associates, 27 Chandler Lane, Hudson, Quebec J0P 1H0, Canada. E-mail: rrreeves@total.net

1.   INTRODUCTION

The global climate is changing, at least in part because of human-generated emissions of ozone-depleting substances (Ashmore and Bell, 1991; United Nations Environmental Program, 1987, 1993; Intergovernmental Panel on Climate Change, 2001a,b, c), world-wide use of hydrocarbons responsible for emissions of greenhouse-effect gases (Flavin, 1992; Wellburn, 1994), and large-scale deforestation and desertification (Barbier *et al.*, 1994). Predictions for the world's oceans range from little or no change to partial melting of Arctic and Antarctic loose ice and shelves, with concomitant rises of sea levels. While many "alarmist" scenarios have been proposed, it is generally accepted that in the next 100 years, sea temperatures will rise by about $1.3\text{-}5.8^0$ C and overall sea level will rise by anywhere from 9 to 88 cm (Intergovernmental Panel on Climate Change 2001a). Because of the overall global change, there are likely to be more local fluctuations in rainfall, storm frequencies, and the incidence of abrupt cold and dry spells, along with other mesoscale changes. Hurricanes and typhoons are known to form with greater frequency when water temperatures are at or above 28^0 C, and even short-term changes in climate (for example, El Niño Southern Oscillation events, ENSOs) can affect the incidence and severity of such storms. This generally grim scenario could be improved if the present decline in ozone-depleting emissions were to continue, or if energy substitutes for fossil fuels were more fully developed and widely adopted (Moore *et al.*, 1996, Intergovernmental Panel on Climate Change, 2001c).

Marine Mammals: Biology and Conservation, edited by
Evans and Raga, Kluwer Academic/Plenum Publishers, 2002

2.   GENERAL POTENTIAL TRENDS

Predictions of impacts of climate change depend heavily on questions of time and geographic scale of the environment, as well as on the longevity, generation time, and geographic distribution of the animals in question. Large but "slow" (in the order of decades or centuries) shifts in climate have occurred throughout Earth's history, and these have helped drive the evolution of adaptive characteristics, clinal variations, population discreteness, and, of course, extinctions. Recent analyses of beach-strewn remains of bowhead whales (*Balaena mysticetus*) demonstrate that these Arctic whales changed their patterns of habitat use several times during the past 11,000 or so years, apparently because of changes in ice conditions, currents, or marine productivity (Dyke *et al.*, 1996; Savelle *et al.*, 2000). The changes in climate that were responsible for major ice ages and the southerly advance and retreat of glaciers in the northern hemisphere appear to have influenced bowhead movements and habitat use, but did not cause the species to go extinct. We suspect that human-caused climate change has caused, and will continue to cause, similar shifts for those animals mobile enough to adapt rapidly. One apparently climate-related example of a recent expansion in range comes from the sub-Antarctic Indian Ocean, where fur seals (*Arctocephalus* sp.) have re-established themselves on Heard Island in the past 50 years, coincident with warmer temperatures, glacier recession, and hypothesized improved food supplies (Shaughnessy *et al.,* 1998, Budd, 2000). Nevertheless, we worry about the impacts of rapid changes on those species that depend on limited patches of specific types of habitat, such as the land-breeding pinnipeds, the coastal and freshwater cetaceans, and the sirenians.

Tynan and DeMaster (1997) present a cogent, if speculative, analysis of the likely effects of Arctic climate change on carnivores and cetaceans that feed in the marine environment, including the polar bear (*Ursus maritimus*). They predict, with caution, that sympagic ("ice-living") amphipods, copepods, and Arctic cod (*Boreogadus saida*), major constituents of Arctic food webs, will undergo highly variable shifts in distribution and biomass in different areas. Some areas will become wetter, with more local snow cover, snow-derived ice, and melt waters that, together, will have the effect of reducing ocean salinities in spring and summer. Other areas will become drier, and so forth. Pinnipeds that need ice as a platform for giving birth, nursing and resting are likely to be affected particularly strongly, but the more mobile bowhead whales, white whales (*Delphinapterus leucas*), and narwhals (*Monodon monoceros*) also may be strongly affected by changes in the distribution or amount of their prey. One important possibility is that more fresh water from melting snow and ice will be released into the North

Atlantic, through the Fram Strait between northeastern Greenland and Svalbard. This could exert a strong influence on salinity in the North Atlantic, shift the Gulf Stream current, and even affect upwelling related to the Great Ocean Conveyor Belt current system (Tynan and DeMaster, 1997; see also Marotzke, 2000). Another predicted effect of Arctic warming is that the Northwest Passage across North America will become more easily navigable, facilitating increased tanker and other traffic and thus increasing the risks of oil and other toxin spills, worsening acoustic pollution, and generally degrading this still relatively "pristine" area (Burns, 2001).

Figure 2.   White whales, *Delphinapterus leucas*, migrating through an early spring ice lead. Such highly mobile animals may not be as susceptible to climate change as more resident or sedentary species. However, an overall lowering of productivity could affect even them. Photo: B. Würsig.

3.   CLASSIFYING EFFECTS

In an earlier forum, Reeves (1990) presented a heuristic model for classifying potential effects of global climate change (see also Würsig and Ortega-Ortiz, 1999). *Primary* effects would be those that debilitate or cause death at the individual level. For example, the water becomes too warm and an individual Galápagos fur seal (*Arctocephalus galapagoensis*) fails to reproduce or dies. This would likely be due to an inadequate food supply, perhaps because sufficient prey of fur seals moved farther away from the area due to sea temperature change. At the primary level, individuals are likely to be affected when already at the edge of their capabilities, such as

Global climate change need not be regarded entirely as "gloom and doom" for biodiversity, but great uncertainty is sufficient cause for great caution to be taken in predicting effects. This was noted by participants in a 1996 Workshop on Climate Change and Cetaceans sponsored by the International Whaling Commission (International Whaling Commission, 1997) as well as in other summaries relating to climate change and cetaceans (MacGarvin and Simmonds, 1996; Burns, 2001). Uncertainties about the nature and degree of future climate change make it impossible to know exactly how weather, ocean circulation, and biological productivity will be affected (but see Weaver and Zwiers, 2000, for ever-more sophisticated models). Predictions of effects for species and populations are therefore highly speculative. Nevertheless, we consider it worthwhile to present several broad scenarios, to alert us to at least major potential problems. In doing so, we hope to inform research initiatives that will address the scientific uncertainties, and give support to lobbying efforts in political arenas for more precautionary management of human activities. We caution that we are not climate-change experts, nor do we have new and compelling information on the actual or potential effects of climate change on marine mammals. In the present chapter, we simply try to summarize some of the known and postulated correlations and forecasts, with the caveat that these vignettes represent a far-from-complete assessment of a rapidly emerging conservation concern.

Figure 1.   The bowhead whale, *Balaena mysticetus*, lives near ice just about wherever it travels. Its foraging efficiency is intricately tied to the Arctic ecosystem, by changes in ice cover, spring breakup, algal blooms, and its food the calanoid copepods and euphausiid crustaceans. Photo: B. Würsig.

youngsters, pregnant females, old animals, or those already stressed in some other manner, such as from lowered immune response due to toxin loads or disease (see Geraci and Lounsbury, this volume). *Secondary* effects would be those pervasive enough, or sufficiently encompassing ecologically, to affect a substantial part or all of a community. For example, the water warms and mullet (*Mugil* sp.) die off or experience reproductive failure, and this in turn affects the plane of nutrition for inshore bottlenose dolphins (*Tursiops truncatus*). As a result, the health status of the dolphins worsens and their population declines. *Tertiary* effects would also manifest at the community level, but involve a feedback loop that includes the initiator of the problem (humans in the present scenario of global warming). For example, the water becomes too warm and fish die. These fish, which once contributed to the food supply of humans, now become scarce or unavailable. Humans therefore may begin to target dusky dolphins (*Lagenorhynchus obscurus*) as an ersatz fishery, for bait to enhance fishing, as direct food, or because of perceived or real competition for scarce marine resources (Van Waerebeek and Reyes, 1994). Other potential tertiary-effect scenarios could be invoked, but in general we have insufficient evidence to support predictions of primary and secondary effects, and the even more complicated "chain reactions" leading to tertiary effects take us far beyond the supporting data. In practice, the primary, secondary, and tertiary effects are inter-related and can be difficult to separate.

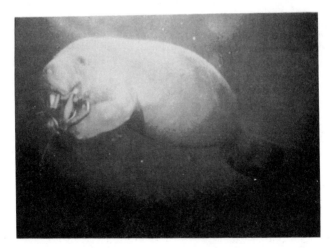

Figure 3.   Manatees are hunted for food in Africa, Central and South America, and in the Amazon basin. Dugongs also suffer from human predation and habitat perturbations. Their nearshore and inshore habitats make sirenians especially vulnerable to changes in ocean edge and riverine ecosystems. Photo: K. Norris Library.

Reeves (1990) also outlined several potential effects for the broad categories of sirenians (manatees and dugongs), cetaceans (whales, dolphins, and porpoises), and carnivores (pinnipeds, sea and marine otters, and the polar bear). Global warming, rise in sea levels, potential increases in numbers and intensities of storms, and changes in circulation and productivity could have the following *primary* effects: (1) The tropical sirenians experience fewer deaths from seasonal thermal stress, but more deaths caused by periodic local cold snaps and storms (cf. Marsh, 1989), or by increased toxic algal or dinoflagellate blooms, such as "red tide" (for example, *Ptychodiscus brevis*) (cf. Bossart *et al.*, 1998). (2) Cetaceans may be less affected at this level, except perhaps for the effects of increases in toxic blooms (cf. Geraci *et al.*, 1989, 1999). However, those species that have very limited distributions or inhabit "pockets" of water surrounded by land (such as the critically endangered vaquita, *Phocoena sinus*, of the northern Gulf of California; or the three small cetacean species in the Black Sea) could suffer direct harm from the ecological degradation of warmer water. (3) Carnivores will probably be most vulnerable. There is the possibility of increased deaths of pups on unstable ice, for example due to collapse of maternity dens of ringed seals (*Pusa* sp.), and an increase in drowning of otariid pups on beaches where storm surges are more frequent or severe. There may well be an increase in deaths due to epizootics such as morbillivirus implicated in the recent die-off of Mediterranean monk seals

Figure 4.   Southern sea lions, *Otaria flavescens*, resting and basking on a small island off the rocky shores of Patagonia, Argentina. The need for resting, mating, and pupping areas represents a limiting factor for many pinnipeds. Rapidly rising sea levels have the potential to change availability patterns of haul-out areas for sea lions, fur seals, true seals, and walruses. Photo: B. Würsig

(*Monachus monachus*; Van de Bildt *et al.*, 1999; but also see Hernández *et al.*, 1998) and various cetacean and other seal populations (Geraci *et al.*, 1999); and increased deaths related to biotoxins, such as the recent die-off of California sea lions (*Zalophus californianus*; Scholin *et al.*, 2000).

At the *secondary* level, all primary effects can, of course, act on the entire community, population, or species. Beyond that, we predict problems for sirenians caused by changes in near-shore and freshwater habitats (for example, Short and Neckles, 1999), and for cetaceans that live in shallow waters (such as, again, the vaquita) or that feed on the sea bottom (such as the gray whale, *Eschrichtius robustus*) caused by the outright loss of habitat. Secondary effects are likely to be severe for the already-beleaguered river cetaceans, especially those in southern Asia (*Platanista gangetica* and *Orcaella brevirostris*) and eastern Asia (*Lipotes vexillifer* and *Neophocaena phocaenoides*) (Reeves *et al.*, 2000). River cetaceans are likely to continue to decline as a result of wholesale changes to their habitat resulting from human activities, many of which are driven or shaped by the exigencies of poverty, which will certainly be exacerbated by climate change. River dolphins need unpolluted waters to survive, and such waters are becoming a scarce commodity in today's world. Carnivores will be affected at the secondary level by the loss of sea ice and the diminution of terrestrial haul-out sites. For example, it is difficult to imagine where tens of thousands of walruses (*Odobenus rosmarus*) will rest when the beaches surrounding mountainous islands disappear due to rapidly rising sea level. These beaches may take tens of years, or even centuries, to be formed; rising waters could outstrip their ability to re-generate. Polar bears will, of course, follow the trends of their pinniped prey, and even slight changes in the timing of near-shore ice formation and break-up could reduce their hunting efficiency (Stirling and Derocher, 1993). This has recently been corroborated, as polar bear physical condition has deteriorated in the past few decades in western Hudson Bay, apparently due to the earlier spring break-up of ice (Stirling *et al.*, 1999). Overall, secondary effects can include lowered immune capabilities due to compromised feeding or other climate-induced environmental stressors. Therefore, epizootics that may or may not be caused by the climate change, *per se*, have a greater potential to create epidemics, associated mass die-offs, and possibly mass strandings of cetaceans (Harvell *et al.*, 1999; Geraci and Lounsbury, this volume).

At the *tertiary* level, predictions become more tenuous. There may be: (1) more fragmentation of sirenian and river cetacean populations as the demand for construction projects to manage flooding increases (cf. Smith *et al.*, 1998, 2000); (2) more boat collisions with manatees and dugongs as recreational activity increases during the longer summers in subtropical and warm temperate areas (cf. Reynolds, 1999); and (3) more hunting pressure on manatees in Africa and the Amazon basin (cf. Reeves *et al.*, 1996), and on near-shore dolphins and whales off Asia, Latin America, Africa, and certain protein-poor island areas (cf. Reeves and Leatherwood, 1994), all related to a general increase of human poverty, famine, and civil conflict. Overall, climate-induced impoverishment of human communities could well lead to a decline in conservation efforts, park designations, and resource management, opening the way to further degradation of coastal and riverine ecosystems, with distressing implications for marine mammals. Table 1 presents speculations on the possible effects of climate change on six almost randomly selected species or species groups. It is, of course, merely illustrative and not intended to be definitive in any way.

Table 1. Speculations regarding some potential effects of global warming on selected marine mammal species.

Taxon	Distribution	Key habitat features	Potential impacts of global warming
Ganges river dolphin, *Platanista gangetica*	Indus, Ganges, Brahmaputra/ Meghna and Karnaphuli river systems of South Asian Subcontinent	Fresh water; access to deep pools and hydraulic refuge for resting and possibly foraging	Glacial melting could increase river flows, but benefits (e.g. more area with adequate water to support dolphins) could be offset by higher evaporation rates on the plains as well as decreased precipitation in the Himalayas; increased flooding on the plains could increase pressure for further river regulation (impoundment dams) and thus more fragmentation of dolphin populations; drier, hotter weather on the plains would increase the already enormous pressure to withdraw more water from the rivers for irrigation and other human uses
Polar bear, *Ursus maritimus*	Circumpolar in Arctic and Subarctic	Sea ice needed as hunting platform; abundant supplies of phocid seals (mainly ringed and bearded); undisturbed coastal maternity denning areas	Reduced duration of pack ice season could affect nutritional condition of females and in turn the survival of offspring; any climate-related reduction in ringed seal abundance would affect polar bears indirectly; milder conditions leading to more industrial and other human activities could affect denning sites and lead to more frequent polar bear-human conflict
Florida manatee, *Trichechus manatus latirostris*	Coastal and inland waters of Florida and adjacent states, USA	Relatively sheltered and shallow water with frequent access to fresh water and regular (nearly	Increase in range made possible by latitudinal extension of 20° C isotherm, but incidence and duration of "cold snaps," if increased, could offset this benefit; warming-

	(including lagoons, rivers, the Everglades and artificial waterways)	constant) access to vegetation (including marsh and sea grasses as well as leafy plants, whether submerged, floating or emergent); lower limit of (prolonged) thermal tolerance ca. 20° C	associated increase in incidence of toxic blooms ("red tides") could cause more frequent mass die-offs (cf. Bossart *et al.* 1998); flooding caused by sea-level rise could increase pressure for further water regulation, resulting in higher manatee mortality from crushing and drowning in flood-control gates
Monk seals, *Monachus schauinslandi* and *M. monachus.*	Hawaiian Islands (mainly Leeward Chain), eastern Mediterranean (mainly Greek Islands and Turkey), Madeira, Saharan coast of NW Africa	Undisturbed beaches or caves for pupping, pup nurturing and resting; adequate prey resources within close proximity to haul-out sites; Hawaiian monk seals forage at least to some extent on coral reefs	Rising sea levels could eliminate already-scarce haul-out habitat, especially by flooding caves that provide the only refuges for some groups of Mediterranean monk seals; climate-driven changes in oceanic productivity could be decisive in reducing (or enhancing) monk seal reproduction and survival (e.g. Polovina *et al.* 1994); degradation or destruction of coral reefs could have negative effects on foraging, especially for Hawaiian monk seals; vulnerability exaggerated by smallness and extremely restricted distribution of extant populations
Bottlenose dolphin, *Tursiops truncatus*	Circumglobal in tropical and temperate marine waters, including open coasts, lagoons, estuaries, oceanic island coasts, and pelagic waters	Exceptionally unspecialised and adaptable as a species, if not as individuals (Shane 1990)	Presumably among the more resilient species to primary and secondary effects, and may benefit from global warming; already, however, targeted for culling and hunted as a source of food and bait - which could increase as a tertiary effect of climate change
Bowhead whale, *Balaena mysticetus*	Arctic circumpolar in five separate geographical populations	Associated with pack ice edge or polynyas in winter, productive zones in high latitudes in summer and autumn	Net effects are difficult to predict, but the species has a demonstrated ability to adapt to large-scale changes in ice and nutrient dynamics; vulnerability is exaggerated, however, by smallness and extremely restricted distribution of four of the five extant populations

4. LESSONS FROM SHORT-TERM PHENOMENA

Most discussions about the effects of climate change on animals have concerned terrestrial rather than marine species. One exception, however, is the examination of reproductive failure in marine birds and pinnipeds subjected to ENSO events. During recent El Niño events, reproductive failure (especially in the form of high juvenile mortality) has affected many colonies of seabirds (Guinet *et al.*, 1998) and some colonies of seals (Guinet *et al.*, 1994; Domingo *et al.*, this volume). For example, during the major El Niño year of 1982, all female Galápagos fur seals (*Arctocephalus galapagoensis*) lost their pups. This massive recruitment failure was attributed to shifts in prey distribution, as at least some lethal and sub-lethal effects were linked to starvation (Trillmich and Dellinger, 1991). Of course, a slower, more prolonged ocean warming than that related to the "natural" occurrence of El Niño may cause shifts in habitat use that simply manifest as alterations in the site fidelity patterns of some marine mammals. However, we believe that land-breeding pinnipeds are particularly vulnerable to such changes, as they utterly depend on access to relatively undisturbed haul-out sites for mating, pupping, resting, and often, moulting, that are within reasonable swimming distance of abundant prey. Sirenians that rely on restricted near-shore habitat may also be affected by changes in physical habitat involving their diet of sea grass and other vegetation. However, all but the least mobile of cetaceans may be less affected by long-term climate warming.

Figure 5.   Fur seals, *Arctocephalus spp.,* of the tropics and sub-Antarctic islands may be especially vulnerable to climate change: they generally do not travel far from their resting beaches, and changes in productivity or current patterns will affect the availability of food. Photo: B. Würsig.

Several examples of odontocete distributional shifts have been described: During the 1982-83 El Niño, near-bottom spawning market squid (*Loligo opalescens*) left the southern California area, and so did the short-finned pilot whales (*Globicephala macrorhynchus*) that normally fed on them (Shane, 1994). It is unknown whether this apparently climate-induced shift resulted in whale deaths or health impairment due to the loss of access to energy-rich prey. Interestingly, the absence of pilot whales was followed several years later by an influx of Risso's dolphins (*Grampus griseus*), also feeding on (the returned) market squid. Risso's dolphins may have taken advantage of a temporarily vacant niche left so by the pilot whales, and there was a shift in cetacean species composition, apparently as a result of the El Niño warming episode (Shane, 1995). During that same El Niño event, nearshore bottlenose dolphins expanded their range from southern to central California, and stayed in the new northern range after the warming event subsided in the mid-1980s (Wells *et al.*, 1990). They are still there. Sperm whales (*Physeter macrocephalus*) of the eastern tropical Pacific had reduced fecundity or calf survival during and after an El Niño event of the late 1980s (Whitehead, 1997). We can surmise that overall climate change will have similarly far-reaching effects. That said, it is important to acknowledge that the time scales of event onset and duration of El Niño episodes are much different from those of global climate change, and therefore that the effects of the latter would be, by comparison, more gradual and potentially more insidious. Nevertheless, it is possible that the severity of El Niño events is exacerbated by present climate change trends, as indicated (for example) by recent studies of coral growth throughout 130,000 years of glacial-interglacial cycles (Tudhope *et al.*, 2001).

5.   RESEARCH IMPERATIVES

The need for better information leads us to consider how it might be obtained. Efforts to gather enough data for large-scale and mesoscale predictions about effects on marine mammals must be made at appropriate climate and ecosystem levels. Certainly, better information on habitat use and habitat requirements can be obtained for almost all species. Even when a species is relatively well known, such as the bottlenose dolphin, there are populations about which virtually nothing is known. Ocean current models with climate-prediction capabilities are constantly being refined (Acoustic Thermometry of Ocean Climate, 1998), and there are relatively recent attempts at multi-level global integrated assessment models, such as TARGETS (Tool to Assess Regional and Global Environmental and Health Targets for Sustainability - Rotmans and deVries, 1997; Weaver and Zwiers,

2000). For marine mammalogists, the best strategy may be to pursue our science in a way that seeks to emphasise the environmental context whenever possible (*e.g.* Forney, 2000; Moore, 2000). We need to integrate knowledge of the animals with knowledge of the ecosystem at multiple levels (Mangel and Hofman, 1999; Schell, 2000). This translates into more than the usual collaborative approach of working together with, for example, behavioural ecologists, physiological ecologists, physical oceanographers, biological oceanographers, and toxicologists to integrate results in the discussion section of a final paper. It means inter-disciplinary involvement from the early design phase and through to the end of an ecology-based research project. The recent GulfCet (Cetaceans of the Gulf of Mexico) project that links cetacean distribution with moving oceanic cyclonic and anticyclonic current systems may be a good model for such an approach (Davis *et al.*, 1998, 1999, in press, Baumgartner *et al.*, 2001). Croll *et al.* (1998) present an excellent integrated approach to the study of marine birds and mammals.

6.   MANAGING FOR ROBUSTNESS

Perhaps more than anything, we need a global change in attitude about how much so-called biodiversity is "enough." It seems that we are in a permanent crisis mode while seeking to reverse the loss of biodiversity, and our (generally meagre) energies are devoted almost exclusively to the most degraded environments and species. This imbalance is, at least in part, a necessary result of there being too few resources available for the conservation cause. However, we argue that at least a portion of those resources should be invested in management for abundance, or in maintaining the "robustness" of ecosystems and species. For example, Asian river cetaceans are undeniably in great peril, and it is appropriate to seek maximal protection for their few remaining pockets of occurrence. They and other already-beleaguered species are suffering from the effects of habitat degradation, including over-fishing, net and line entanglements, toxin pollution, noise, and physical habitat destruction. Global climate change represents one additional insult, and the cumulative impact of these forces may tip the scales against survival of some populations and species.

At the same time, the South American river dolphins, the boto (*Inia geoffrensis*) and tucuxi (*Sotalia fluviatilis*), are relatively numerous and widespread (International Whaling Commission, in press). The greatest challenge with respect to Amazon dolphins (and other presently common or abundant species) may be to avoid complacency. The best, and perhaps only, insurance against the impacts of regional and global environmental

Figure 6. This boto, or Amazon river dolphin (top) belongs to a relatively robust species. However, the Amazon region is rapidly being transformed by human population growth and economic activities; and the status of some other river dolphin species like the baiji or Chinese river dolphin (bottom) prove that there is no room for complacency. Photos: E. Zúñiga and B. Würsig respectively.

perturbations is to have large populations occupying large and diverse spaces, a condition that we call "species robustness." Managing Amazon river dolphins to achieve this kind of robustness, rather than merely to ensure that at least one "viable" population of each species survives in a relict of its historical range, can represent a hedge against the effects of global climate change, and possibly also against non-point source pollution and other unforeseen or unquantifiable threats to species survival. Island and oceanic spinner dolphins (*Stenella longirostris*) and near-shore bottlenose dolphins, the latter perhaps the most widely distributed small cetaceans on Earth, are more common examples. We need to protect them and their environments today, and not wait for an impoverished future before beginning to pay attention to them.

7.   CONCLUSIONS

Finally, for such a pervasive problem as global climate change, we need strong political leadership that cares about the environment. Such leadership can emerge from any part of the globe, but a special responsibility falls on those of us in over-developed countries who can most afford to curb our huge, disproportionate per-capita consumption of energy and output of toxins and other environmental degraders. In the circumstances, it is unconscionable that the U.S. government, for example, has not signed key elements of the Kyoto Protocol (Kyoto Protocol to the United Nations Framework Convention on Climate Change, 1997) to curb environmental emissions of fossil fuels. We also cannot simply call for further studies while hiding behind the smokescreen of: "We do not really know for sure." We know enough to recognise that a large problem exists. Curiously, we even know how to reverse it, at least to some degree. It appears that to forestall or at least to mitigate against wholesale environmental degradation due to climate change, we need a concerted effort to change energy policies, with emphasis on those nations and cultures that are most responsible for the degradation (Intergovernmental Panel on Climate Change, 2001c).

8.   SUMMARY

It is generally accepted that the global climate is changing, fuelled especially by greenhouse gas emissions, ozone depletion, and large-scale deforestation. However, it has not been possible to predict the severity of

Figure 7.   Two marine mammal species with restricted ranges that could be affected by habitat change brought on as a result of global warming are the Hector's dolphin (top) of New Zealand, and the walrus (bottom) of the arctic.

climate change for the coming century, nor the consequences on global or even local scales. Global average temperatures are expected to rise by one to several degrees Celsius within the next 100 years, and a concomitant decrease in Arctic and Antarctic ice thickness and extent will almost certainly take place. These changes are likely to have direct effects on ice-breeding seals and ice-associated cetaceans. Ice conditions and mesoscale climate changes will probably affect circulation, upwelling, salinity, and other variables that in turn will affect primary productivity and other trophic levels. It is uncertain how this will alter migration patterns, feeding locales, and mating/calving areas of marine mammals. In the tropics and mid-latitudes, increases in sea level, changes in salinity, and disturbance/modification (if not complete destruction) of reef assemblages may alter food availability patterns of the warm-water pinnipeds, sirenians, and cetaceans. Hints of the potential future effects have been given during El Niño Southern Oscillation events (or ENSOs), when migration extent and paths of large whales have changed, and when there has been massive reproductive failure in some island-living populations of otariids (eared pinnipeds). Beyond these broad strokes, however, detailed predictions are likely to prove inadequate or downright false. Climate-change scenarios may become manageable (or at least partially understandable) if viewed in terms of three inter-related levels: 1) *primary* level, at which ecological changes act on individuals, perhaps especially the very young, old, or infirm; 2) *secondary* level, at which the entire population or species is affected; and 3) *tertiary* level, at which the results of climate change feed back to the initiators of the problem, humans in this instance, leading to further impacts on marine mammals. An example of a tertiary effect would be the reduction of valued fish resources due to climate change, leading humans to kill dolphins or seals as perceived competitors, or as a replacement food. We suggest that researchers design and conduct interdisciplinary studies that better link, for example, physical oceanography with the health status and productivity of marine mammals. A paradigm shift, from management for mere species survival to management for conservation of species robustness, may be the best long-term strategy for addressing the threat of global climate change. Global climate change is but one threat among many, including over-fishing, net entanglements, toxin pollution, noise, and habitat destruction; and the cumulative effects of these may spell extirpation for those populations and species already depleted or restricted to small geographic areas. We reiterate what has become axiomatic among conservation biologists, that a precautionary stance should be adopted in the face of uncertainty. Finally, it is incumbent on scientists to communicate their knowledge concerning the known and potential effects of climate change in order to influence government policies.

ACKNOWLEDGEMENTS

We thank Steven Katona for providing the germ for this synopsis over 10 years ago, four anonymous reviewers for suggesting ways to update and enhance an earlier version, Melany Würsig for last-minute editing and formatting, and Peter Evans, Antonio Raga, and an anonymous reviewer for thoughtful comments and advice.

REFERENCES

Ashmore, M.R. and Bell, J.N.B. (1991) The role of ozone in climate change. *Annals of Botany*, **67**, 39-48.

Acoustic Thermometry of Ocean Climate Consortium. (1998) Ocean climate change: comparison of acoustic tomography, satellite altimetry, and modeling. *Science*, **281**, 1327-1332.

Barbier, E.B., Burgess, J.C. and Folke, C. (1994) *Paradise lost: the ecological economics of biodiversity*. Earthscan Press, London.

Baumgartner, M.F., Mullin, K.D., May, L.N. and Leming, T.D. (2001) Cetacean habitats in the northern Gulf of Mexico. *Fishery Bulletin*, **99**, 219-239.

Bossart, G.D., Baden, D.G., Ewing, R.Y, Roberts, B. and Wright, S.D. (1998) Brevetoxicosis in manatees (*Trichechus manatus latirostris*) from the 1996 epizootic: gross, histologic, and immuno-histochemical features. *Toxicological Pathology*, **26**, 276-82.

Budd, G.M. (2000) Changes in Heard Island glaciers, king penguins and fur seals since 1947. *Papers and Proceedings of the Royal Society of Tasmania*, **133**, 47-60.

Burns, C.J. (2001) From the harpoon to the heat: climate change and the International Whaling Commission in the 21st Century. *The Georgetown International Environmental Law Review*, **13**, 335-359.

Croll, D.A., Tershy, B.R., Hewitt, R., Demer, D., Hayes, S., Fiedler, P., Popp, J. and Lopez, V.L. (1998). An integrated approach to the foraging ecology of marine birds and mammals. *Deep-Sea Research*, **45**, 1353-1371.

Davis, R.W., Evans, W.E. and Würsig, B. (1999) Distribution and abundance of cetaceans in the northern Gulf of Mexico. *Volume II: Technical Report. U.S. Minerals Management Service*, Gulf of Mexico OCS Region, New Orleans, LA.

Davis, R.W., Fargion, G.S., May, N., Leming, T.D., Baumgartner, M., Evans, W.E., Hansen, L.J. and Mullin, K. (1998) Physical habitat of cetaceans along the continental slope in the north-central and western Gulf of Mexico. *Marine Mammal Science*, **14**, 490-507.

Davis, R.W., Ortega-Ortiz, J.G., Ribic, C.A., Evans, W.E., Biggs, D.C., Ressler, P.H., Cady, R.B., Leben, R.R., Mullin, K.D. and Würsig, B. (In press) Cetacean habitat in the northern oceanic Gulf of Mexico. *Deep Sea Research Part I*: Oceanographic Research Papers.

Dyke, A.S., Hooper, J. and Savelle, J.M. (1996) A history of sea ice in the Canadian Arctic archipelago based on postglacial remains of the bowhead whale (*Balaena mysticetus*). *Arctic*, **49**, 235-255.

Flavin, C. (1992) Natural gas production climbs. In: *Vital signs* (Ed. by L.R. Brown, C. Flavin, and H. Kane), Earthscan Press, London.

Forney, K.A. (2000) Environmental models of cetacean abundance: Reducing uncertainty in population trends. *Conservation Biology*, **14**, 1271-1286.

Geraci, J.R., . Anderson, D.M., Timperi, R.J., St. Aubin, D.J., Early, G.A., Prescott, J.H. and Mayo, C.A. (1989) Humpback whales (*Megaptera novaeangliae*) fatally poisoned by dinoflagellate toxin. *Canadian Journal of Fisheries and Aquatic Sciences,* **46**, 1895-1898.

Geraci, J.R., Harwood, J. and Lounsbury, V.J. (1999) Marine mammal die-offs: causes, investigations, and issures. In: *Conservation and Management of Marine Mammals* (Ed. by J.R. Twiss, Jr. and R.R. Reeves), pp. 367-395. Smithsonian Institution Press, Washington, D.C.

Guinet C., Chastel, O., Koudil, M., Durbec, J.P. and Jouventin, P. (1998) Effects of warm sea-surface temperature anomalies on the blue petrel at the Kerguelen Islands. *Proceedings of the Royal Society of London - Series B: Biological Sciences,* **265**, 1001-1006.

Guinet C., Jouventin, P. and Georges, J.Y. (1994) Long term population changes of fur seals *Arctocephalus gazella* and *Arctocephalus tropicalis* on subantarctic (Crozet) and subtropical (St. Paul and Amsterdam) islands and their possible relationship to El Niño Southern Oscillation. *Antarctic Science* , **6**, 473-478.

Harvell, C.D., Kim, K., Burkholder, J.M., Colwell, R.R., Epstein, P.R., Grimes, D.J., Hofman, E.E., Lipp, E.K., Osterhaus, A.D.M., Overstreet, R.M., Porter, J.W., Smith, G.W. and Vasta, G.R. (1999) Emerging marine diseases - Climate links and anthropogenic factors. *Science,* **285**, 1505-1510.

Hernández, M., Robinson, I., Aguilar, A., González, L.M., López-Jurado, L.F., Reyero, M.I., Cacho, E., Franco, J., López-Rodas V. and Costas, E. (1998) Did algal toxins cause monk seal mortality? *Nature,* (London), **393**, 28-29.

Intergovernmental Panel on Climate Change. (2001a). Climate Change 2001: The Scientific Basis. *Third Assessment Report of the IPCC Working Group* I. Available from IPCC Secretariat, C/O World Meteorological Organization, 7bis Avenue de la Paix, C.P. 2300, CH- 1211 Geneva. http://www.ipcc.ch/

Intergovernmental Panel on Climate Change. (2001b) Climate Change 2001: Impacts, Adaptation and Vulnerability. *Third Assessment Report of the IPCC Working Group II.* Available from IPCC Secretariat, C/O World Meteorological Organization, 7bis Avenue de la Paix, C.P. 2300, CH- 1211 Geneva. http://www.ipcc.ch/

Intergovernmental Panel on Climate Change. (2001c) Climate Change 2001: Mitigation. *Third Assessment Report of the IPCC Working Group III.* Available from IPCC Secretariat, C/O World Meteorological Organization, 7bis Avenue de la Paix, C.P. 2300, CH- 1211 Geneva. http://www.ipcc.ch/

International Whaling Commission. (1997) Report of the IWC workshop on climate change and cetaceans. *Report of the International Whaling Commission,* **47**, 293-313.

International Whaling Commission. (In press) Report of the standing sub-committee on small cetaceans. *Journal of Cetacean Research and Management,* **3** (Suppl.).

Kyoto Protocol to the United Nations Framework Convention on Climate Change. (1997) Obtain as MS #FCCC/CP/1997/L.7/Add.1, from United Nations Publications, Rm. DC2-853, 2 UN Plaza, New York, NY 100017, USA; or as Web Address: www.unfccc.de.

MacGarvin, M. and Simmonds, M. (1996) Whales and climate change. In: *The conservation of whales and dolphins* (Ed. by M.P. Simmonds and J.D. Hutchinson), pp. 321-332. John Wiley and Sons, Chichester, U.K.

Mangel, M. and Hofman, R.J. (1999) Ecosystems: patterns, processes, and paradigms. In *Conservation and management of marine mammals* (Ed. by J. R. Twiss Jr. and R. R. Reeves), pp. 87-98. Smithsonian Institution Press, Washington D.C.

Marotzke, J. (2000) Abrupt climate change and thermohaline circulation: Mechanisms and predictability. *Proceedings of the National Academy of Sciences of the United States of America,* **97**, 1347-1350.

Marsh, H. (1989) Mass stranding of dugongs by a tropical cyclone in northern Australia. *Marine Mammal Science*, **5**, 78-84.

Moore, P.D., Chaloner, B. and Stott, P. (1996) *Global environmental change.* Blackwell Science Press, Oxford.

Moore, S.E. (2000) Variability of cetacean distribution and habitat selection in the Alaskan Arctic, autumn 1982-91. *Arctic*, **53**, 448-460.

Polovina, J.J., Mitchum, G.T., Graham, N.E., Craig, M.P., DeMartini, E.E. and Flint, E.N. (1994) Physical and biological consequences of a climate event in the central North Pacific. *Fisheries Oceanography*, **3**, 15-21.

Reeves, R.R. (1990) Speculations on the impact of global warming on aquatic mammals. *Proceedings of the American Cetacean Society*, Monterey, CA, November 9-11, 1990. American Cetacean Society, San Pedro.

Reeves, R.R. and Leatherwood, S. (1994) Dolphins, porpoises, and whales: 1994-1998 action plan for the conservation of cetaceans. *IUCN*, Gland, Switzerland.

Reeves, R.R., Leatherwood, S., Jefferson, T.A., Curry, B.E. and Henningsen, T. (1996) Amazonian manatees, *Trichechus inunguis*, in Peru: distribution, exploitation, and conservation status. *Interciencia*, **21**, 246-254.

Reeves, R.R., Smith, B.D. and Kasuya, T. (eds.). (2000) *Biology and conservation of freshwater cetaceans in Asia.* Occasional Paper of the IUCN Species Survival Commission, No. 23. IUCN, Gland, Switzerland.

Reynolds, J.E., III. (1999) Efforts to conserve the manatees. In: *Conservation and Management of Marine Mammals* (Ed. by J.R. Twiss, Jr. and R.R. Reeves), pp. 267-295. Smithsonian Institution Press, Wash., D.C.

Rotmans, T. and de Vries, B. (1997) *Perspectives on global change: the targets approach.* Cambridge University Press, Cambridge.

Savelle, J.M., Dyke A.S. and. McCartney, A.P. (2000) Holocene bowhead whale (*Balaena mysticetus*) mortality patterns in the Canadian Arctic Archipelago. *Arctic*, **53**, 414-421.

Schell, D.M. (2000) Declining carrying capacity in the Bering Sea: Isotopic evidence from whale baleen. *Limnological* Oceanograph,y **45**, 459-462.

Scholin, C.A., Gulland, F., Doucette, G.J., Benson, S., Busman, M., Chavez, F.P., Cordaro, J., DeLong, R., DeVogelaere, A., Harvey, J., Haulena, M., LeFebvre, K., Lipscomb, T., Loscutoff, S., Lowenstine, L.J., Marin III, R., Miller, P.E., McLellan, W.A., Moeller, P.D.R., Powell, C.L., Rowles, T., Silvagni, P., Silver, M., Spraker, T., Trainer, V. and VanDolah.F.M. (2000) Mortality of sea lions along the central California coast linked to a toxic diatom bloom. *Nature*, (London), **403**, 80-84.

Shane, S.H. (1990) Comparison of bottlenose dolphin behavior in Texas and Florida, with a critique of methods for studying dolphin behavior. In: *The bottlenose dolphin* (Ed. by S. Leatherwood and R. R. Reeves), pp. 541-558. Academic Press, San Diego.

Shane, S.H. (1994) Occurrence and habitat use of marine mammals at Santa Catalina Island, California from 1983-91. *Bulletin Southern California Academy of Science* **93**, 13-29.

Shane, S.H. (1995) Relationship between pilot whales and Risso's dolphins at Santa Catalina Island, California, USA. *Marine Ecology Progress Series*, **123**, 5-11.

Shaughnessy, P. D., Erb, E., and Green, K. (1998) Continued increase in the population of Antarctic fur seals, *Arctocephalus gazella*, at Heard Island, Southern Ocean. *Marine Mammal Science*, **14**, 384-389.

Short, F.T. and Neckles, H.A. (1999) The effects of global climate change on seagrasses. *Aquatic Botany*, **63**, 169-196.

Smith, B.D.,. Aminul Haque, A.K.M., Hossain M.S. and Khan, A. (1998) River dolphins in Bangladesh: conservation and the effects of water development. *Environmental Management*, **22**, 323-335.

Smith, B.D., Sinha, R.K., Zhou, K., Chaudhry, A.A., Liu, R.,Wang Ding, Ahmed, B., Aminul Haque, A.K.M., Mohan R.S.L. and Sapkota, K. (2000) Register of water development projects affecting river cetaceans in Asia. In: *Biology and Conservation of Freshwater Cetaceans in Asia* (Ed. by R.R. Reeves, B.D. Smith and T. Kasuya), pp. 22-39. Occasional Papers of the IUCN Species Survival Commission **23**.

Stirling, I. and Derocher, A.E. (1993) Possible impacts of climatic warming on polar bears. *Arctic*, **46**, 240-45.

Stirling, I., Lunn, N.J. and. Iacozza, J. (1999) Long-term trends in the population ecology of polar bears in western Hudson Bay in relation to climatic change. *Arctic*, **52**, 294-306.

Trillmich, F. and Dellinger, T. (1991) The effect of El Niño on Galapagos pinnipeds. In: *Pinnipeds and El Niño: responses to environmental stress* (Ed. by F. Trillmich and K. A. Ono). pp. 66-74.. Springer Verlag, Berlin.

Tudhope, A.W., Chilcott, C.P., McCulloch, M.T., Cook, E.R., Chappell, J., Ellam, R.M., Lea, D.W., Lough, J.M. and Shimmield, G.B. (2001) Variability in the El Niño--Southern Oscillation through a glacial-interglacial cycle. *Science*, **291**, 1511-1517.

Tynan, C.T. and Demaster, D.P. (1997) Observations and predictions of Arctic climatic change: Potential effects on marine mammals. *Arctic*, **50**, 308-322.

United Nation Environment Programme. (1987) The ozone layer. *UNEP/GEMS Environmental* Library, **2**, 1-36.

United Nation Environment Programme. (1993) *Environmental data report 1993-94*. Blackwell Press, Oxford.

Van de Bildt, M.W.G., Vedder, E.J., Martina, B.E.E., Sidi, B.A., Jiddou, A.B., Barham, M.E.O., Androukaki, E., Komnenou, A., Niesters, H.G.M. and Osterhaus, A.D.M.E. (1999) Morbilliviruses in Mediterranean monk seals. *Veterinary Microbiology*, **69**, 19-21.

Van Waerebeek, K. and Reyes, J.C. (1994) Interactions between small cetaceans and Peruvian fisheries in 1988/89 and analysis of trends. *Reports of the International Whaling Commission* (special issue **15**), 495-502.

Weaver, A.J. and Zwiers, F.W. (2000) Climatology: Uncertainty in climate change. *Nature* (London), **407**, 571-572.

Wellburn, A. (1994) *Air pollution and climate change: the biological impact,* 2nd edition. Longman Press, New York.

Wells, R.S., Hansen, L.J., Baldridge, A., Dohl, T.P., Kelly, D.L. and Defran, R.H. (1990) Northward extension of the range of bottlenose dolphins along the California coast. In: *The bottlenose dolphin* (Ed. by S. Leatherwood and R. R. Reeves), pp. 421-431. Academic Press, San Diego.

Whitehead, H. (1997) Sea surface temperature and the abundance of sperm whale calves off the Galápagos Islands: Implications for the effects of global warming. *Report of the International Whaling Commission*, **47**, 941-944.

Würsig, B. and Ortega-Ortiz, J.G. (1999) Global climate change and marine mammals In: *European Research on Cetaceans – 13*. Proceedings of the Thirteenth Annual Conference of the European Cetacean Society (Ed. by P.G.H. Evans, J. Cruz and J. A. Raga), pp. 351-356. European Cetacean Society, Valencia, Spain.

SYSTEMATIC LIST OF MARINE MAMMALS

ORDER CARNIVORA

SUB-ORDER PINNIPEDIA

Family Otariidae, the Eared Seals

Subfamily Arctocephalinae (fur seals)

Arctocephalus australis	Falkland fur seal (*A. a. australis*)
	S. American fur seal (*A. a. gracilis*)
A. forsteri	New Zealand fur seal
A. galapagoensis	Galapagos fur seal
A. gazella	Antarctic fur seal
A. philippi	Juan Fernandez fur seal
A. pusillus	Australian fur seal (*A. p. doriferus*)
	S. African fur seal (*A. p. pusillus*)
A. townsendi	Guadalupe fur seal
A. tropicalis	Subantarctic fur seal
Callorhinus ursinus	Northern fur seal

Subfamily Otariinae (sea lions)

Eumetopias jubatus	Steller sea lion, Northern sea lion
Neophoca cinerea	Australian sea lion
Otaria byronia	Southern sea lion
Phocarctos hookeri	Hooker's sea lion, New Zealand sea lion
Zalophus californianus	California sea lion (*Z. c. californianus*)
	Japanese sea lion (*Z. c. japonicus*) - extinct
	Galapagos sea lion (*Z. c. wollebaeki*)

Family Odobenidae (walruses)

Odobenus rosmarus	Pacific walrus (*O. r. divergens*)
	Laptev walrus (*O. r. laptevi*)
	Atlantic walrus (*O. r. rosmarus*)

Family Phocidae, the Earless or True Seals

Subfamily Monachinae (southern phocids)

Hydrurga leptonyx	Leopard seal
Leptonychotes weddellii	Weddell seal
Lobodon carcinophagus	Crabeater seal
Mirounga angustirostris	Northern elephant seal
M. leonina	Southern elephant seal
Monachus monachus	Mediterranean monk seal
M. schauinslandi	Hawaiian monk seal
M. tropicalis	West Indian monk seal (extinct)
Omnatophoca rossii	Ross seal

Subfamily Phocinae (northern phocids)

Cystophora cristata	Hooded seal
Erignathus barbatus	Atlantic bearded seal (*E. b. barbatus*)
	Pacific bearded seal (*E. b. nauticus*)
Halichoerus grypus	Grey or Atlantic seal
Phoca fasciata	Ribbon seal
P. groenlandica	Harp seal
P. hispida	Baltic seal (*P. h. botnica*)
	Arctic ringed seal (*P. h. hispida*)
	Ladoga seal (*P. h. ladogensis*)
	Okhotsk Sea ringed seal (*P. h. ochotensis*)
	Saiman seal (*P. h. saimensis*)
P. largha	Spotted seal, Largha seal
P. vitulina	Western Atlantic harbour seal *(P. v. concolor)*
	Ungava seal (*P. v. mellonae*)
	Eastern Pacific harbour seal (*P. v. richardsi*)
	Western Pacific harbour seal (*P. v. stejnegeri*)
	Eastern Atlantic harbour seal (*P. v. vitulina*)
Pusa caspica	Caspian seal
Pusa sibirica	Baikal seal

SUB-ORDER FISSIPEDIA
Family Ursidae

Ursus maritimus	Polar bear

Family Mustelidae

Enhydra lutris	Sea otter (*E. l. gracilis, E. l. lutris, E. l. nereis*)
Lutra felina	Marine otter

ORDER CETACEA

SUB-ORDER MYSTICETI, the Baleen Whales

Family Balaenidae (right whales)

Balaena mysticetus	Bowhead whale
Eubalaena australis	Southern right whale
E. glacialis	North Atlantic right whale
E. japonica	North Pacific right whale

Family Eschrichtidae

Eschrichtius robustus	Gray whale

Family Balaenopteridae (rorquals)

Balaenoptera acutorostrata	Minke whale
B. bonaerensis	Antarctic minke whale
B. borealis	Sei whale
B. edeni	Bryde's whale
B. musculus	Blue whale
B. physalus	Fin whale
Megaptera novaeangliae	Humpback whale

Family Neobalaenidae
Caperea marginata   Pygmy right whale

SUB-ORDER ODONTOCETI, the Toothed Whales

Family Physeteridae
Physeter macrocephalus*   Sperm whale (* = *P. catodon*)

Family Kogiidae
Kogia breviceps   Pygmy sperm whale
K. simus   Dwarf sperm whale

Family Ziphiidae
Berardius arnuxii   Arnoux's beaked whale
B. bairdii   Baird's beaked whale
Hyperoodon ampullatus   Northern bottlenose whale
H. planifrons   Southern bottlenose whale
Indopacetus pacificus*   Longman's beaked whale
(*previously *Mesoplodon pacificus*)
Mesoplodon bahamondi   Bahamonde's beaked whale
M. bidens   Sowerby's beaked whale
M. bowdoini   Andrews' beaked whale
M. carlhubbsi   Hubbs' beaked whale
M. densirostris   Blainville's beaked whale, dense-beaked whale
M. europaeus   Gervais' beaked whale
M. ginkgodens   Ginkgo-toothed beaked whale
M. grayi   Gray's beaked whale
M. hectori   Hector's beaked whale
M. layardii   Strap-toothed whale
M. mirus   True's beaked whale
M. peruvianus   Peruvian beaked whale, pygmy beaked whale
M. stejnegeri   Stejneger's beaked whale
Tasmacetus shepherdi   Tasman or Shepherd's beaked whale
Ziphius cavirostris   Cuvier's beaked whale

Family Monodontidae
Delphinapterus leucas   White whale, beluga
Monodon monoceros   Narwhal

Family Delphinidae
Cephalorhynchus commersonii   Commerson's dolphin
C. eutropia   Black dolphin
C. heavisidii   Heaviside's dolphin
C. hectori   Hector's dolphin
Delphinus capensis   Long-beaked common dolphin
D. delphis   Common or short-beaked dolphin
D. tropicalis   Arabian common dolphin
Feresa attenuata   Pygmy killer whale
Globicephala macrorhynchus   Short-finned pilot whale
G. melas (= melaena)   Long-finned pilot whale
Grampus griseus   Risso's dolphin
Lagenodelphis hosei   Fraser's dolphin

Family Delphinidae (cont.)

Lagenorhynchus acutus*	Atlantic white-sided dolphin (* = *Leucopleurus*)
L. albirostris	White-beaked dolphin
L. australis	Peale's dolphin
L. cruciger	Hourglass dolphin
L. obliquidens	Pacific white-sided dolphin
L. obscurus	Dusky dolphin
Lissodelphis borealis	Northern right whale dolphin
L. peronii	Southern right whale dolphin
Orcaella brevirostris	Irrawaddy dolphin
Orcinus orca	Killer whale
Peponocephala electra	Melon-headed whale
Pseudorca crassidens	False killer whale
Sotalia fluviatilis	Tucuxi
Sousa chinensis	Indo-Pacific hump-backed dolphin
Sousa plumbea	Indian hump-backed dolphin
S. teuszii	Atlantic hump-backed dolphin
Stenella attenuata	Pantropical spotted dolphin
S. clymene	Clymene dolphin, short-snouted spinner dolphin
S. coeruleoalba	Striped dolphin
S. frontalis	Atlantic spotted dolphin
S. longirostris	Spinner or long-snouted spinner dolphin
Steno bredanensis	Rough-toothed dolphin
Tursiops aduncus	Indo-Pacifc bottlenose dolphin
T. truncatus	Bottlenose dolphin

Family Phocoenidae (porpoises)

Neophocaena phocaenoides	Finless porpoise
Phocoena* dioptrica	Spectacled porpoise (* = *Austraphocoena*)
P. phocoena	Harbour porpoise
P. sinus	Vaquita
P. spinipinnis	Burmeister's porpoise
Phocoenoides dalli	Dall's porpoise

Family Platanistidae

Platanista gangetica	South Asian river dolphin; includes Ganges Susu (*P. g. gangetica*) and Indus Susu (*P. g. minor*)

Family Iniidae

Inia geoffrensis	Boto, boutu or Amazon river dolphin (*I. g. humboldtiana, I. g. geoffrensis*) [Note: Rice (1998) recognises a separate species, Bolivian river dolphin *Inia boliviensis*]

Family Lipotidae

Lipotes vexillifer	Baiji

Family Pontoporiidae

Pontoporia blainvillei	Franciscana, La Plata dolphin

ORDER SIRENIA, the Sea Cows

Family Dugongidae
Dugong dugon   Dugong
Hydrodamalis gigas   Steller's sea cow (extinct)

Family Trichechidae
Trichechus inunguis   Amazonian manatee
T. manatus   West Indian manatee
                                  Antillean manatee (*T. m. manatus*)
                                  Florida manatee (*T. m. latirostris*)
T. senegalensis   West African manatee

Taxonomic Reference Sources: Reynolds, J.E. III & Odell, D.K. 1991. *Manatees and Dugongs*. Facts on File, New York; Jefferson, T.A., Leatherwood, S., and Webber, M.A. 1993. *Marine Mammals of the World*. FAO, Rome; Reijnders, P., Brasseur, J., van der Toorn, P., Boyd, I., Harwood, J., Lavigne, D., and Lloyd, L. 1993. *Seals, Fur Seals, Sea Lions, and Walrus*. IUCN/SSC Seal Specialist Group, IUCN, Gland; Heyning, J.E. & Perrin, W.F. 1994. *Contribs. in Science, Nat. Hist. Museum of Los Angeles County, CA*, 442: 1-35; Rice, D. 1998. *Marine Mammals of the World*. SMM Special Publication No. 4. Allen Press, Kansas; International Whaling Commission. 2001. Annex U. Report of the Working Group on Nomenclature. Report of the Scientific Committee. *Journal of Cetacean Research and Management*, 3, in press.

Index